AF411180

Scientific Advances in STEM:
From Professor to Students

Scientific Advances in STEM: From Professor to Students

Editors

Yadir Torres Hernández
Manuel Felix Angel
Ana María Beltrán Custodio
Francisco Garcia Moreno

MDPI • Basel • Beijing • Wuhan • Barcelona • Belgrade • Manchester • Tokyo • Cluj • Tianjin

Editors
Yadir Torres Hernández
University of Seville (US)
Spain

Manuel Felix Angel
University of Sevilla (US)
Spain

Ana María Beltrán Custodio
University of Seville (US)
Spain

Francisco Garcia Moreno
Helmholtz-Centre Berlin
Germany

Editorial Office
MDPI
St. Alban-Anlage 66
4052 Basel, Switzerland

This is a reprint of articles from the Topical Collection published online in the open access journals *Metals* (ISSN 2075-4701) (available at: https://www.mdpi.com/journal/metals), *Polymers* (ISSN 2073-4360) (available at: https://www.mdpi.com/journal/polymers), *Foods* (ISSN 2304-8158) (available at: https://www.mdpi.com/journal/foods), *Sustainability* (ISSN 2071-1050) (available at: https://www.mdpi.com/journal/sustainability) and *Sensors* (ISSN 1424-8220) (available at: https://www.mdpi.com/journal/sensors).

For citation purposes, cite each article independently as indicated on the article page online and as indicated below:

LastName, A.A.; LastName, B.B.; LastName, C.C. Article Title. *Journal Name* **Year**, *Volume Number*, Page Range.

ISBN 978-3-0365-1775-9 (Hbk)
ISBN 978-3-0365-1776-6 (PDF)

Cover image courtesy of Sergio Muñoz Moreno (University of Seville (Spain)).

Contents

About the Editors

Yadir Torres Hernández: Interests: design and manufacture of porous materials; surface modification (physical and chemical); bio-functional (osseointegration, cells, and bacterial response) and tribo-mechanical (instrumented micro-indentation, fracture, fatigue, scratch resistance, and wear) behavior; biomaterials; tool materials (cemented carbides, cermet's, and multi-layered: alumina-zirconia, WC-Co/WC-Co, and Cermet/WC-Co); powder metallurgy (conventional and space-holder technique).

Manuel Félix Ángel: Interests: active packaging; bioplastics; electrospinning; interfaces; injection moulding, mixing; proteins; rheology; food by-products.

Ana María Beltrán Custodio: Interests: design; nano-structure and chemical studies by scanning-transmission electron microscopy techniques; biomaterials.

Francisco Garcia Moreno: Interests: Process imaging, tomography, metal foams.

Preface to "Scientific Advances in STEM: From Professor to Students"

Universities should generate and relay knowledge and, therefore, placing it at the service of society. In this context, the Higher Polytechnic School (Escuela Politécnica Superior, EPS) of the University of Seville (US), Spain, with a great tradition in the industrial work, aims to train the students both as professional and researcher, promoting knowledge in Science, Technology, Engineering, and Mathematics (STEM) areas. STEM research has a noticeable impact on the industrial sector, where a multidisciplinary approach is required, involving teams formed by people from diverse fields and different level of experience, from full-professor to students at the last steps of their degrees or masters, including Ph.D. students, as well as lab-staff, which allows students to be closer to high-quality research. Advances in STEM education includes innovative approaches and perspectives in promoting and improving STEM-based education and the processes of STEM instructions and further application of knowledge. Thus, the continuous advances in technology not only change the way students learn but also the way they connect and interact every day. These changes have implications in the academia field. In this sense, the interdisciplinary skills developed through STEM provide a network where research can be applied in a fruitful collaboration. This book collects the articles published in special topic Scientific Advances in STEM: From Professor to Students, which includes publications at the MPDI Journals Metals, Polymers, Foods, Sustainability, and Sensors. It covers examples of cutting-edge research hosted at the EPS-US and presented at the 7th Symposium on Research, Development and Innovation of this academic centre, also in collaboration with other national and international institutions. Editors would like to disseminate this multidisciplinary research and contribute to establish new collaborations among groups. This book collects 20 works from academic researchers, reporting novel results in the STEM field. Material science covers 6 contributions, from protein-based materials exhibiting absorbent properties to titanium-based materials for the development of biomaterials. Thus, Alonso et al. [1] analyzed the effect of mould temperature on the mechanical and microstructural properties and water absorption capacity of rice bran-based bioplastics obtained by injection moulding. In this sense, Jiménez-Rosado et al. [2] analyzed the effect of ZnO nanoparticles on soy protein-based bioplastics at three different mould temperature, improving their functional properties, Moreover, in this context, Alvarez-Castillo et al. [3] strengthened the superabsorbent properties of porcine plasma protein through a solubilization-freeze-drying process. These authors proposed the replacement of common acrylic derivatives by biodegradable materials. They analyzed the effect of freeze-drying carried out either directly on the protein flour or after its solubilization in deionized water. Li et al. [4] used test methods and finite element numerical analysis to study the flexural load-bearing performance. These authors concluded that the steel–wood composites had good load-bearing and deformation performance. In the field of biomedical applications, Fernandez-Ponce et al. [5] analyzed the use of hybrid gold-magnetic-iron-oxide nanoparticles synthetized by a novel method suitable for in-vitro biomedical applications. Additionally, Mesa-Restrepo et al. [6] employed titanium for the fabrication of implants superficially modified by direct irradiation synthesis with enhanced osseointegration as it is demonstrated by the cell adhesion, differentiation, and mineralization. On the other hand, the use of cutting edge-computing platforms would play important roles for medical image analysis to aid the detection of illnesses, as presented Duran-Lopez et al. [7] for a quick detection of prostate cancer. Additionally, thinking on healthcare, Luna-Perejón et al. [8] presented the AnkFALL dataset to detect falling risk and used this information to integrate a fall detector in the footwear. Material science is a broad field. Thus, Ancio et al. [9] analyzed the main aspects which affect bonding in stainless steel rebars embedded in a hydraulic medium. These authors based their findings in several finite element simulations for the identification of the relationships between pull-out forces in several situations. Finite elements simulations were also employed by Wang et al. [10] among other tools, to confirm the physical, thermophysical properties, temperature-phase transition, and stress-strain curves of AlFeSi phase for Al6061. Moreover, Tagliabue et al. [11] generated magneto-rheological elastomers based on styrene-butadiene-styrene block copolymers. these authors found that the SiO2 and phosphate coating of the affected the saturation magnetization and the shear modulus of MRE composites. Thus, a combined silica phosphate coating resulted in a higher shear modulus, and, therefore, the magnetorheological effect decreased, As for food-based products, An et al. [12] evaluated the extraction of saponins isolated from Corni Fructus (CT) via ultrasonic microwave. These authors isolated saponins from CF (TSCF) using ultrasonic microwave-assisted extraction, using response surface methodology to optimize the study variables. Moreover, the food-based products in this book also covers the results from Linares et al. [13], where nano-emulsions stabilized by fennel oil and further microfluidization together with the addition of advanced performance xanthan gum were analyzed. Novel design engineering was applied to contribute to the development of children with autism, through a multifunctional smart toy to reinforce the multiple needs and adaptable to each patient, as proposed by Cañete et al. [14]. Regarding to learning, García-Villena et al. [15] evaluated the level of satisfaction of graduates in relation to different online postgraduate programs in the environmental area. It was found that there were significant values of low satisfaction in graduates from the Eurasian area, mainly in terms of organizational issues and academic expectations. On the other hand, two indices to control the stability of power systems were proposed by Borras-Talavera et al. [16] to quantify the effects of non-stationary disturbances with high resolution and precision and contributing to the power quality in electrical systems and equipment. Related to renewable generation systems and mobile load the existence of a good distribution network is required. The optimization of the energy use, balance between consumption and generation, is key for increasing their use. The study presented by Parejo et al. [17] analyzed the implementation and architecture of a demand response solution based on OpenADR standard and its potential integration with a building management system. It also verified compliance with the consumption reductions agreements. Related to the field of sensors, Zhao et al. [18] brought insights about curved text detectors. This field is complemented by the analysis of Ou et al. [19], where weakly supervised instance segmentation was used in region based convolutional neural networks (R-CNN). The use of an algorithm to optimize different procedures is key for the development. For example, chemical and pharmaceutical industries require high measurement accuracy and product selectivity, among other requirements. Choudhary et al. [20] applied their technology for the analyses of Raman spectrometer to demonstrate the enhanced accuracy of detection of their proposal which could be extended to other analytical techniques, in which continuous monitoring is required.

References

1. Alonso-González, M.; Felix, M.; Guerrero, A.; Romero, A. Effects of mould temperature on rice bran-based bioplastics obtained by injection moulding. *Polymers* **2021**, *13*, 398, doi:10.3390/polym13030398.
2. Jiménez-Rosado, M.; Perez-Puyana, V.; Sánchez-Cid, P.; Guerrero, A.; Romero, A. Incorporation of ZnO Nanoparticles into Soy Protein-Based Bioplastics to Improve Their Functional Properties.

Polymers **2021**, *13*, 486, doi:10.3390/polym13040486.

3. Álvarez-Castillo, E.; Bengoechea, C.; Guerrero, A. Strengthening of Porcine Plasma Protein Superabsorbent Materials through a Solubilization-Freeze-Drying Process. *Polymers* **2021**, *13*, 772, doi:10.3390/polym13050772.

4. Li, G.; Liu, Z.; Tang, W.; He, D.; Shan, W. Experimental and Numerical Study on the Flexural Performance of Assembled Steel-Wood Composite Slab. *Sustainability* **2021**, *13*, 3814, doi:10.3390/su13073814.

5. Fernández-Ponce, C.; Mánuel, J.M.; Fernández-Cisnal, R.; Félix, E.; Beato-López, J.; Muñoz-Miranda, J.P.; Beltrán, A.M.; Santos, A.J.; Morales, F.M.; Yeste, M.P.; et al. Superficial Characteristics and Functionalization Effectiveness of Non-Toxic Glutathione-Capped Magnetic, Fluorescent, Metallic and Hybrid Nanoparticles for Biomedical Applications. *Metals* **2021**, *11*, 383, doi:10.3390/met11030383.

6. Mesa-Restrepo, A.; Civantos, A.; Allain, J.P.; Patiño, E.; Alzate, J.F.; Balcázar, N.; Montes, R.; Pavón, J.J.; Rodríguez-Ortiz, J.A.; Torres, Y. Synergistic Effect of rhBMP-2 Protein and Nanotextured Titanium Alloy Surface to Improve Osteogenic Implant Properties. *Metals* **2021**, *11*, 464, doi:10.3390/met11030464.

7. Duran-Lopez, L.; Dominguez-Morales, J.P.; Rios-Navarro, A.; Gutierrez-Galan, D.; Jimenez-Fernandez, A.; Vicente-Diaz, S.; Linares-Barranco, A. Performance Evaluation of Deep Learning-Based Prostate Cancer Screening Methods in Histopathological Images: Measuring the Impact of the Model's Complexity on Its Processing Speed. *Sensors* **2021**, *21*, 1122, doi:10.3390/s21041122.

8. Luna-Perejón, F.; Muñoz-Saavedra, L.; Civit-Masot, J.; Civit, A.; Domínguez-Morales, M. AnkFall—Falls, Falling Risks and Daily-Life Activities Dataset with an Ankle-Placed Accelerometer and Training Using Recurrent Neural Networks. *Sensors* **2021**, *21*, 1889, doi:10.3390/s21051889.

9. Ancio, F.; Rodriguez-Mayorga, E.; Hortigon, B. Analysis of the Main Aspects Affecting Bonding in Stainless Steel Rebars Embedded in a Hydraulic Medium. *Metals* **2021**, *11*, 786, doi:10.3390/met11050786.

10. Wang, H.; Deng, W.; Zhang, T.; Yao, J.; Wang, S. Development of Elastoplastic-Damage Model of AlFeSi Phase for Aluminum Alloy 6061. *Metals*. **2021**, *11*, 954, doi:10.3390/met11060954.

11. Tagliabue, A.; Eblagon, F.; Clemens, F. Analysis of Styrene-Butadiene Based Thermoplastic Magnetorheological Elastomers with Surface-Treated Iron Particles. *Polymers* **2021**, *13*, 1597, doi:10.3390/polym13101597.

12. An, S.; Niu, D.; Wang, T.; Han, B.; He, C.; Yang, X.; Sun, H.; Zhao, K.; Kang, J.; Xue, X. Total Saponins Isolated from Corni Fructus via Ultrasonic Microwave-Assisted Extraction Attenuate Diabetes in Mice. *Foods* **2021**, *10*, 670, doi:10.3390/foods10030670.

13. Llinares, R.; Ramírez, P.; Carmona, J.A.; Trujillo-Cayado, L.A.; Muñoz, J. Assessment of Fennel Oil Microfluidized Nanoemulsions Stabilization by Advanced Performance Xanthan Gum. *Foods* **2021**, *10*, 693, doi:10.3390/foods10040693.

14. Cañete, R.; López, S.; Peralta, M.E. KEYme: Multifunctional Smart Toy for Children with Autism Spectrum Disorder. *Sustainability* **2021**, *13*, 4010, doi:10.3390/su13074010.

15. García Villena, E.; Pueyo-Villa, S.; Delgado Noya, I.; Tutusaus Pifarré, K.; Ruíz Salces, R.; Pascual Barrera, A. Instrumentalization of a Model for the Evaluation of the Level of Satisfaction of Graduates under an E-Learning Methodology: A Case Analysis Oriented to Postgraduate Studies in the Environmental Field. *Sustainability* **2021**, *13*, 5112, doi:10.3390/su13095112.

16. Borrás-Talavera, M.D.; Bravo, J.C.; Álvarez-Arroyo, C. Instantaneous Disturbance Index for Power Distribution Networks. *Sensors* **2021**, *21*, 1348, doi:10.3390/s21041348.

17. Parejo, A.; García, S.; Personal, E.; Guerrero, J.I.; García, A.; Leon, C. OpenADR and Agreement Audit Architecture for a Complete Cycle of a Flexibility Solution. *Sensors* **2021**, *21*, 1204, doi:10.3390/s21041204.

18. Zhao, F.; Shao, S.; Zhang, L.; Wen, Z. A Straightforward and Efficient Instance-Aware Curved Text Detector. *Sensors* **2021**, *21*, 1945, doi:10.3390/s21061945.

19. Ou, J.-R.; Deng, S.-L.; Yu, J.-G. WS-RCNN: Learning to Score Proposals for Weakly Supervised Instance Segmentation. *Sensors* **2021**, *21*, 3475, doi:10.3390/s21103475.

20. Choudhary, S.; Herdt, D.; Spoor, E.; Molina, J.F.G.; Nachtmann, M.; Rädle, M. Incremental Learning in Modelling Process Analysis Technology (PAT)—An Important Tool in the Measuring and Control Circuit on the Way to the Smart Factory. *Sensors* **2021**, *21*, 3144, doi:10.3390/s21093144.

Yadir Torres Hernández, Manuel Felix Angel, Ana María Beltrán Custodio,
Francisco Garcia Moreno
Editors

Article

Performance Evaluation of Deep Learning-Based Prostate Cancer Screening Methods in Histopathological Images: Measuring the Impact of the Model's Complexity on Its Processing Speed

Lourdes Duran-Lopez [1,2,3,4,*], **Juan P. Dominguez-Morales** [1,2,4], **Antonio Rios-Navarro** [1,2,4], **Daniel Gutierrez-Galan** [1,2,4], **Angel Jimenez-Fernandez** [1,2,4], **Saturnino Vicente-Diaz** [1,2,4] and **Alejandro Linares-Barranco** [1,2,3,4]

1   Robotics and Tech. of Computers Lab, Universidad de Sevilla, 41012 Seville, Spain; jpdominguez@atc.us.es (J.P.D.-M.); arios@atc.us.es (A.R.-N.); dgutierrez@atc.us.es (D.G.-G.); ajimenez@atc.us.es (A.J.-F.); satur@us.es (S.V.-D.); alinares@atc.us.es (A.L.-B.)
2   Escuela Técnica Superior de Ingeniería Informática (ETSII), Universidad de Sevilla, 41012 Seville, Spain
3   Escuela Politécnica Superior, Universidad de Sevilla, 41012 Seville, Spain
4   Smart Computer Systems Research and Engineering Lab (SCORE), Research Institute of Computer Engineering (I3US), Universidad de Sevilla, 41012 Seville, Spain
*   Correspondence: lduran@atc.us.es

Citation: Duran-Lopez, L.; Dominguez-Morales, J.P.; Rios-Navarro, A.; Gutierrez-Galan, D.; Jimenez-Fernandez, A.; Vicente-Diaz, S.; Linares-Barranco, A. Performance Evaluation of Deep Learning-Based Prostate Cancer Screening Methods in Histopathological Images: Measuring the Impact of the Model's Complexity on Its Processing Speed. *Sensors* **2021**, *21*, 1122. https://doi.org/10.3390/s21041122

Academic Editors: Yadir Torres Hernández, Manuel Félix Ángel, Ana María Beltrán Custodio and Francisco Garcia Moreno

Received: 13 January 2021
Accepted: 1 February 2021
Published: 5 February 2021

Publisher's Note: MDPI stays neutral with regard to jurisdictional claims in published maps and institutional affiliations.

Abstract: Prostate cancer (PCa) is the second most frequently diagnosed cancer among men worldwide, with almost 1.3 million new cases and 360,000 deaths in 2018. As it has been estimated, its mortality will double by 2040, mostly in countries with limited resources. These numbers suggest that recent trends in deep learning-based computer-aided diagnosis could play an important role, serving as screening methods for PCa detection. These algorithms have already been used with histopathological images in many works, in which authors tend to focus on achieving high accuracy results for classifying between malignant and normal cases. These results are commonly obtained by training very deep and complex convolutional neural networks, which require high computing power and resources not only in this process, but also in the inference step. As the number of cases rises in regions with limited resources, reducing prediction time becomes more important. In this work, we measured the performance of current state-of-the-art models for PCa detection with a novel benchmark and compared the results with PROMETEO, a custom architecture that we proposed. The results of the comprehensive comparison show that using dedicated models for specific applications could be of great importance in the future.

Keywords: deep learning; convolutional neural networks; artificial intelligence; prostate cancer; performance evaluation; benchmark

1. Introduction

Prostate cancer (PCa) is the second most common cancer and the fifth leading cause of cancer death in men (GLOBOCAN [1]). In 2018, almost 1.3 million cases and around 360,000 deaths worldwide were registered due to this malignancy. According to the World Health Organization (WHO), there will be an increase of prostate cancer (PCa) cases worldwide, with 1,017,712 new cases being estimated for 2040. Most of these cases will be registered in Africa, Latin America, the Caribbean and Asia, and appear to be related to an increased life expectancy [2].

To diagnose PCa, digital rectal examination (DRE) is the primary test for the initial clinical assessment of the prostate. Then, prostate-specific antigen (PSA) is used in a screening method for the investigation of an abnormal prostatic nodule found in a digital rectal examination (DRE). Finally, in the case of abnormal DRE and elevated PSA results, trans-rectal ultrasound-guided biopsy is performed to obtain samples of the prostate tissue [3]. Then, these tissue samples are scanned, resulting on gigapixel-resolution images called whole-slide images (WSIs), which are then analyzed and diagnosed by pathologists.

Due to the high increment of new cases, and thanks to the impacts of artificial intelligence (AI) in recent years [4,5], several computer-aided diagnosis (CAD) systems have been developed to speed up the process of PCa diagnosis. A computer-aided diagnosis (CAD) system is an automatic or semi-automatic algorithm whose purpose is to assist doctors in the interpretation of medical images in order to provide a second opinion in the diagnosis. Among the different AI algorithms, deep learning (DL) has become very popular in recent years, and convolutional neural networks (CNNs) particularly [6]. They have been applied in several fields in medical image analysis, such as in disorder classification [7], lesion/tumor classification [8], disease recognition [9] and image construction/enhancement [10], among others.

Deep learning (DL) algorithms have also been applied to other medical image analysis fields such as histopathology, in which whole-slide images (WSIs) are used. Since it is not possible for a convolutional neural network (CNN) to work with a whole WSI as input due to its large size, a common approach is to divide this image into small subimages called patches. This procedure has been widely used in order to develop CAD systems in this field.

Recently, many researchers have investigated the application of CAD systems to the diagnosis of PCa in WSIs. Ström et al. [11] developed a deep learning (DL)-based CAD system to perform a binary classification distinguishing between malignant and normal tissue. The classification was performed using an ensemble of 30 widely used InceptionV3 models [12] pretrained on ImageNet. They achieved areas under the curve (AUC) of 0.997 and 0.986 on the validation and test subsets, respectively. For areas detected as malignant, the authors trained another ensemble of 30 InceptionV3 CNNs in order to discriminate between different PCa Gleason grading system (GGS) scores, achieving a mean pairwise kappa of 0.62 at slide level. Campanella et al. [13] presented a CAD system to detect malignant areas in WSIs. The classification was performed with the well-known ResNet34 model [14] together with a recurrent neural network (RNN) for tumor/normal classification. achieving an area under curve (AUC) of 0.986 at slide level. In a previous study [15], we proposed a CAD system, in which we focused on performing a patch-level classification of histopathological images between normal and malignant tissue. The proposed architecture, called PROMETEO, consisted of four convolution stages (convolution, batch normalization, activation and pooling layers) and three fully connected layers. The network achieved 99.98% accuracy, 99.98% F1 score and 0.999 AUC on a separate test set at patch level after training the network with a 3-fold cross-validation method.

These previous works achieved competitive results in terms of accuracy, precision and other commonly-used evaluation metrics. However, to the best of our knowledge, most state-of-the-art works do not focus on prioritizing the speed of the CAD system as an important factor. Many of them used very complex, well-known networks to train and test, without taking into account the computational cost and the time required to perform the whole process. Since these algorithms are not intended to replace pathologists but to assist them in their task, in some cases it is better to prioritize the speed of the analysis, sacrificing some precision so that the expert has a faster and more dynamic response from the system.

In this paper, a novel benchmark was designed in order to measure the processing and prediction time of a CNN architecture for a PCa screening task. First, the proposed benchmark was run for the PROMETEO architecture on different computing platforms in order to measure the impacts that their hardware components have on the WSI processing time. Then, using the personal computer (PC) configuration that achieved the best performance, the benchmark was run with different state-of-the-art CNN models, comparing them in terms of average prediction time both at patch level and at slide level, and also reporting the slowdown when compared to PROMETEO.

The rest of the paper is structured as follows: Section 2 introduces the materials and methods used in this work, including the dataset (Section 2.1), the CNN models (Section 2.2) and the benchmark proposed (Section 2.3). Then, the results obtained are presented in Section 3, which are divided in two different experiments: first, the performance of a proposed CNN model is evaluated in different platforms, and then it is compared to state-of-the-art, widely-known CNN architectures. Sections 4 and 5 present the discussion and the conclusions of this work, respectively.

2. Materials and Methods

2.1. Dataset

In this work, a dataset with WSIs obtained from three different hospitals was used. These cases consisted of different Hematoxylin and Eosin-stained slides globally diagnosed as either normal or malignant.

From Virgen de Valme Hospital (Sevilla, Spain), 27 normal and 70 malignant cases obtained by means of needle core biopsy were digitized into WSIs. Clínic Barcelona Hospital (Barcelona, Spain) provided 100 normal and 129 malignant WSIs, also obtained by means of needle core biopsy. Finally, from Puerta del mar Hospital (Cádiz, Spain), 65 malignant (26 obtained from needle core biopsy and 39 from incisional biopsy) and 79 (33 obtained from needle core biopsy and 46 from incisional biopsy) WSIs were obtained. Table 1 summarizes the WSIs considered in the dataset.

Table 1. Dataset summary.

Hospital	No. of WSIs		
	Normal	**Malignant**	**Total**
Virgen de Valme Hospital	27	70	97
Clínic Hospital	100	129	229
Puerta del Mar Hospital	79	65	144

2.2. CNN Models

Different CNNs models were considered in this work in order to compare their performance by using the benchmark proposed in Section 2.3. Three different architectures from state-of-the-art DL-based PCa detection works were compared, along with other well-known CNN architectures. The first one is the custom CNN model, called PROMETEO, which we proposed in [15], where we also demonstrated that applying stain-normalization algorithms to the patches in order to reduce color variability could improve the generalization of the model when predicting new unseen images from different hospitals and scanners. The second CNN architecture that was considered in this work is the well-known ResNet34 model [14], which was used by Campanella et al. in [13]. The third one is InceptionV3, introduced in [12], which was used by Ström et al. [11].

Apart from these three CNN models, other widely-known architectures were evaluated with the same benchmark, comparing their performance in terms of execution time with the rest of the networks for the same task. These were VGG16 and VGG19 [16], MobileNet [17], DenseNet121 [18], Xception [19] and ResNet101 [14].

2.3. Benchmark

In this work, a novel benchmark was designed in order to measure and compare the performances of different CNN models and platforms on a PCa screening task. In order to make the benchmark feasible to be shared with other researchers so that it could be run in different computers, a reduced set of WSIs were chosen from the dataset presented in Section 2.1. Since the total amount of WSIs of the dataset represent more than 300 gigabytes (GB) hard drive space, only 40 of them were considered, building up a benchmark of around 50 GB, which is much more shareable. These 40 WSIs were randomly selected, considering all the three different hospitals and scanners, and thus representing well the diversity of the dataset in this benchmark.

The benchmark performs a set of processing steps which are detailed next (see Figure 1). First, as it was introduced in Section 1, since it is not possible for a CNN to use a whole WSI as input due to its large size, these images are divided into small subimages called patches (100×100 pixels at $10\times$ magnification in this case), which are read from each WSI. This process is called "read," and apart from extracting the patches from the input WSI, those corresponding to background are discarded (identified as D in the figure). Then, in the scoring step, a score is given to each patch depending on three factors: the amount of tissue that it contains, the percentage of pixels that are within Hematoxylin and Eosin's hue range and the dispersion of the saturation and brightness channels. This score allows discarding patches corresponding to unwanted areas, such as pen marks, external agents and patches with small amounts of tissue, among others. In Figure 1, discarded patches in this step are highlighted in red, while those that pass the scoring filter are highlighted in green. The third step, called stain normalization, performs a color normalization of the patch based on Reinhard's stain-normalization algorithm [20,21] in order to reduce color variability between samples. In prediction, which is the last step of the process, each of the patches are used as input to a trained CNN, which classifies them as either malignant or normal tissue. Deeper insights into these steps are given in [15]. When the execution of the benchmark finishes, it reports both the hardware and system information of the computer used to run the benchmark, and the results of the execution. These results consist of the mean execution time and standard deviation for each of the four processes (read, scoring, stain normalization and prediction) shown in Figure 1 and presented in [15], both at patch level and at WSI level.

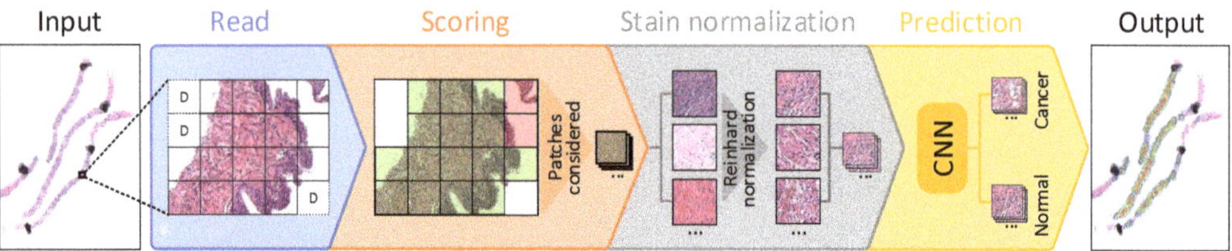

Figure 1. Block diagram detailing each of the steps considered for processing a whole-slide image (WSI) in the proposed benchmark.

3. Results

The CNN-based PROMETEO architecture described in Section 2.2 was proposed and evaluated in terms of accuracy and many other evaluation metrics in [15]. In this work, we evaluated that model in terms of performance and execution time per patch and WSI.

First, the same architecture was tested in different platforms using the benchmark proposed in Section 2.3. These results allow us to measure and quantify the impacts of different components in the whole processing and prediction process, which is useful for designing an edge-computing prostate cancer detection system. Then, the benchmark was used to evaluate the performances of different state-of-the-art CNN architectures on the computing platform that achieved the best results on the first experiment.

Fourteen different PC configurations were used to evaluate the performance of the PROMETEO architecture introduced in Section 2.2. The hardware specifications (central processing unit (CPU) and graphics processing unit (GPU)) of these computers are listed in Table A1 of Appendix A. In Figure 2, the average patch processing time is shown for each of the fourteen configurations, where the mean time for the steps performed when processing a patch (see Section 2.3) is reported. As it can be seen, the step that requires more time is the prediction in most of the cases, but it is highly reduced in configurations consisting of a GPU.

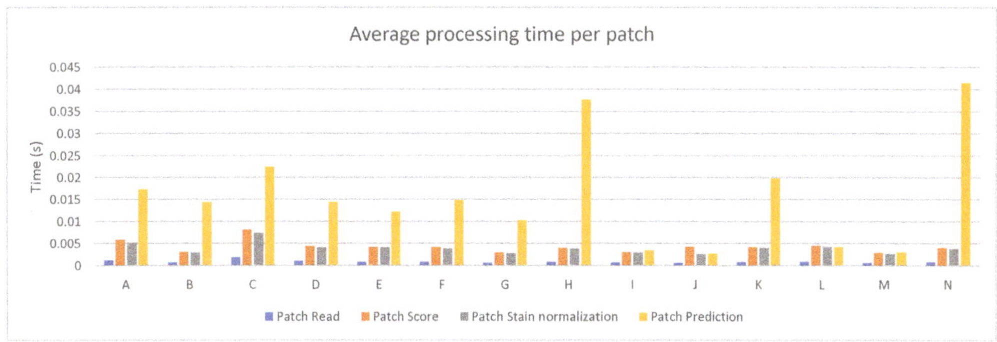

Figure 2. PROMETEO average patch processing time (in seconds) per step for each of the hardware configurations detailed in Table A1.

Figure 3 depicts the average and standard deviation of the execution time needed per WSI when running the benchmark on the fourteen different PC configurations. As in Figure 2, each of the steps considered in the whole process is shown. As it can be seen, reading the whole WSI patch by patch is the step that involves the longest amount of time in most of the devices (mainly in those configurations with no GPU). This might seem contradictory considering Figure 2, but it is important to mention that, in that step, all patches from a WSI are read and analyzed, but not all of them are processed in the following steps. Unwanted areas, such as background regions with no tissue, are discarded before being scored. Then, only those which are not background and pass the scoring step are stain normalized and predicted by the CNN.

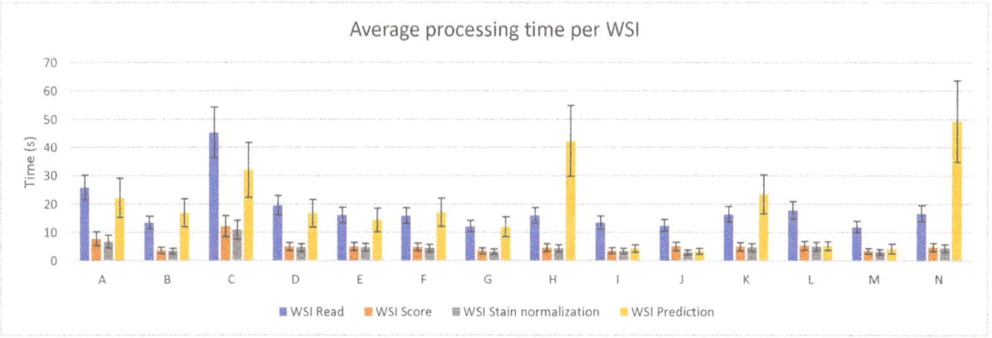

Figure 3. PROMETEO average WSI processing time (in seconds) and standard deviation per step for each of the hardware configurations detailed in Table A1.

3.1. PROMETEO Evaluation

The sum of the average execution time of the four preprocessing steps for each WSI was computed and it can be seen in Figure 4. The best case (device M) takes 22.56 ± 5.67 s on average to perform the whole process per WSI, where the prediction step only represents 4.20 ± 1.73 s.

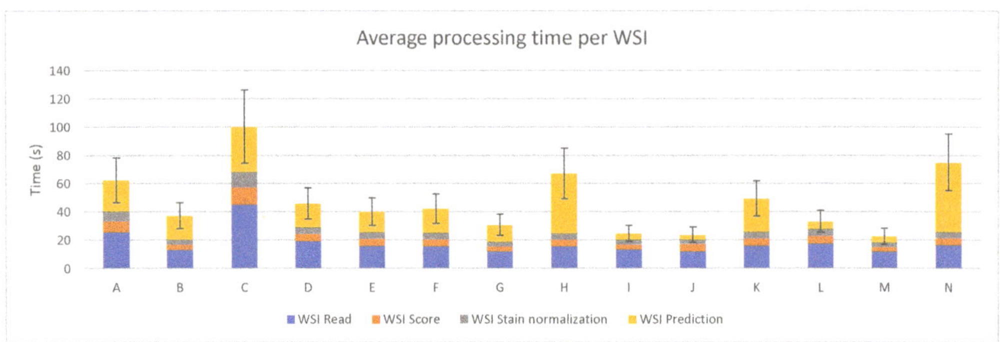

Figure 4. PROMETEO average WSI processing time (in seconds) and standard deviation of the hardware configurations detailed in Table A1.

The execution times obtained and used for generating the plots presented in this subsection are detailed in Table A2 of Appendix A.

3.2. Performance Comparison for Different State-of-the-Art Models

After evaluating the PROMETEO architecture using the benchmark designed for this work with different PCs, the same network was compared to other widely-known architectures. For this purpose, the same computer (device M) was used in order to perform a fair comparison. The same benchmark that was used in the previous evaluation (see Section 3.1) was executed in computer M (see Table A1) for each of the CNN architectures mentioned in Section 2.2. The CNNs considered are PROMETEO [15], ResNet34 and ResNet101 [14], InceptionV3 [12], VGG16 and VGG19 [16], MobileNet [17], DenseNet121 [18] and Xception [19].

The average patch processing time per preprocessing step can be seen in Figure 5 for each of the architectures mentioned. Since the architecture does not have an effect on the first three steps (reading the patch from the WSI, scoring it in order to discard unwanted patches, and normalizing it), the times needed to process them are similar across all the different cases reported in the figure. This does not happen with the prediction time, which directly depends on the complexity of the network.

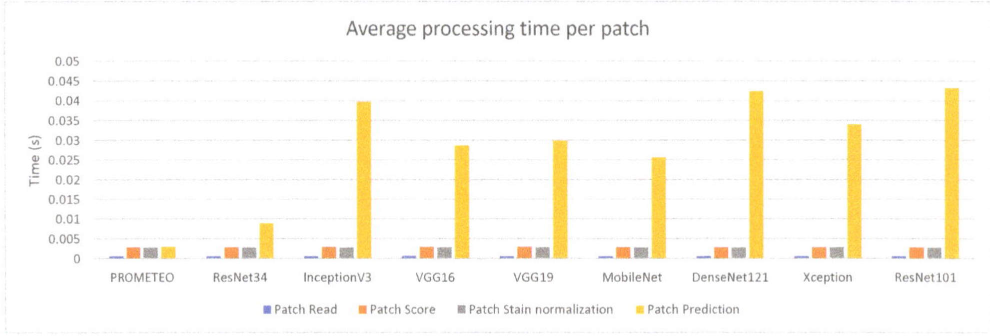

Figure 5. Average patch processing time (in seconds) per step for each of the CNN architectures using computer M (see Table A1).

Figure 6 reports the combined processing time that device M takes to compute a WSI on average, together with its corresponding standard deviation. The same case explained in Section 3.1, where the WSI reading step takes much longer than the patch reading step in relation to the rest of the subprocesses, can also be observed in this figure. It is important to mention that the model proposed by the authors is faster than the rest in terms of prediction time, with a total of 22.56 ± 5.67 s per WSI on average.

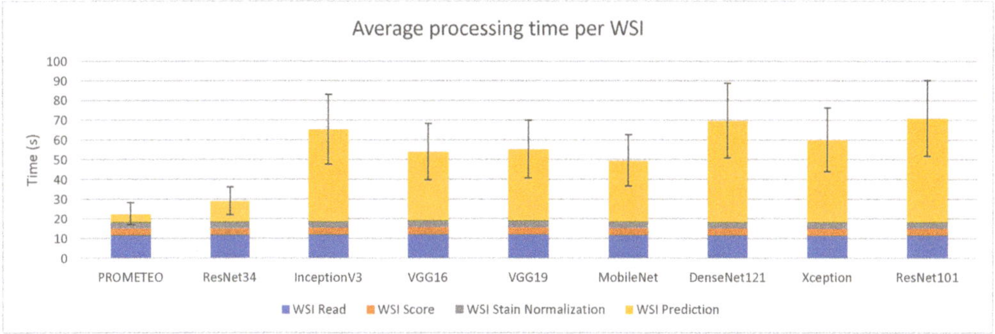

Figure 6. Average WSI processing time (in seconds) and standard deviation for each of the CNN architectures using computer M (see Table A1).

Table 2 presents a summary of the results obtained for each architecture, focusing on the prediction process, which is the only one affected when changing the CNN architecture. Moreover, the number of trainable parameters and the slowdown are also reported. The latter is calculated by dividing the average prediction time per WSI of the corresponding CNN by that obtained with PROMETEO. This way, the improvement in terms of prediction time between PROMETEO and the rest of the architectures considered can be clearly seen. The proposed model predicts $2.55\times$ faster than the CNN used in [13] and $11.68\times$ faster than the one used in [11]. It is also important to mention that, in the latter, the authors did not use only an InceptionV3 model, but an ensemble of 30 of them. In this case, the figures and tables only report the execution times for a single network. When compared to other different widely-known architectures, PROMETEO is between $7.41\times$ and $12.50\times$ faster.

Table 2. Average patch and WSI prediction time, slowdown and number of trainable parameters for each of the CNN architectures considered in this work.

Model	Avg. Prediction Time (patch)	Avg. Prediction Time (WSI)	Slowdown *	Trainable Parameters
PROMETEO	3.054 ± 4.845 ms	4.201 ± 1.739 s	$1\times$	1,107,010
ResNet34	8.982 ± 10.086 ms	10.712 ± 3.134 s	$2.55\times$	21,800,107
InceptionV3	41.301 ± 44.282 ms	49.076 ± 14.353 s	$11.68\times$	23,851,784
VGG16	28.664 ± 9.241 ms	34.921 ± 10.160 s	$8.31\times$	138,357,544
VGG19	29.931 ± 9.305 ms	36.250 ± 10.536 s	$8.63\times$	143,667,240
MobileNet	25.689 ± 10.986 ms	31.110 ± 9.030 s	$7.41\times$	4,253,864
DenseNet121	42.489 ± 16.859 ms	51.483 ± 14.945 s	$12.25\times$	8,062,504
Xception	34.050 ± 11.789 ms	41.764 ± 12.175 s	$9.94\times$	22,910,480
ResNet101	43.287 ± 14.679 ms	52.517 ± 15.266 s	$12.50\times$	44,707,176

* Calculated by using the average prediction time per WSI and taking the PROMETEO architecture as reference. A slowdown of $A\times$ means that model B is A times slower than PROMETEO.

The execution times obtained and used for generating the plots presented in this subsection are detailed in Table A3 of Appendix B.

4. Discussion

In order to design a fast edge-computing platform for PCa detection, an evaluation of a proposed CNN was performed. This allowed us to compare different hardware components and configurations and measure the impacts of them when processing WSIs. Apart from the figures presented in Section 3.1, two specific cases are highlighted in Figure 7. Figure 7a shows the impact that the frequency of the CPU has on the whole process when using the same computer. As it can be seen, the four processing steps clearly benefit when a faster CPU is used. On the other hand, Figure 7b compares two cases where the same configuration is used, except for the GPU, which was removed in one of them. As expected, the GPU highly accelerated the prediction time (by around three times in this case). Therefore, in order to build a low-cost edge-computing platform for PCa diagnosis, this analysis could be useful and should be taken into account in order to prioritize in which component the funds should be invested. As it was explained, all patches from a WSI have to be read, but not all of them have to be predicted, since the majority of them correspond to background and are discarded first. Therefore, the CPU has a higher impact than the GPU in the whole process.

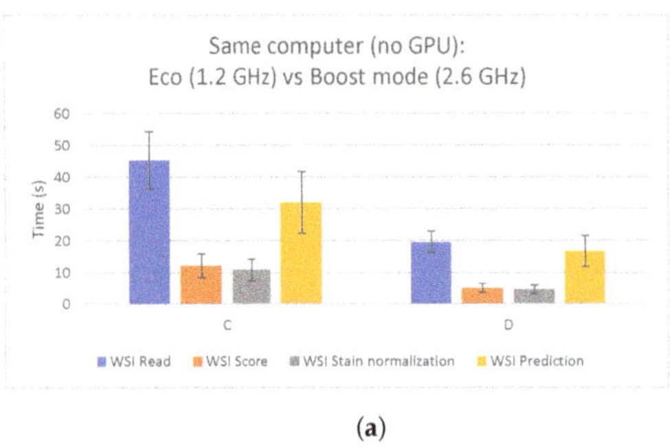

(a)

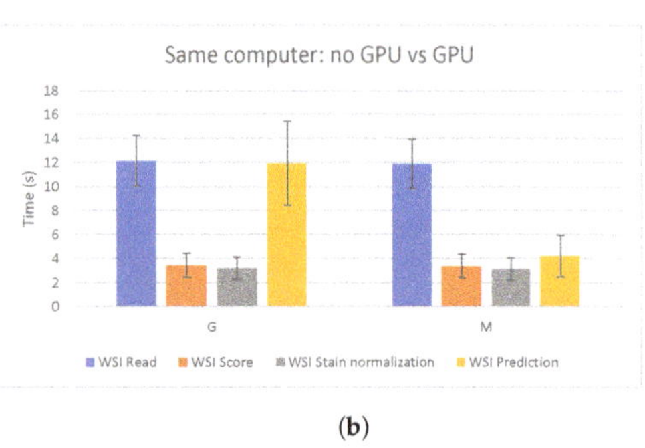

(b)

Figure 7. Impacts of the CPU and the GPU in the different WSI processing steps. (**a**) Same PC, different CPU frequency. Left: 1.2 GHz; right: 2.6 GHz. (**b**) Same PC. Left: without using GPU; right: using GPU.

When comparing PROMETEO to other state-of-the-art CNN models, the former achieved the fastest prediction time, being from 2.55 times up to 12.50 times faster than any of the rest. Although the results in terms of accuracy and other commonly-used metrics in DL algorithms cannot be compared since the authors in [11,13,15] used different datasets, all of them reported state-of-the-art results for PCa detection. In [15], the authors compared PROMETEO to many of the models used in this work in terms of accuracy when using the same dataset for training and testing the CNN, showing that similar results were obtained.

The use of transfer learning in CNNs for medical image analysis has become a commonplace technique, and most of the current research focuses on using this approach for avoiding the problem of having to design, train and validate a custom CNN model for a specific task. This has proved to achieve state-of-the-art results in many different fields and has also accelerated the process of training a custom CNN from scratch [22]. However, when using this technique, very deep CNNs are commonly considered, which, as presented in this work, leads to a higher computational cost when predicting an input image, and therefore, a slower processing time. Some specific tasks could benefit from designing shallower custom CNN models from scratch, such as DL-based PCa screening, providing a faster response to the pathologists in order to help them in this laborious process. With the increases in the number of cases and the mortality produced by PCa, this factor could become even more relevant in the future.

As an alternative, cloud computing has provided powerful computational resources to big data processing and machine learning models [23]. Recent works have focused on accelerating CNN-based medical image processing tasks by using cloud solutions. While it is true that processing images using GPUs and tensor processing units (TPUs) in the cloud is faster than in any local edge-computing device, there is an aspect that is not commonly taken into account when stating this fact: the time required to upload the image to the cloud. This depends on many factors and it is not easy to predict. Moreover, when digitizing histological images, scanners store them in a local hard drive using around 1 GB for each of them. As an example, with an upload speed of 300 Mbps, it would take more than 27 s in ideal conditions just for uploading the WSI to the cloud, which is more than the time it would take to fully process the image on a local platform.

To design a fast, low-cost, edge-computing platform, both the hardware components considered and the CNN model design have to be taken into account. Optimizing these two aspects led to achieving a very short WSI processing time when compared to current DL-based solutions without penalizing the performance of the system in terms of accuracy. In the next future, the authors would like to build a custom bare-bones approach based on the evaluations achieved in this work and test it in some of the hospitals that collaborated with us in this project.

5. Conclusions

In this work, we have presented a comprehensive evaluation of the performance of PROMETEO, a previously-proposed DL-based CNN architecture for PCa detection in histopathological images, which achieved 99.98% accuracy, 99.98% F1 score and 0.999 AUC on a separate test set at patch level.

Our proposed model outperforms other widely-used state-of-the-art CNN architectures such as ResNet34, InceptionV3, VGG16, VGG19, MobileNet, DenseNet121, Xception and ResNet101 in terms of prediction time. PROMETEO takes 22.56 s to predict a WSI on average, including the preprocessing steps needed, using an Intel® Core™ i7-8700K (Intel, Santa Clara, CA, USA) and an NVIDIA® GeForce™ GTX 1080 Ti (NVIDIA, Santa Clara, CA, USA). If we focus only on the prediction time, PROMETEO is between 2.55 and 12.50 times faster than any of the other architectures considered.

The promising results obtained suggest that edge-computing platforms and custom CNN designs could play important roles in the future for AI-based medical image analysis, being able to aid pathologists in their laborious tasks speed-wise.

Author Contributions: Conceptualization, L.D.-L. and J.P.D.-M.; methodology, L.D.-L.; software, L.D.-L. and J.P.D.-M.; validation, L.D.-L., J.P.D.-M., A.R.-N., D.G.-G. and A.J.-F.; formal analysis, L.D.-L. and J.P.D.-M.; investigation, L.D.-L. and J.P.D.-M.; resources, L.D.-L., J.P.D.-M., A.R.-N., D.G.-G., A.J.-F., S.V.-D. and A.L.-B.; data curation, L.D.-L. and J.P.D.-M.; writing—original draft preparation, L.D.-L. and J.P.D.-M.; writing—review and editing, L.D.-L., J.P.D.-M., A.R.-N., D.G.-G., A.J.-F., S.V.-D. and A.L.-B.; visualization, L.D.-L.; supervision, J.P.D.-M., S.V.-D. and A.L.-B.; project administration, S.V.-D. and A.L.-B.; funding acquisition, S.V.-D. and A.L.-B. All authors have read and agreed to the published version of the manuscript.

Funding: This work was partially supported by the Spanish grant (with support from the European Regional Development Fund) MIND-ROB (PID2019-105556GB-C33), the EU H2020 project CHIST-ERA SMALL (PCI2019-111841-2) and by the Andalusian Regional Project PAIDI2020 (with FEDER support) PROMETEO (AT17_5410_USE).

Institutional Review Board Statement: Not applicable.

Informed Consent Statement: Not applicable.

Data Availability Statement: Not applicable.

Acknowledgments: We would like to than Gabriel Jimenez-Moreno and Luis Muñoz-Saavedra for executing the benchmark on their computers and reporting to us its performance. We would also like to thank Antonio Felix Conde-Martin and the Pathological Anatomy Unit of Virgen de Valme Hospital in Seville (Spain) for their support in the PROMETEO project, together with VITRO S.A., along with providing us with annotated WSIs from the same hospital. We would finally like to thank Puerta del Mar Hospital (Cádiz, Spain) and Clínic Barcelona Hospital (Barcelona, Spain) for providing us with diagnosed WSIs from different patients.

Conflicts of Interest: The authors declare no conflict of interest. The funders had no role in the design of the study; in the collection, analyses or interpretation of data; in the writing of the manuscript, or in the decision to publish the results.

Abbreviations

The following abbreviations are used in this manuscript:

AI	Artificial Intelligence
AUC	Area Under Curve
CAD	Computer-Aided Diagnosis
CNN	Convolutional Neural Network
CPU	Central Processing Unit
DL	Deep Learning
DRE	Digital Rectal Examination
GB	Gigabyte
GGS	Gleason Grading System
GPU	Graphic Processing Unit
H&E	Hematoxylin and Eosin
PC	Personal Computer
PCa	Prostate Cancer
PSA	Prostate-Specific Antigen
RNN	Recurrent Neural Network
TPU	Tensor Processing Unit
WHO	World Health Organization
WSI	Whole-Slide Image

Appendix A. PROMETEO Evaluation

Table A1. Hardware specifications (CPU and GPU) of the different computers used in the PROMETEO evaluation.

Device	CPU	GPU
A	Intel® Core™ i7-8850U @ 1.80 GHz 4 cores, 8 threads	-
B	Intel® Core™ i9-7900X @ 3.30 GHz 10 cores, 20 threads	-
C	Intel® Core™ i7-6700HQ @ 1.20 GHz 4 cores, 8 threads	-
D	Intel® Core™ i7-6700HQ @ 2.60 GHz 4 cores, 8 threads	-
E	Intel® Core™ i5-6500 @ 3.20 GHz 4 cores, 4 threads	-
F	Intel® Core™ i7-4770K @ 3.50 GHz 4 cores, 8 threads	-
G	Intel® Core™ i7-8700K @ 3.70 GHz 6 cores, 12 threads	-
H	Intel® Core™ i7-4970 @ 3.60 GHz 4 cores, 8 threads	-
I	Intel® Core™ i9-7900X @ 3.30 GHz 10 cores, 20 threads	NVIDIA® GeForce™ GTX 1080 Ti 11 GB GDDR5X
J	AMD® Ryzen™ 9 3900X @ 4.20 GHz 12 cores, 24 threads	NVIDIA® GeForce™ GTX 1080 Ti 11 GB GDDR5X
K	Intel® Core™ i5-6500 @ 3.20 GHz 4 cores, 4 threads	NVIDIA® GeForce™ GT 730 2 GB GDDR5
L	Intel® Core™ i7-4770K @ 3.50 GHz 4 cores, 8 threads	NVIDIA® GeForce™ GTX 1080 Ti 11 GB GDDR5X
M	Intel® Core™ i7-8700K @ 3.70 GHz 6 cores, 12 threads	NVIDIA® GeForce™ GTX 1080 Ti 11 GB GDDR5X
N	Intel® Core™ i7-4970 @ 3.60 GHz 4 cores, 8 threads	NVIDIA® GeForce™ RTX 2060 6 GB GDDR6

Table A2. PROMETEO evaluation results. The average (Avg) and standard deviation (Std) of the execution times (in seconds) are shown for each of the four processes presented in Section 2.3 (Figure 1), both at patch level and at slide (WSI) level.

Device	Patch								WSI							
	Read		Score		Stain Normalization		Prediction		Read		Score		Stain Normalization		Prediction	
	Avg	Std	Avg	Std	Avg	Std	Avg	Std	Avg	Std	Avg	Std	Avg	Std	Avg	Std
A	0.00120757	0.00311363	0.00585298	0.00502059	0.00512905	0.00418173	0.01730045	0.00850617	25.8035576	4.33829335	7.68691321	2.50054828	6.74114185	2.1823579	22.1192591	6.90573801
B	0.00068973	0.00150109	0.00306787	0.0037015	0.00288733	0.00062438	0.01441587	0.00175146	13.3892811	2.29629633	3.59321811	1.04956324	3.38756321	0.98944353	16.8213335	4.91918301
C	0.00182337	0.00503892	0.00807697	0.00628436	0.00729532	0.00634226	0.02249318	0.0100416	45.2693313	8.99897452	12.1239614	3.77223369	10.8517522	3.36864663	31.9982855	9.74059672
D	0.00103901	0.00228084	0.00437608	0.00080403	0.00404258	0.00098854	0.01435914	0.00211298	19.5256697	3.39723828	4.94245166	1.44889791	4.59097219	1.34596044	16.6959336	4.87815493
E	0.00082695	0.00167953	0.00421429	0.00068222	0.00400594	0.00090984	0.01223286	0.00216942	16.0484505	2.76278471	4.94652829	1.4439207	4.70010431	1.37218657	14.3399619	4.1861481
F	0.00083281	0.00172759	0.00413688	0.00075105	0.0038354	0.00094666	0.01479934	0.00293383	15.8376234	2.74964356	4.7714866	1.3976735	4.42655776	1.29581397	16.9997911	4.98414625
G	0.00062777	0.00133961	0.00292148	0.00038463	0.00270451	0.00067159	0.0102172	0.00150291	12.1361434	2.0832663	3.42516114	1.00071198	3.17680098	0.92726679	11.9159923	3.48494303
H	0.00084291	0.00175864	0.00398322	0.00065638	0.0037566	0.00093803	0.03768491	0.00818776	15.8986914	2.81638821	4.5458395	1.34199731	4.29495389	1.26743003	42.2879135	12.595224
I	0.00069517	0.0015282	0.00299673	0.0003936	0.00285997	0.00066557	0.00345098	0.01023957	13.4663605	2.32710305	3.51854302	1.02719857	3.35297541	0.97965636	4.29145549	1.29811287
J	0.00062976	0.00137508	0.00428461	0.00027188	0.00246943	0.00060632	0.00275394	0.00906872	12.3392324	2.12166155	5.039479	1.47022453	2.91945068	0.85082711	3.35472003	1.00963885
K	0.00084153	0.00173514	0.00417708	0.00059019	0.00397734	0.00091936	0.01973523	0.01713999	16.462836	2.81694585	4.89897847	1.430121	4.67706547	1.36512005	23.4278429	6.84671918
L	0.00091354	0.00194007	0.00449805	0.00163909	0.00421455	0.00178876	0.0420623	0.01316229	17.743488	3.08314193	5.29275112	1.55927849	4.97026951	1.46232287	5.16441015	1.56004368
M	0.0006116	0.00129119	0.00286777	0.00039014	0.00265452	0.00068521	0.0030545	0.03559015	11.8833887	2.0393166	3.36354507	0.98195641	3.11514971	0.90982144	4.20128196	1.7395392
N	0.0008502	0.00498445	0.00405446	0.0007176	0.00375849	0.00097299	0.04150535	0.01602984	16.6332854	2.85486699	4.75020676	1.39576506	4.41236681	1.2946882	49.1934053	14.4530511

Appendix B. Comparison between Different CNN Architectures

Table A3. Execution time comparison between different architectures. The average (Avg) and standard deviation (Std) of the execution times (in seconds) are shown for each of the four processes presented in Section 2.3 (Figure 1), both at patch level and at slide (WSI) level.

| | Patch | | | | | | | | WSI | | | | | | | |
| | Read | | Score | | Stain Normalization | | Prediction | | Read | | Score | | Stain Normalization | | Prediction | |
Architecture	Avg	Std	Avg	Std	Avg	Std	Avg	Std	Avg	Std	Avg	Std	Avg	Std	Avg	Std
PROMETEO	0.000612	0.001291	0.002868	0.00039	0.002655	0.000685	0.003054	0.004845	11.88339	2.039317	3.363545	0.981956	3.11515	0.909821	4.201282	1.739539
ResNet34	0.000612	0.001279	0.002844	0.000405	0.002676	0.000686	0.008982	0.010086	11.92095	2.045551	3.333316	0.973793	3.148634	0.918751	10.71205	3.134596
InceptionV3	0.000621	0.001317	0.002915	0.00041	0.002691	0.001136	0.039772	0.013828	12.10135	2.076446	3.415152	0.997178	3.168544	0.925316	46.72138	13.64997
VGG16	0.000635	0.001341	0.002931	0.000427	0.002785	0.000691	0.028664	0.009241	12.37371	2.140448	3.45475	1.007557	3.280768	0.957531	34.92197	10.16074
VGG19	0.000628	0.001313	0.002931	0.000425	0.00278	0.000682	0.029931	0.009305	12.34846	2.116793	3.44631	1.005834	3.266729	0.953449	36.25006	10.5361
MobileNet	0.000612	0.001278	0.00285	0.00042	0.002688	0.001111	0.025689	0.010986	11.96025	2.044115	3.383497	0.986745	3.208441	0.936362	31.11017	9.030854
DenseNet121	0.000611	0.001284	0.002879	0.000389	0.002687	0.000683	0.042489	0.01686	11.82392	2.035413	3.373127	0.985149	3.148681	0.919557	51.48291	14.94588
Xception	0.0006	0.001261	0.00288	0.000374	0.002758	0.000655	0.03405	0.011789	11.68313	1.987459	3.375536	0.986863	3.235289	0.945187	41.76486	12.17527
ResNet101	0.000607	0.001265	0.002839	0.000398	0.002701	0.00067	0.043287	0.014679	11.84637	2.035472	3.327598	0.971357	3.171104	0.925785	52.51713	15.26661

References

1. Bray, F.; Ferlay, J.; Soerjomataram, I.; Siegel, R.L.; Torre, L.A.; Jemal, A. Global cancer statistics 2018: GLOBOCAN estimates of incidence and mortality worldwide for 36 cancers in 185 countries. *CA Cancer J. Clin.* **2018**, *68*, 394–424. [CrossRef] [PubMed]
2. Rawla, P. Epidemiology of prostate cancer. *World J. Oncol.* **2019**, *10*, 63. [CrossRef] [PubMed]
3. Borley, N.; Feneley, M.R. Prostate cancer: Diagnosis and staging. *Asian J. Androl.* **2009**, *11*, 74. [CrossRef] [PubMed]
4. Hamet, P.; Tremblay, J. Artificial intelligence in medicine. *Metabolism* **2017**, *69*, S36–S40. [CrossRef] [PubMed]
5. Ahuja, A.S. The impact of artificial intelligence in medicine on the future role of the physician. *PeerJ* **2019**, *7*, e7702. [CrossRef] [PubMed]
6. Shen, D.; Wu, G.; Suk, H.I. Deep learning in medical image analysis. *Annu. Rev. Biomed. Eng.* **2017**, *19*, 221–248. [CrossRef] [PubMed]
7. Shi, J.; Zheng, X.; Li, Y.; Zhang, Q.; Ying, S. Multimodal neuroimaging feature learning with multimodal stacked deep polynomial networks for diagnosis of Alzheimer's disease. *IEEE J. Biomed. Health Inform.* **2017**, *22*, 173–183. [CrossRef] [PubMed]
8. Dou, Q.; Chen, H.; Yu, L.; Zhao, L.; Qin, J.; Wang, D.; Mok, V.C.; Shi, L.; Heng, P.A. Automatic detection of cerebral microbleeds from MR images via 3D convolutional neural networks. *IEEE Trans. Med. Imaging* **2016**, *35*, 1182–1195. [CrossRef] [PubMed]
9. Duran-Lopez, L.; Dominguez-Morales, J.P.; Corral-Jaime, J.; Vicente-Diaz, S.; Linares-Barranco, A. COVID-XNet: A custom deep learning system to diagnose and locate COVID-19 in chest X-ray images. *Appl. Sci.* **2020**, *10*, 5683. [CrossRef]
10. Bahrami, K.; Shi, F.; Rekik, I.; Shen, D. Convolutional neural network for reconstruction of 7T-like images from 3T MRI using appearance and anatomical features. In *Deep Learning and Data Labeling for Medical Applications*; Springer: Berlin/Heidelberg, Germany, 2016; pp. 39–47.
11. Ström, P.; Kartasalo, K.; Olsson, H.; Solorzano, L.; Delahunt, B.; Berney, D.M.; Bostwick, D.G.; Evans, A.J.; Grignon, D.J.; Humphrey, P.A.; et al. Artificial intelligence for diagnosis and grading of prostate cancer in biopsies: A population-based, diagnostic study. *Lancet Oncol.* **2020**, *21*, 222–232. [CrossRef]
12. Szegedy, C.; Vanhoucke, V.; Ioffe, S.; Shlens, J.; Wojna, Z. Rethinking the inception architecture for computer vision. In Proceedings of the 2016 IEEE Conference on Computer Vision and Pattern Recognition, Las Vegas, NV, USA, 27–30 June 2016; pp. 2818–2826.
13. Campanella, G.; Hanna, M.G.; Geneslaw, L.; Miraflor, A.; Silva, V.W.K.; Busam, K.J.; Brogi, E.; Reuter, V.E.; Klimstra, D.S.; Fuchs, T.J. Clinical-grade computational pathology using weakly supervised deep learning on whole slide images. *Nat. Med.* **2019**, *25*, 1301–1309. [CrossRef] [PubMed]
14. He, K.; Zhang, X.; Ren, S.; Sun, J. Deep residual learning for image recognition. In Proceedings of the 2016 IEEE Conference on Computer Vision and Pattern Recognition, Las Vegas, NV, USA, 27–30 June 2016; pp. 770–778.
15. Duran-Lopez, L.; Dominguez-Morales, J.P.; Conde-Martin, A.F.; Vicente-Diaz, S.; Linares-Barranco, A. PROMETEO: A CNN-Based Computer-Aided Diagnosis System for WSI Prostate Cancer Detection. *IEEE Access* **2020**, *8*, 128613–128628. [CrossRef]
16. Simonyan, K.; Zisserman, A. Very deep convolutional networks for large-scale image recognition. *arXiv* **2014**, arXiv:1409.1556.
17. Howard, A.G.; Zhu, M.; Chen, B.; Kalenichenko, D.; Wang, W.; Weyand, T.; Andreetto, M.; Adam, H. Mobilenets: Efficient convolutional neural networks for mobile vision applications. *arXiv* **2017**, arXiv:1704.04861.
18. Huang, G.; Liu, Z.; Van Der Maaten, L.; Weinberger, K.Q. Densely connected convolutional networks. In Proceedings of the 2017 IEEE Conference on Computer Vision and Pattern Recognition, Honolulu, HI, USA, 21–26 July 2017; pp. 4700–4708.
19. Chollet, F. Xception: Deep learning with depthwise separable convolutions. In Proceedings of the 2017 IEEE Conference on Computer Vision and Pattern Recognition, Honolulu, HI, USA, 21–26 July 2017; pp. 1251–1258.
20. Reinhard, E.; Adhikhmin, M.; Gooch, B.; Shirley, P. Color transfer between images. *IEEE Comput. Graph. Appl.* **2001**, *21*, 34–41. [CrossRef]
21. Magee, D.; Treanor, D.; Crellin, D.; Shires, M.; Smith, K.; Mohee, K.; Quirke, P. Colour normalisation in digital histopathology images. In Proceedings of the Optical Tissue Image analysis in Microscopy, Histopathology and Endoscopy (MICCAI Workshop), London, UK, 24 September 2009; pp. 100–111.
22. Zhuang, F.; Qi, Z.; Duan, K.; Xi, D.; Zhu, Y.; Zhu, H.; Xiong, H.; He, Q. A comprehensive survey on transfer learning. *arXiv* **2019**, arXiv:1911.02685.
23. Zhang, Q.; Bai, C.; Chen, Z.; Li, P.; Yu, H.; Wang, S.; Gao, H. Deep learning models for diagnosing spleen and stomach diseases in smart Chinese medicine with cloud computing. *Concurr. Comput. Pract. Exp.* **2019**, e5252. [CrossRef]

Article

OpenADR and Agreement Audit Architecture for a Complete Cycle of a Flexibility Solution

Antonio Parejo *, Sebastián García , Enrique Personal, Juan Ignacio Guerrero, Antonio García and Carlos Leon

Department of Electronic Technology, Escuela Politécnica Superior, Universidad de Sevilla, 41011 Sevilla, Spain; sgarcia15@us.es (S.G.); epersonal@us.es (E.P.); juaguealo@us.es (J.I.G.); antgar@us.es (A.G.); cleon@us.es (C.L.)
* Correspondence: aparejo@us.es

Abstract: Nowadays, the presence of renewable generation systems and mobile loads (i.e., electric vehicle) spread throughout the distribution network is increasing. The problem is that this type of system introduces an added difficulty since they present a strong dependence on the meteorology and the mobility needs of the users. This problem forces the distribution system operators to seek tools that make it possible to balance the relationship between consumption and generation. In this sense, automated demand response systems are an appropriate solution that allow the operator to request specific reductions in customers' consumption, offering a discount to the customer and avoiding network congestion. This paper analyzes the implementation and architecture of a demand response solution based on OpenADR standard and its possible integration with a building management system through a use case. As will be analyzed, a key part of the architecture is the measurement system based on smart meters acting as sensors. This is the base of the auditing system which makes it possible to verify compliance with the consumption reduction agreements. Additionally, this study is completed with a parallel auditing system which makes it possible to verify compliance with the consumption reduction agreements. All of the proposed demand response cycle is implemented as a proof of concept in a classroom in the *Escuela Politécnica Superior* at the University of Seville, which makes it possible to identify the advantages of this architecture in the ambit of connection between distribution network and buildings.

Keywords: demand side management; flexibility; demand response; advanced metering infrastructure; smart grid

Citation: Parejo, A.; García, S.; Personal, E.; Guerrero, J.I.; García, A.; Leon, C. OpenADR and Agreement Audit Architecture for a Complete Cycle of a Flexibility Solution. *Sensors* **2021**, *21*, 1204. https://doi.org/10.3390/s21041204

Academic Editors: Yadir Torres Hernández, Manuel Félix Ángel and Ana María Beltrán Custodio
Received: 13 January 2021
Accepted: 5 February 2021
Published: 9 February 2021

Publisher's Note: MDPI stays neutral with regard to jurisdictional claims in published maps and institutional affiliations.

1. Introduction

The latest advances in the power system are mainly marked by the progressive inclusion of renewable generation, which has significantly increased in the daily energy mix of many countries [1]. Renewable generation is also finding places inside customer facilities, allowing them to produce energy. These customers are known as prosumers (that are producers or consumers according to the moment), and these clients are adopting a more active role within the power system. Moreover, the use of electric vehicles is recently getting more attention, as their autonomy has improved a lot from the first protypes that were seen in the previous decade.

The problem with these advances is that this type of system introduces an added difficulty in the power system management since they have a strong dependence on the meteorology [2,3] and the mobility needs of the users [4,5]. This uncertainty and variability, together with its increasing deployment of this kind of system, is creating problems due to the overage of the distribution line power capacity. Traditionally, the solutions to this problem consists of a reinforcement of the infrastructure (known as wire solution). However, the required investment would be too large to be assumed by the companies.

In parallel, this problem forces the system operators to seek tools that make it possible to balance the relationship between generation and consumption, i.e., maintaining the unit commitment (the generation must be equal to the consumption). In this sense, the

15

Distribution System Operators (DSOs) are responsible for maintaining the network stability, orchestrating these assets locally as Distributed Energy Resources (DER). This task is critical in sectors of the distribution network that are weakly connected, where congestion management tasks are delicate.

In this context, Demand Side Management (DSM) and Demand Response (DR) techniques can be a solution to the problem [6,7], allowing customers to provide services of power adjustment when needed. Thus, they can help solve congestion problems as a non-wire solution, reducing their peak consumption, and avoiding oversizing in the distribution grid. Furthermore, they bring the possibility of investment deferral, as they can reduce the overcharge in the most congested power sectors.

In this sense, one of the major benefits is that customer-side resources can transact with the electric grid [8]. In any case, the implication of the prosumer for their participation in DSM and DR programs requires that their facilities become sensorized and connected. Fortunately, after some years, the sensorization of buildings and homes has become relatively easy. Lots of technology and protocols like Wi-Fi smart devices, Z-Wave, ZigBee, Modbus, and advances in computation [9] have facilitated the automation and monitoring at a wide range of levels [10]. These technologies stimulate the extended use of distributed generation, like home solar panels and batteries, which increase the control complexity [11,12]. The users can improve not only the greening of their actions, but also the other high level of the power system, that is, the facilities beyond their homes: the distribution network.

In the European Union (EU), which entity would make this "integration" with prosumers is still being discussed, because it could be done by the Transmission System Operator or TSO (a total centralization), by the DSO with the help of aggregators (in order to respect the mandatory unbundling of the power system), or it could be done in a coordinated way [13,14]. The discussion of these paradigms is not the objective of this paper, but the second one (DSO integration). It will be used for the description of the use case, due to its "lower level" approach, more focused on the possibilities of improving the network at the distribution level.

Following this second class of integration, a prosumer can provide flexibility services to the DSO, helping them make better use of the resources, integrating (in some manner) their resources at DSO network level.

This philosophy is being more and more extended, as can be observed in the new changes in the EU power system regulations [15], where the use of distributed resources is being included under the concept "flexibility". This concept means the ability to respond in a flexible manner to changes and needs of the power system, using a set of available resources (generation or power reduction). In practice, this is implemented with what is called "flexibility programs", i.e., agreements between the operator and the prosumers (or even third parties as the aggregators, when it is desirable or required by the regulations [15]) where those prosumers receive incentives (which are processed separately from their contracted electricity tariff) when they successfully respond to the received request (reducing peaks, generation or consumption during a certain period). The mentioned regulation establishes that all customer groups (industrial, commercial, and households) should have access to the electricity markets to trade their flexibility and self-generated electricity, making it possible for aggregators to play a role as intermediaries between these customers and the market. How these flexibility programs and the incentives are handled is still under discussion. As an example, Ref. [16] proposes that the registry and payment of such services could be done using a local market place at the DSO level using technologies such as Blockchain to manage the incentive payments.

In any case, it is necessary to be prepared for these emergent paradigms, where the users and the rest of the power system need to be perfectly balanced and linked, this being an advantage for both parts not only in Europe, but also in the rest of the world [17].

Likewise, there must be a liquidation and review stage later where the compliances with the agreements are verified. Fortunately, within the development that Smart Grids have experienced in recent years, it is worth highlighting the use of Advanced Metering Infrastructure (AMI) [18,19] and in particular the deployment by the DSOs of customers' smart meters (SM), which following the European directives are fully deployed in most of the European countries [20]. The AMI is composed by two main functions, which are the Automatic Metering Reading (AMR) and Automatic Metering Management (AMM). Both functions are usually integrated in the Smart Metering architecture, which is used for the monitoring of customer consumption and other elements of the power system. In Europe, the deployment of smart metering for gas and electricity started with the publication of the European Directive 2009/72/EC, by pointing out the advantages of such systems, and recommending their deployment [21].

Finally, communication and control of the "internal" elements of a building, as was mentioned before, is solved. There are many technologies and possibilities to integrate sensors, meters, and central control for facilities (even in the case of industrial wide areas, university campuses, etc.). Notwithstanding, communication between the DSO and the user is currently under development and research, mainly since this user-DSO integration is not still fully extended. This final connection between buildings and the distribution network is sometimes referred as "the last mile" of connection, as can be seen in Ref. [22].

Specifically, the work described in this paper is focused on this "last mile" environment. The objective is to define a complete system architecture for prosumers, providing a fluid communication channel with the aggregator/operator and capability to participate in flexibility services, including the audit mechanism. This proposal has been deployed on a testbed, which is called "DERBis lab" here, which is made up of a DR architecture, a Building Management System (BMS) to manage and operate the different loads, and an AMI solution to validate the compliance with the agreements between the different parties. The communication channel with the DSO/aggregator uses the OpenADR standard. Additionally, the smart metering infrastructure is based on the PRIME standard.

In this paper, the whole process of DR planning is analyzed from the point of view of the customer, including the use of their own sensing and control systems to check the flexibility services availability. The architecture and process steps are described, analyzed, and finally tested using real data and information from the BMS and the smart metering devices. It is remarkable that the proposed architecture has been designed considering the same type of smart meters that are already deployed in most countries, which perform hourly consumption measurements. This restriction has been applied in order to avoid the need of deploying new metering systems for the present application, which would lead to higher costs.

Before describing the proposed solution, a brief explanation about the state of the art in the field of flexibility and DSM implementation and AMI solutions is carried out in Section 2. Section 3 contains the description of architecture and its parts. Then, Section 4 details the implementation of the proposed architecture for the DERBis lab. Section 5 shows three study cases of flexibility services. Finally, the conclusions are shown in Section 6.

2. State of the Art

The previous section outlines how the inclusion of flexibility programs could potentially bring multiple benefits for the management of the power system. The next step is to discuss which of the possible architectures is the most convenient one for the inclusion of prosumers, and how their possible actions should be categorized for their scheduling and dispatching.

The tools for power management in the level of unit commitment are classified in different ways depending on the zone. In Ref. [23], J. Ortega, D. Watts, and H. Ren review how the operating reserves are categorized in USA, Chilean, and German markets. They conclude that five categories are used in the case of America, while three of them are established in Europe.

From the list of European categories, the one that can be considered closer to the flexibility from the customer side is the Tertiary Reserve. This reserve, which operates a maximum number of times in 15 min, is manually activated. Its use is a response to a disbalance in the control area and/or bigger congestion problems [23]. This description corresponds to their use in the context of system operations, but the same idea can be applied in congestions at the DSO level, even considering longer times (hours) for dispatched DSM events. In the context of American power markets, these services can fit into the category of Supplemental Reserves.

There are, in fact, some examples of North American distribution companies that have already included DSM collaboration for their customers. This is the case of the companies San Diego Gas & Electric (SD-G&E) [24] or Pacific Gas & Electric (PG&E) [25], whose DR programs are focused on what is called a "Capacity Bidding Program", where a customer commits, under contract, to reduce their power a certain quantity when required. The maximum number of requests is also established in the agreement. Other approaches propose the organization of flexibility and ancillary services under a specific market where TSOs and DSOs could buy those services that they need from prosumers and aggregators [26].

According to [6], DSM techniques can be categorized according to their objective. The categories are direct load reduction, load scheduling, pricing schemes, based on optimization types, and home energy management.

Fortunately, in recent years some technology and protocols have emerged to solve these connections. Not only for DR, of course, but for managing the global Smart Grid environment, where DR is usually included. This has been a natural response to the rising environmental concern around the world, particularly in the U.S., Europe, and Asia [27].

A review of DR is presented in Ref. [28]. Specifically, this last paper studies the advantages of Automatic Demand Response (ADR) for DSM. ADR is the implementation of DR using Information and Communication Technologies (ICTs) without needing manual interventions. Samad et al. states that ADR is critical technology in DR. Additionally, in order to implement ADR, standard communication protocols such as OpenADR [29,30] are essential to allow interoperability as the authors state.

OpenADR is a protocol to allow remote ADR procedures. Despite this protocol being very "general", it offers many possibilities for running DSM and DR programs [31] in a variety of possible architectures. Some examples of these options are exposed in Refs. [32,33], putting special emphasis on the different architectures and connections that could be used. Another good point of this protocol is that it includes functions for dynamic pricing, which are used for example in Ref. [34] as part of more complex Smart Grid managing system.

This is not the only service that can be included in the DR category. Another example of an ancillary service is peak saving [35]. The use of DR is not limited to the domestic/residential level, and examples at the industrial level can be found. An example is [36], where a service of DR is applied to refrigerated warehouses.

The participation of a customer in DR implies two steps: the event dispatching, and the audit of the actuation during such an event. This means that the distribution company must check if the customer accomplished the requirements of consumption during the established time.

Thanks to the network modernization drive by international directives (i.e., European directive 2009/72/EC [21] in which the energy smart meter deployment program was established for the first time), this audit can be done using the same systems that are usually used to establish the power bill, which are the customer' smart meters. As [28] states, although ADR can be conducted without them, smart meters can enhance the implementation of a DR program, as they enable more elaborate means of compensating for DR participation by establishing baselines and comparing demand-side performance against those baselines. An example of use of OpenADR and smart metering can be seen in Ref. [37].

There are many types of solutions for the deployment of smart metering infrastructures, one of the most important characteristics being the method of communication. While non-wired solutions have many advantages, the Power Line Communication (PLC) solutions are most extended for electricity metering, as they use the power lines as communication channels. A general structure of a PLC smart metering infrastructure in the power system can be seen in Figure 1.

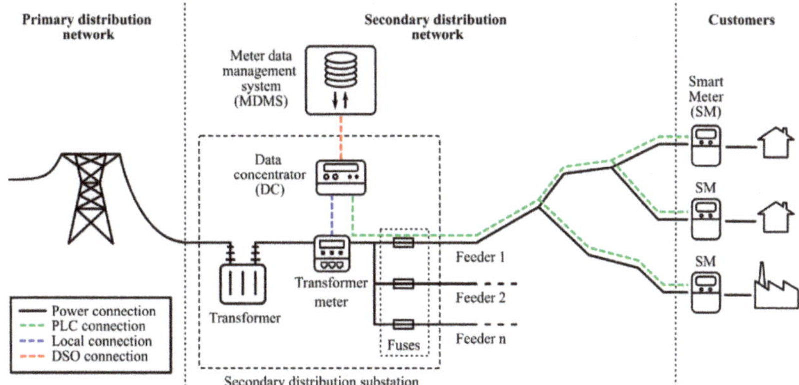

Figure 1. PLC (Power Line Communication)-based smart metering infrastructure [38].

As mentioned in the previous section, the deployment of smart metering is relatively advanced in Europe thanks to the impulse provided by legislative powers. This is very convenient for the future of ADR, as it can serve as auditory infrastructure. In fact, the smart metering deployment is very advanced not only for the electric sector, but also for water and gas utilities [39].

However, currently, the smart meter architectures are traditionally operating with aggregated measurements of one hour. Therefore, if DSOs want to use this system for auditing compliance with flexibility agreements, the offers and actions must match these time intervals. This limitation would be reduced if this granularity became a quarter of an hour, a possible period for this type of network, and more in line with flexibility services.

The conditions of the audit should be based on the expected or estimated consumption of the customer during the period of the event, which will be compared to the actual one to check if the power was effectively reduced or not. In the case of buildings, this estimation can be done using multiple techniques, based on statistical, or even artificial, intelligent methods [40,41].

The review of the state of the art regarding the functionalities and applications of DSM, DR, and Smart Metering has shown that their use can bring multiple benefits to the power system. Nevertheless, their deployment brings about various technical complications, such as the integration and coordination of the different systems and actors. The next section proposes an architecture pointing out the requirements of each of its parts for DSM/DR implementation.

3. Proposed Architecture

In this section, the proposed architecture is described. This architecture includes the complete cycle of DSM, including dispatching and auditing. The general structure is the one shown in Figure 2.

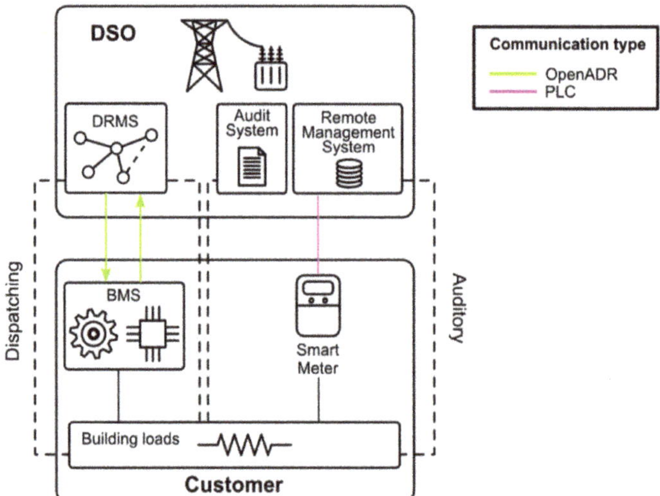

Figure 2. Structure and communication of DSO and customer systems for DSM/DR and auditing.

As can be seen in the Figure 2, the DSM cycle is split into two parts: dispatching and auditing. The dispatching is done through the OpenADR protocol while the auditing is done through the AMI based on PRIME standard. The system in charge of managing the customer flexibility services is the BMS.

The proposed architecture aims to take advantage of the already existing metering system for the audit process, not requiring any remarkable economic effort. The automation systems would require an update to fulfill the requirements, but there is currently a rising tendency to install BMSs in big and medium size customers and domestic automatization systems, so it is expected that these systems will become relatively common and its integration with the proposed approach will be easy. Finally, it is true that it is the DSMs architecture which requires bigger changes for its integration in the DSO control systems. Notwithstanding, it is a clear interest by the DSO. Proof of that is that various DSOs are already exploring the use of this kind of techniques, citing as examples the initiatives of San Diego Gas & Electric Company [24] and Pacific Gas & Electric Company [25].

Therefore, three main parts can be distinguished from the customer side, the DSO-user communications (DRMS-BMS) through the OpenADR protocol, the BMS in the customer side, and the smart metering system to audit the DSM operations. This section attempts to analyze, in detail, the required characteristics of the three parts. Moreover, the Section 3.4 explains how the whole process of DSM dispatching and auditing works.

3.1. OpenADR Standard

As previously stated, a crucial aspect of flexibility (and, more specifically, DSM) and, therefore, congestion management is the communication between service providers (or customers) and the DSO. In this sense, the development of OpenADR standard is becoming an important tool to communicate grid flexibility actions.

OpenADR came up in 2002 at the Demand Response Research Center, which is part of the Lawrence Berkeley National Laboratory (LBNL) in California [30]. In 2009, OpenADR specification (v1.0) specification was donated to different standardization organizations including the Organization for the Advancement of Structured Information Standards (OASIS) turning into an open-based specification. In 2010, the OpenADR Alliance was created by industry stakeholders to support the development, testing, deployment, and standardization of this protocol. The OpenADR 2.0 standard, which was released in its first version (2.0a) in 2011, was created using a profile from the Energy Interoperation version 1.0 standard from OASIS, which is an information model to enable collaborative and transactive use of energy. Currently, it is in version 2.0b since 2015. The main difference of this version with the previous one (2.0a) is that the current version includes four services inherited from the Energy Interoperation (see below for the description of the services) while the previous one only includes one service (the EiEvent) and it is limited to just one signal.

OpenADR is an open specification designed to facilitate and automate DR communication [42]. It is a communication data model, but it also defines transport and security mechanisms. The protocol is based on eXtensible Mark-up Language (XML). As transport mechanisms, OpenADR uses Extensible Messaging and Presence Protocol (XMPP) or Hypertext Transfer Protocol (HTTP) using REpresentational State Transfer (REST).

The protocol defines two kinds of actors or nodes inherited from the OASIS Energy Interoperation, the Virtual Top Node (VTN), and the Virtual End Node (VEN). Communication is hierarchical, always between a VTN and one or more VENs. Thus, no communication between the same kind of nodes could take place on an OpenADR schema. Nevertheless, a node can be at the same time a VEN and a VTN of another hierarchically lower OpenADR stage. In an OpenADR schema, the VTN is responsible for communicating the DSM conditions and other relevant data to other entities (VENs). In this sense, the VEN is responsible for controlling and managing the electrical resources at the final consumer level or aggregating different resources and/or clients. Thus, the VTN will normally be at the DSO side and the VEN will be at the customer side.

The VENs can be either physical devices or a cloud-based platform. Physical devices can have direct control of the customer assets based on the flexibility commands coming from the DSO. Cloud-based platforms can be used by Flexibility Service Providers (FSP) or Aggregators to gather different assets from different locations and owners in one single endpoint making the interaction easier for the DSO. As previously mentioned, the current regulation does not allow the DSO to own DER assets or to communicate flexibility signals to final customers in Europe. Nevertheless, flexibility signals to a customer using a third party like an aggregator are allowed.

The current version of OpenADR defines four main communication services (see Figure 3) inherited from OASIS Energy Interoperation:

- EiRegisterParty: Used to identify and enable communications between a VTN and a VEN.
- EiEvent: Used by the VTN to request DR operations to one or more VENs. In the OpenADR terminology, the process of requesting a DSM is called event. Events can be accepted or rejected by the VEN.
- EiReport: Used to share data. Both the VTN and the VEN are able to report information to each other, e.g., the VEN could report metering information to the VTN.
- EiOpt: Used to communicate availability or unavailability, e.g., a VEN can communicate the period of time in which it will not accept any events coming from the VTN.

In HTTP, VENs can communicate either in pull mode or push mode. In push mode the VEN acts as a client and the VTN acts as a server, the VEN periodically polls the VTN to see if it wants to communicate anything. In push mode, both the VTN and the VEN can act either as server or as client, thus, communication can be asynchronous, and no polling is needed. In a XMPP schema, due to its nature, communication is asynchronous, and no polling is needed.

Additionally, OpenADR defines a fifth service: the oadrPoll, which is only used by the VEN in the HTTP pull mode to periodically poll the VTN.

VTN Development

There exist multiple implementations of VTN software [43], but they do not have an external interface to enable interoperability with other Smart Grid systems like a Demand Response Management System (DRMS) or a Distributed Energy Resources Management System (DERMS). Therefore, a VTN has been developed by the authors to fulfil these requirements. The architecture is shown in Figure 4. The VTN architecture is divided in four main modules. From bottom to top:

- Application Layer and Schema Validation: It handles the OpenADR HTTP services. The VTN is pull and push mode compatible. In addition, the compliance of the messages with the OpenADR protocol is also checked.
- OpenADR Core: It implements the OpenADR 2.0b services.
- Memory Manager: The memory manager is split into a Cache Memory manager which stores frequently requested data in RAM and a Persistence Memory Manager, which manages the connection with a relational database. This split allows the reduction of the database access, increasing the response times of the system.
- API: It provides an interface to other systems to use the VTN services. It is based on REST and JavaScript Object Notation (JSON).

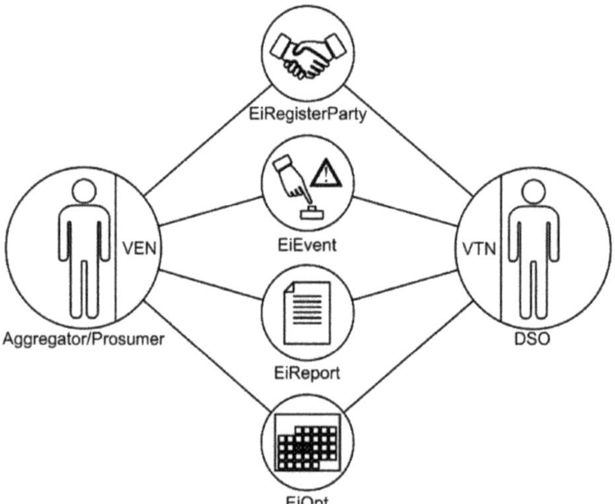

Figure 3. OpenADR communication services.

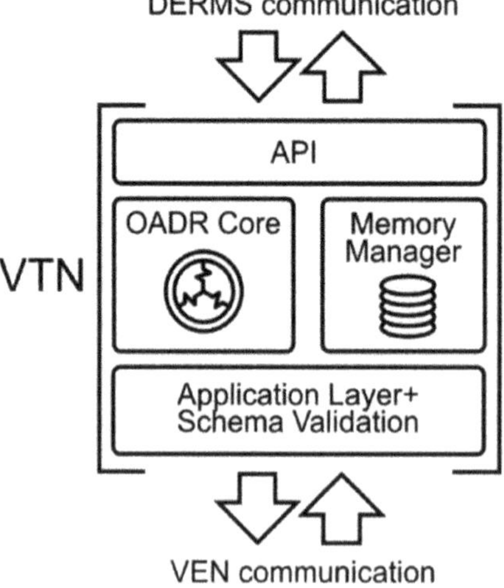

Figure 4. Architecture of the developed VTN with the different layers.

The data models used in the VTN infrastructure, besides OpenADR, also obey the Common Information Model (CIM) from the International Electrotechnical Commission (IEC). OpenADR is harmonized with this standard, which means that it is interoperable with Distribution Management (IEC 61968) and Energy Management (IEC 61970). This, together with the API, makes the VTN interoperable with other Smart Grid systems.

Thanks to the API, the VTN acts as a gateway for other Smart Grid systems. In this sense, the VTN does not take any real actions by itself. These actions are done by means of the API interface. Additionally, the layer-based architecture makes the system easily upgradable to other OpenADR versions, even the EI.

3.2. Building Management System

A BMS is a system that centralizes the information and control of a building or area. Their possibilities go from the mere monitorization of the consumption or comfort conditions to the advanced control algorithms and automation (e.g., [44] that propose a local Energy Consumption Scheduling for HVAC control that which compensates for possible errors between weather forecast and real local conditions). In this sense, most of the BMS control approaches are focusing on getting improvements in energy efficiency [45], obtaining the optimal use of renewable resources [46], or are based on static pricing schemes [47].

Beyond these local or static approaches, in the context of their inclusion as part of a DSM cycle, they should also accomplish certain tasks to permit the participation of the customer resources in flexibility services.

Firstly, it must be aware of "when" and "how much" the resources can provide power reduction (or extra generation) if required.

Secondly, it should include any type of consumption prediction, at least in a horizon of one day, to calculate which is the effective "margin of flexibility" that the building can offer.

Finally, it must count with the communication systems that are used for the flexibility market and the DSO. This point is related to what was previously exposed regarding OpenADR. Thus, the BMS needs a communication endpoint with the DSO as means of an OpenADR VEN.

3.3. Smart Metering and Audit Process

A smart metering system is made up of smart meters (usually one for each client) and a data concentrator, which receives the measurements and sends the information to the DSO central systems. The communication can be performed in multiple ways, but one of the most popular ones is the PLC [48]. This communication is done through the power lines, so it does not require extra elements for the deployment of smart metering.

About the protocol for the PLC, some of the most extended ones in Europe are:

- PRIME: Powerline Intelligent Metering Evolution (PRIME). Defined by the "Prime Alliance".
- Meters&More: Defined by the "Meters and More Association".
- G3PLC: Supported by the G3PLC Alliance.

PRIME will be the chosen protocol for the present proposal, it being one of the most extensive in Spain. Notwithstanding, any of them could be valid from the point of view of required functionality. Thus, as was commented before, the consumption measurement period for billing is typically hourly in Europe. This measurement interval could initially be considered adequate for this application, although it would be advisable to increase the resolution to a quarter of an hour, in order to better adapt to the requirements of flexibility services.

3.4. Complete Flexibility Cycle Process

Figure 5 presents a resume of the whole process of flexibility estimation, participation in the market, event receiving and execution, and the audit.

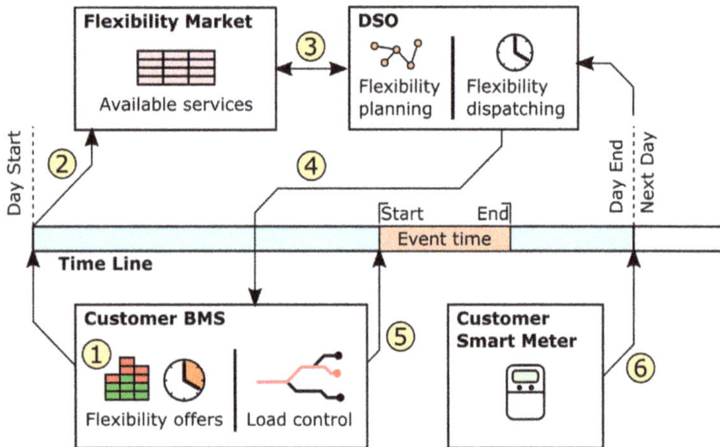

Figure 5. Complete flexibility cycle. (1) Flexibility offers calculation [BMS]; (2) Send offers to market [BMS-Market]; (3) Services selection [DSO-Market]; (4) Service request [DSO-BMS]; (5) Load control for requested service [BMS-Loads]; (6) Auditory of the requested service [DSO].

At the start of the day (or before, depending on the flexibility program), the BMS performs an estimation of their own consumption (and generation, if any), and also considers which devices are curtailable (Stage 1 of the Figure 5). The flexibility offer is derived from these values, which is the description of the possible load reduction for each of the hours of the day (Stage 2). This offer is sent to the flexibility market, where all the offers of all the customers under the market are listed (Stage 3). The DSO selects those offers that are needed for the flexibility planning, and their selection is communicated to the affected customers using an OpenADR event (Stage 4). These events can be sent along the day but should be done some time before the moment when the requested event will start (e.g., at least one hour before). This restriction is usually described in the flexibility program. When the time of the event starts, the BMS will manage the loads to keep the requested power constraints according to the event characteristics, until the event finishes (Stage 5). At the end of the current day (or at the end of the settlement cycle), the DSO will have retrieved the data from the smart meters and make the audit for the billing adjustment, depending on whether the customer had correctly achieved the request of the event (Stage 6).

In the next Section, the implementation of these systems in a real laboratory is detailed.

4. Laboratory Systems

To demonstrate the correct operation of the system and the benefits of OpenADR in flexibility operations and congestion management a use case has been conducted. This use case has been conducted using the accommodations at the "DERBis" located at the *Escuela Politécnica Superior* of the University of Seville. The DERBis is a demonstrator of Smart Grid, DERs and Building Management technologies.

4.1. DERBis Description

The classroom has 23 seats (22 for students and 1 for the professor), each one with a PC. This space is located at the *Escuela Politécnica Superior* and it is mainly used for teaching purposes for the MSc degree in Intelligent Systems, Energy, and Transport.

Nevertheless, this classroom has become a living lab for different Smart Grid and Building Management technologies. This facility makes it possible to train students in the different possibilities of these technologies and how to manage them. Additionally, this space is used by researchers to test different scenarios and use cases related to flexibility.

The complete overview of the different devices and systems deployed in the lab is shown in Figure 6.

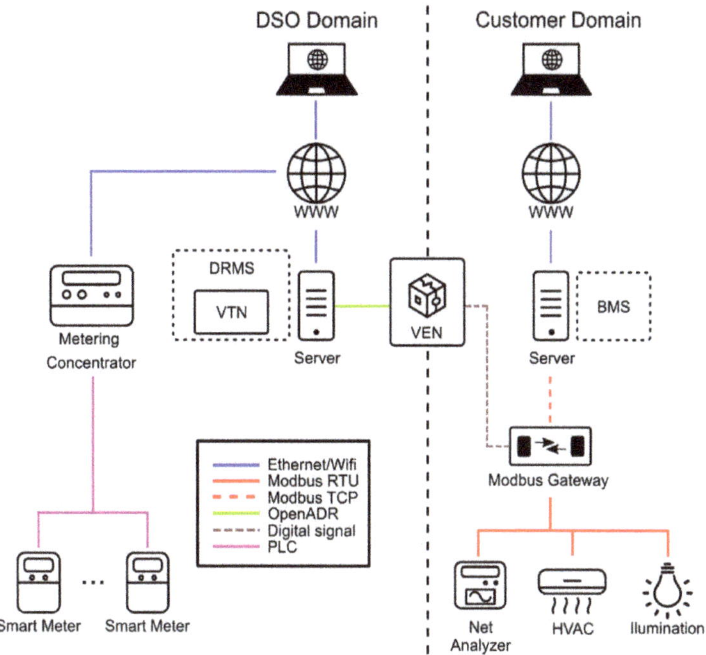

Figure 6. DERBis systems overview. DRMS: Demand Response Management System; VTN: Virtual Top Node; VEN: Virtual End Node; BMS: Building Management System.

On the one hand, the current Smart Grid deployments in the lab are an AMI based on smart meters and a physical OpenADR VEN. On the other hand, the Building Management deployments are based on different controllers with Modbus communications, it being possible to control different systems, such as: The Heat, Ventilation and Air Conditioning (HVAC) modules, the light circuits, and the user's loads (wall plugs). Additionally, a centralized system manages all the resources of the lab. This system acts as a BMS that controls the room assets based on different rules and constraints. Figure 7 shows the devices inside the room.

Thanks to the controllable loads, and based on the energy demand prediction, the lab can provide flexibility services despite not counting with generation resources, simply reducing the load when it is required.

4.1.1. Smart Grid Technologies on the DERBis Living Lab

Additionally, a physically OpenADR VEN has been installed: an EISSBox 3.0 [49] (see Figure 7). This VEN is fully OpenADR 2.0b compatible and it is currently connected to the VTN developed by the authors. Through this VEN, OpenADR commands can be sent to the lab simulating DSO actions. Also, the consumption of the lab can be obtained from the VEN by means of the report service of the OpenADR protocol.

This VEN has four relay outputs that can be configured to respond to OpenADR events. Also, the VEN has four pulse inputs for energy metering purposes. Two of the relays of the VEN are connected to an input Modbus device in order to read their state. Two of the pulse inputs of the VEN are connected to the pulse outputs of the Modbus energy meter.

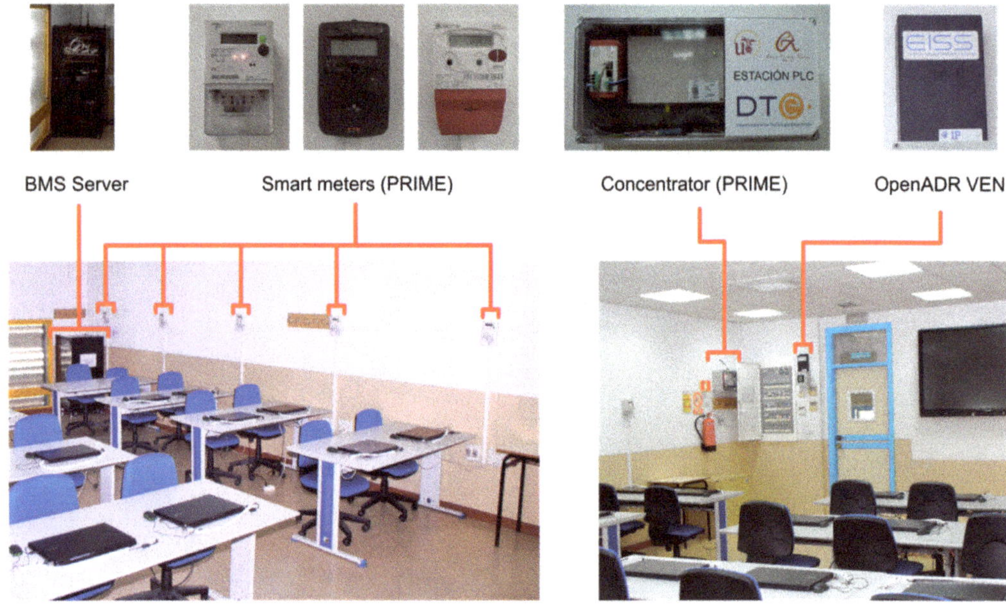

Figure 7. DERBis control and measurement devices.

The smart meters of the DERBis use PLC communications based on PRIME protocol. All smart meters communicate with a PLC concentrator (CIRCUTOR SGE-PL1000) which receives the measurements and can send different commands to the meters. These commands are related to the management and configuration of each smart meter, which can be done separately. A more detailed description of the AMI devices can be found in [38], where their characteristics, structure, functions, and possibilities for teaching purposes are pointed out.

The Smart Meters are distributed through different loads and locations of the lab (student benches, HVAC devices, etc.). In addition, one Smart Meter is installed at the head of the lab, measuring the overall room consumption. The schema of the smart metering system can be seen in the Figure 8.

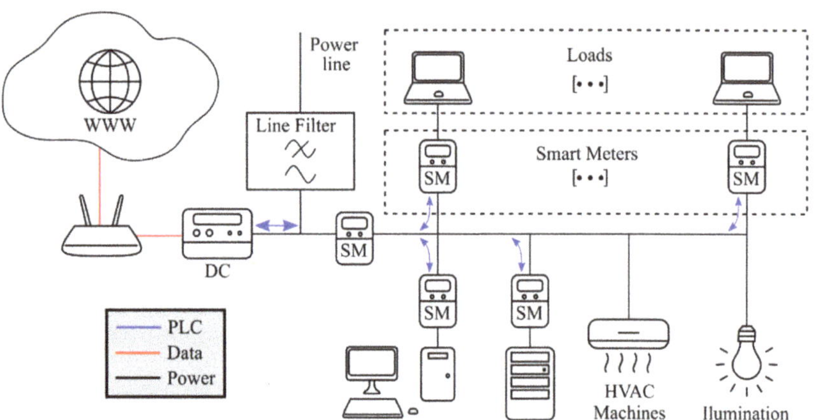

Figure 8. Smart metering system.

4.1.2. Building Management Technologies on the Living Lab

From the Building Management point of view, the lab has different Modbus RTU devices for control and monitoring HVAC, lights and loads. To integrate all these devices with the BMS, a Modbus TCP/RTU gateway has been installed.

Specifically, the lab has three HVAC controllable machines. These machines have a proprietary remote configuration protocol. In this sense, one of these machines is fully controlled by means of a gateway between Modbus RTU and its proprietary protocol. The other two only have an on/off control. The lights and users' loads also can be monitored and have an on/off control. Additionally, different electrical parameters (voltages, currents, active and reactive power, Total Harmonic Distortion (THD), etc.) of the lab can also be obtained by means of an energy analyzer. Ambient conditions are also monitored by a temperature and humidity sensor.

All these systems are controlled by the BMS which integrates all the devices and systems. The BMS offers monitoring and historical services of the Lab and also implements different control schemas to improve and adjust the energy consumption. Additionally, it is the responsibility to set the appropriate adjustments when requested by the DSO by means of the VEN. A capture of the BMS interface can be seen in the Figure 9.

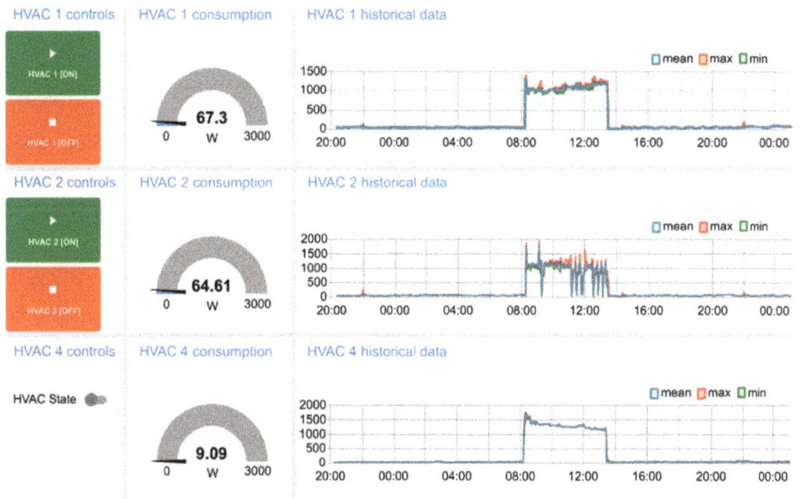

Figure 9. BMS HVAC control interface.

As can be seen, this location offers a great integration of technologies from both Smart Grid and Building Management. In this sense, this space provides a great opportunity to test the interaction between different systems and technologies as well as to test different flexibility operations in the scope of the congestion management in the Smart Grid.

5. Use Cases

Thanks to the great opportunity that this living lab offers in terms of technologies and integration, several use cases have been carried out. These use cases expect to test the living lab capabilities, the OpenADR protocol schema in a real scenario, the auditory process through the AMI as well as the developed systems for grid flexibility operations, especially in congestion management.

The proposed schema will involve a DSO operation request for a customer (or an FSP/Aggregator) to mitigate a congestion in the grid. In this sense, the developed DRMS/VTN will act as the DSO system, and the actions that are supposed to be done by the rest of the DSO systems (Distribution Management Systems (DMS), Energy Management Systems (EMS), etc.) in the DRMS/VTN will be simulated. Although, thanks to the interoperability with different standards that the VTN offers, the connection with the DSO could be easily done.

The DERBis living lab will act as the assets on the customer side. Thus, the OpenADR will be the communication link between the DSO and the customer. In this sense, the VEN located at the lab will be the endpoint on the customer side. Then, the BMS receives the information from the VEN and performs the required management tasks regarding the events and load management.

Furthermore, the DSO could also get the metering of the lab consumption by means of the AMI deployed with the Smart Meters. Even though OpenADR allows one to send metering through the report service, the DSO also has the metering through its AMI infrastructure with which it could audit the fulfilment of the flexibility command. This is an important point in flexibility for the DSO, since with the smart meter infrastructure has an external way to audit the operation.

It must be remarked that the Smart Meter system is considered here the most effective and adequate system to realize the auditory process. The VEN metering capacity constitutes just an additional auxiliary method which can be used for the measurement of specific parts of the consumption/generation, achieving a more complete understanding of the state of the customer systems, but should never be used for audits.

A general schema of the tasks to be performed by the DERBis BMS is:

1. Start of the day:
   - Estimate power (using occupational model).
   - Include controllable loads (if any).
   - This result in the maximum and minimum values for each of the hours of the day.
   - Calculate baseline (mean of consumptions of previous days for each hour according to the DSO rules).
   - Calculate flexibility margin (baseline minus maximum expected consumption).
   - Send participation offer to the flexibility market.
   - The DSO evaluates if they are interested in sending any event and informs the customer.
   - The BSM receives the event information.

2. Event start (if any):
   - The BMS control the loads (if necessary) to accomplish with the event.

3. End of the day and in advance:
   - The audit of the service requested by the DSO will be done. If the audit power minus the actual power is greater or equal to the event reduction power, the customer has successfully performed the requisites.

Once the process to be followed for flexibility prediction and planning is exposed, some study cases will be exposed. The first one corresponds to a day where there are no curtailable loads, but some flexibility services can be provided thanks to the expected use of the lab during that day. The second one considers some controllable loads to provide the service of power reduction during various hours. Finally, the third one presents a day when there is a very low flexibility margin during one of the hours, so it is considered non-sufficient for participation in the market. Of course, the consideration of a "minimum" power of reduction will depend on the preferred strategy for the BMS, on the conditions of the market, or on the requirements of the corresponding DSO for each day. In the third example a minimum limit of 500 W was considered specifically for that day. It is not the objective of this paper to discuss the details of the market coordination. Therefore, for the present study the events have been launched using the mentioned VTN (which corresponds to the actions of the DSO).

5.1. Use Case 1. Absence of Curtailable Loads

The BMS must perform some tasks to manage the flexibility capacity of the laboratory. Firstly, it takes the historical data from smart meter telemetry and applies the rules that are considered by the DSO. In this study case, this baseline will be estimated through the mean power for each hour using the two previous days of the same type (weekend, or non-weekend types). The second task is performing a prediction of the expected power consumption according to their own models (not necessarily based only in previous data, but also including other types of available data). In this case, a simple model based on occupancy has been applied. This model gives a margin of maximum and minimum expected value for each hour and the flexibility margin is calculated as the difference between the maximum expected consumption and the forecasted power. As an example, the values associated for the day 5 March 2020 can be observed in the Figure 10a.

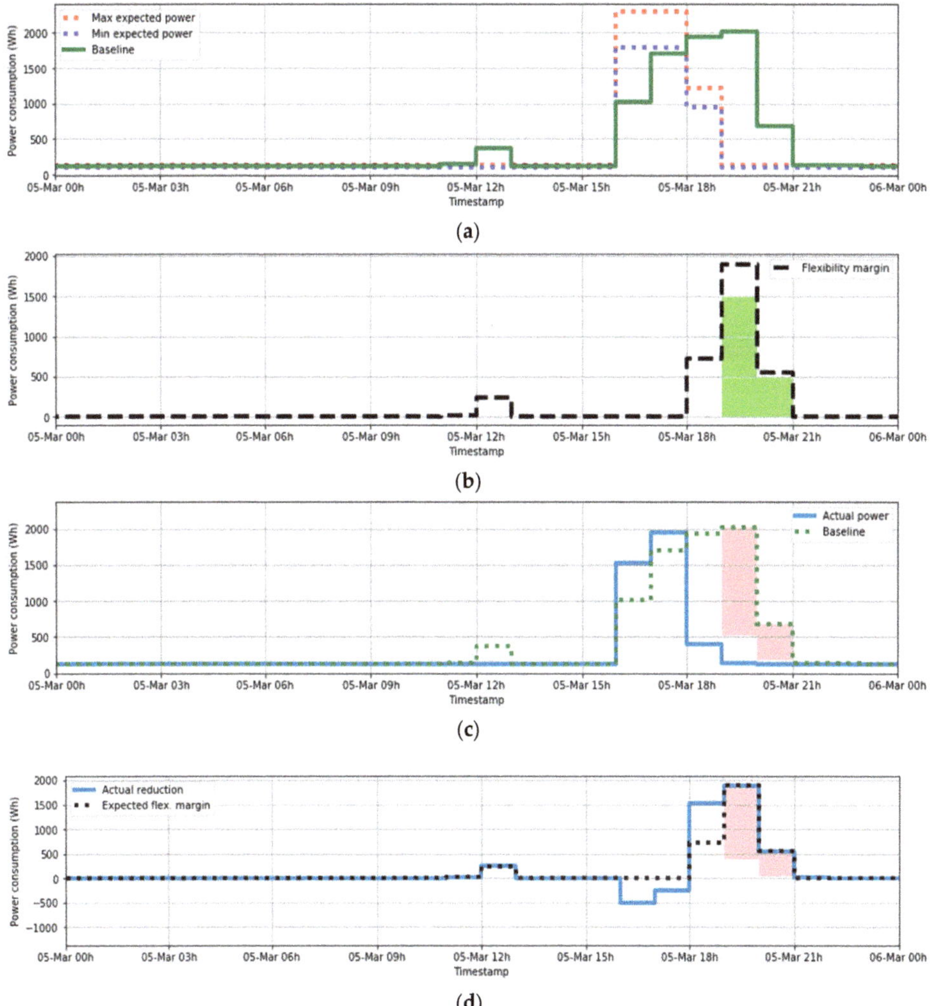

Figure 10. (**a**) Occupancy-based (from BMS) and historical-based (baseline) expected power; (**b**) Expected flexibility Margin (only values >0). Reduction really requested by the DSO (in green); (**c**) Actual power and baseline. Requested power reduction (in red); (**d**) Reduction achieved vs. Expected flexibility margin. Requested power reduction (in red).

In this sense, under atypical laboratory use, the flexibility margin can be positive, providing potential opportunities to participate in any DR event. Based on this information, the BMS will choose their participation or not offer a reduction in that period.

Thus, in the case of a party agreement, it will be necessary to perform the audit. At the end of the day the smart meter data of the real consumption for each hour will be available for the DSO. Therefore, from there (at any moment in advance) the DSO can make the audit to check whether the lab (that acts as customer) has really accomplished the reduction in the events that they promise, in order to apply the corresponding bill adjustment.

For the present example, the flexibility margin can be calculated as the Baseline minus the maximum expected power consumption. Considering the flexibility margin, the DSO could pick up from the market the reduction needed. The Figure 10b shows (in green), which was the request from the DSO. It can be noticed that this request is always equal to or less than the flexibility margin (which was the maximum proposed reduction that the BMS estimated).

Figure 10c shows the comparison between the actual power consumption and the audit power. As can be seen, the reduction that was requested by the DSO from 19:00 to 21:00 was successfully achieved.

The Figure 10d shows the comparison of the actual reduction achieved by the customer and the originally expected flexibility margin. The reduction event requirement appears in red. In those hours where the flexibility margin was expected to be positive, a reduction was achieved. Therefore, the prediction of the possible participation of this customer in DR events was correctly done, showing the best hours for their inclusion in these events.

5.2. Use Case 2. Controllable Loads Available

In the previous section, the exposed example of load reduction was performed without any controllable load action.

In this second example, the flexibility providing is achieved by using some controllable loads, which are the HVAC system of the laboratory. The process of estimating the flexibility margin is very similar to the one performed in the previous case. The only new elements that are introduced here are the parameters of "controllable load availability", i.e., the power that can be put off for each of the hours of the day, estimated in a 24 h horizon (at 00:00 of each day). The next data corresponds to the day 26 February 2020.

The expected power without considering the presence of controllable loads can be seen in the Figure 11a.

In this case, according to the BMS estimations, the HVAC system could be deactivated during the period 18:00–20:00 without affecting the comfort conditions of the laboratory. Therefore, this reduction is included in the Figure 11b. The reduction was about 500 W; therefore, this quantity of reduction is applied to update the maximum and minimum expected power.

Using this estimation and the auditory power (which is the baseline power according to the DSO for the auditory process), it is possible to obtain the estimation of flexibility margin as the difference between the auditory power and the maximum expected power. This is done in Figure 11c.

This flexibility margin marks the maximum flexibility offer of the customer for the day. Let's suppose that the only period where the DSO buys the flexibility services is during the day 26 February from 19:00 to 21:00. The BMS will deactivate the loads during this period (at least). The actual consumption during the day is shown in the Figure 11d.

The comparison between the actual power and the audit power (baseline for auditory process) shows that the reduction was successfully achieved during the period of event can be seen in Figure 11e. The reduction requested appears in red.

Finally, the data of temperature and humidity are shown in Figure 11f. It shows how these two variables start to change their tendency due to the disconnection of the HVAC system, but they maintain adequate values during the reduction event in time of use of the laboratory, keeping good comfort conditions for the users.

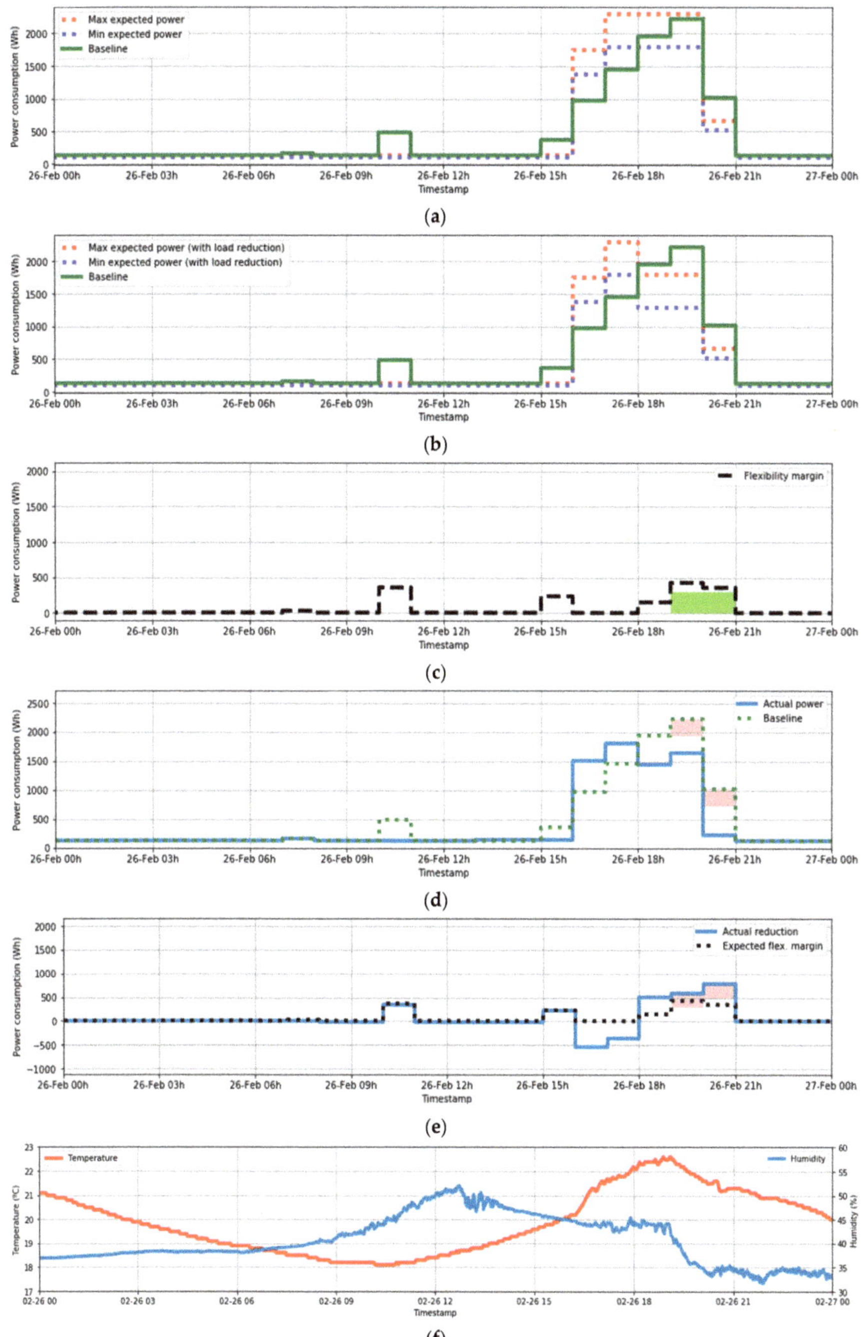

Figure 11. (**a**) Occupancy-based (from BMS) and Historical-based (baseline) expected power; (**b**) Occupancy-based (from BMS) and Historical-based (baseline) expected power—Load reduction considered; (**c**) Expected flexibility Margin (only values > 0). Reduction really requested by the DSO (in green); (**d**) Actual power and baseline. Requested power reduction (in red); (**e**) Reduction achieved vs. Expected Flexibility Margin. Requested power reduction (in red); (**f**) Temperature (red) and humidity (blue).

5.3. Use Case 3. No Significant Flexibility Margin

The third situation that can occur is that the BMS does not find any flexibility margin during the day (or simply that this margin is too low). This means that the customer should not participate in the flexibility market, as they are not expected to be able to offer any load or generation adjustment.

Therefore, the action of the BMS will finish in the calculation of flexibility margin, not offering any reduction or attending any event during the day.

An example corresponding to the day 27 February 2020 is shown in Figure 12a. On this day, there is no hour when the flexibility margin is significant, but only some residual values (see Figure 12a,b). Therefore, participation in the flexibility market is not possible to be considered, so the BMS will not send any offer.

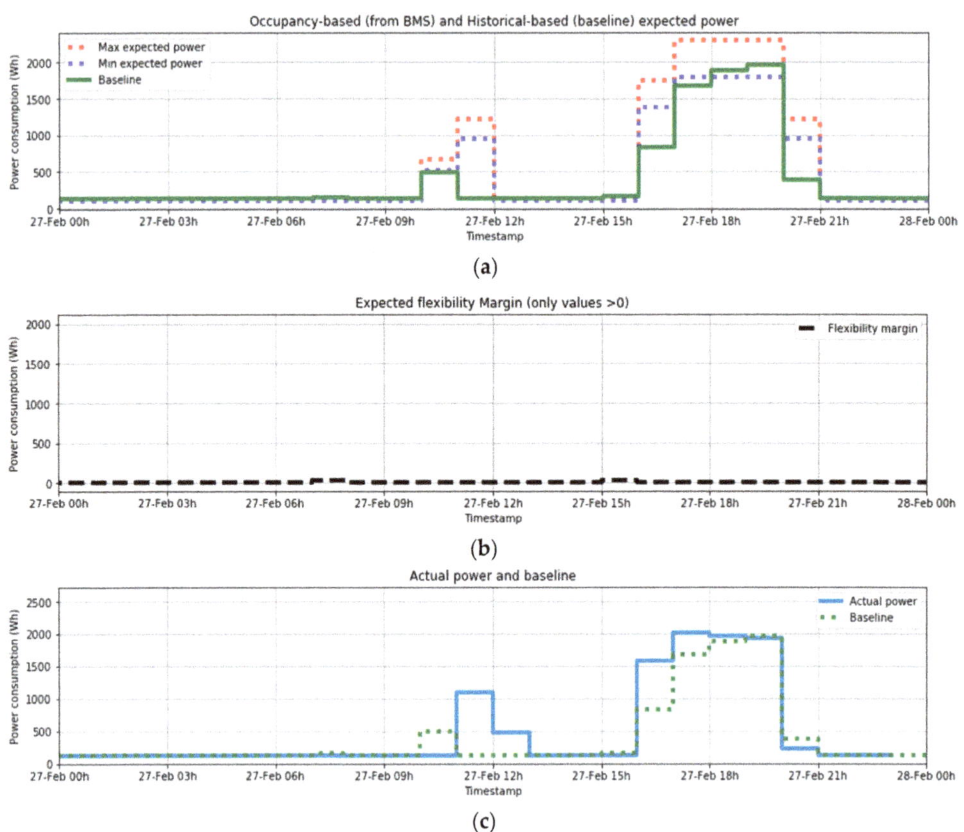

Figure 12. (**a**) Occupancy-based (from BMS) and Historical-based (baseline) expected power; (**b**) Expected flexibility Margin (only values >0); (**c**) Actual power and baseline.

Figure 12c shows the actual power throughout the day and the baseline. Of course, as the customer has not participated in the flexibility market with any offer, there will be no audit process referred to this day.

Throughout the three cases analyzed, how the BMS performs the previsions of flexibility margin in order offer a possible reduction to the DSO can be observed. In some of the days there will not be any possibility of providing any service, while in others it will be possible thanks to the use of controllable loads, or thanks to the analysis of the expected occupation.

6. Conclusions

The inclusion of DSM over the power system undoubtedly constitutes a powerful tool to improve the management capacity and flexibility of distribution networks. However, the use of these kinds of services implies an agreement between the two parts, i.e., the customers and the DSO, which could be performed under multiple existing terms, principles and methods.

In this paper, a possible architecture for DSM dispatching and audit is proposed and tested. In this architecture, all the controllable devices are connected to a BMS. As it was exposed, the BMS established communication to the DSO systems using the standard OpenADR, providing a channel to inform about availability, scheduling, prices or any needed data.

The audit function depends on the DSO smart meter infrastructure that provides remote information about the hourly consumption, this data being used to check the behavior of the loads in question during the period of the dispatched DSM events. The accomplishing or not of such events will therefore be evaluated using information from the smart meters to perform the audit process. Likewise, this hourly information is enough for this purpose, although its extension to a quarter hour is considered interesting for the future to increase network flexibility.

As Proof of Concept the architecture has been tested on a real testbed located in a laboratory of the *Escuela Politécnica Superior* of the University of Seville. In this sense, as can be seen in its results, the proposed architecture clearly shows the possibilities of use of controllable loads under the established DSM contract conditions. The principal difficulty from the customer side would be the requirement of the OpenADR VEN device, which would not be such a big problem in the case of buildings that count with a BMS. Thus, thanks to taking advantage of the available power meter infrastructure that is used for the electricity billing, the audit process can be done by the DSO in a secure and appropriate manner, showing the adequacy of the AMI for this task, as this system makes easier the integration of flexibility services.

It can also be concluded that the described laboratory constitutes a complete platform where some of the most important smart grid related technologies can be evaluated, integrated and taught.

Author Contributions: Conceptualization: E.P., J.I.G., and A.G.; methodology: A.P., S.G. and A.G.; investigation: A.P., S.G. and E.P.; software: S.G., E.P., and J.I.G.; validation: A.P., S.G., E.P. and J.I.G.; formal analysis: A.P., S.G. and E.P.; writing, original draft preparation: A.P., S.G., E.P., and J.I.G.; writing, review and editing: A.G. and C.L.; supervision: A.G. and C.L.; project administration: C.L. All authors have read and agreed to the published version of the manuscript.

Funding: This work was supported by the Enel-Endesa Company, in the project "GRID Flexibility & Resilience Project" (number P020-19/E24), and by the *"Ministerio de Ciencia, Innovación y Universidades"*, Government of Spain under the project *"Bigdata Analitycs e Instrumentación Cyberfísica para Soporte de Operaciones de Distribución en la Smart Grid"*, number RTI2018-094917-B-I00. Sebastián García is also supported by this project. Moreover, Antonio Parejo is supported by the scholarship *"Formación de Profesorado Universitario (FPU)"*, Grant Number FPU16/03522 from the *"Ministerio de Educación y Formación Profesional"*, Government of Spain.

Data Availability Statement: The data presented in this study are available on request from the corresponding author. The data are not publicly available due to project privacy issues.

Acknowledgments: The authors would also like to thank the Enel-Endesa Company, the *"Ministerio de Ciencia, Innovación y Universidades"* and the *"Ministerio de Educación y Formación Profesional"*, Government of Spain, for their funding and support. Furthermore, the authors want to thank Alejandro Gallardo Soto and Julian Rodríguez Galán for their help and support in the implementation of the control system in the laboratory.

Conflicts of Interest: The authors declare no conflict of interest.

References

1. Gerard, H.; Puente, E.I.R.; Six, D. Coordination between transmission and distribution system operators in the electricity sector: A conceptual framework. *Util. Policy* **2018**, *50*, 40–48. [CrossRef]
2. Mohammed, N.A.; Al-Bazi, A. Management of renewable energy production and distribution planning using agent-based modelling. *Renew. Energy* **2021**, *164*, 509–520. [CrossRef]
3. Martinez-Anido, C.B.; Botor, B.; Florita, A.R.; Draxl, C.; Lu, S.; Hamann, H.F.; Hodge, B.-M. The value of day-ahead solar power forecasting improvement. *Solar Energy* **2016**, *129*, 192–203. [CrossRef]
4. Valentine, K.; Temple, W.; Thomas, R.J.; Zhang, K.M. Relationship between wind power, electric vehicles and charger infrastructure in a two-settlement energy market. *Int. J. Electr. Power Energy Syst.* **2016**, *82*, 225–232. [CrossRef]
5. Romero-Ruiz, J.; Pérez-Ruiz, J.; Martin, S.; Aguado, J.A.; la Torre, S.D. Probabilistic congestion management using EVs in a smart grid with intermittent renewable generation. *Electr. Power Syst. Res.* **2016**, *137*, 155–162. [CrossRef]
6. Ahmad, S.; Ahmad, A.; Naeem, M.; Ejaz, W.; Kim, H. A Compendium of Performance Metrics, Pricing Schemes, Optimization Objectives, and Solution Methodologies of Demand Side Management for the Smart Grid. *Energies* **2018**, *11*, 2801. [CrossRef]
7. Malik, S.; Harish, V.S.K.V. Integration of automated Demand Response and Energy Efficiency to enable a smart grid infrastructure. In Proceedings of the 2019 2nd International Conference on Power Energy, Environment and Intelligent Control (PEEIC), Greater Noida, India, 18–19 October 2019; IEEE: New York City, NY, USA, 2019.
8. Ghatikar, G.; Zuber, J.; Koch, E.; Bienert, R. Smart grid and customer transactions: The unrealized benefits of conformance. In Proceedings of the 2014 IEEE Green Energy and Systems Conference (IGESC), Long Beach, CA, USA, 24 November 2014; IEEE: New York City, NY, USA, 2014.
9. Huh, J.-H.; Seo, Y.-S. Understanding Edge Computing: Engineering Evolution with Artificial Intelligence. *IEEE Access* **2019**, *7*, 164229–164245. [CrossRef]
10. Giraldo-Soto, C.; Erkoreka, A.; Mora, L.; Uriarte, I.; Portillo, L.D. Monitoring System Analysis for Evaluating a Building's Envelope Energy Performance through Estimation of Its Heat Loss Coefficient. *Sensors* **2018**, *18*, 2360. [CrossRef] [PubMed]
11. Parejo, A.; Sanchez-Squella, A.; Barraza, R.; Yanine, F.; Barrueto-Guzman, A.; Leon, C. Design and Simulation of an Energy Homeostaticity System for Electric and Thermal Power Management in a Building with Smart Microgrid. *Energies* **2019**, *12*, 1806. [CrossRef]
12. Cheng, Z.; Li, Z.; Liang, J.; Si, J.; Dong, L.; Gao, J. Distributed coordination control strategy for multiple residential solar PV systems in distribution networks. *Int. J. Electr. Power Energy Syst.* **2020**, *117*, 105660. [CrossRef]
13. Papavasiliou, A.; Mezghani, I. Coordination Schemes for the Integration of Transmission and Distribution System Operations. In Proceedings of the 2018 Power Systems Computation Conference (PSCC), Dublin, Ireland, 11–15 June 2018; IEEE: New York City, NY, USA, 2018.
14. Migliavacca, G.; Rossi, M.; Gerard, H.; Dzamarija, M.; Horsmanheimo, S.; Madina, C.; Kockar, I.; Leclecq, G.; Marroquín, M.; Svendsen, H.G. TSO-DSO coordination and market architectures for an integrated ancillary services acquisition: The view of the SmartNet project. In Proceedings of the CIGRE Session 2018, Paris, France, 27–31 August 2018.
15. European Union. *Directive (EU) 2019/944 of The European Parliament and of The Council—of 5 June 2019—on Common Rules for the Internal Market for Electricity and Amending Directive 2012/27/EU.75*; European Commission: Brussels, Belgium, June 2019.
16. Alonso, J.I.G.; Personal, E.; García, S.; Parejo, A.; Rossi, M.; García, A.; Delfino, F.; Pérez, R.; León, C. Flexibility Services Based on OpenADR Protocol for DSO Level. *Sensors* **2020**, *20*, 6266. [CrossRef]
17. Yanine, F.; Sanchez-Squella, A.; Barrueto, A.; Sahoo, S.K.; Parejo, A.; Shah, D.; Cordova, F.M. Homeostaticity of energy systems: How to engineer grid flexibility and why should electric utilities care. *Period. Eng. Nat. Sci.* **2019**, *7*, 474. [CrossRef]
18. Mak, S.T.; So, E. Integration of PMU, SCADA, AMI to accomplish expanded functional capabilities of Smart Grid. In Proceedings of the 29th Conference on Precision Electromagnetic Measurements (CPEM 2014), Rio de Janeiro, Brazil, 24–29 August 2014; IEEE: New York City, NY, USA, 2014.
19. Huh, J.-H. *Smart Grid Test Bed Using OPNET and Power Line Communication*; IGI Global: Hershey, PA, USA, 2018.
20. European Comission. Benchmarking Smart Metering Deployment in the EU-28. Final Report. March 2020. Available online: https://op.europa.eu/en/publication-detail/-/publication/b397ef73-698f-11ea-b735-01aa75ed71a1 (accessed on 28 January 2021).
21. European Parliamen. *Directive 2009/72/EC of the European Parliament and of the Council of 13 July 2009 Concerning Common Rules for the Internal Market in Electricity and Repealing Directive 2003/54/EC 2009*; European Parliamen: Brussels, Belgium, 2009.
22. Cui, T.; Carr, J.; Brissette, A.; Ragaini, E. Connecting the Last Mile: Demand Response in Smart Buildings. *Energy Procedia* **2017**, *111*, 720–729. [CrossRef]
23. Ren, H.; Ortega, J.; Casimis, D.W. Review of Operating Reserves and Day-Ahead Unit Commitment Considering Variable Renewable Energies: International Experience. *IEEE Lat. Am. Trans.* **2017**, *15*, 2126–2136. [CrossRef]
24. San Diego Gas & Electric Company. *Schedule CBP: Capacity Bidding Program*; San Diego Gas & Electric Company: San Diego, CA, USA, April 2019.
25. Pacific Gas & Electric Company. *Electric Schedule E-CBP: Capacity Bidding Program*; Pacific Gas & Electric Company: San Francisco, CA, USA, July 2018.
26. Guerrero, J.I.; Personal, E.; Caro, S.G.; Parejo, A.; Rossi, M.; Garcia, A.; Sanchez, R.P.; Leon, C. Evaluating Distribution System Operators: Automated Demand Response and Distributed Energy Resources in the Flexibility4Chile Project. *IEEE Power Energy Mag.* **2020**, *18*, 64–75. [CrossRef]

27. Samad, T.; Koch, E. Automated demand response for energy efficiency and emissions reduction. In *PES T&D 2012*; IEEE: New York City, NY, USA, 2012.

28. Samad, T.; Koch, E.; Stluka, P. Automated Demand Response for Smart Buildings and Microgrids: The State of the Practice and Research Challenges. *Proc. IEEE* **2016**, *104*, 726–744. [CrossRef]

29. OpenADR Alliance. OpenADR 2.0 Profile Specification B Profile. 2013. Available online: https://www.openadr.org/specification (accessed on 26 November 2020).

30. Kiliccote, S.; Dudley, J.H.; Piette, M.A. *Northwest Open Automated Demand Response Technology Demonstration Project*; (No. LBNL-2573E); Lawrence Berkeley National Lab. (LBNL): Berkeley, CA, USA, 2009.

31. McParland, C. OpenADR open source toolkit: Developing open source software for the Smart Grid. In Proceedings of the 2011 IEEE Power and Energy Society General Meeting, Detroit, MI, USA, 24–28 July 2011; IEEE: New York City, NY, USA, 2011.

32. Herberg, U.; Mashima, D.; Jetcheva, J.G.; Mirzazad-Barijough, S. OpenADR 2.0 deployment architectures: Options and implications. In Proceedings of the 2014 IEEE International Conference on Smart Grid Communications (SmartGridComm), Venice, Italy, 3–6 November 2014; IEEE: New York City, NY, USA, 2014.

33. Jingxi, Z.; Min, C.; Ling, W.; Lixia, Y.; Shixiong, P.; Dong, L. Research on Architecture of Automatic Demand Response System Based on OpenADR. In Proceedings of the 2018 China International Conference on Electricity Distribution (CICED), Tianjin, China, 17–19 September 2018; IEEE: New York City, NY, USA, 2018.

34. Wilcox, J.; Kaleshi, D.; Sooriyabandara, M. DIRECTOR: A distributed communication transport manager for the Smart Grid. In Proceedings of the 2014 IEEE International Conference on Communications (ICC), Sydney, Australia, 10–14 June 2014; IEEE: New York City, NY, USA, 2014.

35. Koch, E.L. Automated demand response—mathsemicolonfrom peak shaving to ancillary services. In Proceedings of the 2012 IEEE PES Innovative Smart Grid Technologies (ISGT), Washington, DC, USA, 16–20 January 2012; IEEE: New York City, NY, USA, 2012.

36. Goli, S.; McKane, A.; Olsen, D. *Demand Response Opportunities in Industrial Refrigerated Warehouses in California*; Lawrence Berkeley National Lab. (LBNL): Berkeley, CA, USA, 2012.

37. Seo, J.; Jin, J.; Kim, J.Y.; Lee, J.-J. Automated Residential Demand Response Based on Advanced Metering Infrastructure Network. *Int. J. Distrib. Sens. Netw.* **2016**, *12*, 4234806. [CrossRef]

38. Parejo, A.; Garcia, S.; Personal, E.; Garcia, A.; Guerrero, J.I.; Leon, C. Living-Lab for Smart Grid technologies teaching. In Proceedings of the 2020 XIV Technologies Applied to Electronics Teaching Conference (TAEE), Porto, Portugal, 8–10 July 2020; IEEE: New York City, NY, USA, 2020.

39. Fagiani, M.; Squartini, S.; Gabrielli, L.; Spinsante, S.; Piazza, F. A review of datasets and load forecasting techniques for smart natural gas and water grids: Analysis and experiments. *Neurocomputing* **2015**, *170*, 448–465. [CrossRef]

40. Daut, M.A.M.; Hassan, M.Y.; Abdullah, H.; Rahman, H.A.; Abdullah, M.P.; Hussin, F. Building electrical energy consumption forecasting analysis using conventional and artificial intelligence methods: A review. *Renew. Sustain. Energy Rev.* **2017**, *70*, 1108–1118. [CrossRef]

41. Deb, C.; Zhang, F.; Yang, J.; Lee, S.E.; Shah, K.W. A review on time series forecasting techniques for building energy consumption. *Renew. Sustain. Energy Rev.* **2017**, *74*, 902–924. [CrossRef]

42. Piette, M.A.; Kiliccote, S.; Ghatikar, G.; McKane, A.T.; Matson, N.; Page, J.; MacDonald, J.S.; Aghajanzadeh, A.; Black, D.R.; Yin, R. *Demand Response Research Center—Final Report*; Berkeley Lab's Energy Technologies Area (ETA): Berkeley, CA, USA, 2015.

43. Electric Power and Research Institute (EPRI). *OpenADR 2.0b Open Source Virtual Top Node (OADR2.0b VTN) Version 0.9.7.0.*; Electric Power and Research Institute (EPRI): Palo Alto, CA, USA, 2017.

44. Du, Y.F.; Jiang, L.; Duan, C.; Li, Y.Z.; Smith, J.S. Energy Consumption Scheduling of HVAC Considering Weather Forecast Error Through the Distributionally Robust Approach. *IEEE Trans. Ind. Inform.* **2018**, *14*, 846–857. [CrossRef]

45. Bianco, V.; Piazza, G.; Scarpa, F.; Tagliafico, L.A. Energy, economic and environmental assessment of the utilization of heat pumps for buildings heating in the Italian residential sector. *Int. J. Heat Technol.* **2017**, *35*, S117–S122. [CrossRef]

46. Hosseinzadeh, M.; Salmasi, F.R. Robust Optimal Power Management System for a Hybrid AC/DC Micro-Grid. *IEEE Trans. Sustain. Energy* **2015**, *6*, 675–687. [CrossRef]

47. Lagrange, A.; de Simón-Martín, M.; González-Martínez, A.; Bracco, S.; Rosales-Asensio, E. Sustainable microgrids with energy storage as a means to increase power resilience in critical facilities: An application to a hospital. *Int. J. Electr. Power Energy Syst.* **2020**, *119*, 105865. [CrossRef]

48. Huh, J.-H.; Otgonchimeg, S.; Seo, K. Advanced metering infrastructure design and test bed experiment using intelligent agents: Focusing on the PLC network base technology for Smart Grid system. *J. Supercomput.* **2016**, *72*, 1862–1877. [CrossRef]

49. IPKeys Technologies. *EISS Box 3.0 System/Device Monitoring and Management*; IPKeys Technologies: Stafford, VA, USA, 2017.

Article

Instantaneous Disturbance Index for Power Distribution Networks

María Dolores Borrás-Talavera *, Juan Carlos Bravo and César Álvarez-Arroyo

Electrical Engineering Department, Escuela Politécnica Superior, Universidad de Sevilla, c/Virgen de África 9, 41011 Sevilla, Spain; carlos_bravo@us.es (J.C.B.); cesaralvarez@us.es (C.Á.-A.)
* Correspondence: borras@us.es

Abstract: The stability of power systems is very sensitive to voltage or current variations caused by the discontinuous supply of renewable power feeders. Moreover, the impact of these anomalies varies depending on the sensitivity/resilience of customer and transmission system equipment to those deviations. From any of these points of view, an instantaneous characterization of power quality (PQ) aspects becomes an important task. For this purpose, a wavelet-based power quality indices (PQIs) are introduced in this paper. An instantaneous disturbance index (ITD(t)) and a Global Disturbance Ratio index (GDR) are defined to integrally reflect the PQ level in Power Distribution Networks (PDN) under steady-state and/or transient conditions. With only these two indices it is possible to quantify the effects of non-stationary disturbances with high resolution and precision. These PQIs offer an advantage over other similar because of the suitable choice of mother wavelet function that permits to minimize leakage errors between wavelet levels. The wavelet-based algorithms which give rise to these PQIs can be implemented in smart sensors and used for monitoring purposes in PDN. The applicability of the proposed indices is validated by using a real-time experimental platform. In this emulated power system, signals are generated and real-time data are analyzed by a specifically designed software. The effectiveness of this method of detection and identification of disturbances has been proven by comparing the proposed PQIs with classical indices. The results confirm that the proposed method efficiently extracts the characteristics of each component from the multi-event test signals and thus clearly indicates the combined effect of these events through an accurate estimation of the PQIs.

Keywords: power quality indices; signal processing; multi-resolution analysis; renewable energy applications

Citation: Borrás-Talavera, M.D.; Bravo, J.C.; Álvarez-Arroyo, C. Instantaneous Disturbance Index for Power Distribution Networks. *Sensors* **2021**, *21*, 1348. https://doi.org/10.3390/s21041348

Academic Editor: Hossam A. Gabbar
Received: 14 January 2021
Accepted: 11 February 2021
Published: 14 February 2021

Publisher's Note: MDPI stays neutral with regard to jurisdictional claims in published maps and institutional affiliations.

1. Introduction

The extensive use of power-switching devices for source conditioning, renewable energy supply and motion control in modern industrial applications has detrimental side-effects on power quality, such as increase potential for unacceptable harmonic levels, poor power factor, or unbalanced currents and voltages in power distribution networks. In addition, transients caused by faults and switching events in power systems significantly affect the power-transfer quality of a supply [1,2]. All these undesirable effects cause huge economic losses [3] and require an effective power quality analysis in the grid and affected facilities. Smart sensors proposals are rapidly increasing in response to these new requirements [4–9]. Although the first applications were almost exclusively limited to billing, they now include new features related to power quality detection and have a common denominator: they require high processing speed to handle large computer data in a shorter time. In this context it is crucial to provide a small number of parameters and indicators that can effectively characterize all these events and derive from a very fast and efficient analysis tool. This task is the main objective of this work.

The power quality index (PQI) is the summary of waveform distortions in voltage and current from the perfect sinusoids. The PQIs are used to characterize the degree of quality degradation in a quantitative manner. They represent the impacts of non-ideal waveforms on electrical power systems in a compact but expressive manner. Existing indices such as Total Harmonic Distortion (THD), Power factor, Flicker factor, etc. reflect the degree of power disturbance in each of these categories individually, but fail to assess most of the phenomena mentioned together in an exhaustive and concise manner by a single value [10].

Several methods have been applied to research the real-time behavior of controllers, test and protection equipment [11] and fault diagnosis devices [12,13]. All of these applications, together with those mentioned above, have a positive and economic impact on the industry. A measuring device integrated into the system offers the additional characteristic of quantifying the various characteristics that affect power quality and thus identifies any specific aspect that needs attention.

Power quality measurement and analysis has typically been divided between steady-state concerns, such as harmonic distortion, and transient concerns, such as those resulting from faults or switching transients. Fourier analysis has been applied to the first class of problems while Joint Time-Frequency Analysis (JTFA), including wavelets, has traditionally been used for the second class [14].

When using the discrete Fourier transform (DFT) in nonstationary situations, the estimation of a time-varying signal during a specific time interval results in a serious deterioration of measurement accuracy. The effect is equivalent to the frequency deviation of the power system from its nominal value when measuring electrical signals [15,16]. To avoid this problem, the Fourier Transform applied in time segments as the Short-Time Fourier Transform (STFT) determines the frequency contents of a signal in each time window. Nevertheless, the size of the window affects the capacity of multi-resolution capability and the cost of the calculations is high. These deficiencies can be solved by using the Discrete Wavelet Transform (DWT) that also avoids the problems of interference of time and frequency distributions [17]. Besides, the DWT method offers a better time and frequency resolutions for high and low-frequency components, respectively. This method is best for locating disturbances in non-stationary conditions. However, the limitation of the DWT uncertain can be a shortcoming that can be minimized by an appropriate choice of the analyzing wavelet basic function. In this article, the Daubechies40 wavelet offers the best results for the analyzed electrical signals present in power distribution networks.

Traditionally, DWT [18–24], the generalized S-transform (GST) [25–30], the Time-Frequency distributions, and the Short-DFT (SDFT) [31] have been used for the analysis of the transient and time-varying nature of disturbance signals in electric power systems. In particular, power quality indices [32–38] have been defined using these transforms. Thus, the estimation of standard time-varying PQIs is done based on a proper JTFA [32], adaptive window-based fast generalized S-transform [33], empirical wavelet transform (EWT)-based time-frequency technique [34], cluster analysis of long-term power quality (PQ) data [35], wavelet packet transform [36], load composition rate and Euclidean norm of total harmonic distortions [37], and global harmonic parameters for phasor measurement units [38]. Most of these works use many PQIs in their proposals, for example, the global PQIs introduced in [35] contain up to seven PQI factors. As will be shown later, only two PQIs are used in the present work. Specific treatment of transient events is done in [39], whereas in [37] only stationary distortion is considered. Thus, it is necessary to conceive a measure of power quality in order to capture simultaneously both the "transient" characteristics of disturbance signals in electrical power systems [39] and the stationary ones. Moreover, it must be done with a minimum computational cost and using a single descriptive indicator. Wavelet-based techniques are good candidates for this purpose. However, a method based on empirical wavelet transform will work properly for off-line data processing but not for real-time analysis due to its computational cost [34,40].

Due to their powerful characteristics, in this work, DWT and multiresolution analysis (MRA) are chosen for the joint analysis of the stationary and transient parts of electrical signals. These parts are extracted from a monitoring window to ensure the correct use of the

DWT. In this way, the fundamental component of the electrical signal is extracted and the window containing the transient disturbance is processed. Both aspects are subsequently used to construct new DWT based power-quality indices that replace existing counterparts. On the other hand, the main drawback of this method is based on its limitation in the analysis of highly noisy signals. In these noisy conditions, the method has been proven to work properly until the minimum value of 34 dB for the signal-to-noise ratio (SNR) [19], which is more than enough for most of practical PQ events detections in low-voltage AC power distribution networks.

Reference [41] presents an interesting overview of power quality in low-voltage DC distribution networks. This study highlights the most relevant disturbances in DC networks and how they can be accurately characterized by means of power quality indices, some of which have been defined based on DFT. Due to the close relationship between the indices defined in AC power system networks and DC networks, a characterization of PQ in DC networks using the current DWT-based method seems to be interesting and deserves future work. In this possible context, the implementation of the algorithms derived from the present work in smart sensors may provide new functionalities for PQ improvement. However, they could increase the power consumption requirements of such sensors. In this way, reference [42] presents a useful study of the power requirement of functional sensors in a traditional PV system. It is based on Neural Network maximum power point tracking with cloud method and estimates the reduction percentage of power consumption in functional sensors.

In this paper we propose two wavelet-based power quality indices, the Instantaneous Disturbance Index (ITD) and the Global Disturbance Ratio (GDR), which comprehensively assess the power transfer quality of a given supply in steady-state and/or transient situations. The proposed ITD instantly shows the evolution of the PQ. It has the advantage of evaluating the PQ in real-time situations under any conditions and extracting the characteristics of the disturbance for any load in the power distribution networks. In addition, GDR has the advantage of assessing PQ by means of a single value and allows distinguishing between different events, so the GDR can be used as an input in a disturbance classifier. Therefore, the new PQ indices can identify properly practical waveform distortions in power networks. They can also be used to evaluate both the effectiveness and dynamic responses of PQ mitigation equipment in practical applications. In this work, a real-time platform is being developed to experimentally validate the feasibility of the proposed PQI measurement and analysis method.

The rest of this document is organized as follows: In Section 2, a simplified outline of wavelet filter selection criteria is presented. Section 3 briefly describes the DWT-based instantaneous indices used to evaluate PQ. Section 4 shows a detailed summary of the proposed measurement process is shown. In Section 5, results are discussed. The last section draws conclusions from the results.

2. Fundamentals of the Proposed Indices

2.1. Components Signal Estimation

The proposed estimation technique uses DWT and MRA for extracting the fundamental component of the input signal $s(t)$, which can be described by wavelet coefficients [14,24]

$$s(t) = \sum_{k=1}^{2^J} a_{J,k}\, \phi_{J,k}(t) + \sum_{j=1}^{J} \sum_{k=1}^{2^j} d_{j,k}\, \psi_{j,k}(t) \tag{1}$$

where j and k are the wavelet frequency scale and wavelet time scale, respectively, J is the highest j scale, i.e., the lowest frequency band; $a_{J,k}$ and $d_{J,k}$ are the wavelet coefficients; and $\psi(t)$ and $\varphi(t)$ are the mother and scale wavelet functions, respectively. Equations (1)–(3) present the basic concepts of MRA. The aim of MRA is to develop representations of a complex signal $f(t)$ in terms of scaling and wavelet functions.

The digitized version of input signal $s(t)$ is a sequence of n samples, $s(n)$, which can be processed in the same way as the DFT. In addition, the number of DWT levels (decomposition) is limited by the number of the original signal samples, which in turn must be a power of two.

Therefore, signal $s(n)$ can be presented in terms of its frequency components, i.e., coefficient $a_{J,k}$, $k = 1, \ldots, 2^J$, is the smoothed version of signal $s(n)$, and coefficients $d_{j,k}$, $k = 1, \ldots, 2^J$, $j = 1, \ldots, J$ are detailed versions of $s(n)$. They contain the lower and higher-frequency components, respectively [14].

$$s(n) = a_J(n) + \sum_{j=1}^{J} d_j(n) \tag{2}$$

With

$$\begin{aligned} a_j(k) &= \sum_n g(n - 2k)a_{j-1}(k) \\ d_j(k) &= \sum_n h(n - 2k)a_{j-1}(k) \end{aligned} \tag{3}$$

where a_j and d_j are the approximation and detail coefficients at level j, $g(k)$ and $h(k)$ are the high pass and low pass filters corresponding to the scaling and wavelet filters. The coefficients have half of the original input data due to the downsampling process.

For extracting the fundamental component of a signal by using MRA, the sampling rate and, so, the number of MRA steps must be specified. If it is assumed that H is the frequency band order with central frequency, $\omega/(2\pi)$, equal to fundamental frequency and fs is the sampling frequency, the number of MRA steps, J, satisfies the following expression:

$$\frac{fs}{2^J} = 2 \times H \tag{4}$$

In this work has been assumed that the most important transients occurring in actual situations of Power Distribution Networks (PDN) are captured into a frequency band of 6400 Hz, for $fs = 1/T_S = 12.8$ kHz. It assures accurate results with $J = 6$.

2.2. Decomposition Structures of MRA Method

MRA method uses a decomposition structure based on Quadrature Mirror Filter (QMF). QMF consists of two complementary filters, one low pass and the other high pass, which disjoin the frequency range into two equal parts. They decompose the input signal into two frequency intervals, the low pass filter output is downsampled and is used as new input of another identical filter pair corresponding to the next decomposition level. This operation is repeated recursively, decomposing the signal into approximation (a) and detail (d) coefficients for various scales.

The coefficients of the scale g and wavelet h filters, and the efficiency of wavelet analysis, are related to the selected mother wavelet. The right selection of the mother wavelet is the main task to obtain the desired results in signal analysis with WMRA. In this work, the selection of the analyzer function is based on the efficiency of the scale and wavelet filters used for this purpose.

The QMF filters frequency response of the first level DWT decomposition is shown in Figure 1. It depicts a comparison of the scale and wavelet filters between the db4, db10, db20, db40, and dmey mother wavelets used on the MRA method. The filters have overlapping frequency bands and energy leakage occurs between the two adjacent bands affecting the obtained coefficients. It is less pronounced with db40 and dmey mother wavelet.

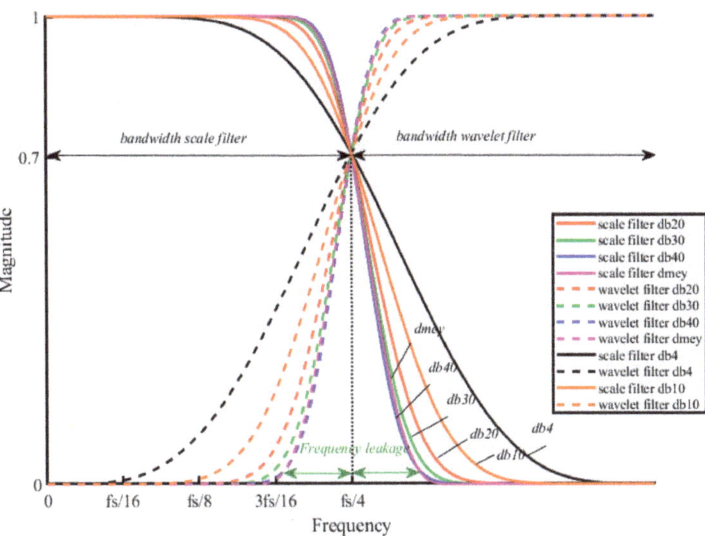

Figure 1. Comparative frequency response of mother wavelets.

2.3. Wavelet Filters Selection

Several mother wavelets have been evaluated in order to select the most suitable for the specific application of event detection and classification methods. In this paper, the selection of an explicit wavelet is based on the following principles:

- Minimum frequency leakage of QMF in the first levels of decomposition.
- Number of filter coefficients.
- Similarity between classical THD (Total Harmonic Distortion Ratio) with the TWD (Total Wavelet Disturbance Ratio) defined with wavelets.

The main selection criterion is the frequency selectivity. It is based on the magnitude transition zone slope of the frequency response of the wavelet filters. The selected mother wavelet will have better characteristics if the slope of both filters has the highest value and the information dispersion related to the frequency content of the signal will be less. In this way, a greater concentration of energy is obtained in single frequency bands.

A comparison between Fourier and Wavelet analysis is made to select the most appropriated. The STFT has good performance with signals disturbed by harmonics. According to the Parseval theorem, the energy of a signal can be decomposed in terms of the energy of a and d coefficients, the selected mother wavelet will be that allocate the signal energy correctly.

Figure 2a shows a signal that contains three harmonics, the 3rd (7% Urms), the 5th (10% Urms), and the 9th (10% Urms). This signal is chosen to support the selection of the best mother wavelet available. Figure 2b shows the FFT spectrum where the fundamental harmonic has 97.570% of the total signal energy, the 3rd has 0.478% and both the 5th and 9th have 0.976%. Figure 2c also shows as a bar graph the energy percentage of this disturbed signal obtained with various mother wavelets. Table 1 shows the percentage of signal energy achieved with the mother wavelets depicted in the legend of Figure 2c.

Table 2 shows the THD and TWD results calculated, the results are similar excluding bior6.8 and db10 ones.

According to Table 1, the mother wavelets with the best frequency response correctly distribute the energy of the signal according to the FFT spectrum (Figure 2). Therefore, the db40 and dmey wavelets are the best mother wavelets. Since the db40 wavelet has fewer coefficients than the dmey wavelet, the db40 is the mother wavelet with the best performance.

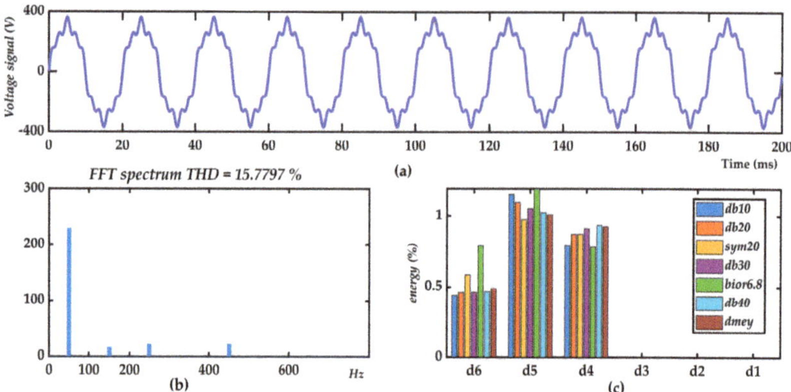

Figure 2. Voltage signal with harmonics: (**a**) Instantaneous representation, (**b**) FFT spectrum, (**c**) distribution of the energy percentage of the signal according to the coefficients of the mother wavelets.

Table 1. Energies (%) of voltage signal decomposition shown in Figure 2.

f(Hz)	a6 0–100	d6 100–200	d5 200–400	d4 400–800	d3 800–1600	d2 100–200	d1 3200–6400
db10	97.6142	0.4378	1.1561	0.7912	0.0007	0.0000	0.0000
db20	97.5719	0.4610	1.0966	0.8706	0.0000	0.0000	0.0000
sym20	97.5705	0.5844	0.9745	0.8706	0.0000	0.0000	0.0000
db30	97.5704	0.4616	1.0550	0.9130	0.0000	0.0000	0.0000
bior6.8	97.2315	0.7903	1.1931	0.7817	0.0035	0.0000	0.0000
db40	97.5699	0.4673	1.0257	0.9371	0.0000	0.0000	0.0000
dmey	97.5705	0.4905	1.0116	0.9274	0.0000	0.0000	0.0000

Table 2. Total Harmonic Distortion (THD) and Total Wavelet Disturbance Ratio (TWD) of voltage signal decomposition shown in Figure 2.

THD	TWD db10	TWD db20	TWD sym20	TWD db30	TWD bior6.8	TWD db40	TWD dmey
15.7797	15.6335	15.7751	15.7798	15.7801	16.8741	15.7818	15.7796

3. Quantitative Formulations of Steady-State and Transient Power Quality Aspects

This formulation is based on the IEC 61000-4-7 standard [43], EN-50160 standard [44] and the IEEE Std 1159™-2009 [45] guidelines, in order to fulfil the essential requirements and to define the characteristics of the supplied voltage and the most common disturbances.

3.1. DWT-Based Disturbance Ratio

The Total Harmonic Distortion ratio (*THD*) for single-phase networks (or polyphase balanced networks) has been defined traditionally as [43,46]

$$THD = \frac{\sqrt{\sum_{h=2}^{H} S_k^2}}{S_1} \cdot 100 \quad H = \frac{f_m}{2 \cdot f_0} \tag{5}$$

$$THD = \frac{\sqrt{S^2 - S_1^2}}{S_1} \cdot 100 \tag{6}$$

where S denotes RMS value of signal $s(n)$, S_1 denotes fundamental component of $s(n)$, and S_k the k component of $s(n)$. Using the MRA tool, S is given by

$$S = \sqrt{Sa_J^2 + \sum_{j \le J} Sd_j^2} \tag{7}$$

Here Sa_J, is the RMS value of the N samples signal $a_J(n)$ in the lowest frequency band J, where the fundamental component S_1 is included. $\{Sd_j\}$ is the set of RMS values of $d_j(n)$ signal in the higher frequency band, or wavelet-level lower than or equal to the scaling level J. Then, the Total Wavelet Disturbance ratio TWD [47] is defined as

$$TWD = \frac{\sqrt{\sum\limits_{j \le J} Sd_j^2}}{Sa_J} \cdot 100 \tag{8}$$

3.2. DWT-Based Instantaneous Disturbance Ratio

The Instantaneous Transient Disturbance ratio ($ITD(n)$), the transient version of TWD, is defined (9) in terms of the time-scale distribution of the MRA components:

$$ITD(n) = \frac{\sqrt{\sum\limits_{j \le J} d_j^2(n)}}{A_J} \cdot 100 \tag{9}$$

where A_J is the fundamental energy component defined as:

$$A_J^2 = \frac{1}{N} \sum_{n=1}^{N} a_J^2(n) \tag{10}$$

It can be seen that Sa_J is identical to A_J.

The definition of the $ITD(n)$ can be interpreted as a "time-varying" power quality evaluation determined by the time-frequency localized energy ratio of the disturbance events to the fundamental frequency energy.

As an instantaneous quantity, the proposed $ITD(n)$ index can reveal the time-varying characteristics of the transient disturbance for assessment purposes.

The time-varying signature can be quantified as a single number, as in the case of steady-state disturbances. Therefore, a "transient-interval average" of the $ITD(n)$, <ITD>, can be defined over a sample interval N as follows:

$$\langle ITD \rangle = \frac{1}{N} \sum_{n=1}^{N} ITD(n) \tag{11}$$

Note that <ITD> is almost identical to THD when only steady-state disturbances are presented in the signal.

The non-stationary events duration is a very relevant parameter to be considered. It can be measured with high precision by the wavelet procedure used in this work. Then a Global Disturbance Ratio, GDR, can be defined,

$$GDR = \left(1 + \frac{T_0}{T}\right) \langle ITD \rangle \tag{12}$$

where T_0 is the duration of the transient disturbance and T is the time interval window used. The selection of the time interval (T_0) can be determined by the time index of the first maximum peak value of the $ITD(n)$, t_0, and the time index of the last maximum peak value of the $ITD(n)$, $t_0 + T_0$. Only if steady-state disturbances are present in the signal, GDR = <ITD>, otherwise GDR is greater than <ITD>.

Consequently, the proposed index GDR is the transient-interval average, <ITD>, plus its magnitude weighted by a term related to the event duration.

A loaded power network with sinusoidal voltages yields an ideal null *GDR*. Conversely, a high value of *GDR* would indicate a high level of steady-state and/or transient disturbances, with the contribution of each event aspect well defined and measured. Note that the proposed index *GDR* presents the advantage over the *THD* of distinguishing transients and stationary events. The duration of the disturbance plays a significant role in the *GDR* index.

4. Measurement Process

4.1. Developed Platform

In this work, a developed platform is used to test the effectiveness of the proposed indices under common real-time working conditions (Figure 3). This procedure permits the emulation of actual power systems. In this context, the proposed instantaneous indices are suitable.

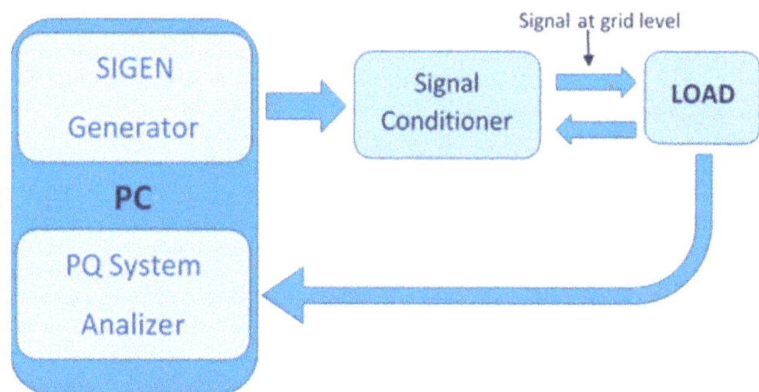

Figure 3. Developed system scheme: Voltage generator with preset disturbances, power amplifier, and signal conditioner, load and Power Quality Analyzer.

The developed system consists of a signal generator that allows the design of any kind of disturbance. This signal is amplified, conditioned and applied to a real load. Finally, the voltage and current are measured on the load side and processed in a power quality analyzer.

Matlab® software has been used to program a virtual signal generator, called *Sigen*, which allows a complete configuration of the signals required for testing. *Sigen* mainly generates a pattern based on the parameters defined by the user; thus, steady-state and/or transient-state disturbances can be modelled. The Graphical user interface of single-phase *Sigen* is shown in Figure 4.

The processes of *Sigen* are performed to complete the effective generation of electric signals, as follows.

- First, electrical inputs signals are defined and the user sets their parameters.
- Second, signals according to these specifications are built.
- Finally, the designed signals are sent to a file or to the data acquisition board (DAQ).

Sigen is designed to program disturbances as described in IEEE standard 1159-09 for monitoring electric power quality [45].

The design is based on generating single or three-phase voltages and line currents. Generated data sets are obtained from a host PC in data files with the American Standard Code for Information Interchange format compatible with the most popular data analysis tools.

The host PC is equipped with the NI USB-6259; it is a 16-Bit High-Speed M Series Multifunction DAQ for USB. It acquires eight differential inputs. Analog inputs are converted with 16 bits of resolution sampled at 1.25 MS/s. Voltage and current sensors are built with Hall Effect voltage and current transducers, type LV25-P and LA25-NP, respectively. Low-level voltage signals proportional to the phase-neutral voltages are available.

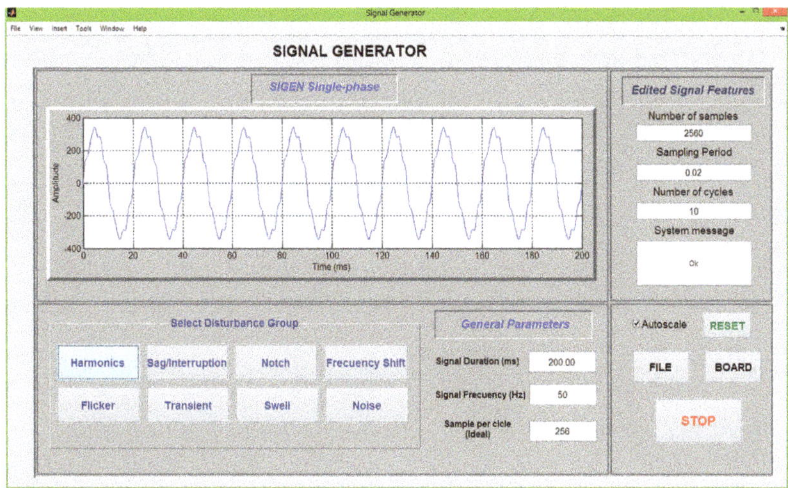

Figure 4. Graphical user interface of the *Sigen* system.

At the final stage, an amplifier section increases the voltage signals up to the grid voltage level. A decisive request remains on the amplifier section since it must ensure accurate and constant gain and phase shift overall bandwidth required. A Pacific Power Source Model 320 is used as a power amplifier to fulfill all the proposed requirements (output voltage up to ±600 peak volts; maximum output power: 1.2 kVA; bandwidth (30–5 kHz) at full power; THD < 0.2%).

Disturbed voltage signals at the grid level are generated for studying Power Quality Events in several types of loads.

The generated voltage signals are used to simulate a power system with actual voltage sources and arbitrary loads. It can process polyphase sinusoidal voltages added with simple or multiple disturbances.

The power-quality analyzer is a virtual device that processes the signal data file from an A/D converted connected to the signal conditioner (Figure 5). A control program developed in MATLAB® diagnoses quality aspects of the input signals, such as frequency stability, distortion level, symmetry of three-phase signals (balance between phases R, S and T), and others that can be inferred from the graphical user interface of Figure 4.

In order to carry out this diagnosis, the system can measure and present/display the graphs (with its time evolution):

- Instantaneous network frequency, following its changes at intervals of measurement of one cycle, considering deformed signals and with adding noise.
- Harmonics, represented in phasor form using two bar charts, one for magnitudes and another for phases.
- Instantaneous PQI and coefficients of power quality indices (percent), and a presentation of the data corresponding to the signals.
- DWT coefficients with a representation of the wavelet level selected by the user.
- Three-phase signals of voltage and current, or in its place, their respective fundamental symmetrical components, i.e., the fundamental components of positive, negative and zero sequences.

Furthermore, the PQ System analyzer computes power quantities in Wavelet and Fourier domain specified by IEEE standard 1459–2010 [46].

The developed platform and the proposed indices can be further used in real-time for both monitoring and detecting faults in power networks [20] and electrical machinery [48]. By means of a signal-based fault diagnosis method, the proposed novel indices can be applied to loads such as induction motors, power converters, and mechanical components.

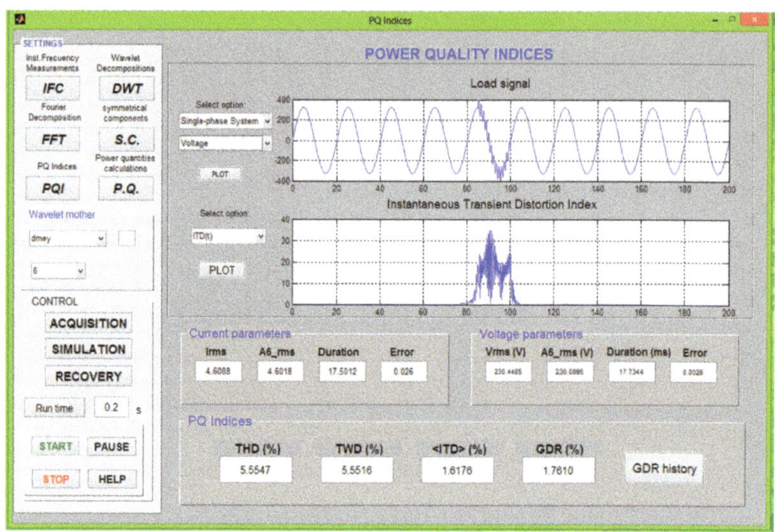

Figure 5. Graphical user interface of the power quality (PQ) System Analyzer.

4.2. PQI Measurements

Different kinds of power quality disturbances are applied through the power voltage amplifier to linear and nonlinear loads. In these loads, voltages and currents are taken by the PQ System Analyzer.

For the voltage quality assessment, instantaneous frequency measurement is performed which enables synchronization between the signal period and the sampling sequence.

For the considered voltage and current windows specified by IEC standard 61000-4-30 [49], time-frequency-based quality aspects are calculated by the DWT. For the case of 12.8-kHz sampling rate in the *Sigen* system, Table 3 summarizes the frequency band information for different wavelet analysis levels. The db40 mother wavelet is applied in MRA of the voltage and current signals.

Table 3. Frequency bands and harmonics of six levels of the Discrete Wavelet Transform (DWT).

Level	Freq. Band (Hz)	Odd Band Harmonics
7 (d1)	3200–6400	63rd–127rd (odd num.)
6 (d2)	1600–3200	33rd–63rd (odd num.)
5 (d3)	800–1600	17th–31st (odd num.)
4 (d4)	400–800	9th, 11th, 13th, 15th
3 (d5)	200–400	5th, 7th
2 (d6)	100–200	3rd
1 (a6)	DC-100	1st

According to Equations (5), (6), (8), (9), (11) and (12) the PQIs are computed. The PQ indices have been tested on a variety of stationary, non-stationary single-phase signals as well as balanced or unbalanced three-phase signals.

To measure properly disturbances with a higher duration than the considered window, the PQ system analyzer provides a *GDR* history tool to save stored data of the *GDR* index.

The fundamental component of the grid voltage presented in all the used test signals is distorted by harmonics and/or non-stationary events, in accordance with both, the IEEE standard 1159-2009 [45] and the European standard EN-50160 [44].

5. Illustrative Results

To show the effectiveness of *GDR* and *<ITD>*, *TWD*, and *THD* indices are also calculated and the result obtained has been compared in two examples sets. In all of them, the grid voltage waveform contains a fundamental component of 230 Vrms, 50 Hz of nominal frequency, and stationary and/or transient disturbances. In the first set, a group of signals with single disturbances and almost identical *THD* index are considered. In the second set, complex signals with more significant and combined disturbances are studied.

5.1. Voltages Waveform with Similar THD

Several disturbed voltage waveforms extracted from the developed platform are analyzed. In particular, the considered signals are three voltage sags and two swells with different duration and amplitude, three oscillatory transients with distinct frequency and duration, a steady-state distorted voltage with three harmonics (5th, 7th, and 9th with relative amplitudes depicted in the FFT spectrum of Figure 6) and a flicker disturbance. Some of them are depicted in Figure 7 and they all only have in common a similar THD value.

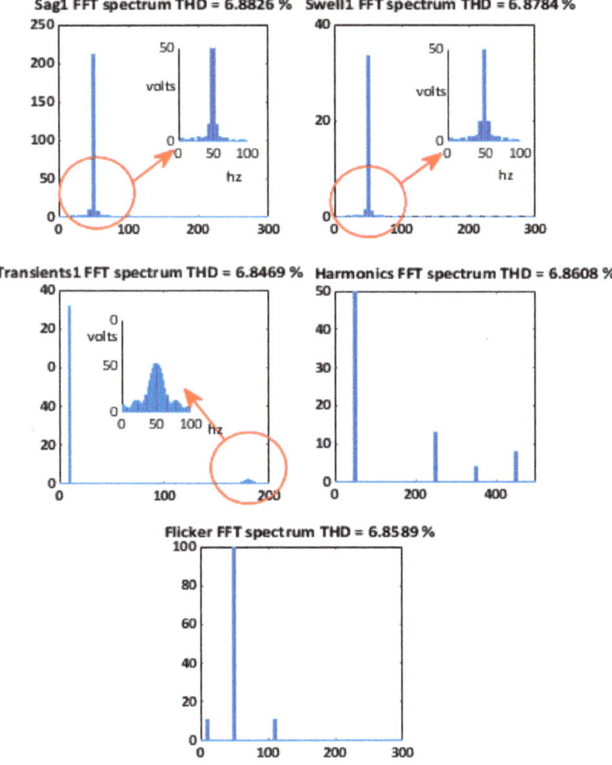

Figure 6. FFT spectra for several signals used on case A.

These voltage signals are applied to linear and nonlinear loads, and the resulting load currents are analyzed in our PQ System Analyzer too. The corresponding voltage FFT spectra are shown in Figure 7. The $ITD(t)$ of the signals shown in Figure 6 are respectively depicted in Figure 8. The first $ITD(t)$ at the left top of Figure 8 shows two peak values: 3.1% at 31.1 ms and 2.7% at 146.2 ms. The second $ITD(t)$ at the right top shows two peak values: 3.3% at 21.2 ms and 2.5% at 136 ms. The peak values in both signals are associated with the depth/crest of the corresponding sag/swell. The $ITD(t)$ attaching to oscillatory transient of Figure 8 shows a peak value of 44.27% at 53 ms. In this context, the peak values indicate that the oscillatory transient is a more severe event than the harmonic distortion. However, it presents the shortest time duration and the highest frequency content of the five disturbances (FFT spectra of Figure 7). The $ITD(t)$ for steady-state conditions do not offer any new relevant information over the THD, nevertheless $<ITD>$ is almost identical to THD (Table 4). In this case, the error between both quantities is less than 0.5%.

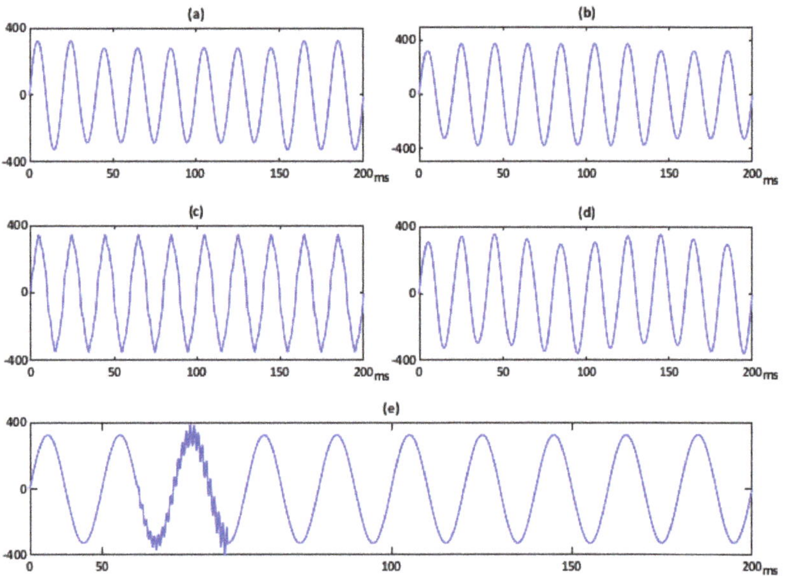

Figure 7. Single-phase voltage disturbances with similar THD. (**a**) Voltage sag1. (**b**) Voltage swell1. (**c**) Harmonic disturbances. (**d**) Flicker disturbance. (**e**) Transient1 disturbance.

Table 4 provides the quantities obtained for all the examples discussed in terms of their respective duration (T_0), Vrms value and time-frequency based transient power quality indices. It can be seen that THD is almost the same for all disturbed signals even if they correspond to different classes of disturbances. TWD is similar to THD in those signals in which the frequency bands are far from the fundamental energy component. On the contrary, in those with the frequency components near to the fundamental, the values are very different and the duration of the disturbance does not make any substantial difference. Instead, GDR and $<ITD>$ assessments are more consistent with the energy content of these disturbance signals.

Table 4. Summary of Power Quality Indices for the case A.

	Sag1	Sag2	Sag3	Harmonics	Transient1	Transient2	Transient3	Flicker	Swell1	Swell2
T_0 (ms)	120.62	80.31	60.22	0	82.89	57.42	28.12	0	120.62	80.62
Vrms (V)	212.67	218.35	220.65	230.54	230.53	230.56	230.42	230.54	251.63	244.18
THD	6.8826	6.8857	6.8821	6.8608	6.8410	6.8852	6.8469	6.8589	6.8784	6.8742
TWD	0.6491	0.6515	0.7078	6.8607	6.8407	6.8848	6.8425	4.6067	0.6564	0.6610
$<ITD>$	0.3279	0.3329	0.3552	6.4397	3.8688	3.3018	2.2395	4.0494	0.3284	0.3409
GDR	0.5257	0.4765	0.4628	6.4397	5.4723	4.2498	2.5544	4.0494	0.5265	0.4783

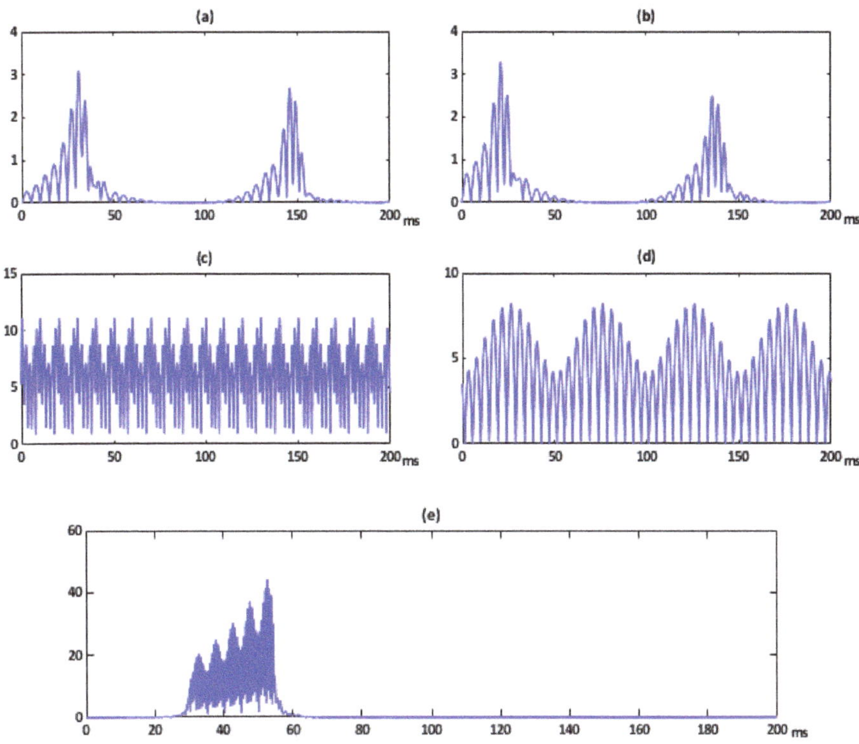

Figure 8. Instantaneous Transient Disturbance ratio index *ITD*(*t*) of voltage disturbances corresponding to Figure 5.

The *ITD*(*t*) (Figure 8) shows the instantaneous character of the disturbance noting the relevance of any suddenly introduced change. Instead, its averaged energy over the observation window is considered by the *GDR*. Furthermore, the *GDR* permits to distinguish between different disturbances and can be used for classification purposes. In particular, the *GDR* differentiates the proposed disturbances, with almost the same *THD*, by giving importance to both, the amplitude and the duration of the disturbance. This index offers the advantage to indicate clearly the most relevant of them, as can be seen in Table 4 for any kind of the disturbances numbered as "1". So transient events are perfectly characterized by this procedure. Nevertheless, the studied sag and swell signals have the same frequency contents and are undistinguished by all the indices, so the RMS value has to be considered as an assisting index.

5.2. Disturbances Combination in Voltage Waveforms

In practical situations, a PQ event usually consists of a combination of singular disturbances, most of them treated in case A. Therefore, a set of complex disturbed signals with more severe and combined effect are studied.

Figure 9a depicts the *ITD*(*t*) corresponding to a pronounced sag. It shows two peak value: 5.5% at 56.25 ms and 3.3% at 147.25 ms. Figure 9b shows this index for a combination of the mentioned sag and a steady-state distorted voltage with five harmonics (3rd, 5th, 7th, 9th, and 11th and Vrms equal to 10 V, 17 V, 7 V 13 V, and 3 V, respectively). The *ITD*(*t*) resulting from a steady-state distorted voltage only with the five harmonics is provided in Figure 9c. In addition, Figure 9d shows this index for a combination of the same steady-state voltage and an important transient oscillation. The *ITD*(*t*) of a voltage disturbed only with this single transient is shown in Figure 9e. It has a peak value of 30% at 127 ms.

In this case, the waveform offers more relevant changes than the previous set of examples because more substantial disturbances are present. When the signal is disturbed with two events, both are evinced in $ITD(t)$ graph.

The PQ indices obtained are presented in Table 5. By comparing the values obtained, the GDR index remarks the effect of the disturbances combination. As it can be observed from Table 5, the proposed index is the only one capable to indicate clearly the accumulative effect of the combined events in complex signals. Although the GDR index has not the additivity property, it performs the indication by significantly increasing its relative value when distinct disturbances are present.

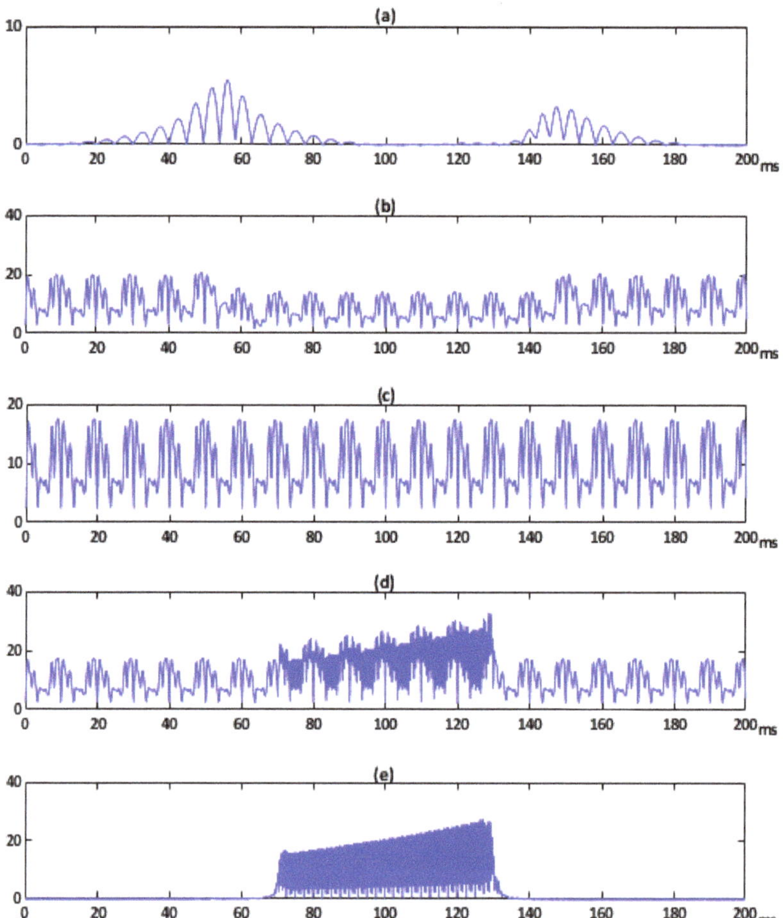

Figure 9. Instantaneous Transient Disturbance ratio index $ITD(t)$ of voltage disturbances combination.

Table 5. Summary of Power Quality Indices for case B.

	Sag	Sag + Harmonics	Harmonics	Harmonics + Transients	Transients
T_0 (ms)	100.62	100.62	0	62.81	62.81
Vrms	200.30	201.51	231.36	232.14	230.79
THD	16.8966	20.2710	10.8804	13.6716	8.2782
TWD	1.2597	11.0999	10.8804	13.6705	8.2766
<ITD>	0.7170	10.0536	10.1025	12.2901	4.1376
GDR	1.0778	15.1118	10.1025	16.1499	5.4370

6. Conclusions

The motivation for this work stemmed from the growing need for a more effective analysis of power quality in electrical systems and equipment. In this way, the main contribution of this research is the introduction of a novel wavelet-based single indicator, designated global disturbance ratio (GDR) that has been tested on real signals to address an integral assessment of the electrical network PQ. The objective of these test is to guarantee the applicability of such index to smart network sensors in particular, and to PQ monitoring in general. To this end, a PQ System Analyzer based on wavelet techniques is developed, turning out to be an effective device to verify the behavior of the proposed indices.

The GDR is based on an instantaneous index $ITD(t)$, which is also introduced, and considers two quality aspects of the electrical signal: steady-state power quality relative to harmonic level, and the non-stationary index relative to oscillatory transients or sudden amplitude changes in the signal.

In the initial theoretical stage of signal analysis it is remarkable the mother wavelet selection procedure used to guarantee the most accurate frequency decomposition of the voltage signal under steady-state and non-stationary events.

The applicability of the GDR and $ITD(t)$ is illustrated for typical source-load configurations. The proposed PQIs have been tested on single-phase stationary and non-stationary complex signals, as well as three-phase balanced and unbalanced signals.

Finally, the $ITD(t)$ and GDR indices are a powerful tool for detecting and monitoring non-stationary signal components and they can extract relevant characteristics. In particular, the GDR index may be applied as the input of a classifier. In this sense, the authors are working on the physical implementation of an intelligent sensor based on a DSP, which provides the proposed PQIs. This guideline deserves future research.

Author Contributions: M.D.B.-T. Conceived and developed the idea of this research, generated the data, performed the simulations, designed the whole structure of the experimental device, and wrote most of the paper; J.C.B. Contributed to the methodology and wrote some of the paper; C.Á.-A. contributed to the development of the experimental device and the simulations and wrote some of the paper. All authors have read and agreed to the published version of the manuscript.

Funding: This research was funded by the Universidad de Sevilla (VI Plan Propio de Investigación y Transferencia) under grant 2020/00000596.

Institutional Review Board Statement: Not applicable.

Informed Consent Statement: Not applicable.

Data Availability Statement: Not applicable.

Conflicts of Interest: The authors declare no conflict of interest.

References

1. Montaño, J.-C.; Bravo, J.-C.; Borrás, M.-D. Joint Time–Frequency Analysis of the Electrical Signal. In *Power Quality: Mitigation Technologies in a Distributed Environment*; Moreno-Muñoz, A., Ed.; Springer London: London, UK, 2007; pp. 41–72, ISBN 978-1-84628-772-5.
2. Chawda, G.S.; Shaik, A.G.; Shaik, M.; Padmanaban, S.; Holm-Nielsen, J.B.; Mahela, O.P.; Kaliannan, P. Comprehensive Review on Detection and Classification of Power Quality Disturbances in Utility Grid with Renewable Energy Penetration. *IEEE Access* **2020**, *8*, 146807–146830. [CrossRef]
3. Sharma, A.; Rajpurohit, B.S.; Singh, S.N. A review on economics of power quality: Impact, assessment and mitigation. *Renew. Sustain. Energy Rev.* **2018**, *88*, 363–372. [CrossRef]
4. Viciana, E.; Alcayde, A.; Montoya, F.; Baños, R.; Arrabal-Campos, F.; Manzano-Agugliaro, F. An Open Hardware Design for Internet of Things Power Quality and Energy Saving Solutions. *Sensors* **2019**, *19*, 627. [CrossRef]
5. Artale, G.; Caravello, G.; Cataliotti, A.; Cosentino, V.; Di Cara, D.; Dipaola, N.; Guaiana, S.; Panzavecchia, N.; Sambataro, M.G.; Tinè, G. PQ and Harmonic Assessment Issues on Low-Cost Smart Metering Platforms: A Case Study. *Sensors* **2020**, *20*, 6361. [CrossRef] [PubMed]
6. Thongkhao, Y.; Pora, W. A low-cost Wi-Fi smart plug with on-off and Energy Metering functions. In Proceedings of the 2016 13th International Conference on Electrical Engineering/Electronics, Computer, Telecommunications and Information Technology (ECTI-CON), Chiang Mai, Thailand, 28 June–1 July 2016; IEEE: Piscataway, NJ, USA; pp. 1–5.
7. Yaemprayoon, S.; Boonplian, V.; Srinonchat, J. Developing an innovation smart meter based on CS5490. In Proceedings of the 2016 13th International Conference on Electrical Engineering/Electronics, Computer, Telecommunications and Information Technology (ECTI-CON), Chiang Mai, Thailand, 28 June–1 July 2016; IEEE: Piscataway, NJ, USA, 2016; pp. 1–4.
8. Sun, Q.; Li, H.; Ma, Z.; Wang, C.; Campillo, J.; Zhang, Q.; Wallin, F.; Guo, J. A comprehensive review of smart energy meters in intelligent energy networks. *IEEE Internet Things J.* **2015**, *3*, 464–479. [CrossRef]
9. Albu, M.M.; Sănduleac, M.; Stănescu, C. Syncretic use of smart meters for power quality monitoring in emerging networks. *IEEE Trans. Smart Grid* **2016**, *8*, 485–492. [CrossRef]
10. Sharon, D.; Montano, J.; Lopez, A.; Castilla, M.; Borras, D.; Gutierrez, J. Power Quality Factor for Networks Supplying Unbalanced Nonlinear Loads. *IEEE Trans. Instrum. Meas.* **2008**, *57*, 1268–1274. [CrossRef]
11. Montano, J.; Leon, C.; Garcia, A.; Lopez, A.; Monedero, I.; Personal, E. Random Generation of Arbitrary Waveforms for Emulating Three-Phase Systems. *IEEE Trans. Ind. Electron.* **2012**, *59*, 4032–4040. [CrossRef]
12. Jin, H.; Titus, A.; Liu, Y.; Wang, Y.; Han, Z. Fault Diagnosis of Rotary Parts of a Heavy-Duty Horizontal Lathe Based on Wavelet Packet Transform and Support Vector Machine. *Sensors* **2019**, *19*, 4069. [CrossRef]
13. Gao, Z.; Cecati, C.; Ding, S.X. A survey of fault diagnosis and fault-tolerant techniques-part I: Fault diagnosis with model-based and signal-based approaches. *IEEE Trans. Ind. Electron.* **2015**, *62*, 3757–3767. [CrossRef]
14. Borras, D.; Castilla, M.; Moreno, N.; Montano, J.C. Wavelet and neural structure: A new tool for diagnostic of power system disturbances. *IEEE Trans. Ind. Appl.* **2001**, *37*, 184–190. [CrossRef]
15. Wen, H.; Guo, S.; Teng, Z.; Li, F.; Yang, Y. Frequency Estimation of Distorted and Noisy Signals in Power Systems by FFT-Based Approach. *IEEE Trans. Power Syst.* **2014**, *29*, 765–774. [CrossRef]
16. Lopez, A.; Montano, J.; Castilla, M.; Gutierrez, J.; Borras, M.D.; Bravo, J.C. Power System Frequency Measurement Under Nonstationary Situations. *IEEE Trans. Power Deliv.* **2008**, *23*, 562–567. [CrossRef]
17. Jeong, J.; Williams, W.J. Kernel design for reduced interference distributions. *IEEE Trans. Signal Process.* **1992**, *40*, 402–412. [CrossRef]
18. Borrás, M.D.; Montaño, J.C.; Castilla, M.; López, A.; Gutiérrez, J.; Bravo, J.C. Voltage index for stationary and transient states. In Proceedings of the Mediterranean Electrotechnical Conference-MELECON, Valletta, Malta, 26–28 April 2010; University of Seville: Sevilla, Spain, 2010; pp. 679–684.
19. Borrás, M.D.; Bravo, J.C.; Montaño, J.C. Disturbance Ratio for Optimal Multi-Event Classification in Power Distribution Networks. *IEEE Trans. Ind. Electron.* **2016**, *63*, 3117–3124. [CrossRef]
20. Bravo, J.C.; Borras, M.D.; Torres, F.J. Development of a smart wavelet-based power quality monitoring system. In Proceedings of the 2018 International Conference on Smart Energy Systems and Technologies (SEST), Sevilla, Spain, 10–12 September 2018; IEEE: Piscataway, NJ, USA, 2018; pp. 1–6.
21. Lu, S.-D.; Sian, H.-W.; Wang, M.-H.; Liao, R.-M. Application of Extension Neural Network with Discrete Wavelet Transform and Parseval's Theorem for Power Quality Analysis. *Appl. Sci.* **2019**, *9*, 2228. [CrossRef]
22. Khokhar, S.; Asuhaimi, A.; Zin, M.; Memon, A.P.; Safawi Mokhtar, A. A new optimal feature selection algorithm for classification of power quality disturbances using discrete wavelet transform and probabilistic neural network. *Measurement* **2017**, *95*, 246–259. [CrossRef]
23. Wang, J.; Xu, Z.; Che, Y. Power Quality Disturbance Classification Based on DWT and Multilayer Perceptron Extreme Learning Machine. *Appl. Sci.* **2019**, *9*, 2315. [CrossRef]
24. Costa, F.B. Boundary Wavelet Coefficients for Real-Time Detection of Transients Induced by Faults and Power-Quality Disturbances. *IEEE Trans. Power Deliv.* **2014**, *29*, 2674–2687. [CrossRef]
25. Bravo-Rodríguez, J.C.; Torres, F.J.; Borrás, M.D. Hybrid Machine Learning Models for Classifying Power Quality Disturbances: A Comparative Study. *Energies* **2020**, *13*, 2761. [CrossRef]

26. Kumar, R.; Singh, B.; Shahani, D.T.; Chandra, A.; Al-Haddad, K. Recognition of Power-Quality Disturbances Using S-Transform-Based ANN Classifier and Rule-Based Decision Tree. *IEEE Trans. Ind. Appl.* **2015**, *51*, 1249–1258. [CrossRef]
27. Shafiullah, M.; Abido, M.A.; Abdel-Fattah, T. Distribution grids fault location employing ST based optimized machine learning approach. *Energies* **2018**, *11*, 2328. [CrossRef]
28. Wang, H.; Wang, P.; Liu, T. Power quality disturbance classification using the S-transform and probabilistic neural network. *Energies* **2017**, *10*, 107. [CrossRef]
29. Reddy, M.V.; Sodhi, R. A modified S-transform and random forests-based power quality assessment framework. *IEEE Trans. Instrum. Meas.* **2018**, *67*, 78–89. [CrossRef]
30. Tang, Q.; Qiu, W.; Zhou, Y. Classification of Complex Power Quality Disturbances Using Optimized S-Transform and Kernel SVM. *IEEE Trans. Ind. Electron.* **2020**, *67*, 9715–9723. [CrossRef]
31. Heydt, G.T.; Fjeld, P.S.; Liu, C.C.; Pierce, D.; Tu, L.; Hensley, G. Applications of the windowed FFT to electric power quality assessment. *IEEE Trans. Power Deliv.* **1999**, *14*, 1411–1416. [CrossRef]
32. Lin, T.; Domijan, A. On power quality indices and real time measurement. *IEEE Trans. Power Deliv.* **2005**, *20*, 2552–2562. [CrossRef]
33. Biswal, M.; Dash, P.K. Estimation of time-varying power quality indices with an adaptive window-based fast generalised S-transform. *IET Sci. Meas. Technol.* **2012**, *6*, 189. [CrossRef]
34. Thirumala, K.; Umarikar, A.C.; Jain, T. Estimation of Single-Phase and Three-Phase Power-Quality Indices Using Empirical Wavelet Transform. *IEEE Trans. Power Deliv.* **2015**, *30*, 445–454. [CrossRef]
35. Jasiński, M.; Sikorski, T.; Kostyła, P.; Leonowicz, Z.; Borkowski, K. Combined cluster analysis and global power quality indices for the qualitative assessment of the time-varying condition of power quality in an electrical power network with distributed generation. *Energies* **2020**, *13*, 2050. [CrossRef]
36. Morsi, W.G.; El-Hawary, M.E. Wavelet Packet Transform-Based Power Quality Indices for Balanced and Unbalanced Three-Phase Systems Under Stationary or Nonstationary Operating Conditions. *IEEE Trans. Power Deliv.* **2009**, *24*, 2300–2310. [CrossRef]
37. Jo, S.; Son, S.; Park, J. On Improving Distortion Power Quality Index in Distributed Power Grids. *IEEE Trans. Smart Grid* **2013**, *4*, 586–595. [CrossRef]
38. Granados-Lieberman, D. Global harmonic parameters for estimation of power quality indices: An approach for PMUs. *Energies* **2020**, *13*, 2337. [CrossRef]
39. Shin, Y.; Powers, E.J.; Grady, M.; Arapostathis, A. Power quality indices for transient disturbances. *IEEE Trans. Power Deliv.* **2006**, *21*, 253–261. [CrossRef]
40. Sahani, M.; Dash, P.K.; Samal, D. A real-time power quality events recognition using variational mode decomposition and online-sequential extreme learning machine. *Meas. J. Int. Meas. Confed.* **2020**, *157*, 107597. [CrossRef]
41. Barros, J.; de Apráiz, M.; Diego, R.I. Power quality in DC distribution networks. *Energies* **2019**, *12*, 848. [CrossRef]
42. Chang, S.; Wang, Q.; Hu, H.; Ding, Z.; Guo, H. An NNwC MPPT-Based energy supply solution for sensor nodes in buildings and its feasibility study. *Energies* **2019**, *12*, 101. [CrossRef]
43. IEC 61000-4-7. *Electromagnetic Compatibility (EMC)—Part 4-7: Testing and Measurement Techniques—General Guide on Harmonics and Interharmonics Measurements and Instrumentation, for Power Supply Systems and Equipment Connected Thereto*; International Electrotechnical Commission: Geneva, Switzerland, 2008.
44. *EN 50160: Voltage Characteristics of Electricity Supplied by Public Distribution Network*; British Standards: London, UK, 2010.
45. *IEEE Recommended Practice for Monitoring Electric Power Quality*; IEEE Std 1159-2019 (Revision of IEEE Std 1159-2009); Institute of Electrical and Electronics Engineers: Piscataway, NJ, USA, 2019; pp. 1–98.
46. *IEEE Standard Definitions for the Measurement of Electric Power Quantities under Sinusoidal, Nonsinusoidal, Balanced, or Unbalanced Conditions*; IEEE Std 1459-2010 (Revision of IEEE Std 1459-2000); Institute of Electrical and Electronics Engineers: Piscataway, NJ, USA, 2010; pp. 1–50.
47. Morsi, W.G.; El-Hawary, M.E. Reformulating Power Components Definitions Contained in the IEEE Standard 1459–2000 Using Discrete Wavelet Transform. *IEEE Trans. Power Deliv.* **2007**, *22*, 1910–1916. [CrossRef]
48. Gritli, Y.; Zarri, L.; Rossi, C.; Filippetti, F.; Capolino, G.; Casadei, D. Advanced Diagnosis of Electrical Faults in Wound-Rotor Induction Machines. *IEEE Trans. Ind. Electron.* **2013**, *60*, 4012–4024. [CrossRef]
49. IEC 61000-4-30:2015. *Electromagnetic Compatibility (EMC)—Part 4-30: Testing and Measurement Techniques—Power Quality Measurement Methods*; International Electrotechnical Commission: Geneva, Switzerland, 2015.

Article

AnkFall—Falls, Falling Risks and Daily-Life Activities Dataset with an Ankle-Placed Accelerometer and Training Using Recurrent Neural Networks

Francisco Luna-Perejón [1,2,*], Luis Muñoz-Saavedra [1,2], Javier Civit-Masot [1,2], Anton Civit [1,2,3] and Manuel Domínguez-Morales [1,2,3]

1   Architecture and Computer Technology Department, ETSII-EPS, University of Seville, 41004 Sevilla, Spain; lmsaavedra@us.es (L.M.-S.); jcivit@atc.us.es (J.C.-M.); civit@us.es (A.C.); mjdominguez@us.es (M.D.-M.)
2   Robotics and Technology of Computers Laboratory, University of Seville, 41004 Sevilla, Spain
3   Research Institute of Computer Engineering (I3US), University of Seville, 41004 Sevilla, Spain
*   Correspondence: fluna1@us.es

Abstract: Falls are one of the leading causes of permanent injury and/or disability among the elderly. When these people live alone, it is convenient that a caregiver or family member visits them periodically. However, these visits do not prevent falls when the elderly person is alone. Furthermore, in exceptional circumstances, such as a pandemic, we must avoid unnecessary mobility. This is why remote monitoring systems are currently on the rise, and several commercial solutions can be found. However, current solutions use devices attached to the waist or wrist, causing discomfort in the people who wear them. The users also tend to forget to wear the devices carried in these positions. Therefore, in order to prevent these problems, the main objective of this work is designing and recollecting a new dataset about falls, falling risks and activities of daily living using an ankle-placed device obtaining a good balance between the different activity types. This dataset will be a useful tool for researchers who want to integrate the fall detector in the footwear. Thus, in this work we design the fall-detection device, study the suitable activities to be collected, collect the dataset from 21 users performing the studied activities and evaluate the quality of the collected dataset. As an additional and secondary study, we implement a simple Deep Learning classifier based on this data to prove the system's feasibility.

Keywords: accelerometer; deep learning; embedded system; fall detection; wearable; recurrent neural networks

Citation: Luna-Perejón, F.; Muñoz-Saavedra, L.; Civit-Masot, J.; Civit, A.; Domínguez-Morales, M. AnkFall—Falls, Falling Risks and Daily-Life Activities Dataset with an Ankle-Placed Accelerometer and Training Using Recurrent Neural Networks. *Sensors* **2021**, *21*, 1889. https://doi.org/10.3390/s21051889

Academic Editors: Ki H. Chon and Marco Iosa

Received: 25 January 2021
Accepted: 5 March 2021
Published: 8 March 2021

Publisher's Note: MDPI stays neutral with regard to jurisdictional claims in published maps and institutional affiliations.

1. Introduction

Among most events related to the gait study, fall detection clearly stands out. These events can lead to severe injuries and sometimes chronic problems or even death. Its importance grows as users age, becoming one of the most important causes of morbidity and mortality worldwide among the elderly according to the World Health Organization (WHO) [1]. The obtained results estimate that around 28% to 35% of people over 65 fall at least one time per year. Additionally, this rate increments in people who suffered falls in the past. Key factors regarding falls include not only physical variables but also psychological aspects, such as fear of falling again. These aspects condition the gait of these persons, leading to higher falling risks. All these factors are correlated with the person's way of walking.

After a fall event, it is very important to have a quick response to avoid increasing the consequences of the fall. However, in many occasions, elderly people live alone or spend too much time alone. In addition, in exceptional situations, such as the current global pandemic caused by COVID-19, we must significantly reduce mobility and contact with this population sector, since it is the most sensitive to these diseases. This is one of the reasons why automatic fall detection systems have become very important in recent years and especially in the current pandemic situation.

There are multiple fall detection system types that achieve good results in these situations—solutions integrated into the user's home (within the "smart home" domain) using visual sensors, telecare solutions with user interaction, mobile applications or wearable systems, among others.

As has already been explained, the responses of these systems occur only after the user falls, acting as a rapid intervention mechanism when these events occur. However, in order to continuously monitor and prevent these falls, it is very interesting to detect possible risk of fall situations, although these events are much more complex to detect. Thus, it is very important to analyze the efficiency of the system based on these three states—fall, falling risk and activity of daily-living (ADL).

In order to design a new fall-detection system, it is important to collect a large amount of data to be able to use it during the system testing phase. For this purpose, researchers usually use public datasets to save time and resources in the design and implementation of the data gathering device as this eliminates or greatly reduces the workload for the collection, filtering and labeling of the required data. Also, the more data we have, the more reliable the system results will be.

As an important point, it should be noted that there are several sensors and/or systems that can be used to detect falls. However, the most widely used sensor that is integrated in almost all fall detection systems is the accelerometer. Some works add other complementary sensors, like gyroscopes (Gyr) or magnetometers (Mag) to filter the information, but the accelerometer (Accel) is the most widely used sensor (sometimes integrated in a wearable independent device and others as part of a smartphone).

In previous works, information about fall events was collected using several users, and the processed and labelled data were made publicly available for researchers as datasets. Most of these datasets are linked with published papers that describe the collecting process. Several of these datasets are summarized in Table 1. As can be observed, there are two main tendencies—using a smartphone or a wearable device. However, almost all published works use an accelerometer.

Another important point shown in Table 1 is the sensor (or device) location. Of these works, 31% place the device in the pocket (usually those that use a smartphone), another 31% place it on the waist, 6% on the chest and 6% on the wrist. Apart from this, two works place several devices in different locations and only one places it on the lower limb. With this summary, it can be observed that 87.5% of the analyzed datasets locate the device in the upper half of the body. Only one work locates the sensor at the bottom, and another uses multiple sensors (locating one of them on the ankle).

The long-term objective of this work is to implement an integrated system to analyze events associated with gait, merging the fall detection system with other gait anomaly detection systems. To achieve this goal, the fall detection system must be near the foot, in order to be able to integrate this new system with the one developed in previous works [2,3], where the pressure sensors that measured the gait trend were located in the footwear insole. For this reason, in order to study the viability of this type of systems, we need a dataset that collects the information about falls and activities of daily living (ADL) from the foot or from the ankle.

Moreover, as detailed previously, it is very important to detect not only falls, but also falling risks. As observed in Table 1, only two of the works include falling risk activities due to the difficulties regarding data collection. But, for our case, it would be very important to include these events. So, attending to the aforementioned problems and the detailed future integration work, we need a dataset obtained from the ankle or from the foot using an accelerometer and including three types of events: Falls, Falling risks and Activities of daily living (ADL).

This dataset creation would set a precedence in the field of fall detectors and could be of great help to other researchers who work in this field. This dataset could be combined in future works with classic Machine Learning (ML) or Deep Learning (DL) techniques to automatically classify the events related to the collected data. These techniques come from the field of Artificial Intelligence (AI), and they have had (and continue to have) great success when automatically extracting the meta-characteristics of large amounts of data in order to perform these classifications correctly. Not only in this area, but also in many others, both ML and DL are used assiduously to solve this type of problem; and as a result of this fact, several studies have emerged with remarkable results [4–8].

Table 1. Datasets summary.

Dataset	Year	Participants	Sensors	Location	Classes	#Activities
Frank et al. [9]	2010	16	Accel	Waist	Pasive, ADL, Risk, Fall	5
Kerdegari et al. [10]	2012	50	Accel	Waist	ADL, Fall	20
Anguita et al. [11]	2013	30	Smartphone (Accel)	Waist	Pasive, ADL	8
Medrano et al. [12]	2014	20	Smartphone (Accel)	Pocket	ADL, Fall	8
Ojetola et al. [13]	2015	42	Accel, Gyr	Chest	ADL, Risk, Fall	15
Fall-MobileGuard [14]	2015	20	Shimmer2R (Accel)	Pocket	ADL, Fall	29
Vilarinho et al. [15]	2015	3	Smartwatch (Accel)	Hand	ADL, Fall	19
Wertner et al. [16]	2015	5	Smartwatch (Accel, Gyr)	Pocket	ADL, Fall	14
MobiAct [17]	2016	57	Smartwatch (Accel, Gyr)	Pocket	ADL, Fall	13
UniMiB SHAR [18]	2017	30	Smartwatch (Accel)	Pocket	ADL, Fall	17
UMAFall [19]	2017	17	Smartwatch, Tags (Accel, Gyr, Mag)	Ankle, Waist, Wrist, Chest	ADL, Fall	11
SisFall [20]	2017	38	Accel, Gyr	Waist	ADL, Fall	33
Rescio et al. [21]	2018	15	EMG	Lower limb muscles	ADL, Fall	6
Quadros et al. [22]	2018	22	Accel, Gyr and Mag	Wrist	ADL, Fall	12
FallDroid [23]	2018	20	Smartphone (Accel)	Waist	ADL, Fall	19
UP-Fall [24]	2019	17	Wearables (Accel), EEG, Cameras, Context-aware	Wrist, Neck, Pocket, Waist, Ankle	ADL, Fall	11

Clearly, the application of these technologies in the field of e-Health, both in the analysis of physiological signals and as diagnostic aid systems with medical images, has an enormous impact and helps to significantly reduce the workload of healthcare professionals. Several works related to this area can be found [25–28], and currently we can even find some interesting applications related to the detection of COVID-19 [29,30]. Due to the rise of these systems, it is interesting to carry out a preliminary dataset study using DL techniques in order to evaluate its quality.

Thus, summarizing the aforementioned, in this work one main fundamental objective is established: designing and developing a dataset for the detection of fall events, falling risk events and activities of daily living (ADL) using a wearable device as data collector (based on low-power microcontroller, accelerometer and Bluetooth BLE transmission), located on the user's ankle.

As a complementary objective, but not one of the main objectives, we include a study regarding the other published datasets (16 in total) in order to extract their main characteristics and their most common activities. This information will be used to determine the activities used for our own dataset. It is important to mention that, although this is not one of the main objectives, this initial study will help to determine the most suitable activities to include in the dataset; thus, this study will help to accomplish the main objective.

Finally, as an additional objective, a dataset classification study is carried out using DL techniques based on Recurrent Neural Networks (RNN). This final objective will determine the dataset quality and, for this reason, a deep study about the different parameters used for classification purposes is carried out. However, it is important to clarify that, as this is not one of our main objectives, we are not looking for the best classification result, but just for a good classifier that justifies the correct samples' labelling. The results that will be shown regarding this classification can be (and will be) optimized using more complex RNNs (like in [31]) in future works after certifying the correctness of the dataset.

About the device location, the ankle was chosen instead of the foot for the ease and comfort when integrating the designed prototype (larger than the final device). However, the final and future device will be miniaturized and located in the footwear in order to be combined with the gait pathology detection system implemented and presented in [3].

The rest of the paper is divided as follows—first, in the 'Materials and Methods' Section, the characteristics of the collected dataset and the device used for this purpose are presented, as well as the recurrent neural network implemented to test the dataset quality. Next, the results obtained after training and testing the dataset collected are detailed and explained in the Results and Discussion Section. Finally, conclusions are presented.

2. Materials and Methods

In this section, we will present in depth the collected dataset—we will discuss the users who have participated in the data collection as well as detail the activities selected to be carried out by all of them based mainly on the analysis of previous public datasets. In addition, we will emphasize the device designed for this task—we will detail the hardware design, its firmware and the user application that receives and stores the information.

Last, but not least, we will detail the selected Recurrent Neural Network architectures used to test the quality of the dataset, as well as all the parameter configurations that have been tested.

2.1. Ankfall Dataset

The dataset developed for this work has been called "AnkFALL" due to the location of the data recorder device (ankle), and it is publicly available to any researcher (https://github.com/mjdominguez/AnkFall) taking into account the limitations indicated in the information web page (accessed on 8 March 2021) .

This dataset is stored with version control. In this way, any modification of its content would provoke the creation of a new version; so that, when researchers download the dataset, they will be able to know which version they are using and thus be able to compare themselves with other works. At the time of publication of this work, the version of the dataset is "v1", and it will be increased numerically as future changes are made.

This section will detail the users who have participated in the data collection, the activities carried out by each one of them and the device designed for data collection.

2.1.1. Activity Set and Participants

In order to consider the best activities for our dataset, it is important to analyze the sixteen previous studies related to the actual datasets that contain activities related to gait. All these studies were detailed in the Introduction Section and collect activities of daily living (ADL), fall simulations and, some of them, falling risk activities (as detailed in Table 1).

We analyzed the three activity types recorded by all of them and included only those which are more common among the sixteen datasets. However, it was established that activities which are too dangerous to carry out or too difficult to simulate would be discarded.

- Activities of Daily Living (ADL): in this kind of activity, most of the datasets include the most common ones like walking, sitting down on a chair, getting up from a chair, crouching down, getting upstairs and getting down stairs. Those ADL activities that are only present in one or two datasets were not taken into account. So, based on this study, only those common activities were included in our activity set.
- Falling Risks: regarding these activities, it is important to mention that only two of the sixteen datasets include this kind of events due to the difficult of simulating them. These two datasets include activities like recovering after trying to sit down into the void, trying to get up, stepping down from a platform, trying to dodge an obstacle on the ground and sitting with an imaginary wall. These situations were discarded due to the danger involved when performing a simulation or the difficulties about conducting a realistic simulation. However we wanted to include falling risk situations and, according to the most common situations among elderly, the risk activities included in our study were: tripping over an obstacle and walking with an improper weight change due to dizziness.
- Falls: the simulation of these activities is carried out by all the previous studies in very different ways, so a deeper analysis is needed in order to select the best activities for our dataset. A summary about fall simulation activities used in the sixteen analyzed datasets can be found in Table 2. According to this review, the most frequent fall events used in the datasets are dropping down (used in eleven of the sixteen datasets: almost 69%), tripping (used in four of the sixteen datasets: 25%), sitting into the void (used in four of the sixteen datasets: 25%) and fainting (used in four of the sixteen datasets: 25%). Thus, these four fall events are included in our dataset.

Table 2. Fall types considered in previous datasets. Reference [11] is empty as it does not consider any fall event.

	[9]	[10]	[11]	[12]	[13]	[14]	[15]	[16]	[17]	[18]	[19]	[20]	[21]	[22]	[23]	[24]
With flexed knees		✓					✓		✓							
Sitting on empty		✓		✓			✓									✓
Base on wall					✓		✓									
Fall as subject prefers		✓													✓	
From standing (drop down)	✓			✓	✓	✓	✓		✓	✓	✓			✓	✓	✓
Pushing															✓	
Improper weight shift															✓	
Trip							✓	✓				✓			✓	
Walking with improper weight shift															✓	
Sitting with imaginary wall															✓	
Walking and slip								✓				✓			✓	
Syncope/fainting/falling asleep				✓		✓						✓			✓	

Table 2. *Cont.*

	[9]	[10]	[11]	[12]	[13]	[14]	[15]	[16]	[17]	[18]	[19]	[20]	[21]	[22]	[23]	[24]
Trying to get up (seat)								✓		✓						
Trying to sit down (seat)												✓				
Lying down on a bed													✓			
Lying initially on the knees						✓			✓							✓
Rolling out of bed						✓										
With compensation strategies to prevent the impact				✓						✓						
With contact with an obstacle before hitting the ground				✓						✓						
Steing down from platform							✓									

After evaluating all the datasets, we obtain a total amount of twelve activities. Another important problem we detected after evaluating the previous datasets is that the different kind of activities are unbalanced—in some datasets more than 75% of the activities are ADL; and the great majority of them do not have risk events. These problems affect the classifier system severely, as it needs balanced data in order to extract the meta-characteristics of the dataset. Thus, ADL activities were reduced in our dataset to achieve this goal. Table 3 shows the final list of activities considered for the preparation of the dataset. In this table a short description is included, as well as a detailed preparation process to record them. As can be observed, our dataset has five ADL activities, three Falling risk activities and four Fall simulations.

Table 3. List of recorded activities. Last column indicates the mean execution time (in seconds) for each activity.

# Activity	Activity Description	Activity Steps	Time
Act. 1	The subject walks.	0. Standing; 1. Walk; 2. Stop	17
Act. 2	The subject sits in a chair.	0. Standing in front of a chair; 1. Turn around; 2. Sit	7
Act. 3	The subject gets up from a chair.	0. Sitting on a chair; 1. Get up; 2. Stand	6
Act. 4	The subject crouches down with the intention of touching the ground.	0. Standing; 1. Crouch down; 2. Return to upright position; 3. Stand	6
Act. 5	The subject goes up and down stairs.	0. Standing; 1. Go up stairs; 2. Stop; 3. Turn around; 4. Go down stairs; 5. Stand	X
Act. 6	The user trips over with the left foot.	0. Standing; 1. Walk; 2. Trip over a creased carpet with the left foot; 3. Regain balance; 4. Walk; 5. Stand	8
Act. 7	The user trips over with the right foot.	0. Standing; 1. Walk; 2. Trip over a creased carpet with the right foot; 3. Regain balance; 4. Walk; 5. Stand	8
Act. 8	The user walks while dizzy.	0. Standing; 1. Turn on itself several turns; 2. Walk; 3. Stop	19
Act. 9	The user falls backwards while sitting on the void.	0. Stand; 1. Try to sit; 2. Fall; 3. Stay still	6
Act. 10	The subject trips over and falls forward.	0. Stand; 1. Walk; 2. Trip over a creased carpet with the left foot; 3. Fall; 4. Stay still	10
Act. 11	The subject falls to the left (faint).	0. Stand; 1. Fall without resisting to the left; 2. Stay still	5
Act. 12	The subject falls to the right (faint).	0. Stand; 1. Fall without resisting to the right; 2. Stay still	5

All these activities were performed by 21 volunteers who were previously informed and trained about the way to carry them out. Due to the difficulty to perform some of the falling risk events, the volunteers had the possibility to avoid some of them. The participants were 16 males and 5 females aged between 21 and 60, with heights between 1.60 and 1.95 m, weights between 70 and 110 kg and none of them presented any gait limitations.

2.1.2. Acquisition Process

As described before, the long-term goal after this study is to develop an integrated system that records and analyzes all types gait related events. These events include not only fall events, falling risk events and ADL events; but also problems related to user balance, way of walking, plantar problems like pronation or supination, and so forth. Regarding these problems, some works have been previously developed [2,3], and the device implemented for those other works was integrated in the footwear. Thus, in order to fuse both systems (fall detector and gait analyzer) in the near future, the acquisition device was located in the ankle (as the prototype is currently too big to fit in the footwear).

Moreover, we have consulted two sets of people—the first set was composed by people who are using a similar device; and the second set includes some of the participants of the collected dataset. However, because of the difficulties of making an extensive consultation due to the current pandemic, the first set was formed by relatives and acquaintances of the authors of this manuscript.

Finally, the first set includes five persons (four of them older than 65); and the device location used by them varied according to each user—two of them wore it on the waist, one on the wrist and two on the chest. The two main problems that they commented on were discomfort when wearing it (3/5; those who wore the device on their waist did not notice discomfort), and forgetfulness (4/5; only one indicated that he did not forget it because he wears a bracelet all day). For the second set, four of the participants in the data collection were able to provide a comparative view with other devices they had used. They all agreed that the new device was more comfortable to wear. Moreover, some of the participants, at the end of the collecting process, did not remember which ankle they had the device on (as it was put over the sock).

So, according to these consultations and taking into account the proposed future works detailed in the Introduction Section, there are two main reasons why we focus the study on acquiring data on the footwear:

- This area of the body is directly influenced by risk events and falls; however, the vast majority of published datasets do not record from this position.
- This location opens the opportunity to combine the information with other sensors, such as pressure sensors, to analyze pathologies or other gait-related problems.

Regarding the forgetfulness problem, we need to analyze the focus population of the device:

- Users with many pathologies are used to wear insoles. Forgetting to use the device would be unlikely, especially if the sensors were integrated into the orthotics devices.
- The elderly usually wear a small number of shoes and, thus, it would be less common to forget the device.
- Athletes usually wear sports shoes, and therefore sensors could be installed on them.

Thus, according to this extensive study, locating the device in the footwear in a near future could be an interesting solution to avoid the inconveniences indicated before. For now, our device will be located in the ankle; but the information recorded by the accelerometer will be very similar to what will be obtained by a device located in the footwear. For this purpose, the full recording system is made up of a sensing and transmission device (placed on the ankle) and an application for monitoring and saving the received information.

The wearable device is managed by a 32-bit low-power microcontroller. The full list of used components is described next (see Figure 1a):

- STMicroelectronics microcontroller (STM32L432KCU6): 32-bits ARM Cortex-M4 with floating-point unit (FPU), 80 Mhz CLK, 256 KB FLASH, 64 KB SRAM, 12-bit resolution ADC (analog-to-digital converter) up to 10 channels and up to 22 GPIOs (general purpose pins).
- ADXL345: triaxial and analog accelerometer with a resolution up to ±16 g for each axis.
- HM-10: low-energy bluetooth modem controlled by a serial port. This modem is used to transmit the collected data to the monitoring application.
- Power supply: as the main objective is the dataset collection, we used a comercial powerbank to feed the device. In a near future, with the final implementation after testing the dataset and integrating the RNN inside the embedded system, a 150–200 mAh lipo battery will be used; however, at this time, no power-consumption studies have been performed.

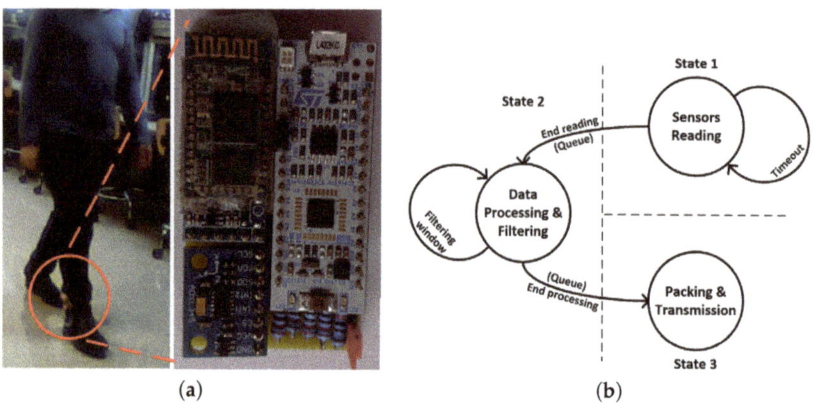

Figure 1. Wearable acquisition device: (**a**) wearable device placed in the ankle; (**b**) firmware's state machine.

Regarding the device's firmware, a real-time operating system (FreeRTOS) is used to correctly process the information (see Figure 1b) and transmit it without losing data, configuring the accelerometer with a ±16 g accuracy and positioning it in the following way—x axis was aligned with the horizontal line, y axis with the vertical, and the z axis was aligned with the march direction. The information collected from each accelerometer axis is filtered using the mean with a 10-sample window, and the resulting values are encapsulated into a frame (which contains a starting sequence, the three accelerometer filtered axis values, a checksum value to check the correct reception of the information in the computer, and an ending sequence), and transmitted at a 50 Hz rate to the computer (see Figure 1b).

On the other hand, the monitoring system plots the information received from the wearable device by the computer in real time (at 50 Hz as commented before), allowing us to visualize whether the information has been correctly captured. In the meanwhile, this data is stored locally in a csv file.

After the collecting step, the information stored for all the activities are accelerometer values (3 values, one for each axis). At this stage the different classes are not indicated yet, and we need to precisely discriminate the correct label to tag each sample individually.

In order to perform the labelling task, a webcam records the activities while the monitoring system is plotting the information. A local script launches the monitoring application and the webcam recording tasks at the same time. Thus, the csv file (that contains the accelerometer information) and the clip recorded by the webcam share the timestamps, easing the subsequent labelling task. Therefore, we store the accelerometer information and the videoclip for every task for all users. However, for confidentiality reasons these videos are not publicly available. A recording sequence from both the monitoring application and the videoclip is shown in Figure 2.

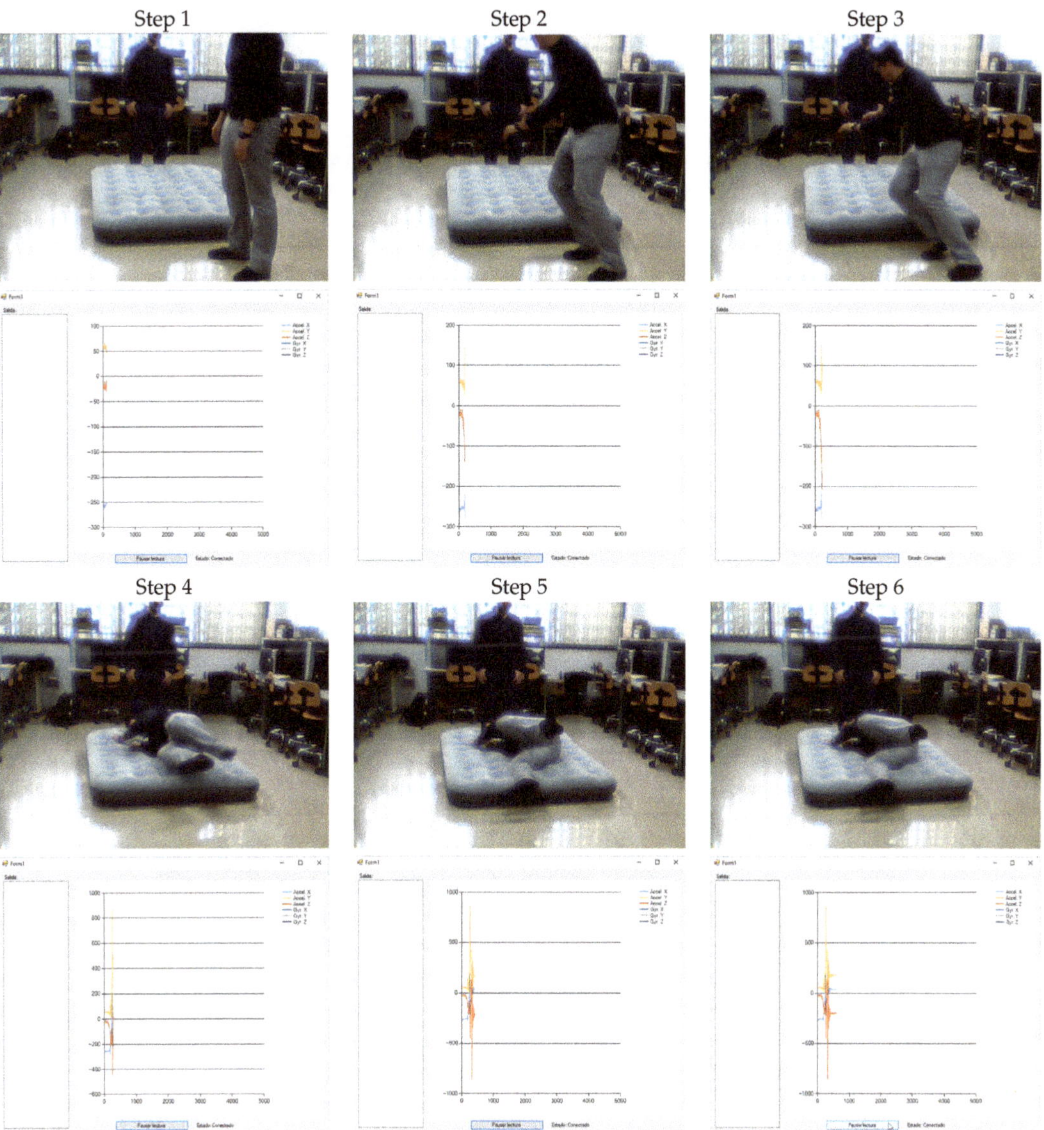

Figure 2. Monitoring application and videoclip during a Fall event recording sequence (in order, from step 1 to 6).

Viewing the video in slow motion, we labelled the entire dataset sample by sample between the available classes, that is, fall, falling risk and ADL. However, those periods in which none of the three previous possibilities occurred (a section where the user does not move), were labelled with an auxiliary category called background (BKG) that corresponds to the class of other datasets named "Passive" (see Table 1). So, finally, our dataset distinguishes between four classes—BKG, ADL, FALL and RISK. The reasons why we use these classes are detailed below:

- Most studies group events to simplify the final problem. In fact, there are many studies that only classify between fall events and ADL, without addressing risk events.
- Knowing the specific user activity is not as interesting for healthcare systems as detecting dangerous anomalies like falls or falling risks.
- Using more classes requires more complex models and the systems would loose the ability to alert in real time.

2.2. Recurrent Neural Network Classifier

As detailed previously, the main goal of this work is the elaboration of one dataset that discretizes between falls, falling risks and activities of daily living (ADL), and the way in which the activities are selected as well as the methodology to used to perform them was described before. However, in order to evaluate the quality of the collected dataset, a DL based classification will be carried out using recurrent neural networks (RNN) in the same way as it was done in previous works with other datasets [31]. It is important to emphasize that our objective is not to optimize the final classification accuracy of the system; however, we study the different parameter combinations for the network in order to evaluate if it is possible to classify the different activities and, if this is the case, the adequacy of the collected dataset will be demonstrated. For this purpose, we need to evaluate the specificity metric—this metric determines measures of the proportion of values not belonging to a class that are correctly identified as such. So, a good result in this metric means that the different classes can be differentiated easily and, thus, the dataset collecting and labeling processes can be verified using this metric.

In order to perform this task, we used Gated Recurrent Neural Network algorithms. These algorithms can automatically extract the appropriate characteristics from a temporal sequence (like our dataset) for performing the final classification. Long Short-Term Memory (LSTM) and Gated Recurrent Units (GRU) are the most used gated recurrent layers (see Figure 3), which have been demonstrated to obtain an acceptable performance in signal classification problems [32,33].

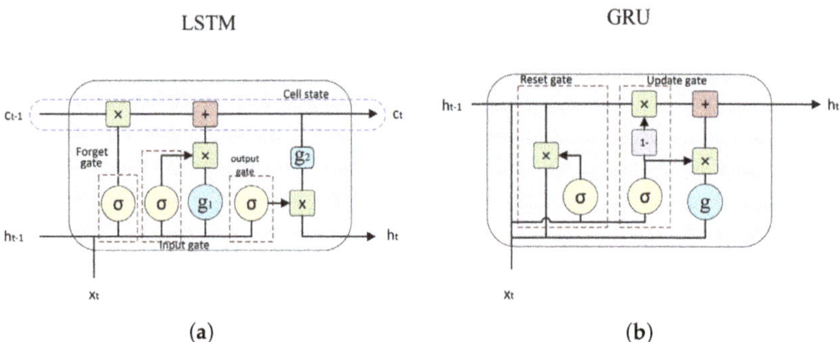

(a) (b)

Figure 3. Long Short-Term Memory (LSTM) (**a**) and Gated Recurrent Units (GRU) (**b**) units.

If we generalize the problem, the gait-related events require a study of acquired signals sequential/temporal characteristics, as detailed before. To contemplate this temporality there are two main approaches; by using frequency analysis (using FFT or DWT), extracting features and combining them with a classic algorithm such as an MLP network; or by using deep learning algorithms such as RNNs. As the final future objective of his work is to create real-time systems, the feature extraction process needed in the frequency analysis method requires more computational time than working directly with RNNs (with not too many layers); therefore, for a real-time system, it is more consistent to work with RNNs.

In previous studies, we used these recurrent layers with accelerometer signals taken at the waist (using the information of other datasets), obtaining high effectiveness results and real-time classifications using low-power microcontrollers [3]. These layers implement a memory cell that retains relevant information from the analyzed sequence section, to use it in the analysis of the full sequence and, finally, to provide appropriate information for the classification problem. In summary, their operation is detailed below:

- LSTM cell: this cell consists of an input gate, a forget gate and an output gate. The first gate adds new information to the cell from the sequence sample in the current instant of time t and the outputs of the same layer processing for the previous instant $t - 1$. With the same inputs, the forget gate determines the non-necessary data for the analysis of rest of the sequence. Finally, the output gate passes information from the memory that considers relevant as input to the layer processing the next instant $t + 1$.
- GRU cell: it has only two gates (no forget gate is used), so it can add (update gate) and remove (reset gate) information from the cell, allowing all stored information to be used by the neural network throughout the sequence analysis.

In this study two main architectures were used (see Figure 4). Both of them contain a single recurrent layer and a dense layer with four nodes. A softmax activation function is used to normalize the results in a probability distribution; and, thanks to this, both the loss calculation and the 4-class classification results can be performed. Additionally, we use a batch normalization layer during the training step to normalize the input data, allowing a faster convergence of the system. The two architectures differ in the type of RNN layer used—the first one is based on LSTM cells while the other one uses GRU cells.

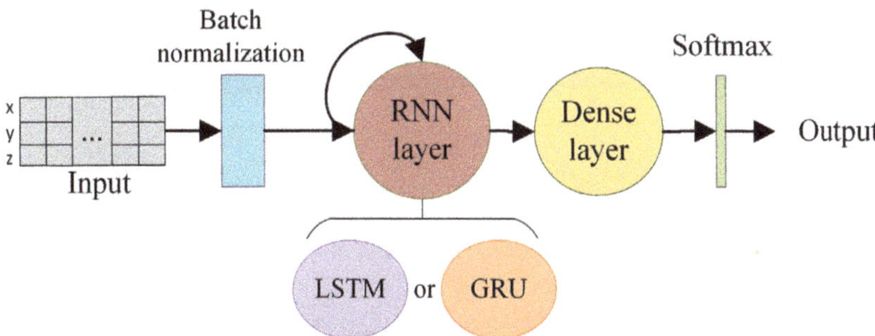

Figure 4. Diagram of the Gated RNN architectures assessed.

Data Segmentation and Labelling Criteria

In order to adapt each registered activity to the RNN models, we established a fixed temporal window of 64 samples, which, considering an acquisition frequency of 50 Hz, is equivalent to 1.28 s. This window is appropriate to contemplate a fall event or even risk situations [34]. Each activity was split into blocks with the same window length of 64 and was labelled with a unique category. The criteria used for labelling each block consisted of assigning it to the most relevant class whose occurrence percentage exceeds an established threshold (see Figure 5). If none of the thresholds is reached, the segment is classified as background (BKG).

In the first step of this study, we consider different threshold values for each relevant category and analyze the training results to obtain the best threshold values. In a second step, grid search is used for model optimization. The parameters considered were the number of nodes and dropout value for the recurrent layer, as well as learning rate and batch size used during training. All these tests and the obtained results will be detailed in the next section.

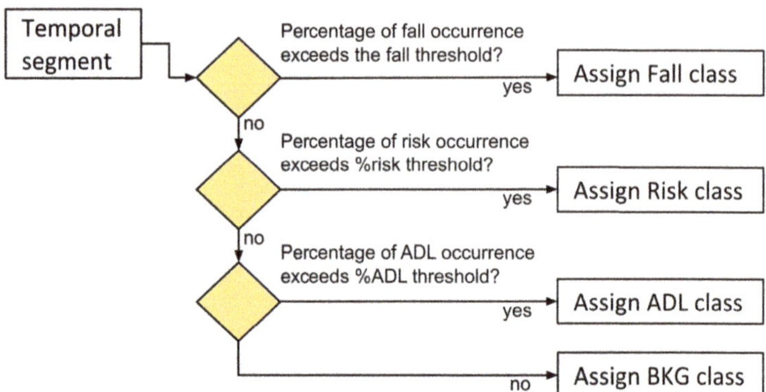

Figure 5. Labelling criteria based on occurrence percentage in each block.

3. Results and Discussion

After analyzing all the datasets developed in recent years related to fall events and selecting the activities that must be collected for the new dataset, its final distribution is shown below.

In addition, as already indicated above, we consider that it is necessary to test our dataset using a DL classification system. In this way, we will assess whether the collected dataset is useful to future fall detection systems studies.

3.1. Collected Dataset

While recording the dataset, three repetitions of each activity were performed. However, due to the COVID-19 pandemic lockout, some activities could not be registered. It mainly affected activities 5 (going up and down stairs) and 8 (walking while dizzy) because they needed special considerations during the preparation. Thus, the data recollection for these activities was delayed until the other activities were recorded. While activity 8 (walking while dizzy) was performed by 10 users, activity 5 (going up and down stairs) could not be performed by any user.

So, finally, eleven activities were recorded. Moreover, as activity 5 is considered an ADL (activity of daily living), the final dataset is more balanced than our initial expectations as it contains 4 ADL activities, 4 fall activities and 3 falling risk activities.

Additionally, some participants did not carry out some activities for security reasons or at their own request and, in some cases, some errors were detected after some recordings (in particular with two activities for participant seventeen); all these reasons reduced the final amount of records. Figure 6 illustrates the process to obtain the final dataset. Finally, the total set contains 615 records.

So, after recording the AnkFALL dataset, it is important to compare it with the datasets presented in the Introduction Section. However, as there are too many differences among them, only datasets that use a wearable device with an accelerometer (not a smartphone or smartwatch) and that include fall events are taken into account. The comparison is presented on Table 4.

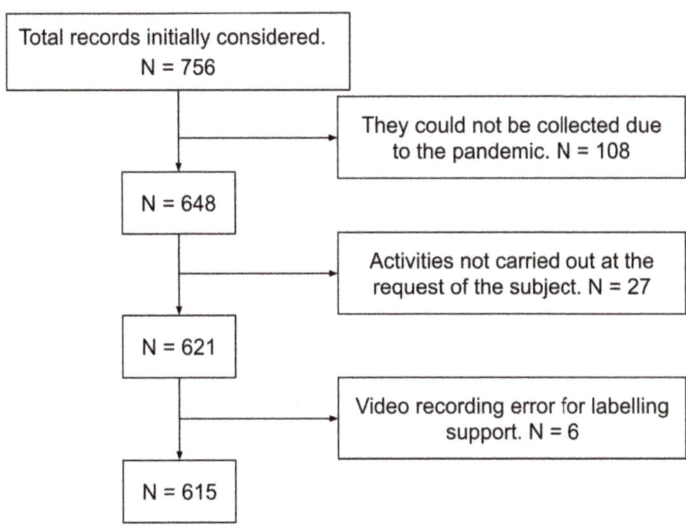

Figure 6. Resulting dataset.

Table 4. Comparison between AnkFALL and the most similar datasets (from Table 1). The last column indicates the number of activities from each type (in order: activities of daily living (ADL), Risk, Fall).

Dataset	Year	#Users	Sensors	Location	Classes	#Activities	Balance
Frank et al. [9]	2010	16	Accel	Waist	Pasive, ADL, Risk, Fall	5	3-1-1
Kerdegari et al. [10]	2012	50	Accel	Waist	ADL, Fall	20	17-0-3
Ojetola et al. [13]	2015	42	Accel, Gyr	Chest	ADL, Risk, Fall	15	8-1-6
SisFall [20]	2017	38	Accel, Gyr	Waist	ADL, Fall	33	29-0-4
Quadros et al. [22]	2018	22	Accel, Gyr and Mag	Wrist	ADL, Falls	12	11-0-1
AnkFALL	2020	21	Accel	Ankle	BKG, ADL, Risk, Fall	11	4-3-4

As can be observed in Table 4, AnkFALL registers a total number of participants and a number of activities slightly lower than average (around 30 and 19, respectively). As detailed before, the dataset is continuously growing and, in the near future, the total number of participants will increase; however, due to the pandemic lockout, no participant record has been registered since March 2020. Regarding the number of activities, AnkFALL seems to have very few, but the dataset it was designed to be as balanced as possible—as it is shown in the last column of Table 4, we can observe some goodness about AnkFALL:

- AnkFALL is the most balanced dataset among the existing ones according to the number of activities: around 36%, 28% and 36% of ADL, Risks and Falls, respectively. If we do not take into consideration AnkFall, the more balanced dataset is the one presented by Ojetola et al. [13] with 53%, 7% and 40% of ADL, Risks and Falls, respectively.
- Only two datasets from Table 1 register more Fall events than AnkFALL—the one from Ojetola et al. [13] with six different Fall activities and FallDroid [23], which has no Risk activities and uses a smartwatch.

- No other dataset has more Risk activities than AnkFALL—the datasets that have Risk activities only register one or two types.

Thus, the two most important things that make AnkFALL useful for researchers is that it has several Risk activities records, and it is clearly the most balanced available dataset. These two characteristics are very useful when implementing ML classification systems due to the importance of balancing the training data for obtaining good classification results without overtraining one of the classes. The internal data structure of AnkFall dataset can be observed in Figure 7.

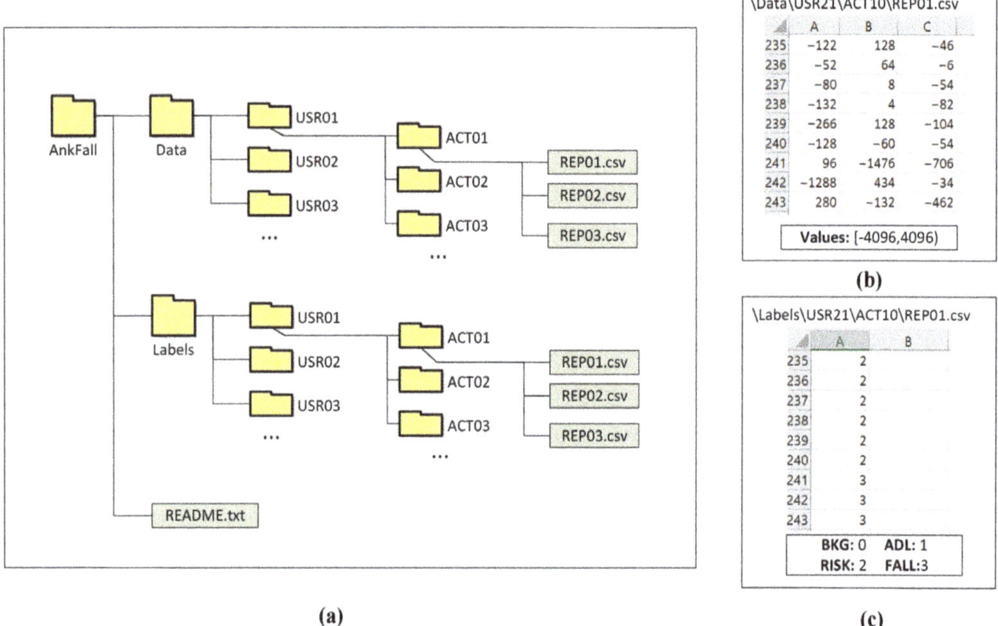

Figure 7. AnkFall internal structure: (**a**) Folder structure; (**b**) Data collected from user 21 during activity 10, repetition 1; (**c**) Labels for each sample of data collected during activity "(**b**)".

3.2. Testing the Dataset

Although the main goal of this work is already achieved, it is very interesting to perform a first classification study in order to check that the information recorded and the applied pre-processing techniques are correct.

For this purpose, the Hold-Out technique has been applied, dividing the dataset in two subsets—one subset was used for training and the remaining was used for evaluation. The distribution was carried out in such a way that each subset contained data from different users, to avoid bias. Data from 5 users were randomly selected, avoiding choosing the users who had carried out the least number of activities. In the dataset segmentation process, we used a 64-sample temp window, as mentioned above, and a 25% displacement of the sample size, which corresponds to 16 temporal samples and applied this approach as a data augmentation technique: this means that there is an overlapping of 75% of the information between two consecutive temporal windows. Data distribution is shown in Table 5.

Table 5. Participant distribution for each subset using the hold-Out method.

Subsets	Users	# Activities
Train	1, 2, 3, 4, 6, 7, 11, 12, 13, 14, 15, 16, 17, 18, 19, 20	463
Validation	5, 8, 9, 10, 21	152

We compared the effectiveness of the classification system using different and well-known metrics: sensitivity (also known as recall), specificity, precision and F1-score [35]. This last metric is the harmonic mean of precision and sensitivity. Those metrics are presented in the next equations:

$$\text{Specificity} = \sum_c \frac{TN_c}{TN_c + FP_c}, c \in classes \tag{1}$$

$$\text{Precision} = \sum_c \frac{TP_c}{TP_c + FP_c}, c \in classes \tag{2}$$

$$\text{Sensitivity} = \sum_c \frac{TP_c}{TP_c + FN_c}, c \in classes \tag{3}$$

$$\text{F1-score} = 2 * \frac{\text{precision} * \text{sensitivity}}{\text{precision} + \text{sensitivity}}. \tag{4}$$

About those metrics:

- Specificity: proportion of "true negative" values in all cases that don't belong to this class (see Equation (1)).
- Precision: proportion of "true positive" values in all cases that have been classified as it (see Equation (2)).
- Sensitivity (or Recall): proportion of "true positive" values in all the cases that belong to this class (see Equation (3)).
- F1: It considers both the precision and the sensitivity (recall) of the test to compute the score. It is the harmonic mean of both parameters (see Equation (4)).

In the first stage of the study, the best threshold values for block labelling were analysed. We considered the values shown in Table 6.

Table 6. Occurrence thresholds values analyzed for each class.

Threshold	Set of Values
ADL	0.3, 0.4, 0.5, 0.6, 0.7
RISK	0.2, 0.3, 0.4, 0.5
FALL	0.1, 0.2, 0.3, 0.4

For training our classification system, we used the best architecture considered in a previous work [31]—this architecture consists, as detailed in the previous section, of a recurrent layer followed by a dense layer and a softmax activation function. The resulting best values for thresholds were 30% for Fall events, 20% for Falling risk events and 30% for ADL, with a mean F1-score up to 0.75 and a standard deviation 0.08 after three training repetitions. The dataset distribution using the considered thresholds can be consulted in Table 7. It can be seen how, despite trying to homogenize the number of samples for each type of event, due to the short duration of risk situations and fall events, an imbalance in the number of dataset samples can be observed.

Table 7. Dataset distribution for each subset with best labelling thresholds.

		Samples Per Class			
Subsets	**Total**	**ADL**	**BKG**	**Risk**	**Fall**
Train	9099	2064	4587	1165	1283
Validation	3172	837	1500	331	504

The model optimization was carried out with a grid search considering the parameter values shown in Table 8. The parameters consisted of the number of nodes of the recurrent layer, dropout rate, learning rate and batch size. This range of values was selected based on the parameters considered in previous studies [31,34].

Table 8. Grid search values for exhaustive parameters optimization.

Parameters	**Set of Values for Grid Search**
Learning rate	0.0005, 0.001, 0.0015, 0.002
Batch size	32, 48, 64
Number of nodes	24, 32, 40
Dropout	0, 0.2, 0.35

After carrying out all the training studies with all the possibilities for each parameter, the best trained models were obtained and are shown in Table 9. In order to check the quality of these models, this table shows the values of the metrics described in Equations (1)–(4). As can be observed, two possibilities are shown in this table based on the two network types considered in this work—the first one contains LSTM units, and the second one contains GRU units.

Table 9. Best results obtained after grid search optimization.

RNN Architecture	**RNN Nodes**	**Learn. Rate**	**Batch Size**	**RNN Drop.**	**Precision**	**Specificity**	**Sensitivity**	**F1-Score**
One LSTM layer	32	0.002	32	0.2	0.780	0.928	0.768	0.774
One GRU layer	40	0.002	32	0.2	0.762	0.924	0.771	0.766

If we analyze the results shown in Table 9, we can obtain the next conclusions for each metric:

- Specificity: we can see values over 92% that denote a high rate of true negatives among the total amount of true negatives and false positives. So, in general, this system classifies correctly the values that do not belong to each class.
- Precision and Sensitivity: these values are lower than the specificity, however both have similar values around 77–78%. These results are not bad at all, but the important difference between the specificity and these two metrics indicate that the system does not behave in the same way with all classes. Maybe one of the four classes has quite worse results than the others. In fact, if we study the system deeply, we can check that the class RISK has much worse results than the others due to two main reasons: the difficulty of distinguishing it and the alternation between RISK, ADL and BKG during the Falling risk activities. This aspect will be studied later.
- F1-score: As this metric is obtained from precision and sensitivity, it is normal that its value is similar to them. The main conclusion obtained with this value is the same that was already obtained for the previous metrics.

After presenting all the metrics, we can observe that the obtained "specificity" values are high (92–93%). This fact indicates that the system classifies the values that not belong to each class correctly. This means that the different classes can be easily differentiated and, thus, the dataset collecting and labeling processes are proven to be correct. So, this is a good metric to evaluate the quality of the collected dataset. Regarding the "precision" metric, it depends on each class independently. But, thanks to the good specificity results, we can assume that, using more complex RNNs the classification results can be improved.

So, in order to corroborate the results predicted previously by evaluating the metrics, it is important to study each class independently for the two neural networks (LSTM and GRU from Table 9) selected after our initial study. For this purpose, the confusion matrices are shown in Figure 8.

This figure reveals the predicted main obstacle to be faced in the detection problem—the correct distinction of falling risk activities from activities of daily living (ADL). However, if we focus our attention on the other classes, their results are high, obtaining an 87% success rate when classifying falls. These results demonstrate that the information recorded for the AnkFALL dataset is very useful for being used in fall detection studies. The problem concerning the RISK class is not new as, in a previous work [31], a RISK class was included in the SisFALL dataset, and the conclusions were similar—the falling risk events are much more difficult to distinguish than any other event type.

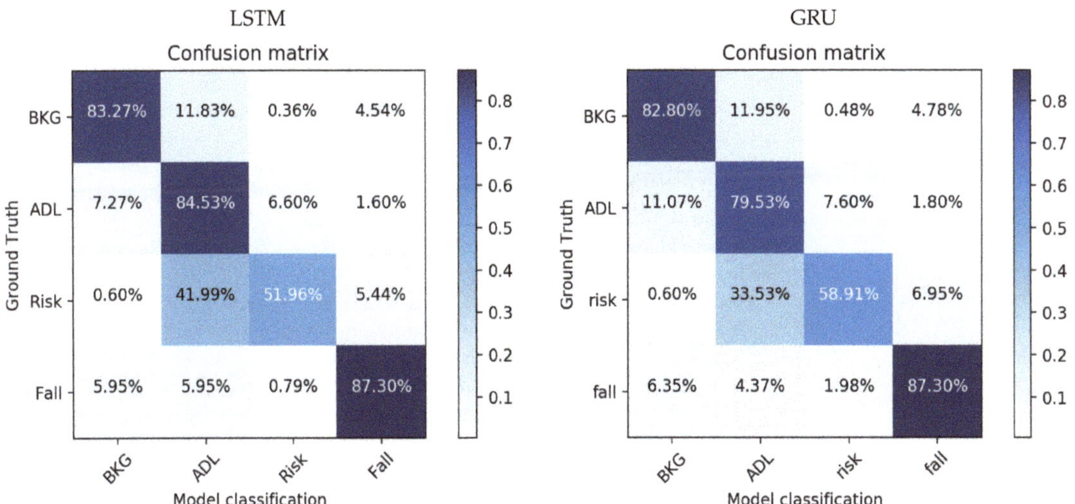

Figure 8. Confusion matrices for best LSTM and GRU models. Each specific box in the confusion matrix represents the percentage of samples of the class indicated by the row that have been classified as the class indicated in the column.

Concerning the system ROC curves (see Figure 9), they reveal the same problem again: RISK class obtains worse results than the others. However, the area under the curve is high, revealing that, in binary classifications of the One-vs-All type, the system has the ability to distinguish each event class individually.

The curves corresponding to ADL and RISK are those with the lowest area under the curve, 90% and 80% respectively, which is correctly correlated with the results shown in the confusion matrices. Regarding the RISK class, a stagnation can be seen from a sensitivity value of 0.75, which reveals that there are samples of this class which are very complex to identify. These samples are mainly those corresponding to activity 8 (walking dizzy); so, it would be interesting to eliminate this activity from a future training study in order to corroborate these preliminary conclusions.

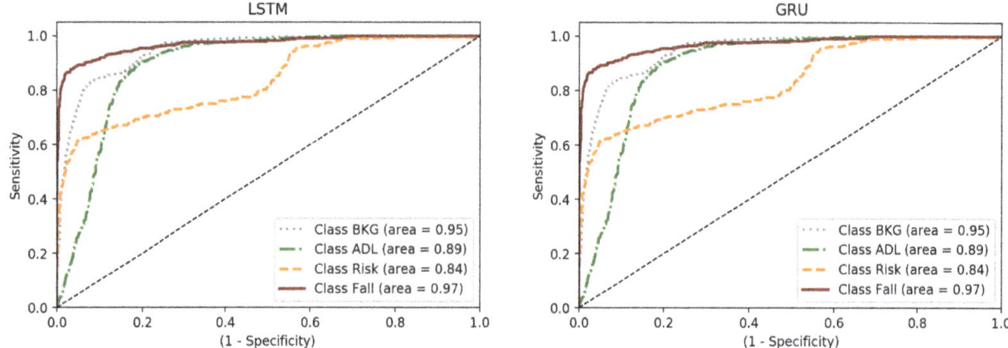

Figure 9. ROC curves for best LSTM and GRU hmodels.

4. Conclusions

In this work, a new dataset called AnkFALL is presented. It contains labelled information from 21 participants performing 11 activities, including activities of daily-living (ADL), Falls and Falling risk activities. These activities have been carefully selected by performing a study of the different available datasets and taking into account that one of the main objectives is to obtain a dataset that contains all types of activities and that is as balanced as possible.

According to the explanations given in the introduction, the results of this work can be used in healthcare and fitness areas. Our main focus area is healthcare, and more precisely telecare services for older people who use devices to alert about emergency situations (mainly falls). Using a device placed in the footwear may prevent forgetfulness and improve wearing comfort and, thanks to this dataset, researchers can start to develop footwear-placed fall detectors. Although this is our main interest area, these principles can be easily applied to the fitness field too.

To record the dataset, a personalized wearable device, based on a low-power microcontroller, an accelerometer and a Bluetooth low energy interface, has been designed and implemented to transmit the information from the user's ankle; on the other hand, a computer application has been implemented for the reception, visualization and storage of the received data. The labelling task has been thoroughly carried out by checking the received information sample by sample with the synchronized videoclips recorded while performing the activities.

The final result is the first dataset of its kind that collects information from the ankle, and it is also the most balanced fall-detection dataset among of all the datasets currently described in the literature. Also, this dataset includes a type of activity (falling risk) that is only available in a small subset of the available datasets.

To verify the quality of the dataset, a deep learning classifier system based on recurrent neural networks has been designed and implemented for the classification of the four classes in this dataset—BKG, ADL, Fall and Risk. A detailed study is carried out with multiple variations of thresholds, learning rates, batch sizes, number of nodes and dropout; obtaining, in combination, more than 600 different neural networks that have been trained for this purpose. The best result for each parameter is presented and the final classification results are shown. The obtained specificity values (between 92% and 93% depending on the implementation) indicate that the system classifies correctly the values that do not belong to each class. This means that the different classes can be differentiated easily, so the quality of the dataset is proven by this result. Regarding the precision, it depends on each class but, thanks to the specificity results, we can assume that, using more complex RNNs the classification results can be improved.

Moreover, the classification results show that the main difficulty is detecting falling risk events—this is mainly due to the speed at which they occur and the combination with other ADL activities; not surprisingly, most of the datasets do not include this type of activities and most of the research studies do not include them either.

For future versions this dataset needs to be expanded with a larger number of participants. Moreover, after testing the dataset, we have observed that the information collected for one of the activities has several errors and this affects the classification results; thus, in the next versions of the dataset we need to evaluate whether the instructions given for the participants were clear enough and if this activity is essential or not. Finally, in order to obtain a better adequacy of the activities collected for the dataset, we have started to apply "Explainable Deep Learning" techniques to the collected data; we hope that these techniques will help us improve the next versions of the dataset.

Author Contributions: Conceptualization, F.L.-P. and M.D.-M.; methodology, F.L.-P. and M.D.-M.; software, F.L.-P. and L.M.-S.; validation, F.L.-P., L.M.-S. and J.C.-M.; formal analysis, F.L.-P. and J.C.-M.; investigation, F.L.-P., L.M.-S. and M.D.-M.; resources, F.L.-P. and M.D.-M.; data curation, L.M.-S. and J.C.-M.; writing—original draft preparation, F.L.-P. and M.D.-M.; writing—review and editing, L.M.-S., J.C.-M. and A.C.; visualization, F.L.-P., L.M.-S. and J.C.-M.; supervision, M.D.-M. and A.C.; project administration, M.D.-M. and A.C.; funding acquisition, M.D.-M. and A.C. All authors have read and agreed to the published version of the manuscript.

Funding: This research was funded by the Spanish Agencia Estatal de Investigación (AEI) project MINDROB: "Percepción y Cognición Neuromórfica para Actuación Robótica de Alta Velocidad" (PID2019-105556GB-C33/AEI/10.13039/501100011033).

Institutional Review Board Statement: Ethical review and approval were exempted for this study, because all the participants are part of the research group or relatives of some of its members, being aware of the risk and participating completely voluntarily.

Informed Consent Statement: Informed consent was obtained from all subjects involved in the study.

Data Availability Statement: Public available **AnkFall Dataset** on https://github.com/mjdominguez/AnkFall.

Acknowledgments: We want to thank the Telefonica Chair "Intelligence in Network", but also the Research Group "TEP-108: Robotics and Computer Technology" from Universidad de Sevilla (Spain) for using its resources.

Conflicts of Interest: The authors declare no conflict of interest.

References

1. World Health Organization: Ageing and Life Course Unit. *WHO Global Report on Falls Prevention in Older Age*; World Health Organization: Geneva, Switzerland, 2008.
2. Domínguez-Morales, M.J.; Luna-Perejón, F.; Miró-Amarante, L.; Hernández-Velázquez, M.; Sevillano-Ramos, J.L. Smart Footwear Insole for Recognition of Foot Pronation and Supination Using Neural Networks. *Appl. Sci.* **2019**, *9*, 3970. [CrossRef]
3. Luna-Perejón, F.; Domínguez-Morales, M.; Gutiérrez-Galán, D.; Civit-Balcells, A. Low-Power Embedded System for Gait Classification Using Neural Networks. *J. Low Power Electron. Appl.* **2020**, *10*, 14. [CrossRef]
4. Muñoz-Saavedra, L.; Luna-Perejón, F.; Civit-Masot, J.; Miró-Amarante, L.; Civit, A.; Domínguez-Morales, M. Affective State Assistant for Helping Users with Cognition Disabilities Using Neural Networks. *Electronics* **2020**, *9*, 1843. [CrossRef]
5. Gutierrez-Galan, D.; Dominguez-Morales, J.P.; Cerezuela-Escudero, E.; Rios-Navarro, A.; Tapiador-Morales, R.; Rivas-Perez, M.; Dominguez-Morales, M.; Jimenez-Fernandez, A.; Linares-Barranco, A. Embedded neural network for real-time animal behavior classification. *Neurocomputing* **2018**, *272*, 17–26. [CrossRef]
6. Saleem, M.H.; Potgieter, J.; Arif, K.M. Plant Disease Classification: A Comparative Evaluation of Convolutional Neural Networks and Deep Learning Optimizers. *Plants* **2020**, *9*, 1319. [CrossRef]
7. Chen, C.H.; Kung, H.Y.; Hwang, F.J. Deep learning techniques for agronomy applications. *Agronomy* **2019**, *9*, 142. [CrossRef]
8. Wang, J.; Ma, Y.; Zhang, L.; Gao, R.X.; Wu, D. Deep learning for smart manufacturing: Methods and applications. *J. Manuf. Syst.* **2018**, *48*, 144–156. [CrossRef]
9. Frank, K.; Vera Nadales, M.J.; Robertson, P.; Pfeifer, T. Bayesian recognition of motion related activities with inertial sensors. In Proceedings of the 12th ACM International Conference Adjunct Papers on Ubiquitous Computing-Adjunct, Copenhagen, Denmark, 26–29 September 2010; pp. 445–446.

10. Kerdegari, H.; Samsudin, K.; Ramli, A.R.; Mokaram, S. Evaluation of fall detection classification approaches. In Proceedings of the 2012 4th International Conference on Intelligent and Advanced Systems (ICIAS2012), Kuala Lumpur, Malaysia, 12–14 June 2012; Volume 1, pp. 131–136.

11. Anguita, D.; Ghio, A.; Oneto, L.; Parra, X.; Reyes-Ortiz, J.L. *A Public Domain Dataset for Human Activity Recognition Using Smartphones*; Esann: Midlothian, VA, USA, 2013.

12. Medrano, C.; Igual, R.; Plaza, I.; Castro, M. Detecting falls as novelties in acceleration patterns acquired with smartphones. *PLoS ONE* **2014**, *9*, e94811. [CrossRef]

13. Ojetola, O.; Gaura, E.; Brusey, J. Data set for fall events and daily activities from inertial sensors. In Proceedings of the 6th ACM Multimedia Systems Conference, Portland, OR, USA, 18–20 March 2015; pp. 243–248.

14. Fortino, G.; Gravina, R. Fall-MobileGuard: A smart real-time fall detection system. In Proceedings of the 10th EAI International Conference on Body Area Networks, Sydney, Australia, 28–30 September 2015; ICST (Institute for Computer Sciences, Social-Informatics and Telecommunications Engineering: Brussels, Belgium, 2015; pp. 44–50.

15. Vilarinho, T.; Farshchian, B.; Bajer, D.G.; Dahl, O.H.; Egge, I.; Hegdal, S.S.; Lønes, A.; Slettevold, J.N.; Weggersen, S.M. A combined smartphone and smartwatch fall detection system. In Proceedings of the 2015 IEEE International Conference on Computer and Information Technology, Ubiquitous Computing and Communications, Dependable, Autonomous and Secure Computing, Pervasive Intelligence and Computing, Liverpool, UK, 26–28 October 2015; pp. 1443–1448.

16. Wertner, A.; Czech, P.; Pammer-Schindler, V. An open labelled dataset for mobile phone sensing based fall detection. In Proceedings of the 12th EAI International Conference on Mobile and Ubiquitous Systems: Computing, Networking and Services, Coimbra, Portugal, 22–24 July 2015; pp. 277–278.

17. Vavoulas, G.; Chatzaki, C.; Malliotakis, T.; Pediaditis, M.; Tsiknakis, M. The MobiAct Dataset: Recognition of Activities of Daily Living using Smartphones. In Proceedings of the ICT4AgeingWell, Rome, Italy, 21–22 April 2016; pp. 143–151.

18. Micucci, D.; Mobilio, M.; Napoletano, P. Unimib shar: A dataset for human activity recognition using acceleration data from smartphones. *Appl. Sci.* **2017**, *7*, 1101. [CrossRef]

19. Casilari, E.; Santoyo-Ramón, J.A.; Cano-García, J.M. Umafall: A multisensor dataset for the research on automatic fall detection. *Procedia Comput. Sci.* **2017**, *110*, 32–39. [CrossRef]

20. Sucerquia, A.; López, J.; Vargas-Bonilla, J. SisFall: A fall and movement dataset. *Sensors* **2017**, *17*, 198. [CrossRef]

21. Rescio, G.; Leone, A.; Siciliano, P. Supervised machine learning scheme for electromyography-based pre-fall detection system. *Expert Syst. Appl.* **2018**, *100*, 95–105. [CrossRef]

22. de Quadros, T.; Lazzaretti, A.E.; Schneider, F.K. A movement decomposition and machine learning-based fall detection system using wrist wearable device. *IEEE Sens. J.* **2018**, *18*, 5082–5089. [CrossRef]

23. Shahzad, A.; Kim, K. FallDroid: An automated smart-phone-based fall detection system using multiple kernel learning. *IEEE Trans. Ind. Inform.* **2018**, *15*, 35–44. [CrossRef]

24. Martínez-Villaseñor, L.; Ponce, H.; Brieva, J.; Moya-Albor, E.; Núñez-Martínez, J.; Peñafort-Asturiano, C. UP-fall detection dataset: A multimodal approach. *Sensors* **2019**, *19*, 1988. [CrossRef]

25. Civit-Masot, J.; Domínguez-Morales, M.J.; Vicente-Díaz, S.; Civit, A. Dual Machine-Learning System to Aid Glaucoma Diagnosis Using Disc and Cup Feature Extraction. *IEEE Access* **2020**, *8*, 127519–127529. [CrossRef]

26. Dominguez-Morales, J.P.; Jimenez-Fernandez, A.F.; Dominguez-Morales, M.J.; Jimenez-Moreno, G. Deep neural networks for the recognition and classification of heart murmurs using neuromorphic auditory sensors. *IEEE Trans. Biomed. Circuits Syst.* **2017**, *12*, 24–34. [CrossRef]

27. Sahoo, A.K.; Pradhan, C.; Das, H. Performance evaluation of different machine learning methods and deep-learning based convolutional neural network for health decision making. In *Nature Inspired Computing for Data Science*; Springer: Berlin/Heidelberg, Germany, 2020; pp. 201–212.

28. Sahoo, A.K.; Pradhan, C.; Barik, R.K.; Dubey, H. DeepReco: Deep learning based health recommender system using collaborative filtering. *Computation* **2019**, *7*, 25. [CrossRef]

29. Civit-Masot, J.; Luna-Perejón, F.; Domínguez Morales, M.; Civit, A. Deep learning system for COVID-19 diagnosis aid using X-ray pulmonary images. *Appl. Sci.* **2020**, *10*, 4640. [CrossRef]

30. Duran-Lopez, L.; Dominguez-Morales, J.P.; Corral-Jaime, J.; Vicente-Diaz, S.; Linares-Barranco, A. COVID-XNet: A custom deep learning system to diagnose and locate COVID-19 in chest X-ray images. *Appl. Sci.* **2020**, *10*, 5683. [CrossRef]

31. Luna-Perejón, F.; Domínguez-Morales, M.J.; Civit-Balcells, A. Wearable Fall Detector Using Recurrent Neural Networks. *Sensors* **2019**, *19*, 4885. [CrossRef] [PubMed]

32. Gao, C.; Neil, D.; Ceolini, E.; Liu, S.C.; Delbruck, T. DeltaRNN: A Power-efficient Recurrent Neural Network Accelerator. In Proceedings of the 2018 ACM/SIGDA International Symposium on Field-Programmable Gate Arrays, Monterey, CA, USA, 25–27 February 2018; ACM: New York, NY, USA, 2018; pp. 21–30. [CrossRef]

33. Yu, S. Residual Learning and LSTM Networks for Wearable Human Activity Recognition Problem. In Proceedings of the 2018 37th Chinese Control Conference (CCC), Wuhan, China, 25–27 July 2018; pp. 9440–9447.

34. Musci, M.; De Martini, D.; Blago, N.; Facchinetti, T.; Piastra, M. Online fall detection using recurrent neural networks. *arXiv* **2018**, arXiv:1804.04976.

35. Sokolova, M.; Lapalme, G. A systematic analysis of performance measures for classification tasks. *Inf. Process. Manag.* **2009**, *45*, 427–437. [CrossRef]

Article

A Straightforward and Efficient Instance-Aware Curved Text Detector

Fan Zhao *, Sidi Shao, Lin Zhang and Zhiquan Wen

Department of Information Science, Xi'an University of Technology, Xi'an 710054, China; 2180820019@stu.xaut.edu.cn (S.S.); 2180821086@stu.xaut.edu.cn (L.Z.); 2180820012@stu.xaut.edu.cn (Z.W.)
* Correspondence: vcu@xaut.edu.cn; Tel.: +86-150-9105-0688

Abstract: A challenging aspect of scene text detection is to handle curved texts. In order to avoid the tedious manual annotations for training curve text detector, and to overcome the limitation of regression-based text detectors to irregular text, we introduce straightforward and efficient instance-aware curved scene text detector, namely, look more than twice (LOMT), which makes the regression-based text detection results gradually change from loosely bounded box to compact polygon. LOMT mainly composes of curve text shape approximation module and component merging network. The shape approximation module uses a particle swarm optimization-based text shape approximation method (called PSO-TSA) to fine-tune the quadrilateral text detection results to fit the curved text. The component merging network merges incomplete text sub-parts of text instances into more complete polygon through instance awareness, called ICMN. Experiments on five text datasets demonstrate that our method not only achieves excellent performance but also has relatively high speed. Ablation experiments show that PSO-TSA can solve the text's shape optimization problem efficiently, and ICMN has a satisfactory merger effect.

Keywords: text detection; convolutional neural networks; article swarm optimization; curved text

Citation: Zhao, F.; Shao, S.; Zhang, L.; Wen, Z. A Straightforward and Efficient Instance-Aware Curved Text Detector. *Sensors* **2021**, *21*, 1945. https://doi.org/10.3390/s21061945

Academic Editor: Yadir Torres Hernández

Received: 11 February 2021
Accepted: 7 March 2021
Published: 10 March 2021

Publisher's Note: MDPI stays neutral with regard to jurisdictional claims in published maps and institutional affiliations.

1. Introduction

As a crucial premise of text recognition, scene text detection has largely attracted the attention of many academics and industrial researchers, and many promising results have been achieved in recent decades. However, due to existing large differences in size, aspect ratio, direction and shape, as well as the presence of distortion and occlusion, success detection of scene text is still a very challenging task. To deal with these challenges, traditional text detection pipelines [1,2] usually focused on two subtasks: text detection and non-text removal, however, they are limited by hand-crafted features and usually involve heavy post-processing.

In recent years, with the renaissance of convolutional neural networks (CNNs), many deep learning-based methods [3–34] have achieved remarkable achievements in text detection, and these methods can be divided into top-down and bottom-up methods. The top-down methods [3–20], also commonly referred to as regression-based methods, usually adopt popular object detection pipelines to first detect text on the block level and then break a block into the word or line level if necessary. However, because of the structural limitations of the corresponding CNN models, these methods cannot efficiently handle long text and arbitrarily shaped texts. The bottom-up methods [21–34] first detect text components with a CNN and then group these components into text instances. According to the granularity of the text components, the bottom-up methods are mainly divided into two categories: pixel-level methods and part-level methods. Borrowing from the idea of semantic segmentation, pixel-based methods [21–29] produce a text saliency map for text detection by employing an FCN [30] to perform pixel-level text/nontext prediction. By regarding a text region as a group of text components, part-level methods [31–34] usually first detect individual characters and then group them into words or text lines. Compared

75

with the top-down methods, the bottom-up methods have more flexibility and better detection performance. Unfortunately, however, the training model requires a large number of sample annotations and complex network design, which limits their generalization.

For taking advantage of the high efficiency of top-down methods while avoiding the bottom-up method for large-scale labeling of samples, we introduce an efficient and straightforward curved text detection method, namely, look more than twice (LOMT), which progressively localizes a complete text instance multiple times without heavy learning. First, the text proposals are located in the original image by adopting a direct regression network, and then the candidate characters are extracted from each text proposal by using maximally stable extremal regions (MSER). Next, a particle swarm optimization-based text shape approximator (PSO-TSA) gradually approximates an arbitrary text shape by a specific PSO technique. Finally, one complete text instance is generated by merging two adjacent or intersecting text proposals by an instance-aware component merging network (ICMN).

Our main contributions can be summarized as follows:

(1) We propose a text detector, which can accurately locate curved text appearing in the scene through several consecutive straightforward and effective steps.
(2) To fine-tune the result of the regression-based text detector, we propose a particle swarm optimization-based text shape approximator called PSO-TSA, which can quickly approach text shapes without heavy pre-training or pre-learning in advance.
(3) To improve the text instance completeness, an instance-aware component merging network (ICMN) is designed to merge adjacent text subparts, which can be flexibly adapted to text detection results of any shape.
(4) Although the entire pipeline is not an end-to-end mechanism, the experiments on five text benchmark datasets show that our method not only achieves excellent harmonic mean (H-mean) performance but also has relatively high speed.

The rest of this paper is organized as follows. We briefly review some related works on scene text detection in Section 2, followed by describing the details of the proposed method in Section 3. Then, we present experimental results in Section 4. Finally, we conclude and give some perspectives on future work in Section 5.

2. Related Works

Recently, numerous approaches have been proposed to address different challenges in text detection, such as low resolution, perspective distortion, low contrast, complex background, multi-lingual, multi-view, multi-direction, and arbitrary shape and size. In early works, connected component analysis (CCA) was mainly utilized to detect text in scene images by first extracting candidate text fragments and then filtering non-text candidates. Representative connected-component-based approaches such as the stroke width transform (SWT) [1] and MSER [2] have achieved outstanding performance on various test datasets, particularly for well-focused texts, e.g., ICDAR13 [35]. However, these methods fall short for the more challenging ICDAR15 [36] and MSRA-TD500 [37] due to the limitation of hand-crafted features. With the advance of CNNs, different CNN models have been exploited for scene text detection tasks. In general, CNN-based methods can be roughly formulated into two categories: top-down methods and bottom-up methods.

Top-down methods first detect text on the block level and break a block into the word or line level if necessary, and these methods are often based on general object detection frameworks [3–20]. The Faster Region-based CNN (R-CNN) [3] is the most representative and accurate generic object detection method and has also achieved promising results on text detection tasks [4]. TextBoxes++ [5] adopts irregular 1×5 convolutional filters instead of the standard 3×3 filters and leverages recognition results to refine the detection results. ITN [6], E2E-MLT [11] and FOTS [18] are end-to-end text instance networks. Liu et al. applied feature pyramid networks (FPNs) in the CNN part to extract features of multi-scales, and used the bidirectional long short-term memory (Bi-LSTM) network to generate text proposals [7]. By connecting a region proposal network (RPN) and a regression module (TLOC), Liu et al. [8] proposed a method named the curve text detector (CTD) to detect curved texts. EAST [9], deep regression [20], DeconvNet [10] and E2E-MLT [11] all directly use a fully convolutional network (FCN) to predict the score map for each pixel. Inspired by EAST, Zhang et al. [12] presented a text detector named look more than once (LOMO) that progressively localized text multiple times. SegLink++ [13] detects dense and arbitrarily shaped scene text by instance-aware component grouping. By combining deep learning and histogram-oriented gradient features, NguyenVan et al. [14] designed a pooling-based scene text proposal technique and integrated it into an end-to-end scene text reading system. He et al. [15] proposed multi-scale scene text detection with a scale-based RPN. Zhu et al. [16] presented a shape robust detector with coordinate regression. Wang et al. [17] proposed a progressive scale expansion network (PSENet) to detect text instances with arbitrary shapes. Wang et al. [19] proposed a text region proposal network (text-RPN) and verified and refined it through a refinement network.

The above state-of-the-art regression-based methods have achieved promising results on standard benchmarks. However, most of these methods need a comprehensive design with complex anchors and cumbersome multistage network structure, and suboptimal performance is achieved by exhaustive tuning. Unsatisfactory results often occur when dealing with long text.

Bottom-up methods usually first detect text at the pixel or part level and then segment these components into desired word- or line-level text instances. Borrowing from the idea of semantic segmentation, pixel-based methods [21–29] produce a text saliency map by employing an FCN [30] to perform pixel-level text/non-text prediction. Wu et al. [21] proposed self-organized text detection with minimal post-processing via border learning. He et al. [22] presented an end-to-end scene text detection by combining a multiscale FCN with a cascade-style instance segmentation. PixelLink [23] generates an 8-direction margin to separate text lines. Xue et al. [24] exploited bootstrapping and text border semantic techniques for long text localization. Zhang et al. [25] first used an FCN to extract text blocks and then search for character candidates from these blocks with MSER [2]. By fusing the proposed generator technology with the FCN, a framework named FAST [26] was proposed to reduce the number of text proposals while maintaining a high recall in scene text images. Xu et al. [27] presented a text detector named TextField, which directly regresses the direction field. Pixel-level methods mostly need complex post-processing to extract accurate bounding boxes of words or text lines. Liu et al. propose the Mask Tightness Text Detector (Mask TTD) [29], which uses a tightness prior and text frontier learning to enhance pixel-wise masks prediction.

By regarding a text region as a group of text components, part-level methods [30–34] usually first detected components or characters and then grouped them into words or text lines. SPC-Net [30] uses an instance segmentation framework and context information to detect text in arbitrary shapes while suppressing false positives. Lyu et al. [32] combined position-sensitive segmentation with corner detection to generate text instances. Long et al. [33] used ordered disks to represent curved text. Baek et al. [34] developed a framework, referred to as character region awareness for text detection (CRAFT), which explores each character and the affinity between characters.

This work is inspired by the idea of LOMO [12], which progressively localizes text multiple times. However, the detection results of LOMO are multiple subsections of a word or text line instead of a single complete text instance, which leads to a certain degree of incompleteness of semantics. Similar to LOMO, by combining the backbone of ResNet50 [38] with FPN [39], our method also integrates the advantages of top-down methods and bottom-up methods. In order to overcome the limitation of regression-based detector to locate arbitrary shape text and to make the located text box to fully contain text while fitting its out-line, we first propose a straightforward and efficient text shape approximation method called PSO-TSA, which is based on particle swarm optimization method. Without prior training and heuristic parameters, PSO-TSA achieves very competitive performance on curved and non-curved datasets. Moreover, a lightweight ICMN is proposed to merge two adjacent or intersecting text subsections in a text line, which can further improve the completeness of a text instance. Compared with most deep learning algorithms, the advantages of our method are very simple and straightforward, PSO-TSA is general to any data set and ICMN can be readily plugged into any CNN-based detector.

3. The Proposed Method

3.1. Background

A challenging aspect of scene text detection is to handle curved text which is common in natural scenes. The regression-based methods have achieved promising results on standard benchmarks, and the efficiency is also very high. Using regression-based method for text proposal detection is a premise of our method. Here, we choose the classic EAST detector [9] to detect the text proposals appearing in the scene.

By inputting the text image to EAST detector [9], a dense prediction channel of text/non-text is outputted to indicate the pixel-wise confidence of being text or not. The dense proposals from the network are then filtered based on the text confidence score rp with the threshold value 0.9. For further processing in the next step, the pixel proposals with $rp \geq 0.9$ are labeled as white pixels, and the rest are labeled as black pixels, which are also considered text pixels and non-text pixels, respectively. The image formed by these white/black pixels is called a text confidence map, marked as the symbol I_m. Finally, locality-aware NMS [9] is performed on the filtered proposals to generate candidate quadrilateral text proposals, which are collected into a text proposal set $R = \left\{ r_t \big|_{t \in [1,T]} \right\}$. Here, r_t is the $t - th$ text proposal, and T is the total number of text proposals.

Usually, regression-based text detectors fall quite short of detecting extremely long curved text. In order to overcome the limitation of regression-based detector to irregular text, we introduce an efficient and straightforward mechanism called PSO-TSA, which makes the regression-based detection result gradually change from loosely bounded rectangle box to compact polygon.

In the PSO-TSA method, we need to fine-tune the text proposal detected by EAST. Character extraction is another prerequisite of our method. The accuracy of character extraction in the text proposal largely determines the performance of our method. In order to avoid a large number of sample annotations and complicated network designs, we adopt a simple traditional character extraction method like MSER [2] instead of a deep framework like CRAFT. The reason for choosing MSER is that the MSER algorithm assumes that text components have a similar homogeneous background and foreground and thus are stable with respect to a range of thresholds. Here, an extracted connected component is also considered a candidate character. The candidate characters extracted from the $t - th$ text proposal r_t are then combined into a set C_t, expressed as $C_t = \left\{ c_m^t | m \in [1, M] \right\}$, in which c_m^t is the $m - th$ character in $t - th$ text proposal and M is the total number of candidate characters. It should be noted here that a detected connected component does not refer to an individual character in a strict sense; it may be a single character or consecutive multiple characters in a text proposal.

3.2. LOMT

As illustrated in Figure 1, the pipeline of our approach consists of text proposal detection, connected components (CCs) extraction, shape approximation and component merging, the first two are prerequisites, and the last two are the main strategies of our algorithm.

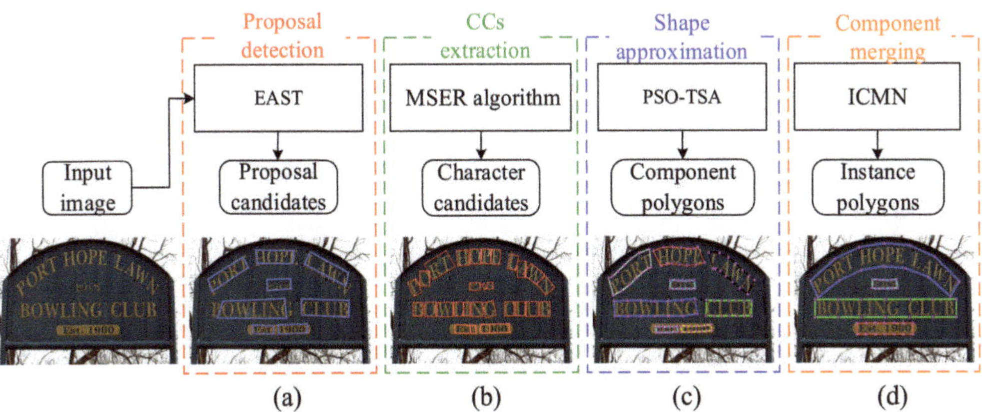

Figure 1. Illustration of the overall pipeline, (**a**) proposal detection module, (**b**) CCs extraction module, (**c**) shape approximation module and (**d**) component merging module.

3.2.1. PSO-TSA

Inspired by the sociological behavior associated with birds flocking, Kennedy and Eberhart proposed a PSO method [40] to solve a global optimization problem. Each individual in the swarm is called a particle, which represents a candidate position or solution to the problem. By constantly adjusting their positions and speeds with shared information or experience, particles fly through the search space influenced by two factors: one is the individual's best position ever found (pbest); the other is the group's best position (gbest). The PSO algorithm possesses the advantages of simplified computing, rather quick convergence speed, global optimization performance and fewer control parameters, so we apply an improved PSO in this paper to solve the optimal approximation problem of an arbitrary text shape. What needs to be pointed out here is that in addition to the position and velocity properties of the particles, we introduce isometric information into the particle in order to better use the improved PSO algorithm to approximate the shape of the text. To the best of our knowledge, this is the first time that the PSO algorithm has been used in curved text detection. A vivid example is shown in Figure 2 to explain the procedure of the proposed PSO-TSA algorithm, in which the text proposals and candidate characters are extracted from EAST and MSER algorithm, respectively. First, a curve is fitted with all the center points of the candidate characters, which is called the fitted character centerline. Because the accuracy of character detection by MSER algorithm is normally not high enough, this fitted curve is most likely not the real centerline of the curved text, which is also a problem that is usually faced by bottom-up algorithms. In response, points are evenly sampled on the fitted character centerline, which serve as the initial points of the particle, and also as the center point of the sampling neighborhood of the particle's corresponding dimension position. Here, the sampling neighborhood is also the region where the particle position changes, and particle positions in all dimensions are fitted to generate the approximated centerline of the text. Because the particle's initial position points are sampled on the fitted character centerline, the first approximated centerline is the same as the fitted character centerline. However, as the particle's position changes when the particle swarm flies in the sampling neighborhood, the approximated centerline will be farther away from the fitted character centerline and closer to the real centerline of the text such that it eventually coincides with the real text centerline. The polygon, connected in series by equidistant control points moving in the direction perpendicular to the approximated centerline, is the approximate polygon for curved text by PSO-TSA algorithm. By combining the shared foraging flight of the particle swarm and the continuous adjustment of the equidistant points of the corresponding particles according to their distance change, the global optimal particle position gradually approaches the true centerline of the text; in the same instant, the polygon gradually approaches the text shape, and finally, both optima are reached.

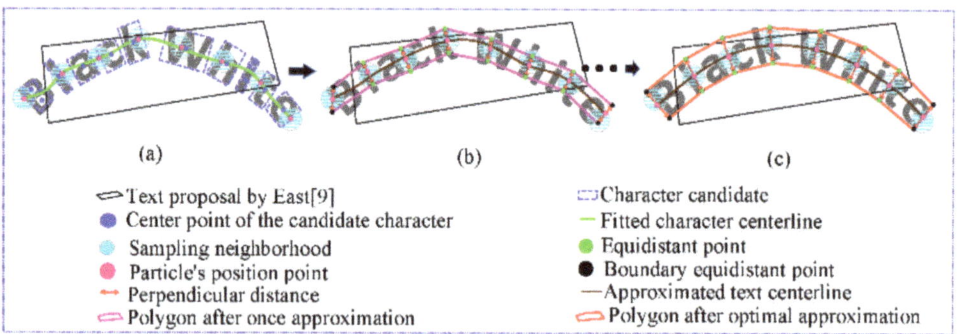

Figure 2. The framework of the PSO-TSA algorithm, (**a**) particle initialization process, (**b**) generating polygon by once approximation, (**c**) generating polygon by optimal approximation.

The specific initialization of particles is described in detail in Algorithm 1. First, we define a particle swarm $X = \{X_1, X_2, \cdots, X_I\}$, in which X_i represents the $i-th$ particle, $i \in [1, I]$, and I is the total number of particles in particle swarm X. Each particle is represented by its position, velocity and distance between its position and equidistant points, namely, $X_i = (X_i.p, X_i.v, X_i.d)^T$. Before initializing the particles, we first need to set the particles' range of activity. Assume that N points are uniformly sampled on the character centerline, which are expressed as $P_1, P_2, \cdots, P_N$. The spatial neighborhood $N_{r1 \times r1}(P_n)$ with center point P_n and radius $r1$ is the variation range of the n-*th* dimension position x_i^n of $X_i.p$ for $i \in [1, I]$ & $n \in [1, N]$. That is, the position of particle X_i is expressed as $X_i.p = \left(x_i^1, x_i^2, \cdots x_i^N\right)^T$, in which $x_i^n \in N_{r1 \times r1}(P_n)$. x_i^n is initialized as a point P_n^s sampled in $N_{r1 \times r1}(P_n)$, whose x and y coordinates are equal to $P_n^s.x$ and $P_n^s.y$, respectively. The $n-th$ dimensional velocity v_i^n of $X_i.v$ is initialized to $(0, 0)^T$. The distance variable d_i^n between the $n-th$ position point x_i^n and its two equidistant points is initialized with a value uniformly sampled in the interval $[d_1, d_2]$.

Algorithm 1. Particle Swarm Initialization.

---

1. Define particle swarm X $=\{X_1, \cdots, X_i, \cdots, X_I\}$, in which $X_i \cdot p = \left(x_i^1, \cdots, x_i^n, \cdots, x_i^N\right)^T$, $X_i.v = \left(v_i^1, \cdots v_i^n, \cdots, v_i^N\right)^T$ and $X_i.d = \left(d_i^1, d_i^2, \cdots d_i^N\right)^T$, respectively.
2. Construct N spatial neighborhoods, each denoted as $N_{r1 \times r1}(P_n)$, $n \in [1, N]$.
3. Set the range of equidistant values of particle position points as $[d_1, d_2]$.
4. *for* each $X_i \in X$ *do*
5.   Randomly sample a point P_n^s in each $N_{r1 \times r1}(P_n)$, and all sampled points form a point set $P^s = \left\{P_1^s, \cdots, P_n^s, \cdots, P_N^s\right\}$.
6.   Randomly sample N values in interval $[d_1, d_2]$ to form a equidistant set $D^S = \left\{d_1^s, \cdots, d_n^s, \cdots, d_N^s\right\}$.
7.   *for* each $n \in [1, N]$ *do*
8. $x_i^n = (P_n^s \cdot x, \ P_n^s \cdot y)^T$, $v_i^n = (0, 0)^T$ and $d_i^n = d_n^s$.
9.   *end for*
10. *end for*

When the number of iterations k does not reach the maximum number of iterations K or the error criterion $\Delta \varepsilon$ does not reach the minimum $\varepsilon_{\min}$, the proposed PSO-TSA algorithm is implemented iteratively to approximate the text shape. Each cyclical process mainly includes the following three steps:

Step 1: Two equidistant points $y_i^{n,1}$ and $y_i^{n,2}$ are calculated in the normal direction at $X_i's$ $n-th$ position point x_i^n according to Equations (1) and (2), respectively.

$$\begin{cases} y_i^{n,1} \cdot x = x_i^n - d_i^n \times \sin \theta_n \\ y_i^{n,1} \cdot y = x_i^n + d_i^n \times \cos \theta_n \end{cases} \tag{1}$$

$$\begin{cases} y_i^{n,2} \cdot x = x_i^n + d_i^n \times \sin \theta_n \\ y_i^{n,2} \cdot y = x_i^n - d_i^n \times \cos \theta_n \end{cases} \tag{2}$$

Here, θ_n is the angle between the tangent line at point x_i^n and the horizontal positive half-axis, that is, $\theta_n = \arctan(k_n)$. k_n is the tangent slope at the n-*th* point x_i^n of particle X_i. The isometric points of all position points are connected head to tail to form the corresponding polygon S_i of particle X_i.

Step 2: The valuating indicator f_{X_i} of particle X_i is calculated as

$$f_{X_i} = 0.5 \times \frac{A_{S_i}}{A_{r_t}} + 0.5 \times \frac{N_{S_i}^c}{N_{r_t}} \tag{3}$$

The fitting value f_{X_i} is expressed as the sum of two parts: one corresponds to the ratio of the text aggregation degree in S_i and r_t, and the other corresponds to the ratio of the number of characters in S_i and r_t. Each part has a weight of 0.5. Here, A_{S_i} and A_{r_t} are the text aggregation degree of S_i and r_t, respectively. $N_{S_i}^C$ and N_{r_t} is the number of characters in S_i and r_t. Here the variable N_{r_t} takes the value M. Given the text confidence map I_m, which is obtained in Section 3.1, the text aggregation degree A_{S_i} is defined as the ratio of the number of white pixels in the polygon S_i within I_m to the area of S_i, and A_{r_t} is the ratio of the number of white pixels in the text proposal r_t to the area of r_t.

The reason for considering the parameter ratio rather than the parameter itself of each part is that doing so can ensure the completeness of the text instance while ensuring the compactness.

Step 3: After $k - th$ single loop ends, the individual particle optima $Pbest_i$ and the particle swarm global optimum $Gbest^k$ are found from the history, and each particle's velocity, position and distance are updated according to Equations (4)–(6), respectively, in which c_1, c_2 represent the acceleration constants, μ_1, μ_2 are uniformly distributed random numbers and $\Delta\delta$ is white Gaussian noise.

$$X_i^k.v = X_i^{k-1}.v + c_1 \times \mu_1 \times (Pbest_i - X_i^k.p) + c_2 \times \mu_2 \times (Gbest^k - X_i^k.p) \tag{4}$$

$$X_i^k.p = X_i^{k-1}.p + X_i^k.v \tag{5}$$

$$X_i^k.d = X_i^{k-1}.d + \Delta\delta \tag{6}$$

When the optimization is completed, the optimal polygon will be obtained, which is composed of 14 equidistant points connected in series after the PSO-TSA optimal approximation.

3.2.2. ICMN

Even if the PSO-TSA algorithm is used to revise the EAST detection result from quadrilateral to polygon, the approximate polygon may not be a complete text instance but only a part of the text instance. To avoid incomplete scene text detection when processing long words or text lines, it is necessary to merge adjacent or intersecting text components that belong to the same text instance. Therefore, ICMN is proposed to solve the instance incomplete detection problem. The intuition behind the ICMN is that humans can easily judge whether two separate text components belong to a complete text instance and naturally perceive its shape without examining the entire instance; similarly, a specially designed neural network might also be able to infer the boundary points of a text instance.

The specific architecture of the ICMN is shown in Figure 3, which consists of an input layer, a pooling layer, two fully connected layers (FC) and an output layer. The input layer first accepts the boundary points of two adjacent polygons as individual features, both of which have a size of 1×28. In the pooling layer, these two features from the input are simply concatenated together to build a 1×56-sized snippet feature. There are two fully connected layers in the ICMN, the sizes of which are 56×48 and 48×30. Using fully connected layers in the ICMN reduces the dimensions of the pooled feature to the desired size of 1×30. The final output consists of two sibling parts: the first one (with $1 \times 2 - sized$) outputs two confidence scores indicating whether the two input components belong to the same text instance or not, and the second one (with $1 \times 28 - sized$) outputs 14 regression boundary points of the instance, half of which are upper boundary points and half of which are lower boundary points. Because each point has x and y coordinates, the ICMN outputs a total of 28 corresponding point coordinates. The salient aspects of the ICMN are that it can predict whether two components merge and can refine the text instance boundary points by spatial coordinate regression.

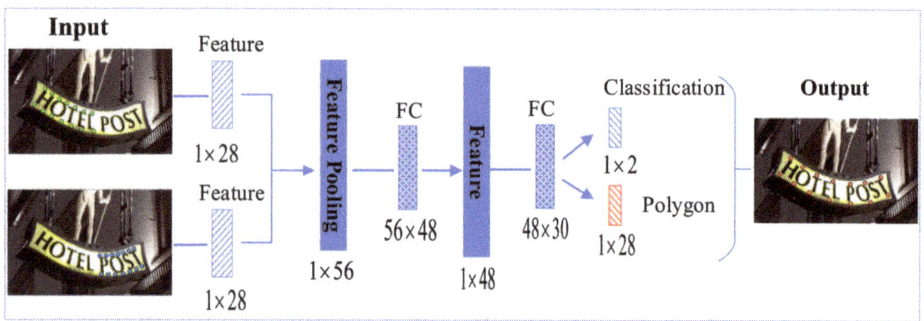

Figure 3. ICMN architecture.

For training the ICMN, we assign a binary class label (whether it belongs to an instance) to each text snippet. A positive label is assigned to a snippet if (i) two subsections in it belong to an instance or (ii) the Intersection over Union (IoU) is greater than or equal to 0.3. We design a multitask loss L to jointly train classification and coordinate regression.

$$L = L_{cls} + \lambda L_{reg} \tag{7}$$

Here, L_{cls} is the loss for merging/non-merging classification, which is a standard softmax loss; L_{reg} is the loss for spatial coordinate regression; and λ is a hyperparameter used to control over fitting. The regression loss is

$$L_{reg} = \frac{1}{N_{pos}} \sum_{i=1}^{N_b} \left(l_i^* \times \frac{1}{14} \sum_{j=1}^{14} \left\| \left(P_j^i.x - P_j^{i*}.x \right) + \left(P_j^i.y - P_j^{i*}.y \right) \right\|_2^2 \right) \tag{8}$$

where L_2 distance is adopted; l_i^* is the label, which is 1 for positive samples and 0 for negative samples; N_{pos} is the number of positive samples in a mini-batch; and N_b is the number of all samples in a mini-batch. Here, $P_j^i.x$ and $P_j^i.y$ denote the x and y coordinates of the predicted $j - th$ point P_j^i of the $i - th$ sample, respectively, and $P_j^{i*}.x$ and $P_j^{i*}.y$ denote the x and y coordinates of the $j - th$ ground truth point of the $i - th$ sample, respectively. Before training, we obtain the real ground truth by the method proposed in [8]. The regression loss is calculated only for positive samples. To assist LOMT in dealing with long curved text, a small and light ICMN is used in conjunction with PSO-TSA. Two adjacent or intersecting polygons may be subparts of the same text instance, so they are input to the ICMN to decide whether to merge them into a more complete text instance. The input of the ICMN is 28 points formed by concatenating each pair of 14 points of adjacent polygons, and the outputs of the network are the merging flag of the two parts and the regressed

14 points of the more complete text polygon. After a series of merging steps, each complete text instance appearing in the scene can be detected.

4. Experiment and Discussion

4.1. Datasets and Implementation Details

4.1.1. Datasets

ICDAR2015 (IC15) [36]: IC15 was built for Challenge 4 of the ICDAR-2015 Robust Reading Competition and consists of 1000 training images and 500 test images. The images are acquired using Google Glass, and the text accidentally appears in the scene. The ground truth is annotated by a word-level quadrangle.

ICDAR2017 (IC17) [41]: IC17 contains 7200 training images, 1800 validation images, and 9000 test images with text in 9 languages for multilingual scene text detection. Similar to IC15, the text regions in IC17 are also annotated by the 4 vertices of quadrilaterals.

MSRA-TD500 (TD500) [37]: TD500 contains 500 natural images, which are split into 300 training images and 200 test images, collected both indoors and outdoors using a pocket camera. The images contain English and Chinese scripts. This dataset is mainly used for multidirectional text detection, in which text regions are annotated by rotated rectangles.

SCUT-CTW1500 (CTW1500) [8]: CTW1500 consists of 1000 training and 500 test images. Every image has curved text instances, which are annotated by polygons with 14 vertices. The annotation is given at the text-line level such that a complete sentence is annotated as a single polygon.

Total-Text (Total-Text) [10]: Total-Text contains 1555 scene images, which are divided into 1255 training images and 300 test images. This dataset mainly provides curved texts, which are annotated at the word level by polygon-shaped bounding boxes.

4.1.2. Implementation Details

In this section, we carry out experiments on curved and non-curved datasets to test our method. In actual experiments, while maintaining the aspect ratio, the longer sides of the images in IC15, IC17, and TD500 are resized to 768, 1024, and 768, respectively, the longer sides of the images within Total-Text and CTW1500 are resized to 1280 and 1024, respectively. All experiments are performed with a single image resolution. The results are evaluated using the standard PASCAL VOC protocol [42], which uses a 0.5 intersection-over union (IoU) threshold to determine true and false positives.

For training the text proposal detector, a union of the IC15 and IC17 dataset is used. The network is trained using an improved adaptive moments (ADAM) [43] optimizer. To speed up learning, we uniformly sample 512×512 crops from images to form a mini-batch of size 24. The learning rate of ADAM starts from 1×10^{-3}, decays to one-tenth every 25,000 mini-batches, and stops at 1×10^{-5}. The network is trained until the performance stops improving.

When executing the PSO-TSA model, the parameters are set as follows. The particle swarm population consists of I particles moving in the N-dimensional search space, in which I and N are set as 20 and 7, respectively. The radius of the sampling neighborhood, $r1$, is a certain proportion of the maximum of the average width and height of characters, which is set to 0.25 in the experiment. The distance range of the particle's position point and its two equidistant points is $[d_1, d_2]$, in which $d_1 = 1.5 \times r1$ and $d_2 = 2.5 \times r2$. The maximum number of iteration K and the minimum error ε_{min} are set 20 and 1×10^{-3} respectively. The particle's acceleration constants C_1 and C_2 are the same, both of which are set to 1.2, and both u_1 and u_2 are random numbers in $(0, 1)$.

In the objective function of ICMN, the hyper-parameter λ was set as 1.5. During training of the ICMN network model, the negative-to-positive sample ratio was set to 10 in a mini-batch, the learning rate was set to 0.005, the batch size was 128, and the number of iterations was set to 30,000. In text detection process, when the IoU of two polygons is greater than or equal to 0.3, the two polygons are input into ICMN network for further merging.

4.2. Comparison with State-of-the-Arts

4.2.1. Experiments on Multi-Oriented Text (IC15, IC17 and TD500)

We first conduct experiments on three multi-oriented text datasets, IC15, IC17, and TD500, to demonstrate that the proposed method performs well for multi-oriented scene text detection. For IC15, annotations are provided at the word level, so the test experiments are performed at the word level. For the IC17 and TD500 datasets, annotations are provided at the line level, without a post-processing step applied to generate word boxes, similar to the method in [34]. We evaluate the performance of the proposed method directly at the line level. To make a fair comparison with end-to-end methods [11,18], we include their detection-only results by referring to the original papers. We also compare the proposed method with other state-of-the-art methods, including five recent multi-oriented text detectors, EAST [9], TextBoxes++ [5], FTPN [7] and the methods in [25,32] and five arbitrarily shaped text detectors: TextSnake [33], PixelLink [23], PSENET-1s [17], CRAFT [34] and ICG (SegLink++) [13]. The comparison is given in Table 1. As depicted in Table 1, LOMT achieves 85.7%, 74.6%, and 82.7% in terms of the harmonic mean (H-mean) on IC15, IC17, and TD500, respectively, outperforming most competitors.

Table 1. Results on quadrilateral-type datasets, such as IC15, IC17 and TD500. P, R and H refer to the precision, recall rate and H-mean, respectively. The best H score is highlighted in bold, and * denotes that the results based on multiscale tests.

Method	Backbone	Year	IC15			IC17			MSRA-TD500		
			P (%)	R (%)	H (%)	P (%)	R (%)	H (%)	P (%)	R (%)	H (%)
Zhang et al. [25]	VGG-16 + FCN	2016	71	43	54	-	-	-	83	67	74
EAST * [9]	VGG-16	2017	80.5	72.8	76.4	-	-	-	81.7	61.6	70.2
TextSnake [33]	VGG-16 + UNet	2018	84.9	80.4	82.6	-	-	-	83.2	73.9	78.3
TextBoxes++ * [5]	VGG-16	2018	87.8	78.5	82.9	-	-	-	-	-	-
PixelLink [23]	VGG-16	2018	85.5	82	83.7	-	-	-	83	73.2	77.8
Lyu et al. [32]	VGG-16 + FCN	2018	89.5	79.7	84.3	74.3	70.6	72.4	87.6	76.2	81.5
E2E-MLT [11]	ResNet-34 + FPN	2018	-	-	-	64.6	53.8	58.7	-	-	-
FOTS [18]	ResNet-50 + FPN	2018	88.8	82.0	85.3	79.5	57.5	66.7	-	-	-
PSENET-1s [17]	ResNet + FPN	2019	86.9	84.5	85.7	75.3	69.2	72.2	-	-	-
CRAFT [34]	VGG-16 + UNet	2019	89.8	84.3	**86.9**	80.6	68.2	73.9	88.2	78.2	**82.9**
FTPN [7]	ResNet + FPN	2019	68.2	78.0	72.8	-	-	-	-	-	-
ICG (SegLink++) [13]	VGG-16	2019	83.7	80.3	82.0	-	-	-	-	-	-
LOMT	ResNet-50 + FPN	2020	86.9	84.6	85.7	79.1	70.6	**74.6**	88.9	77.3	82.7

IC15 is a dataset of complicated backgrounds, and the text size is small. As shown in the fourth to sixth columns of Table 1, the H-mean performance of LOMT is better than all other competitors, except that it is slightly lower than the CRAFT method. In addition, we demonstrate some test examples in Figure 4a, and LOMT can accurately locate text instances in different directions and different sizes.

IC17 is a large-scale multilingual text dataset. We next verify the ability of our method to detect multilingual text for IC17. As shown in columns 7 to 9 of Table 1, our method outperforms all other competing methods in terms of H-mean. Some qualitative detection results on the IC17 dataset are given in Figure 4b, and the results show that our LOMT method can accurately detect multilingual text such as English, Chinese and Korean in different directions and sizes.

On MSRA-TD500, we evaluate the performance of LOMT for detecting long and multilingual text lines. As shown in the results of columns 10 to 12 of Table 1, our method has certain advantages in the detection of long texts, and the H-mean is only 0.2% lower than CRAFT [34]. In addition, we demonstrate some test examples in Figure 4c, and our method can accurately locate long text instances with various orientations.

Figure 4. Some qualitative results of the proposed method on IC15 (**a**), IC17 (**b**), and TD500 (**c**).

4.2.2. Experiments on Two Curve Text Datasets (Total-Text and CTW1500)

To test the ability of LOMT for the detection of curved text, we evaluate the detection results on both the Total-Text and CTW1500 datasets, which mainly contain curved text.

We first conduct text detection experiments on the Total-Text dataset. Total-Text was constructed in late 2017, and curved text was collected from various scenes, including text-like and low-contrast background scenes. Most images in this dataset contain at least one curved text, which is annotated in word-level granularity using a strictly 14-point polygon. To compare LOMT with other algorithms under the same conditions, the number of uniformly sampled points on the character centerline is set to 7 so that the polygon with the same 14 vertices as the ground truth. Columns 3 to 5 of Table 2 show the results of different methods on Total-Text. As shown in Table 2, the proposed method achieves 88.2%, 79.6%, and 83.7% in precision, recall and H-mean on Total-Text, respectively. The results show that our method outperformed recent state-of-the-art methods on Total-Text. Especially, compared with the text detector EAST, the H-mean is improved by 41.7%, and our method is slightly better than the excellent CRAFT algorithm. Examples of the detection results of the proposed method are illustrated in Figure 5. From the results, we can find that the proposed method can work well for text of any shape and length. In addition, we show some test examples of different methods on Total-Text in Figure 6, where the results in columns (a), (b) and (c) correspond to CTD + TLOC [8], CRAFT [34] and the proposed LOMT method, respectively. CTD + TLOC [8] detects a false text in the first row in Figure 6a, and all detection results are inaccurate. The detection results of CRAFT [34] shown in Figure 6b are relatively ideal, but the parts of a single character, such as the letter "C", are not completely contained in the detected text box. Compared to other methods, the proposed method has better completeness of the text instances and smoother boundaries of the polygons.

Table 2. Results on curved text datasets Total-Text and CTW1500. The best H score is highlighted in bold and * denotes the results based on multiscale tests.

Method	Year	Total-Text			CTW1500		
		P (%)	R (%)	H (%)	P (%)	R (%)	H (%)
EAST * [9]	2017	50.0	36.2	42.0	78.7	49.1	60.4
DeconvNet * [10]	2017	33.0	40.0	36.0	-	-	-
TextSnake [33]	2018	82.7	74.5	78.4	67.9	85.3	75.6
PSENet-1s [17]	2019	81.8	75.1	78.3	80.6	75.6	78.0
LOMO-1s [12]	2019	88.6	75.7	81.6	-	-	-
CTD + TLOC [8]	2019	77.4	69.8	73.4	77.4	69.8	73.4
ICG (SegLink++) [13]	2019	82.1	80.9	81.5	82.8	79.8	81.3
CRAFT [34]	2019	87.6	79.9	83.6	86.0	81.1	**83.5**
Wang et al. [19]	2019	80.9	76.2	78.5	80.1	80.2	80.1
Mask-TTD [29]	2020	74.5	79.1	76.7	79.7	79.0	79.4
LOMT	2020	88.2	79.6	**83.7**	85.1	79.9	82.5

Figure 5. Some qualitative results of the proposed method on Total-Text.

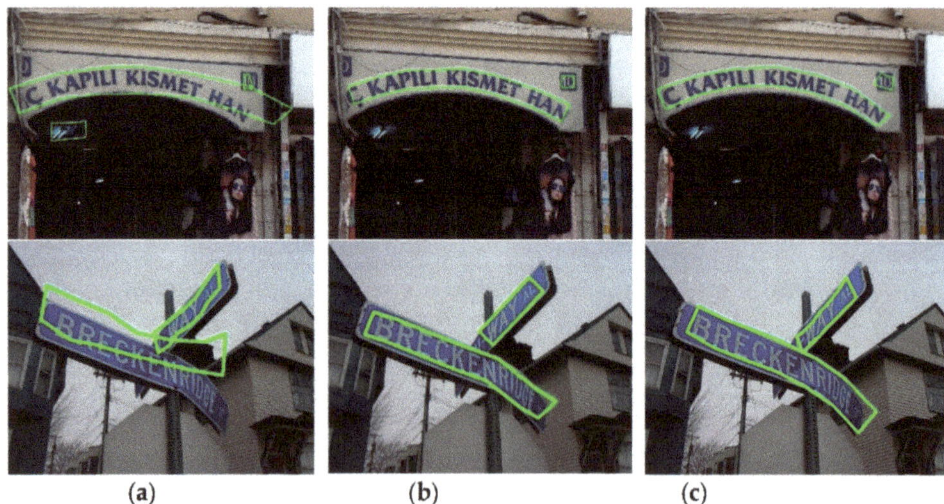

Figure 6. Comparison results of several methods, (**a**) CTD + TOLC [8], (**b**) CRAFT [34], and (**c**) LOMT.

We continue to evaluate the results of LOMT on CTW1500, whose annotation is given at the text-line level such that a complete text instance is annotated as a single 14-point polygon. A comparison with the quantitative evaluations of other methods is depicted in Table 3, indicating that H-mean performance of the proposed LOMT method is 22.1% higher than the EAST [9] algorithm. Compared with CTD + TLOC [8], TextSnake [33], PSENet-1s [17] and Mask-TTD [29], H-mean improvement is 9.1%, 6.9%, 4.5% and 3.1%, respectively. The proposed LOMT is slightly better in H-mean performance than ICG (SegLink++) [13] and the method in [19] and is only 1.0% lower than CRAFT [34]. The results demonstrate that the proposed method can handle arbitrarily shaped line-level text detection. Some qualitative results are shown in Figure 7. The results show that LOMT can handle long text detection well.

Table 3. The ablation experiment results.

Dataset	EAST	MSER Extraction	PSO-TSA	ICMN	H (%)
	√	X	X	X	60.4
CTW1500	√	√	√	X	71.2
	√	√	√	√	81.5
	√	X	X	X	42.0
Total-Text	√	√	√	X	81.7
	√	√	√	√	83.7

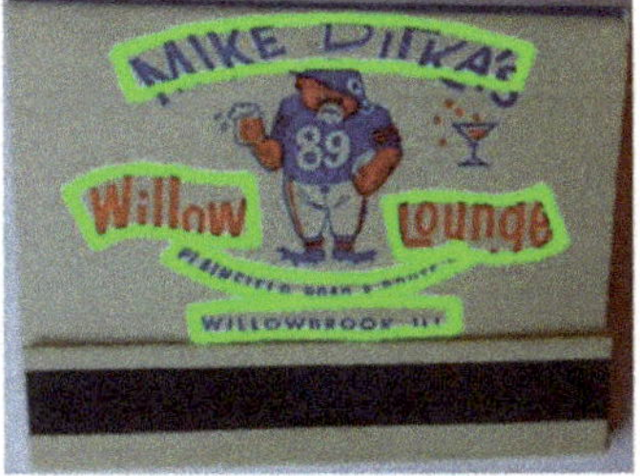

Figure 7. Some qualitative results of the proposed method on CTW1500.

4.2.3. Speed and Ablation Analysis

All the experiments are conducted on a PC using a single NIVIDA GeForce GTX 1080 graphics card with the Pascal architecture and an Intel(R) Core(TM) i7-6700K @ 4.00 GHZ CPU. The speed of PSO-TSA and ICMN are 20.0 FPS and 30.9 FPS, respectively, which can basically achieve real-time running. Although the H-mean performance of the CRAFT algorithm is currently the most competitive compared with other state of the arts, its average frame rate is only 3.3.

The proposed method consists of four modules: text proposal detection, candidate character extraction using MSER, text shape approximations by PSO-TSA, and text subsection merging via ICMN. The first two modules are the prerequisites for PSO-TSA. Without the extracted characters, the PSO-TSA module cannot be executed independently. That is, the contribution of the character extraction module is included in the performance improvement of PSO-TSA. Therefore, we need to verify only the effectiveness of the three modules, namely, text proposal detection, PSO-TSA and ICMN. Table 3 shows the ablation experiment results on two curved text benchmarks. From Table 3, for the CTW1500 and Total-Text datasets, compared to the text proposal detection module, the improvements in the H-mean performance of the PSO-TSA module are 10.8% and 39.7%, respectively. Furthermore, compared with the PSO-TSA module without ICMN, the gain of the H-mean performance of LOMT is 10.3% and 2.0%. In particular, it should be emphasized that compared to the general regression-based detection module, EAST, the overall performance of our algorithm can be improved by at least 21.1% and up to 41.7%. The experimental results demonstrate that each module in Table 3 has a certain contribution to the overall performance, thus verifying the necessity of each part.

4.2.4. Disadvantages and Advantages

The main disadvantage of the proposed method is not an end-to-end framework. The traditional MSER method is used in our character extraction module, and the accuracy rate is 68.3% in CTW1500 data set, while the accuracy rate of CRAFT algorithm is 75.7%. Due to unsatisfactory text detection performance of MSER, the accuracy of the proposed method is not good enough compared to the existing methods, especially CRAFT.

Although the performance of our method is slightly inferior to the CRAFT method, one of the biggest advantages of our method is that it does not depend too much on a specific data set. In order to test the generalization of the algorithm, Table 4 lists the comparison results of the generalization experiment of the two methods, in which CRAFT-IC15-20k.pth and CRAFT-MLT-25k.pth denote the CRAFT models trained on IC15 and IC13 + IC17 data sets, respectively. It can be seen from Table 4 that the model CRAFT-IC15-20k.pth performs well on the IC15 dataset, but its performances on the other four datasets have dropped a lot, and the performances of the model CRAFT-MLT-25k.pth on the IC15 and IC17-MLT data sets are also inferior to our method. In other words, the CRAFT model is strongly based on training samples, and its generalization ability is not particularly ideal. Fortunately, neither our character extraction module nor our PSO-TSA module depends on a specific dataset; consequently, they are highly applicable to any dataset. ICMN module can be readily plugged into to any regression-based text positioning module to complete the text instance detection.

Table 4. Comparison of generalization experimental results.

Data Sets	CRAFT-IC15-20k.pth			CRAFT-MLT-25k.pth			LOMT		
	P (%)	R (%)	H (%)	P (%)	R (%)	H (%)	P (%)	R (%)	H (%)
IC15	89.8	84.3	86.9	81.7	82.5	82.0	86.9	84.6	85.7
IC17-MLT	51.2	47.7	49.4	80.6	68.2	73.9	79.1	70.6	74.6
TD500	16.1	29.6	20.9	88.0	78.1	82.9	88.9	77.3	82.7
CTW1500	69.8	70.6	70.2	86.0	81.1	83.5	85.1	79.9	82.5
Total-Text	72.5	76.3	74.3	87.6	79.9	83.6	88.2	79.6	83.7

5. Conclusions

In this paper, we have proposed a straightforward and efficient instance-aware curved text detector, which is composed of quadrilateral text proposal detection, character extraction, particle swarm optimization-based text shape approximation (PSO-TSA), and an instance-aware component merging network (ICMN). Without training or learning in advance, PSO-TSA can adjust the quadrilateral box obtained by a regression-based text detector to a polygon, thus approaches the shape of the text. ICMN can merge two adjacent or intersecting text components into one single text instance that is more complete. A large number of experiments have been carried out on five scene text detection benchmark datasets, H-mean performance achieve to 85.7%, 74.6%, 82.7%, 82.5% and 83.7% for IC15, IC17, MSRA-TD500, CTW1500 and Total-Text dataset, respectively. As the two core components of the text detector, the average running speeds of PSO-TSA and ICMN modules are 20.0 FPS and 30.9 FPS, respectively. Experimental results demonstrate that the proposed algorithm has competitive advantages in H-mean performance and execution speed.

In the future, we would like to replace the MSER extraction with more elegant character detection method to further improve the efficiency of the algorithm.

Author Contributions: Conceptualization, F.Z.; methodology, F.Z., S.S. and L.Z.; software, S.S. and L.Z.; validation, Z.W.; formal analysis, S.S. and L.Z.; writing—original draft preparation, F.Z.; writing—review and editing, Z.W. All authors have read and agreed to the published version of the manuscript.

Funding: This work was funded by the National Natural Science Foundation of China (NSFC), grant number 61671376 and 61771386.

Institutional Review Board Statement: Not applicable.

Informed Consent Statement: Not applicable.

Data Availability Statement: The datasets IC15 and IC17 in this paper can be obtained from the following link: https://rrc.cvc.uab.es/?ch=4&com=downloads, accessed on 26 August 2015. The dataset TD500 in this paper can be obtained at http://www.iapr-tc11.org/mediawiki/index.php/MSRA_Text_Detection_500_Database_%28MSRA-TD500%29, accessed on 20 June 2012. The dataset CTW1500 in this paper can be obtained at https://www.github.com/Yuliang-Liu/Curve-Text-Detector, accessed on 6 December 2017. The dataset TotalText in this paper can be obtained at https://github.com/cs-chan/Total-Text-Dataset, accessed on 15 November 2017.

Conflicts of Interest: The authors declare no conflict of interest.

References

1. Epshtein, B.; Ofek, E.; Wexler, Y. Detecting Text in Natural Scenes with Stroke Width Transform. In Proceedings of the IEEE Conference on Computer Vision and Pattern Recognition (CVPR), San Francisco, CA, USA, 13–18 June 2010; pp. 2963–2970.
2. Neumann, L.; Matas, J. A method for text localization and recognition in real-world images. In Proceedings of the Asian Conference on Computer Vision, Queenstown, New Zealand, 8–12 November 2010; pp. 770–783.
3. Ren, S.; He, K.; Girshick, R.; Sun, J. Faster R-CNN: Towards real-time object detection with region proposal networks. In Proceedings of the Neural Information Processing Systems, Montreal, QC, Canada, 7–10 December 2015; pp. 91–99.
4. Ma, J.; Shao, W.; Ye, H.; Wang, L.; Wang, H.; Zheng, Y.; Xue, X. Arbitrary-oriented scene text detection via rotation proposals. *IEEE Trans. Multimed.* **2018**, *20*, 3111–3122. [CrossRef]
5. Liao, M.; Shi, B.; Bai, X. Textboxes++: A single-shot oriented scene text detector. *IEEE Trans. Image Process.* **2018**, *27*, 303–338. [CrossRef]
6. Wang, F.; Zhao, L.; Li, X.; Wang, X. Geometry-aware scene text detection with instance transformation network. In Proceedings of the IEEE Conference on Computer Vision and Pattern Recognition (CVPR), Salt Lake City, UT, USA, 18–23 June 2018; pp. 1381–1389.
7. Liu, F.; Chen, C.; Gu, D.; Zheng, J. Ftpn: Scene text detection with feature pyramid based text proposal network. *IEEE Access* **2019**, *7*, 44219–44228. [CrossRef]
8. Liu, Y.; Jin, L.; Zhang, S.; Luo, C.; Zhang, S. Curved scene text detection via transverse and longitudinal sequence connection. *Pattern Recognit.* **2019**, *90*, 337–345. [CrossRef]
9. Zhou, X.; Yao, C.; Wen, H.; Wang, Y.; Zhou, S.; He, W.; Liang, J. East: An efficient and accurate scene text detector. In Proceedings of the IEEE Conference on Computer Vision and Pattern Recognition (CVPR), Honolulu, HI, USA, 21–26 July 2017; pp. 2642–2651.
10. Ch'ng, C.K.; Chan, C.S. Total-text: A comprehensive dataset for scene text detection and recognition. In Proceedings of the 2017 14th IAPR International Conference on Document Analysis and Recognition (ICDAR), Kyoto, Japan, 9–15 November 2017; Volume 1, pp. 935–942.
11. Bušta, M.; Patel, Y.; Matas, J. E2e-mlt-an unconstrained end-to-end method for multi-language scene text. In Proceedings of the Asian Conference on Computer Vision, Perth, WA, Australia, 2–6 December 2018; Springer: Cham, Switzerland; pp. 127–143.
12. Zhang, C.; Liang, B.; Huang, Z.; En, M.; Han, J.; Ding, E.; Ding, X. Look more than once: An accurate detector for text of arbitrary shapes. In Proceedings of the IEEE Conference on Computer Vision and Pattern Recognition (CVPR), Long Beach, CA, USA, 16–20 June 2019; pp. 10552–10561.
13. Tang, J.; Yang, Z.; Wang, Y.; Zheng, Q.; Xu, Y.; Bai, X. Seglink++: Detecting dense and arbitrary-shaped scene text by instance-aware component grouping. *Pattern Recognit.* **2019**, *96*, 106954. [CrossRef]
14. NguyenVan, D.; Lu, S.; Tian, S.; Ouarti, N.; Mokhtari, M. A pooling based scene text proposal technique for scene text reading in the wild. *Pattern Recognit.* **2019**, *87*, 118–129. [CrossRef]
15. He, W.; Zhang, X.-Y.; Yin, F.; Luo, Z.; Ogier, J.-M.; Liu, C.-L. Realtime multi-scale scene text detection with scale-based region proposal network. *Pattern Recognit.* **2020**, *98*, 107026. [CrossRef]
16. Zhu, Y.; Ma, C.; Du, J. Rotated cascade R-CNN: A shape robust detector with coordinate regression. *Pattern Recognit.* **2019**, *96*, 106964. [CrossRef]

17. Wang, W.; Xie, E.; Li, X.; Hou, W.; Lu, T.; Yu, G.; Shao, S. Shape robust text detection with progressive scale expansion network. In Proceedings of the IEEE Conference on Computer Vision and Pattern Recognition (CVPR), Long Beach, CA, USA, 16–20 June 2019; pp. 9336–9345.
18. Liu, X.; Liang, D.; Yan, S.; Chen, D.; Qiao, Y.; Yan, J. Fots: Fast oriented text spotting with a unified network. In Proceedings of the IEEE Conference on Computer Vision and Pattern Recognition (CVPR), Salt Lake City, UT, USA, 18–23 June 2018; pp. 5676–5685.
19. Wang, X.; Jiang, Y.; Luo, Z.; Liu, C.L.; Choi, H.; Kim, S. Arbitrary shape scene text detection with adaptive text region representation. In Proceedings of the IEEE Conference on Computer Vision and Pattern Recognition (CVPR), Long Beach, CA, USA, 16–20 June 2019; pp. 6449–6458.
20. He, W.; Zhang, X.-Y.; Yin, F.; Liu, C.-L. Deep direct regression for multi-oriented scene text detection. In Proceedings of the IEEE International Conference on Computer Vision, Venice, Italy, 22–29 October 2017; pp. 745–753.
21. Wu, Y.; Natarajan, P. Self-organized text detection with minimal post-processing via border learning. In Proceedings of the IEEE Conference on Computer Vision and Pattern Recognition (CVPR), Honolulu, HI, USA, 21–26 July 2017; pp. 5000–5009.
22. He, D.; Yang, X.; Liang, C.; Zhou, Z.; Ororbi, A.G.; Kifer, D.; Giles, C.L. Multi-scale fcn with cascaded instance aware segmentation for arbitrary oriented word spotting in the wild. In Proceedings of the IEEE Conference on Computer Vision and Pattern Recognition (CVPR), Honolulu, HI, USA, 21–26 July 2017; pp. 474–483.
23. Deng, D.; Liu, H.; Li, X.; Cai, D. Pixellink: Detecting scene text via instance segmentation. In Proceedings of the AAAI-18 AAAI Conference on Artificial Intelligence, New Orleans, LA, USA, 2–7 February 2017; pp. 6773–6780.
24. Xue, C.; Lu, S.; Zhan, F. Accurate scene text detection through border semantics awareness and bootstrapping. In Proceedings of the European Conference on Computer Vision, Munich, Germany, 8–14 September 2018; pp. 35–372.
25. Zhang, Z.; Zhang, C.; Shen, W.; Yao, C.; Liu, W.; Bai, X. Multi-oriented text detection with fully convolutional networks. In Proceedings of the IEEE Conference on Computer Vision and Pattern Recognition (CVPR), Las Vegas, NV, USA, 27–30 June 2016; pp. 4159–4167.
26. Bazazian, D.; G'omez, R.; Nicolaou, A.; Gomez, L.; Karatzas, D.; Bagdanov, A.D. Fast: Facilitated and accurate scene text proposals through fcn guided pruning. *Pattern Recognit. Lett.* **2019**, *119*, 112–120. [CrossRef]
27. Xu, Y.; Wang, Y.; Zhou, W.; Wang, Y.; Yang, Z.; Bai, X. Textfield: Learning a deep direction field for irregular scene text detection. *IEEE Trans. Image Process.* **2019**, *28*, 5566–5579. [CrossRef] [PubMed]
28. Li, J.; Zhang, C.; Sun, Y.; Han, J.; Ding, E. Detecting text in the wild with deep character embedding network. In Proceedings of the Asian Conference on Computer Vision, Perth, WA, Australia, 2–6 December 2018; Springer: Cham, Switzerland, 2018.
29. Liu, Y.; Jin, L.; Fang, C. Arbitrarily Shaped Scene Text Detection with a Mask Tightness Text Detector. *IEEE Trans. Image Process.* **2020**, *29*, 2918–2930. [CrossRef] [PubMed]
30. Long, J.; Shelhamer, E.; Darrell, T. Fully convolutional networks for semantic segmentation. In Proceedings of the IEEE Conference on Computer Vision and Pattern Recognition (CVPR), Boston, MA, USA, 7–12 June 2015; pp. 3431–3440.
31. Xie, E.; Zang, Y.; Shao, S.; Yu, G.; Yao, C.; Li, G. Scene text detection with supervised pyramid context network. In Proceedings of the AAAI Conference on Artificial Intelligence, Honolulu, HI, USA, 27 January–1 February 2019; pp. 9038–9045.
32. Lyu, P.; Yao, C.; Wu, W.; Yan, S.; Bai, X. Multi-oriented scene text detection via corner localization and region segmentation. In Proceedings of the IEEE Conference on Computer Vision and Pattern Recognition (CVPR), Salt Lake City, UT, USA, 18–23 June 2018; pp. 7553–7563.
33. Long, S.; Ruan, J.; Zhang, W.; He, X.; Wu, W.; Yao, C. Textsnake: A flexible representation for detecting text of arbitrary shapes. In Proceedings of the European Conference on Computer Vision, Munich, Germany, 8–14 September 2018; pp. 20–36.
34. Baek, Y.; Lee, B.; Han, D.; Yun, S.; Lee, H. Character region awareness for text detection. In Proceedings of the IEEE Conference on Computer Vision and Pattern Recognition (CVPR), Long Beach, CA, USA, 16–20 June 2019; pp. 9365–9374.
35. Karatzas, D.; Shafait, F.; Uchida, S.; Iwamura, M.; Bigorda, L.G.i.; Mestre, S.R.; Mas, J.; Mota, D.F.; Almazan, J.A.; de las Heras, L.P. ICDAR 2013 robust reading competition. Document Analysis and Recognition (ICDAR). In Proceedings of the 2013 12th International Conference on IEEE Computer Society, Niigata, Japan, 16–20 June 2013; pp. 1484–1493.
36. Karatzas, D.; Gomez-Bigorda, L.; Nicolaou, A.; Ghosh, S.; Bagdanov, A.; Iwamura, M.; J Matas, L.N.; Chandrasekhar, V.R.; Lu, S. ICDAR 2015 competition on robust reading. In Proceedings of the 2015 13th International Conference on Document Analysis and Recognition (ICDAR), Tunis, Tunisia, 23–26 August 2015; pp. 1156–1160.
37. Yao, C.; Bai, X.; Liu, W.; Ma, Y.; Tu, Z. Detecting texts of arbitrary orientations in natural images. In Proceedings of the IEEE Conf. on Computer Vision and Pattern Recognition (CVPR), Providence, RI, USA, 18–20 June 2012; pp. 1083–1090.
38. He, K.; Zhang, X.; Ren, S.; Sun, J. Deep residual learning for image recognition. In Proceedings of the IEEE Conference on Computer Vision and Pattern Recognition (CVPR), Las Vegas, NV, USA, 27–30 June 2016; pp. 770–778.
39. Lin, T.-Y.; Dollár, P.; Girshick, R.; He, K.; Hariharan, B.; Belongie, S. Feature pyramid networks for object detection. In Proceedings of the IEEE Conference on Computer Vision and Pattern Recognition (CVPR), Honolulu, HI, USA, 21–26 July 2017; pp. 2117–2125.
40. Kennedy, J.; Eberhar, R. Particle swarm optimization. In Proceedings of the ICNN'95-International Conference on Neural Networks, Perth, WA, Australia, 27 November–1 December 1995; pp. 1942–1948.

41. Nayef, N.; Yin, F.; Bizid, I.; Choi, H.; Feng, Y.; Karatzas, D.; Luo, Z.B.; Pal, U.; Rigaud, C.; Chazalon, J.; et al. ICDAR2017 robust reading challenge on multi-lingual scene text detection and script identification-rrc-mlt. In Proceedings of the 2017 14th IAPR International Conference on Document Analysis and Recognition (ICDAR) IEEE, Kyoto, Japan, 9–15 November 2017; pp. 1454–1459.
42. Everingham, M.; van Gool, L.; Williams, C.K.; Winn, J.; Zisserman, A. The pascal visual object classes (voc) challenge. *Int. J. Comput. Vis.* **2010**, *88*, 303–338. [CrossRef]
43. Zhang, Z. Improved adam optimizer for deep neural networks. In Proceedings of the 2018 IEEE/ACM 26th International Symposium on Quality of Service (IWQoS), Banff, AB, Canada, 4–6 June 2018; pp. 1–2.

Communication

Incremental Learning in Modelling Process Analysis Technology (PAT)—An Important Tool in the Measuring and Control Circuit on the Way to the Smart Factory

Shivani Choudhary [1,*,†], Deborah Herdt [1,*,†], Erik Spoor [1], José Fernando García Molina [2], Marcel Nachtmann [1] and Matthias Rädle [1]

1   Center for Mass Spectrometry and Optical Spectroscopy, Mannheim University of Applied Sciences, Paul-Wittsack-Straße 10, 68163 Mannheim, Germany; e.spoor@hs-mannheim.de (E.S.); m.nachtmann@hs-mannheim.de (M.N.); m.raedle@hs-mannheim.de (M.R.)
2   Institute of Process Control and Innovative Energy Conversion, Mannheim University of Applied Sciences, 68163 Mannheim, Germany; Jose.Fernando.Garcia.Molina@continental-corporation.com
*   Correspondence: s.choudhary@hs-mannheim.de (S.C.); d.herdt@hs-mannheim.de (D.H.)
†   These authors contributed equally to this work.

Citation: Choudhary, S.; Herdt, D.; Spoor, E.; García Molina, J.F.; Nachtmann, M.; Rädle, M. Incremental Learning in Modelling Process Analysis Technology (PAT)—An Important Tool in the Measuring and Control Circuit on the Way to the Smart Factory. *Sensors* **2021**, *21*, 3144. https://doi.org/10.3390/s21093144

Academic Editors: Fabian Khateb, Bernhard Wilhelm Roth, Krzysztof M. Abramski and Guillermo Villanueva

Received: 2 March 2021
Accepted: 28 April 2021
Published: 1 May 2021

Publisher's Note: MDPI stays neutral with regard to jurisdictional claims in published maps and institutional affiliations.

Abstract: To meet the demands of the chemical and pharmaceutical process industry for a combination of high measurement accuracy, product selectivity, and low cost of ownership, the existing measurement and evaluation methods have to be further developed. This paper demonstrates the attempt to combine future Raman photometers with promising evaluation methods. As part of the investigations presented here, a new and easy-to-use evaluation method based on a self-learning algorithm is presented. This method can be applied to various measurement methods and is carried out here using an example of a Raman spectrometer system and an alcohol-water mixture as demonstration fluid. The spectra's chosen bands can be later transformed to low priced and even more robust Raman photometers. The evaluation method gives more precise results than the evaluation through classical methods like one primarily used in the software package Unscrambler. This technique increases the accuracy of detection and proves the concept of Raman process monitoring for determining concentrations. In the example of alcohol/water, the computation time is less, and it can be applied to continuous column monitoring.

Keywords: SVM; incremental learning; Raman spectroscopy; process technology

1. Introduction

Process analysis technology (PAT) is established in many chemical industry plants. It enables the production of the required technical quality in compliance with safety standards. This is made possible with the best possible use of raw materials, systems, and energy. The chemical industry is the most energy-intensive manufacturing industry in Germany, with consumption of 1137.3 petajoules in 2018 [1]. This corresponds to a work or energy of 36.06 gigawatt years. From an economic perspective alone, resource efficiency is of great importance. From 1990 to 2018, the chemical–pharmaceutical industry was able to increase its production by 76%, reduce energy consumption by 17%, and reduce greenhouse gas emissions by 51% [2]. It is of the utmost importance to use sensors that significantly increase process understanding and allow more profound insight into the process. The profit is immense, especially for existing systems. Direct in-line measurement of the current composition for regulation is currently rarely used due to the high costs. Therefore, the typical sensors in-process application focus on process control variables—such as pressure and temperature—instead of in-line-product variables. However, measuring the composition of substances is particularly valuable to optimize processes economically and energetically. Optical processes are growing slowly because knowledge is obtained in-line and no extractions of the process have to be made, but accuracy is too low and the price too high. These systems must be flexible and quick to use to understand the process further. Permanent measuring points must be characterized by high measuring accuracy and reliability with low installation and operating costs. To guarantee process stability and, therefore, a high performance, the process has to be monitored. With high measurement density, digital output of the data, and numerous individual and real-time measurements, the system's state becomes more predictable. Additionally, simultaneous measurements of several indicators of the process stabilize the predicted status, and consequently, rapid intervention, if necessary, is ensured.

A possible and suitable method for those mentioned challenges is Raman technology with high selectivity and detection sensitivity. The integration into the process can be realized with a flange connector mounted on an inspection glass in a pipe. The excitation wavelength from the laser and the scattered light from the sample can pass through the window [3,4]. With a calibration model, the resulting Raman shift intensities can be converted into the substance's concentration. This continuous measurement variable leads to sustainable and safe controlled operations. From an economic point of view, the plant's productivity can be increased or decreased as required. Therefore, this method can be adjusted according to demand, changing energy and raw material prices, leading to enhanced profit. Production efficiency is an increasingly differentiating characteristic for companies. This characteristic is per the sustainability requirements. Worldwide, there is a demand for conserving resources, reducing global warming gases, and, therefore, sustainable energy usage.

In the present work, an incremental learning evaluation model using a Support Vector Machine (SVM) model for Raman spectroscopic data is presented, which ensures user-friendly recalibration during operation. Its performance is determined experimentally and compared with conventional modelling techniques. SVM are robust classifiers, but large datasets lead to long computation times, high memory requirements, and increased complexity of the model. To solve this issue, SVM ensembles, where each SVM sees only a fraction of the data, are a viable solution [5]. Standard methods are used to achieve a high accuracy classifier by computing the best hyperparameters for the SVM model like tenfold cross-validation and grid search [6,7].

Generally, the studies using SVM learning from new data involve discarding the existing classifier, integrating the new data to the old set, and training a new classifier from scratch. The studies do not learn incrementally with the addition of new data, and they result in unlearning of data [5,8]. This means that the system cannot learn new information without forgetting previously learned classifiers. Such a problem is solved with the help of an incremental learning algorithm, defined as one that meets the following criteria [9,10]:

1. Can learn additional information from new data
2. Does not require access to the original data used to train the existing classifier
3. Preserves previously acquired knowledge

Additionally, the proposed incremental learning systems [11–14] suffer from high computation time and complexity. This framework's main contribution is the implementation and evaluation of an incremental learning algorithm based on Garcia et al. [15], which uses parallel computing and helps reduce the computation time with reduced complexity as opposed to previously used learning algorithms [16]. The developed method's relevance in the industrial environment is represented and discussed from a technical perspective. In the example process of rectification, a feed stream of ethanol and water is thermally separated, and ethanol is removed overhead as a distillate. Water leaves the column via the sump stream.

2. Materials and Methods

2.1. Experimental Setup Raman-Spectroscopy

Raman Spectroscopy was performed with a tec5 MultiSpec®Raman system (tec5, Steinbach, Germany), equipped with a coaxial designed RamanProbe II (InPhotonics, Norwood, Massachusetts, USA;fibre configuration: 105 µm excitation fibre; 600 µm collection fibre/working distance 7.5 mm). Raman scattering was excited by a 500 mW temperature-stabilized semiconductor laser (Raman Boxx™, PD-LD500) source at 785 nm. The Raman spectra were collected with a 1 cm^{-1} solution ranging from 300–3215 cm^{-1} using MultiSpec Pro II Raman process software (v1.4.1189.1826, tec5, Steinbach, Germany). The probe was clamped with a laboratory stand in the 50 mL borosilicate beaker (from Schott, Mainz, Germany). Figure 1 shows the setup schematically. Each solution (1–10) was measured ten times with an integration time of 10 s.

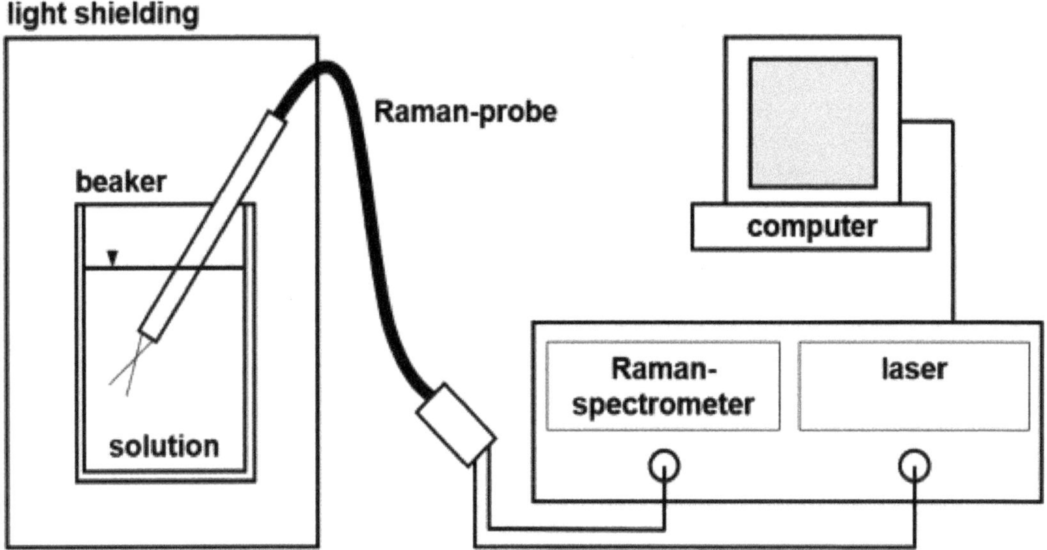

Figure 1. Experimental setup of Raman measurement.

The dilution series was prepared with 96.2 Vol% Ethanol from VWR (Darmstadt, Germany, CAS No. 64-17-5) and distilled water. In total, ten solutions were prepared in 50 mL volumetric flasks (from Schott, Mainz, Germany). Initially, ethanol was transferred, with an Eppendorf Reference pipette (100-1000µL; Eppendorf AG, Hamburg, Germany), into the volumetric flask, and consequently, the remaining volume was filled up with distilled water according to Table 1. Solutions were then transferred into a 50 mL beaker for subsequent measurement.

Table 1. Dedicated samples numbers with calculated ethanol concentration.

Sample Number	Volume H$_2$O/mL	Volume EtOH/mL	Calculated EtOH Concentration/Vol%
0	50.000000000	0.000000000	0.000000000
1	49.500000000	0.500000000	0.962000000
2	49.750000000	0.250000000	0.481000000
3	49.875000000	0.125000000	0.240500000
4	49.937500000	0.062500000	0.120250000
5	49.968750000	0.031250000	0.060125000
6	49.984375000	0.015625000	0.030062500
7	49.992187500	0.007812500	0.015031250
8	49.996093750	0.003906250	0.007515625
9	49.998046875	0.001953125	0.003757813

2.2. Algorithm

In this work, the developed data evaluation method's advantages were presented, which ultimately resulted in lowering the detection limit and increased robustness against outliers or poorly representative data. As a method, a learning algorithm was developed that has particular advantages concerning the learning speed in the continuous expansion of the database when new data were added. This was achieved by taking samples during monitoring as well as analyzing them in the laboratory. In time, they can be integrated into the model. The model improved continuously with accumulating data, which resulted in increased accuracy and a decreased error rate. The mathematical model used in this paper was adapted from the works of Garcia et al. [15].

Current computer-aided tools have made strides towards accurate and efficient detection of chemical concentrations. However, a common drawback observed in these approaches was the use of models that suffer from unlearning [5,8]. In these models, the previously learned knowledge was discarded, and new models had to be trained from scratch in the learning phase as soon as new data became available. In actual process conditions, in which the trained data were presented with time delay over a period of time, the standard gold data had a lack of stability. This was due to the sample taking influences, inhomogeneous product distribution, and impurities. This gave the reason that new emerging areas in machine learning systems must be investigated.

The advantage of an incremental learning algorithm was that it could learn additional information step by step as soon as new data were available. The ensemble learning algorithm used was implemented based on the work of Robi Polikar et al. [9]. The data set was divided into two classes: The first class corresponded to the spectra with a dilution greater than or equal to a specific dilution limit, and the second class contained the dilutions below this limit.

2.3. Mathematics

Based on the dynamically updated distribution of the training data set, the ensemble classifier was trained so that the samples that were more difficult to classify were given an increased probability of increasing their chances of being selected in the following training data set. The algorithm used the database $DB_k, k = 1,\ldots\ldots, K$ where K represents the number of available measurement series, in this work ten. The random samples of all measurement series were first permuted arbitrarily and divided into $K = 10$ stacks of equal size.

The inevitable case wording of SVM was used as the primary classifier, which is referred to below as 1C-SVM. The goal of training with the 1C-SVM was to establish an optimal hypothesis h with which the two classes were separated. The distance between the dividing line or hypothesis and the training data (called support vectors) of each class was maximized, and thus the model was protected against incorrect specifications and increased the robustness of the forecast. The goal was to determine an optimal hyperplane $f(x) = \varphi(x_i)^T w + w_0$ in the feature space. The following formula was used for this:

$$min_{w,w_0,\varepsilon} \frac{1}{2}\|w\|^2 + C\sum_{i-1}^{N}\varepsilon_i \tag{1}$$

Subject to $\varepsilon_i \geq 0$

$$y_i\left(\varphi(x_i)^T w + w_0\right) \geq 1 - \varepsilon_i \; \forall i$$

In it forms $\phi(x_i)$ maps x_i in a higher-dimensional space. w is the weight vector. C is a regulated hyperparameter that added a penalty to the target function in the event of overfitting. ε is a slip variable to weaken the limits and is called an error range or misclassification error. The Equation (1) can be determined using a quadratic approach with a second kind Lagrangian function [17]. By using this function, the kernel trick $K(x_i, x_j)$ can be used. With this it was possible to transfer the data to a higher dimension so that non-linear dependencies between different training data can be considered. The mapping in a higher dimensionality enabled the detection of similarities between data characteristics. The Gaussian radial basis function was used as the kernel, and accordingly, the following Equation (2) was obtained:

$$K(x_i, x_j) = exp\left(-\gamma\|x_i - x_j\|^2\right), \gamma = \frac{1}{2\sigma^2} \tag{2}$$

σ is the variance. γ is a hyperparameter that smoothed the kernel function, and this means a stronger or weaker relationship between the samples can be found depending on the γ values. To estimate the optimal hyperparameters γ and C for 1C-SVM, Nelder-Mead's heuristic method was used [18,19]. The selected criterion to find the optimal classifier was the area under the Receiver Operating Characteristic curve (AUC-ROC), which is increasingly used in machine learning and evaluating more significant amounts of data.

The inputs for the ensemble algorithm were:

1. Training data $S_k = \{(x_i, y_i)|i = 1, \ldots\ldots, N_k$. The data set consisted of N_k training data points with $x_i \in R^d$, where d represents the number of dimensions and $y_i \in \{-1, 1\}$, the associated class. The N_k data points were randomly selected from the kth database (DB_k).
2. A primary classifier to generate hypothesis h. The classifier required that at least 50 percent of the training data record was classified correctly.
3. An integer T_k that specified the number of iteration steps $t = 1, 2, \ldots\ldots, T_k$ for each data set, with $t \in N$. The prediction error could be reduced sufficiently with T_k.

The ensemble algorithm started with the initialization of a series of weights α_t for the training data set S_k and a distribution D_t obtained from α_t [1]. According to D_t, S_k was divided into two subsets, TR_t for training and TE_t for validating during the tth iteration of the algorithm. D_t was initially defined uniformly without deductive information. At each iteration, the weights adjusted at iteration $t - 1$ were divided by the sum of all $W_t - 1$ to ensure a legitimate distribution, and a new D_t was computed. Training and test subsets were drawn randomly according to D_t. A hypothesis h_t was obtained as the tth classifier, whose error ε_t (3) was computed on the entire data set S_k with:

$$\varepsilon_t = \frac{\sum_{i:h_t(x_i)} D_t \cdot |y_i - h_k(x_i)|}{\sum_{i:h_t(x_i)} D_t} \tag{3}$$

ε_t was required to be less than 0.5 to ensure a reasonable performance of h_t. If the condition was satisfied, h_t was accepted, and the error was normalized to calculate the normalization error β_t (4):

$$\beta_t = \frac{\varepsilon_t}{(1 - \varepsilon_t)}, \; 0 < \beta_t < 1 \tag{4}$$

The current h_t was discarded if the condition was not satisfied, and a new training subset was selected. All hypotheses generated so far were then combined using the weighted majority voting to obtain a composite one H_t (5), which allowed efficient incremental learning capability when new classes were introduced. The hypothesis with good performance was awarded a higher voting weight [20].

$$H_t = \text{argmax}_{y \in Y} \sum_{t:h_t(x)=y} \log\left(\frac{1}{\beta_t}\right) \tag{5}$$

The error of H_t was computed with (6) and must have also been less than 0.5 to ensure a reasonable performance of H_t; otherwise, the algorithm discarded that one and returned to select a new TR_t.

$$E_t = \frac{\sum_{i:H_t(x_i)} D_t \cdot |y_i - H_t(x_i)|}{\sum_{i:H_t(x_i)} D_t} \tag{6}$$

Then B_t was computed with (7).

$$B_t = \frac{E_t}{(1 - E_t)},\ 0 < B_t < 1 \qquad (7)$$

The rule of Equation (8) was used to reduce the weights of those data points that were correctly classified by the composite hypothesis H_t. Furthermore, this lowered the probability of being selected in the following training subset.

$$\alpha_{t+1}(i) = \alpha_t(i) \cdot \begin{cases} B_t, & if\ H_t(x_i) = y_i \\ 1, & otherwise \end{cases} \qquad (8)$$

The hypothesis H_F for the training subset and the subset of features could be obtained by combining all hypotheses that had been generated so far using the weighted majority voting rule (see Figure 2).

$$H_F(x) = \text{argmax}_{y \in Y} \sum_{k=1}^{K} \sum_{t:H_t(x)=y} \log\left(\frac{1}{B_t}\right) \qquad (9)$$

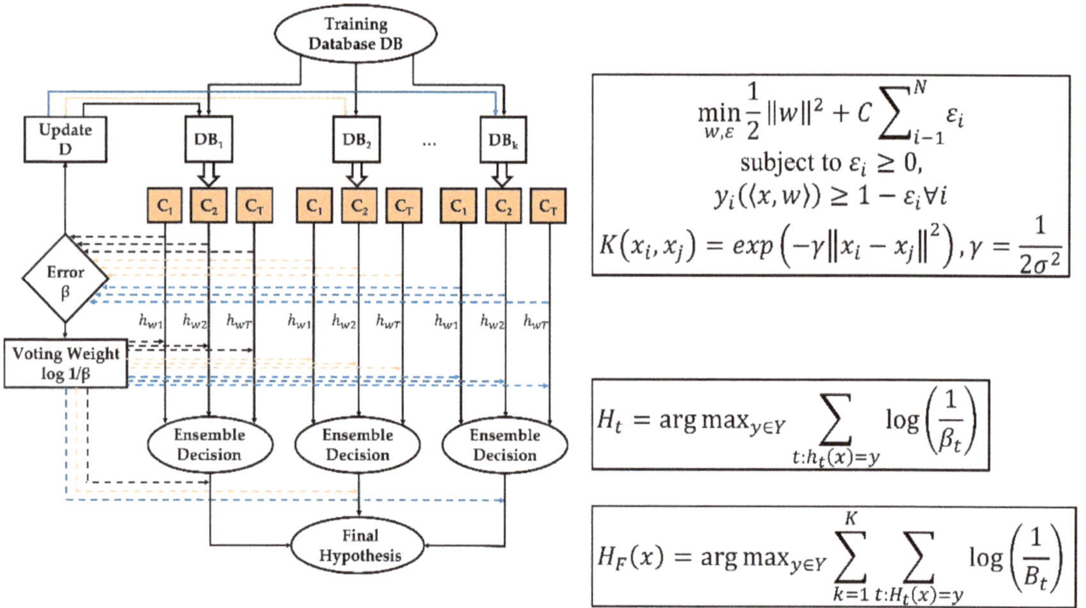

Figure 2. Schematic representation of the incremental learning ensemble classifier (altered from [20]).

2.4. Evaluation

The water spectrum was subtracted from each sample's spectrum. Then, 454 main features from intervals of the spectrum were used to evaluate the data. The intervals contained the spectrum's descriptive peaks, which were between the Raman shifts 850–910, 1010–1130, 1410–1510, and 2840–3010 cm^{-1}. These characteristics corresponded to the value of the derivative and the integral in each interval.

The following methodology was implemented to obtain an objective evaluation. The data set was divided into three subsets: training, validation, and test. The validation subset aimed to make a fair estimation of performance, independency of the test data, and to pick optimal parameters, which in turn provided a more generalized solution. We implemented a nested cross-validation (CV) method for unbiased estimation of prediction error in an independent test data. Leave-one-out (LOO) was used to evaluate the classifier, i.e., nine experiments were used for training and validation and one experiment for independent testing. To estimate the classifier's unknown tuning parameters, tenfold CV was used with data points from the nine experiments selected for training and validation. The data points were randomly permuted and divided in ten parts, one part for the validation and nine for training. Statistical measures such as the sensitivity true positive results (TPR) and precision were also computed (see Equation (10)). A thresholding procedure was used to design a binary classification. The threshold T_h was selected as the concentration level such that a particular concentration was higher than the threshold concentration if the probability $p\left(\frac{y_i}{x_i}\right) \geq T_h$ else it was classified as lower concentration than required, and a ground truth table based on this idea was generated for every concentration. The use of TPR and Positive Predictive Value (PPV) provided additional information about the classifier's performance and could be used to compare results obtained with other methods. Additional parameters like the PPV and TPR were also calculated as follows:

$$\mathrm{TPR} = \frac{\mathrm{TP}}{\mathrm{TP} + \mathrm{FN}} \tag{10}$$

$$\mathrm{PPV} = \frac{\mathrm{TP}}{\mathrm{TP} + \mathrm{FP}} \tag{11}$$

TP and TN denote the number of true positive and true negative data points, respectively. FP and FN denote the number of false-positive and false-negative data points, respectively.

The SVM model in The Unscrambler X version 10.4 (Camo Software, Oslo, Norway) was used to compare the algorithm with a standard program. The ground truth table generated using the MATLAB algorithm was used as input data for binary classification. For each concentration, cross-validation was performed in the same way the MATLAB algorithm does. This resulted in the creation of nine models and predictions per concentration.

3. Results

Figure 3 shows the resulting Raman spectra of ethanol and water in the complete recorded region. The descriptive peaks for ethanol in between 850–910, 1010–1130, 1410–1510, and 2840–3010 cm^{-1} were used in the algorithm. Additionally, two Raman peaks at 435 and 1275 cm^{-1} were also visible in the ethanol spectrum. Water showed typical bands at 500 cm^{-1} (hydrogen bond), 1640 cm^{-1} (OH bending), and 3100–3600 cm^{-1} (OH stretching). Remaining peaks were derived from the glass beaker.

In Figure 4, the spectral data of the dilution series is displayed. For reasons of clarity, the range in between 850–1130 cm^{-1} was enlarged to show the linear concentration dependency more clearly.

The outputs from the predictions obtained through Unscrambler X must be categorised manually into true positive, true negative, false positive, and false negative to calculate the mean value for accuracy, precision, and recall (sensitivity) for each concentration. The results obtained through Unscrambler X are shown in Table 2.

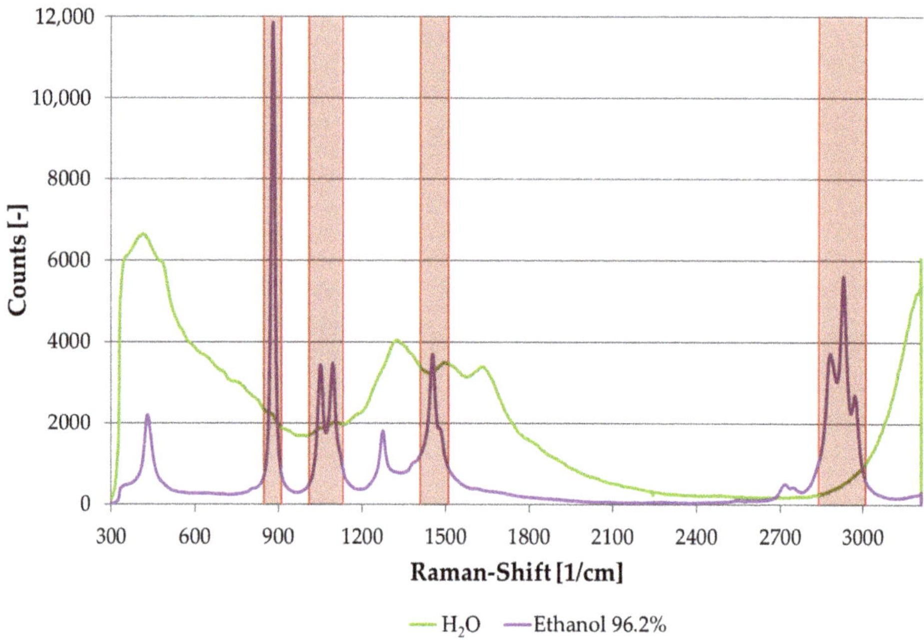

Figure 3. Raman spectra of water and ethanol with evaluated intervals.

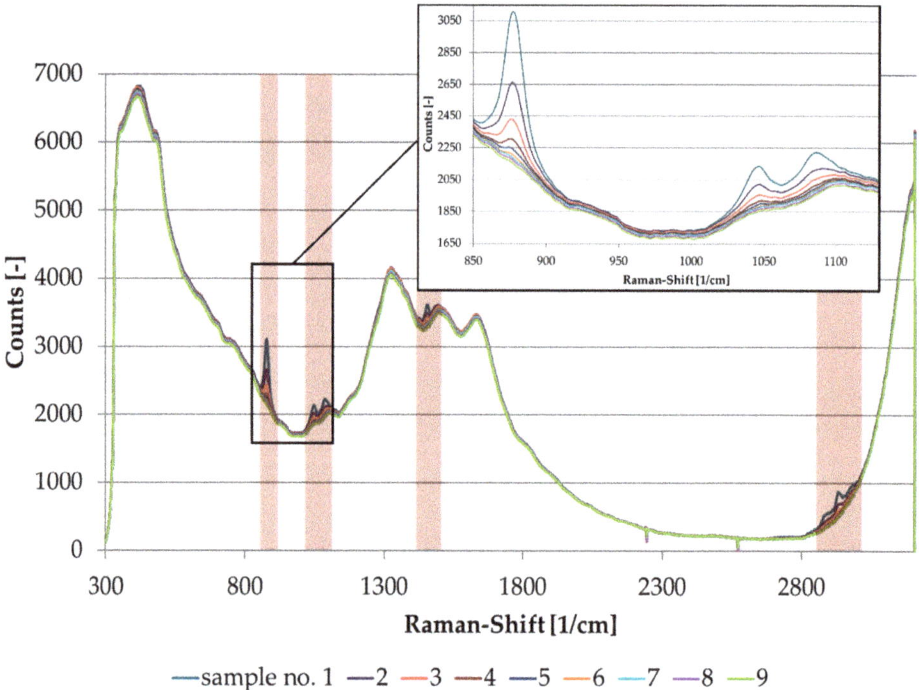

Figure 4. Raman spectra of ethanol dilution series (sample number 1–9).

Table 2. Calculated mean value for accuracy, precision, recall/sensitivity, and computation time for each concentration with the c-value and gamma obtained from the algorithm using Unscrambler X.

Calculated EtOH Concentration (Vol%)	Accuracy Unscrambler (Unscrambler Parameters) (%)	Precision (-)	Recall/Sensitivity (-)
0.9620	100.0	1.00	1.00
0.4810	100.0	1.00	1.00
0.2405	100.0	1.00	1.00
0.1203	98.7	0.99	0.98
0.0601	95.8	0.95	0.97
0.0301	90.9	0.95	0.90
0.0150	83.4	0.90	0.87
0.0075	84.7	0.90	0.92
0.0038	90.7	0.93	0.98

Table 3 displays the results obtained from the MATLAB algorithm. The different parameters for the performance of the classifier were computed using the algorithm. The training time, as well as the classification time, was calculated for each concentration. The time taken for validation and training was also calculated.

Table 3. Calculated mean value for accuracy, precision, recall/sensitivity, and time required for training and validation for each concentration obtained from the MATLAB algorithm.

Calculated EtOH Concentration (Vol%)	Accuracy MATLAB Algorithm (MATLAB Algorithm Parameters) (%)	Precision (-)	Recall/ Sensitivity (-)	Training and Validation Time (s)
0.9620	100.0	1.00	1.00	14.5748
0.4810	100.0	1.00	1.00	17.8897
0.2405	99.9	0.99	1.00	43.3492
0.1203	99.9	0.98	0.98	40.5498
0.0601	99.0	0.97	0.97	148.2467
0.0301	96.2	0.96	0.95	225.5823
0.0150	91.9	0.97	0.93	140.3069
0.0075	89.1	0.98	0.93	92.7786
0.0038	83.7	0.99	0.94	48.4829

As can be observed from the results obtained, when the concentration is higher, both the methods' performances were more or less identical. However, as the concentration decreases, the MATLAB algorithm showed better accuracy, precision, and sensitivity. For concentrations from 0.962 to 0.12025, the performance with respect to precision and sensitivity of both methods was almost the same, but the MATLAB algorithm's accuracy was still higher. As we go to lower concentrations from 0.060125 to 0.007515625, there was a significant difference between the accuracy, precision, and sensitivity from the MATLAB algorithm and Unscrambler X. This shows that the MATLAB algorithm gave better results in terms of accuracy, precision, and sensitivity.

In Figure 5, the accuracy from The Unscrambler X and the accuracy from the MATLAB algorithm are plotted over concentration. For concentrations from sample number 1–3, the Unscrambler X models with the MATLAB parameters showed a significantly lower accuracy, while the MATLAB algorithm and the Unscrambler X model with the new parameters showed an accuracy of 100%. This implied that the Unscrambler X and MATLAB work with different parameters that cannot be taken from each other to create the same results. For calculated ethanol concentrations lower than 0.481 Vol% (sample number 3), the MATLAB algorithm generated higher accuracy and displayed slighter decrease in accuracy over the samples. At a calculated ethanol concentration of 0.0300625 Vol% (sample number 7), the

MATLAB algorithm had the highest difference in accuracy compared to the Unscrambler model with approximately 8%.

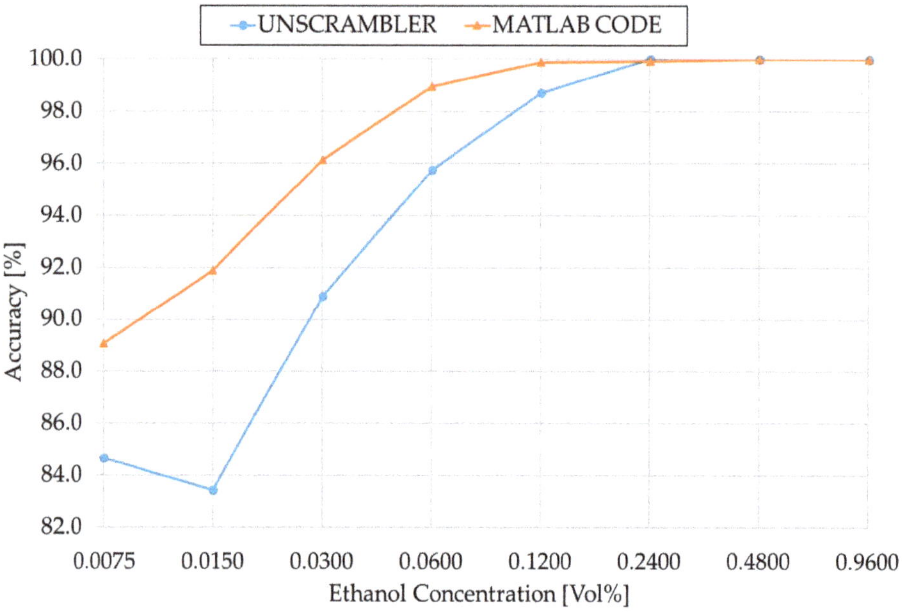

Figure 5. Comparison of the accuracy from The Unscrambler X and the MATLAB code.

4. Discussion

To solve the problem of unlearning in models used in PAT processes, pattern recognition systems that can be better adapted, scalable, and given the ability to learn different data distributions dynamically should be used, but are not available. The method for incremental learning algorithm that uses an ensemble algorithm for classification in a rectification process of ethanol is a possible solution for unlearning. It can learn additional information as soon as new data are available without unlearning the previously acquired knowledge and prevents the unlearning of old Raman spectroscopic data. The system is continuously updated with the current data and significantly increases robustness and accuracy without the loss of any data.

The main advantage of the proposed method over the standard methods available, like Unscrambler X, was that the computation time was less, accuracy was higher, and manual intervention was not involved. The disadvantage in the SVM model created using Unscrambler X was that the data had to be prepared manually, and for each concentration, a model had to be created. The proposed method works without manual intervention saving the users a lot of time and effort. Additionally, a Grid Search had to be performed for each concentration in Unscrambler X. This resulted in a total of 81 models that had to be evaluated manually. Hence, the total time to calculate the accuracy for all concentrations depended on the operator's efficiency and speed.

The algorithm of the developed new model's self-learning follow-up enabled the computing time (batch processing, parallel execution, and distributed data) to be reduced. Additionally, the accuracy and the detection sensitivity of the measuring devices that are used can be increased. In this paper, the spectrum range peaks were selected and used, but the algorithm's application can be extended to any number of peaks and any range in the spectrum that suits the application. The algorithm can be adapted for different data and applications where continuous monitoring is required with the addition of new data. The algorithm performs better than standard methods until the concentration is above 0.0075.

Further investigations need to be performed using different ranges of spectroscopic data and with different chemical processes. The method presented further increases the robustness against outliers or poorly representative data, which do occur in the measured values obtained during operation determined by the system's operating personnel. The fast availability of measurement data increases the system flexibility and thus the constant optimization to changing demands. To the best of our knowledge, the present work is the first to apply the incremental learning SVM model in Raman spectroscopic data used in various chemical processes where continuous monitoring is required.

Author Contributions: Conceptualization: D.H. and S.C.; methodology: D.H. and S.C.; software: J.F.G.M., S.C., and E.S.; validation: D.H. and S.C.; formal analysis: D.H., S.C. and E.S.; investigation: D.H. and M.N.; writing—original draft preparation: D.H. and S.C.; writing—review and editing: D.H. and S.C.; visualization: D.H. and S.C.; supervision: M.R.; funding acquisition: M.R. All authors have read and agreed to the published version of the manuscript.

Funding: This research was funded by the Bundesministerium für Wirtschaft und Energie (BMWi) and the Arbeitsgemeinschaft industrieller Forschungsvereinigungen (AiF, ZF4013934RE7).

Institutional Review Board Statement: Not applicable.

Informed Consent Statement: Not applicable.

Data Availability Statement: Restrictions apply to the availability of these data. Data was obtained from José Fernando García Molina's thesis and is available from the authors with the permission of García Molina.

Acknowledgments: The authors like to thank Frank Braun for his technical support.

Conflicts of Interest: The authors declare no conflict of interest.

References

1. Breitkopf, A. Energieverbrauch in Deutschland Nach Wirtschaftszweig 2018: Energieverbrauch Des Verarbeitenden Gewerbes in Deutschland Nach Ausgewählten Sektoren im Jahr 2018. Available online: https://de.statista.com/statistik/daten/studie/4325 96/umfrage/energieverbrauch-im-verarbeitenden-gewerbe-in-deutschland-nach-sektor/ (accessed on 31 March 2020).
2. Hohmann, M. Entwicklung von Produktion, Energieverbrauch und Treibhausgasemissionen der Chemisch-Pharmazeutischen Industrie in Deutschland im Zeitraum 1990 bis 2018. Available online: https://de.statista.com/statistik/daten/studie/186496 /umfrage/entwicklung-der-chemischen-industrie-in-deutschland/ (accessed on 31 March 2020).
3. Schalk, R.; Braun, F.; Frank, R.; Rädle, M.; Gretz, N.; Methner, F.-J.; Beuermann, T. Non-contact Raman spectroscopy for in-line monitoring of glucose and ethanol during yeast fermentations. *Bioprocess Biosyst. Eng.* **2017**, *40*, 1519–1527. [CrossRef] [PubMed]
4. Braun, F.; Schalk, R.; Brunner, J.; Eckhardt, H.S.; Theuer, M.; Veith, U.; Hennig, S.; Ferstl, W.; Methner, F.-J.; Beuermann, T.; et al. Nicht-invasive Prozesssonde zur Inline-Ramananalyse durch optische Schaugläser. *tm—Tech. Messen* **2016**, *83*. [CrossRef]
5. Ai, H.; Wu, X.; Zhang, L.; Qi, M.; Zhao, Y.; Zhao, Q.; Zhao, J.; Liu, H. QSAR modelling study of the bioconcentration factor and toxicity of organic compounds to aquatic organisms using machine learning and ensemble methods. *Ecotoxicol. Environ. Saf.* **2019**, *179*, 71–78. [CrossRef] [PubMed]
6. Zaidi, S. Novel application of support vector machines to model the two phase boiling heat transfer coefficient in a vertical tube thermosiphon reboiler. *Chem. Eng. Res. Des.* **2015**, *98*, 44–58. [CrossRef]
7. Li, M.; Wei, D.; Liu, T.; Liu, Y.; Yan, L.; Wei, Q.; Du, B.; Xu, W. EDTA functionalized magnetic biochar for Pb(II) removal: Adsorption performance, mechanism and SVM model prediction. *Sep. Purif. Technol.* **2019**, *227*, 115696. [CrossRef]
8. French, R. Catastrophic forgetting in connectionist networks. *Trends Cognit. Sci.* **1999**, *3*, 128–135. [CrossRef]
9. Polikar, R.; Upda, L.; Upda, S.S.; Honavar, V. Learn++: An incremental learning algorithm for supervised neural networks. *IEEE Trans. Syst. Man Cybern. C* **2001**, *31*, 497–508. [CrossRef]
10. Erdem, Z.; Polikar, R.; Gurgen, F.; Yumusak, N. Ensemble of SVMs for Incremental Learning. In *Multiple Classifier Systems, Proceedings of the 6th International Workshop, MCS 2005, Seaside, CA, USA, 13–15 June 2005*; Oza, N.C., Ed.; Springer: Berlin, Germany, 2005; pp. 246–256. ISBN 978-3-540-26306-7.
11. Kho, J.B.; Lee, W.; Choi, H.; Kim, J. An incremental learning method for spoof fingerprint detection. *Expert Syst. Appl.* **2019**, *116*, 52–64. [CrossRef]
12. Li, Y.; Su, B.; Liu, G. An Incremental Learning Algorithm for SVM Based on Combined Reserved Set. *J. Shanghai Jiaotong Univ.* **2016**. [CrossRef]

13. Liu, J.; Pan, T.-J.; Zhang, Z.-Y. Incremental Support Vector Machine Combined with Ultraviolet-Visible Spectroscopy for Rapid Discriminant Analysis of Red Wine. *J. Spectrosc.* **2018**, *2018*, 4230681. [CrossRef]
14. Ardakani, M.; Escudero, G.; Graells, M.; Espuña, A. Incremental Learning Fault Detection Algorithm Based on Hyperplane-Distance. In *26th European Symposium on Computer Aided Process Engineering: Part A*; Kravanja, Z., Ed.; Elsevier Science: Amsterdam, The Netherlands, 2016; pp. 1105–1110. ISBN 9780444634283.
15. García Molina, J.F.; Zheng, L.; Sertdemir, M.; Dinter, D.J.; Schönberg, S.; Rädle, M. Incremental learning with SVM for multimodal classification of prostatic adenocarcinoma. *PLoS ONE* **2014**, *9*, e93600. [CrossRef] [PubMed]
16. Xu, G.; Zhou, H.; Chen, J. CNC internal data based incremental cost-sensitive support vector machine method for tool breakage monitoring in end milling. *Eng. Appl. Artif. Intell.* **2018**, *74*, 90–103. [CrossRef]
17. Hastie, T.; Tibshirani, R.; Friedman, J.H. *The Elements of Statistical Learning: Data Mining, Inference, and Prediction*; Springer: New York, NY, USA, 2009.
18. Cosme, R.D.C.; Krohling, R.A. *Support Vector Machines Applied to Noisy Data Classification Using Differential Evolution with Local Search*; Technical Report; Universidade Federal do Espirito Santo: Vitória, Espírito Santo, Brazil, 2011.
19. Nelder, J.A.; Mead, R. A Simplex Method for Function Minimization. *Comput. J.* **1965**, *7*, 308–313. [CrossRef]
20. García Molina, J.F. Modeling and Analysis of Prostate Cancer Structures Alongside an Incremental and Supervised Learning Algorithm. Ph.D. Thesis, Ruprecht-Karls University of Heidelberg, Heidelberg, Germany, 2014.

Article

WS-RCNN: Learning to Score Proposals for Weakly Supervised Instance Segmentation

Jia-Rong Ou, Shu-Le Deng and Jin-Gang Yu *

School of Automation Science and Engineering, South China University of Technology,
Guangzhou 510641, China; au_jaring@mail.scut.edu.cn (J.-R.O.); audsl@mail.scut.edu.cn (S.-L.D.)
* Correspondence: jingangyu@scut.edu.cn

Abstract: Weakly supervised instance segmentation (WSIS) provides a promising way to address instance segmentation in the absence of sufficient labeled data for training. Previous attempts on WSIS usually follow a proposal-based paradigm, critical to which is the proposal scoring strategy. These works mostly rely on certain heuristic strategies for proposal scoring, which largely hampers the sustainable advances concerning WSIS. Towards this end, this paper introduces a novel framework for weakly supervised instance segmentation, called Weakly Supervised R-CNN (WS-RCNN). The basic idea is to deploy a deep network to learn to score proposals, under the special setting of weak supervision. To tackle the key issue of acquiring proposal-level pseudo labels for model training, we propose a so-called Attention-Guided Pseudo Labeling (AGPL) strategy, which leverages the local maximal (peaks) in image-level attention maps and the spatial relationship among peaks and proposals to infer pseudo labels. We also suggest a novel training loss, called Entropic OpenSet Loss, to handle background proposals more effectively so as to further improve the robustness. Comprehensive experiments on two standard benchmarking datasets demonstrate that the proposed WS-RCNN can outperform the state-of-the-art by a large margin, with an improvement of 11.6% on PASCAL VOC 2012 and 10.7% on MS COCO 2014 in terms of mAP$_{50}$, which indicates that learning-based proposal scoring and the proposed WS-RCNN framework might be a promising way towards WSIS.

Keywords: weakly supervised learning; instance segmentation; proposal scoring network

Citation: Ou, J.-R.; Deng, S.-L.; Yu, J.-G. WS-RCNN: Learning to Score Proposals for Weakly Supervised Instance Segmentation. *Sensors* **2021**, *21*, 3475. https://doi.org/10.3390/s21103475

Academic Editor: Loris Nanni

Received: 26 March 2021
Accepted: 7 May 2021
Published: 17 May 2021

1. Introduction

Instance segmentation [1–4] refers to the task of jointly localizing, categorizing and segmenting the spatial extents of individual visual objects from a given image. Like many other computer vision tasks, remarkable progress has recently been made on instance segmentation driven by the prosperity of convolutional neural networks (CNN) [2,4–7]. Nevertheless, CNN-based solutions to vision tasks commonly suffer from the data-hungry nature, i.e., the necessity of a large amount of annotated data for training. This may be particularly infeasible for instance segmentation since pixel-wise annotations at the instance level are extremely labor-intensive. Weakly supervised instance segmentation (WSIS) [8,9] is one possible way to alleviate the dependency on such strong annotations, which aims to achieve instance segmentation by the use of weaker and thus less labor-intensive annotations [10–12], ideally image-level labels only [8] as we are concerned with in the present work.

There recently emerge a couple of attempts on WSIS with image-level labels in the literature [8,9,13–15], which can be typically outlined as three major steps: (1) proposal generation, i.e., a number of class-agonistic segment proposals are generated from the given image; (2) proposal scoring, i.e., classification scores are assigned to the proposals; (3) postprocessing, i.e., final results are retrieved from the scored proposals by using non-maximal suppression or certain postprocessing procedures (e.g., applying Mask R-CNN for refinement [14]). It has been well-established that CNN classifiers trained globally at the image level have remarkable ability of spatial localization [16,17], and the so-called attention maps of a certain form (e.g., Class Activation Maps [18], Occlusion Maps [19], Saliency Maps [20], Excitation Backprop [17]) are utilized to represent such localization cues, where the intensity stands for the possibility of spatial occurrence of visual objects. Taking advantage of this fact, previous works on WSIS mainly put their efforts on the step of proposal scoring, i.e., how to infer reasonable classification scores for the proposals from global attention maps. As a pioneering work on WSIS, Zhou et al. [8] proposed Peak Response Map (PRM) to boost the instance localization ability of CNNs, which then enables better proposal scoring. Zhu et al. [13] presented the Instance Activation Map which aims to enhance the spatial extents of instances in PRM and consequently improves proposal scoring. Ahn et al. [14] deployed the Inter-pixel Relation Network to learn a high-quality proposal generator, as well as pairwise semantic affinities which guide the proposal scoring.

While these wisdoms of making improvements on proposal scoring (or other particular components) can indeed benefit to an extent, previous works on WSIS share one major limitation that they commonly follow a heuristic way to exploit attention maps for proposal scoring, lacking of a unified framework. Concretely, they utilize certain attention maps and hand-crafted scoring rules to assign classification scores to proposals [8,13]. In this way, the attention maps are expected to be aware of the spatial extents of visual objects, such that they can match the shapes of true objects and consequently assign them with high scores. On the other hand, attention maps can only sparsely highlight some sites discriminative for classification, usually at very coarse spatial solutions and unaware of object extents. Such discrepancy explains why previous works all devote their efforts to enhancing the perception of object extents of attention maps. Nevertheless, this is by nature a rather difficult perceptual grouping task (see Figure 1), which may limit the substantial advances on this topic. These observations motivate us to consider if these exist a simpler but more effective way towards WSIS.

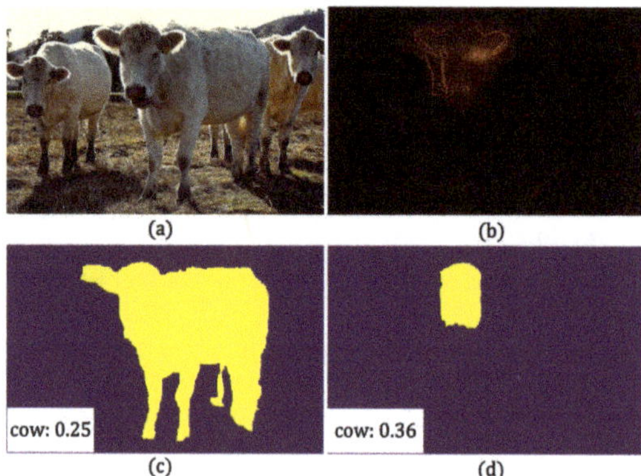

Figure 1. Illustration of the motivation of our work. Existing proposal-based approaches to WSIS commonly follow a heuristic way to exploit attention maps for proposal scoring. In order to assign high scores to the proposals of true objects, the attention maps are expected to be aware of the spatial extents of objects, which is also the main focus of previous efforts on WSIS. But unfortunately, this is by nature a very difficult perceptual grouping task, since attention maps can only have very coarse resolution sparsely highlighting some discriminative sites (unaware of object extents). For intuition, we show some exemplary results obtained by the poineering approach of PRM [8], where (**a**) is the original image, (**b**) is the PRM of the *cow* class, and (**c**,**d**) are two proposals and the obtained classification scores. As can be seen, the more favorable proposal in (**c**) is undesirably assigned with a lower score.

In this paper, we introduce a general framework for weakly supervised instance segmentation, called Weakly Supervised R-CNN (WS-RCNN). The underlying idea is very simple and natural: Instead of relying on heuristic strategies for proposal scoring, we deploy a deep network to learn to score proposals. For this purpose, there exists an inherent challenge, i.e., how to acquire proposal-level pseudo labels to enable the learning since only image-level labels are available in our problem setting (the pseudo labeling issue)? This is actually a key issue for any weakly supervised learning problem including ours. In addition, an appropriate training loss function is also well worthy of exploration in our learning task. We propose specific solutions to these key issues and further design a unified framework for WSIS. Since the resulting framework can be conceptually interpreted as a Fast R-CNN model [21], which is a representative model of general object detection among the R-CNN family [2,21–23], under the particluar setting of weak supervision, we name our framework as Weakly Supervised R-CNN (WS-RCNN).

As another major contribution, we propose an effective solution to this pseudo labeling issue, termed as Attention-Guided Pseudo Labeling (AGPL). AGPL also utilizes attention maps, however it relies only on local maxima (peaks) in attention maps and the spatial relationship among peaks and proposal regions. Specifically, it admits object instances of a target class by inspecting the positional inclusion relationship between proposals and peaks in the attention maps associated with that particular class (see Section 3.3 for details). The key insight lies in that AGPL leverages very weak information, instead of necessitating to take into account object extents like existing approaches [8,9,13,14], which makes the pseudo labeling procedure simple but robust. We stress that such careful design of AGPL is crucial to the success of the whole WS-RCNN framework (see Section 4.3 for experimental validation).

As our third contribution, we suggest a novel loss function for model training to further boost the performance, called Entropic Open-Set (EOS) Loss. As our method is proposal-based, the proposal generator will inevitably yield a portion of background proposals (those belonging to none of the concerned classes) even with the state-of-the-art method. How to make our model conscious of background proposals to avoid misclassification is important to its robustness. This is by nature an open-set recognition problem [24,25], which has been well studied in the context of robust pattern recognition. We propose to adapt the state-of-the-art method for CNN-based open-set recognition [25] to our background handling. To our knowledge, this is the work which considers the open-set issue in object detection/instance segmentation.

We will carry out comprehensive experiments on standard benchmarking datasets to demonstrate that WS-RCNN can push forward the state-of-the-art on WSIS by a remarkable margin, with an improvement of 11.6% on PASCAL VOC 2012 and 10.7% on MS COCO in terms of mAP_{50} (see Section 4.2 for details). Such overwhelming superiority in performance suggests that WS-RCNN might be a more appropriate framework for WSIS worthy of further exploration than existing ones which focus on deducing object extents from attention maps for proposal scoring. To sum up, the main contributions of our work are as below:

- We propose a novel framework called Weakly Supervised R-CNN (WS-RCNN) for weakly supervised instance segmentation. The key insight is to deploy a deep network to learn to score proposals, instead of relying on heuristic proposal scoring strategies, which may provide a new perspective for future exploration on WSIS.
- We propose a simple but effective strategy for inferring pseudo labels from attention maps, called Attention-Guided Pseudo Labeling (AGPL).
- We introduce an Entropic Open-Set (EOS) Loss for handling background proposals in model training to further boost the performance.

The remainder of this paper is organized as follows: Section 2 briefly reviews related literature. Section 3 details the proposed WS-RCNN framework, including the AGPL scheme for pseudo labeling and the EOS loss. Section 4 focuses experiments, including comparison, ablation study and related analysis. In Section 5, we make a conclusion and give some remarks on future work.

2. Related Work

In this section, we present a brief review of the literature that are closely related to ours, including those on weakly supervised instance segmentation, weakly supervised semantic segmentation, weakly supervised object detection and open-set recognition and background handling.

2.1. Weakly Supervised Instance Segmentation

Instance segmentation aims to produce accurate masks that can distinguish between instances of specific object classes. It generally requires pixel-wise object masks to train an instance segmentation model, which are extremely expensive. Weakly supervised instance segmentation (WSIS) devotes to conquering this challenge by using certain forms of weaker annotations, most ideally image-level labels only as concerned in this paper. Along the line of WSIS using image-level annotations, PRM [8] is a pioneering work which treats peaks in class response maps obtained by CNN classifiers as indicators of the existence of object instances, and introduces the so-called Peak Response Map (PRM) to score object proposals so as to identify true object instances. In [26], the results obtained by PRM are used as pseudo annotations to train a Mask R-CNN [2] for further refinement. Zhu et al. [13] proposed the Instance Activation Map to enhance PRM [8] by taking better care of spatial extents of objects. Ahn et al. [14] presented the Inter-pixel Relation Network to learn a good proposal generator, and also inter-pixel connections which can guide proposal scoring. In [9], the authors collaboratively combined weakly supervised object detection with WSIS under a unified framework of course learning. As previously stated, WSIS generally remains at the early stage of exploration, and the few existing approaches mainly focus on figuring out the shapes of object instances from attention maps for better proposal scoring, which faces a very difficult perceptual grouping task.

2.2. Weakly Supervised Semantic Segmentation

The mainstream paradigm for weakly supervised semantic segmentation (WSSS) can be summarized as to derive pseudo semantic masks to enable the training of a model of fully supervised semantic segmentation (most typically FCN [27]), and existing approaches can be distinguished by how they acquire the pseudo masks. In [28], discriminative regions are selected from CAMs [18] based on three principles, called "seed, expand and constrain", as pseudo labels to supervise a segmentation network. Roy et al. [29] utilized a novel deep architecture which fuses bottom-up, top-down, and smoothness cues to acquire pseudo masks. Other representative strategies for enhancing the quality of pseudo masks include the LSE pooling [30], the EM algorithm [31], seeded region growing [32], semantic affinity [33], dilated convolution [34], the anti-erase strategy [35,36], similar region mining [37], the self-erasing strategy [38], visual saliency [39,40], etc. Although WSSS is another task different from WSIS, our work was somewhat inspired by the paradigm of inferring pseudo labels to train fully supervised models.

2.3. Weakly Supervised Object Detection

The task of weakly supervised object detection (WSOD) is similiar to WSIS except for the necessity of yielding instance segmentation masks. Earlier works on WSOD mostly follow the pipeline of multiple instance learning [41–43]. Recently CNN-based approaches have attracted more and more research interest. As a milestone work along this direction, WSDDN [44] adopts a two-branch network where softmax operations are performed over proposals and classes respectively in the two branches, and the obtained classification scores are synthesized into an image-level score so as to establish the training loss. Diba et al. [45] introduced a novel cascaded network for WSOD, which performs ROI pooling at multiple levels to boost the performance. Ref. [46] proposes a weakly supervised region proposal network which is trained using only image-level annotations and the proposed region proposal network can be plugged into a WSOD network easily. In [47], Tang et al. introduced the Proposal Clustering Learning which generates proposal clusters to learn refined instance classifiers by an iterative process. Other successful approaches to WSOD include the Min-Entropy Latent Model [48], the Category-Aware Spatial Constraint [49], etc. While the tasks are essentially different, these works on WSOD are similar to ours in being proposal-based.

2.4. Open-Set Recognition and Background Handling

Traditional recognition systems obey the closed-set assumption, i.e., the training and testing data are drawn from the the same set of semantic classes. However, a more realistic scenario is that data from unseen classes can emerge unexpectedly during testing, which may drastically decrease the robustness of the system. This is well-known as the open-set issue in pattern recognition, which has attracted a lot of research interests [24,25,50]. In two-stage object detectors (like FRCNN [21] and our WS-RCNN), the proposals obtained by the proposal generator will inevitably include a portion of proposals from image background (unseen classes to the model), which is by nature an open-set recognition problem [24]. But surprisingly, no previous authors have addressed this problem from the perspective of open-set recognition to our knowledge. Instead, they usually perform background handling in certain ad-hoc strategies, e.g., adding a dummy background class to the model.

3. Approach

In this section, we start with the problem statement of WSIS, followed by describing the proposed WS-RCNN framework. Then we detail the key components, including the Attention-Guided Pseudo Labeling and the Entropic Open-Set Loss. Finally, we present some remarks.

3.1. Problem Statement

Given a set of classes of interest $\mathcal{C} = \{1, \ldots, C\}$ and a training set $\mathcal{I} = \{(\mathbf{I}_k, \mathbf{y}_k)\}_{k=1}^{K}$, where $\mathbf{I}_k$ is an image and $\mathbf{y}_k \in \{0,1\}^{C \times 1}$ is the corresponding multi-class label vector, the task of weakly supervised instance segmentation (WSIS) in our work can be roughly stated as to segment, for an input testing image, all the object instances belonging to the classes $\mathcal{C}$. Such a problem setting differs intrinsically from general instance segmentation [2,4] in that no pixel-wise instance annotations but only image-level labels are available for model establishment, which makes the task very challenging.

Like general instance segmentation, WSIS can also follow a proposed-based paradigm [8,13,33], which can be epitomized as a three-step pipeline as aforementioned. These approaches can then be viewed as to retrieve true object instances from a pool of proposals according to the assigned scores, central to which is proposal scoring, i.e., how to appropriately assign classification scores to proposals. One commonly-used strategy for proposal scoring is to make use of the well-established localization ability of CNNs [8,16,18]. Specifically, the training set $\mathcal{I}$ with image-level labels are firstly taken to train an image-level CNN classifier, from which a collection of class-specific attention maps are derived to assign classification scores to the proposals. For this purpose, it is desired that these attention maps can preserve object shapes, which is however a difficult perceptual grouping task. In addition, the hand-crafted scoring rules adopted by existing methods are also limited as well. These facts motivates us to propose the WS-RCNN framework.

3.2. The Proposed WS-RCNN Framework

The basic idea of WS-RCNN is to deploy a deep network to learn to score proposals under the special setting of weak supervision, instead of relying on heuristic proposal scoring strategies. To achieve this goal, one major obstacle is the absence of proposal-level labels necessitated for training. To conquer this challenge, we develop an effective strategy, called Attention-Guided Pseudo Labeling (AGPL), to take advantage of the attention maps associated with the image-level CNN classifier to infer proposal-level pseudo labels. Furthermore, we introduce an Entropic Open-Set Loss (EOSL) to handle the background issue in training to further improve the robustness of our framework. In the following, we will first present an overview of WS-RCNN, followed by detailing the AGPL stategy and the EOSL loss.

Network Architecture: The overall network architecture of WS-RCNN is shown in Figure 2. Following the notations above, the input image $\mathbf{I}$ sized by $H_I \times W_I$ is first fed into a proposal generator (using the off-the-shelf method [51,52] in our implementation) to obtain the segment proposals $\{\mathbf{R}_n\}_{n=1}^N$, where each $\mathbf{R}_n$ is an $H_I \times W_I$ binary mask representing a segment proposal with arbitrary shapes (rather than regular bounding-boxes). The image then goes through a backbone CNN for feature extraction, yielding the feature maps $\mathbf{F} \in \mathbb{R}^{H \times W \times M}$, where $H \times W$ is the size and M the number of the feature maps. Afterwards, the network bifurcates into two branches, i.e., the proposal scoring branch and the pseudo labeling branch. Notice that these two branches share the same backbone CNN.

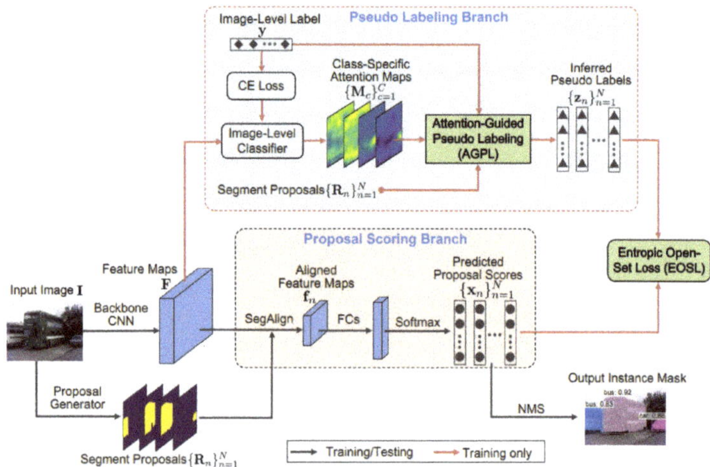

Figure 2. Overview of the proposed Weakly Supervised R-CNN (WS-RCNN) framework for weakly supervised instance segmentation. WS-RCNN adapts Fast R-CNN (FRCNN) [21], a representative model for general object detection, to the particular setting of weak supervision, which can be roughly interpretted as to derive pseudo labels from the image-level CNN classifier to initiate the FRCNN. The network mainly consists of two streams, i.e., the proposal scoring stream and the image-level classification stream. The former learns to score proposals, trained by the Entropic Open-Set Loss, and the latter performs pseudo labeling using the Attention-Guided Pseudo Labeling.

In the proposal scoring branch, the features corresponding to each individual proposal are extracted. A standard operation for this task is RoIAlign [2], widely used in two-stage object detectors, which however cannot be directly applied to our case since it is designed for bounding-box proposals. Therefore, we modify RoIAlign to adapt to segment proposals, resulting in the SegAlign operation (see details below). For each proposal $\mathbf{R}_n$, the corresponding features can be extracted from $\mathbf{F}$ and aligned to a canonical grid via SegAlign, denoted by $\mathbf{f}_n \in \mathbb{R}^{h \times w \times M}$ (we use h = w = 7; M = 512 in this paper), which is followed by three fully-connected layers (FCs, with the node numbers being 4096, 4096 and C respectively) and a softmax layer to get the proposal-level classification score $\mathbf{x}_n \in \mathbb{R}^{C \times 1}$.

The pseudo labeling branch is executed for training only, where the feature maps $\mathbf{F}$ are followed by an image-level classifier. Then, a set of class-specific attention maps, denoted by $\{\mathbf{M}_c\}_{c=1}^C$, are extracted from this classifier, where each $\mathbf{M}_c$ reflects the spatial probability of occurrence of object instances belonging to the class c. Among possible choices of attention maps, we adopt the Class Peak Responses [8] in our implementation due to its excellent localization ability. These attention maps (as well as the image-level label $\mathbf{y}$) are then utilized to infer the proposal-level pseudo class labels $\{\mathbf{z}_n\}_{n=1}^N$, where

$\mathbf{z}_n \in \mathbb{R}^{C \times 1}$ is a one-hot vector $\mathbf{z}_n \in \mathbf{0}^{C \times 1}$ standing for the background class), by the use of AGPL.

Training Strategy: We adopt a two-phase training strategy to train the WS-RCNN model. In the first phase, we train the image-level classifier in the pseudo labeling branch, which is initialized by the model pre-trained on ImageNet. Proposal-level pseudo labels are then inferred from the trained imagelevel classifier using AGPL. In the second phase, we train the proposal scoring branch, where the backbone CNN is reinitialized with the model pre-trained on ImageNet. We will validate the effectiveness of this two-phase training strategy by comparative experiments in Section 4.3. Notice that since there usually exist significantly more background proposals than target-class ones after pseudo labeling, we always make their numbers identical by uniformly sampling background proposals.

Training Loss: For the training of the image-level classifier (the first-phase training), we use the given image-level labels $\{\mathbf{y}_k\}_{k=1}^{K}$ and the conventional cross-entropy loss function for multi-label classification to establish the training loss.

For the training of the proposal scoring branch (the secondphase training), suppose for the image $\mathbf{I}_k$ labeled with $\mathbf{y}_k$ in the training set $\mathcal{I}$, the proposals obtained are $\{\mathbf{R}_n^k\}_{n=1}^{N}$. For each $\mathbf{R}_n^k$, let us denote by $\mathbf{x}_n^k$ the classification score predicted by the proposal scoring branch, and by $\mathbf{z}_n^k$ n the pseudo class label inferred by the pseudo labeling branch using AGPL. Given all these, the training loss for the proposal scoring stream can be established by

$$L_p = \frac{1}{K} \sum_{k=1}^{K} \left[\frac{1}{N} \sum_{n=1}^{N} \ell_{\text{EOSL}}(\mathbf{x}_n^k, \mathbf{z}_n^k) \right], \tag{1}$$

where ℓ_{EOSL} is the proposed Entropic Open-Set Loss (see details in Section 3.4).

SegAlign: As shown in Figure 3, following the notations above, suppose for a segment mask $\mathbf{R}$ in the $W_I \times H_I$ image $\mathbf{I}$, the corresponding receptive field mapped to the feature maps $\mathbf{F} \in \mathbb{R}^{H \times W \times M}$ is $\mathbf{R}_F$. SegAlign extracts from $\mathbf{F}$ the features corresponding to $\mathbf{R}_F$ and maps them to canonical feature maps $\mathbf{f} \in \mathbb{R}^{h \times w \times M}$, which is basically a modified RoIAlign to adapt to segment masks. Concretely, suppose $\mathbf{R}_F$ is bounded by the rectangle $\mathbf{B}$, φ is the bilinear transform from the spatial coordinates $(i, j) \in \mathbf{f}$ to $(i', j') \in \mathbf{B}$, i.e., $(i', j') = \varphi(i, j)$, and g is the bilinear interpolation function over $\mathbf{F}$. The SegAlign operation can then be defined by

$$\mathbf{f}(i,j) = \begin{cases} g(\mathbf{F}, \varphi(i,j)), & \text{if } \varphi(i,j) \in \mathbf{R}_F, \\ 0, & \text{otherwise.} \end{cases} \tag{2}$$

Note we drop the channel dimension of feature maps above without loss of clarity.

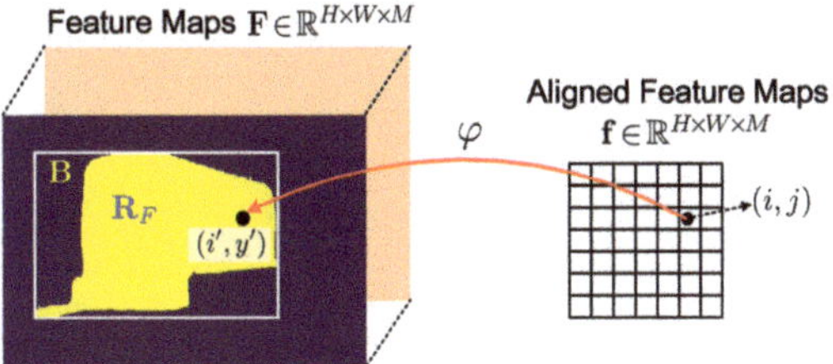

Figure 3. Illustration of the SegAlign operation. SegAlign adapts the widely-used RoIAlign [2] to segment masks.

3.3. Attention-Guided Pseudo Labeling

AGPL leverages the localization ability of CNNs and the spatial relationship among proposals to achieve pseudo labeling. As shown in Figure 4, for the image $\mathbf{I} \in \mathcal{I}$, given the class label vector $\mathbf{y} = [y_1, \ldots, y_C]^T$, the segment proposals $\{\mathbf{R}_n\}_{n=1}^{N}$ and the class-specific attention maps $\{\mathbf{M}_c\}_{c=1}^{C}$, AGPL can be outlined as follows:

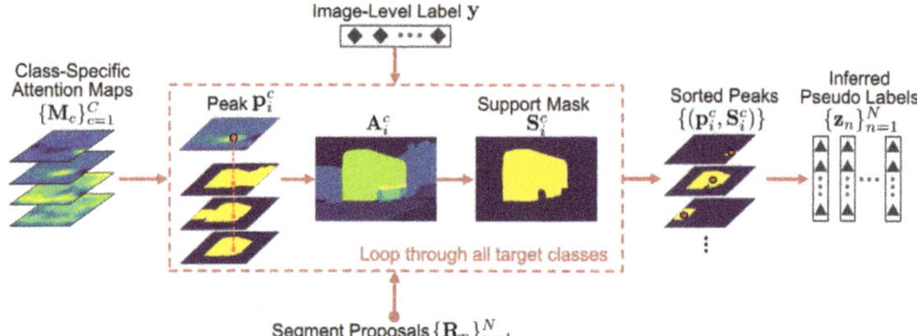

Figure 4. Illustration of Attention-Guided Pseudo Labeling (AGPL). AGPL leverages the localization ability of CNNs and the spatial relationship among proposals to achieve pseudo labeling.

(1) For each target class c (with $y_c = 1$), all the local maxima (peaks) are identified from M_c, denoted as $\{\mathbf{p}_i^c\}_{i=1}^{m_c}$, where $\mathbf{p}_i^c \in \mathbb{R}^2$ stands for pixel coordinates and m_c the number of peaks. For each $\mathbf{p}_i^c$, we pick up all the proposals spatially including this point, which are further averaged and thresholded to get a support mask $\mathbf{S}_i^c$ as follows

$$\mathbf{A}_i^c = \frac{1}{m_i^c} \sum_{\{n|\mathbf{p}_i^c \in \mathbf{R}_n\}} \mathbf{R}_n, \tag{3}$$

$$(\mathbf{S}_i^c)_{pq} = \begin{cases} (\mathbf{A}_i^c)_{pq}, & \text{if } (\mathbf{A}_i^c)_{pq} > \beta, \\ 0, & \text{otherwise.} \end{cases} \tag{4}$$

where m_i^c is the number of picked proposals corresponding to $\mathbf{p}_i^c$, p and q are pixel indices, and the threshold $\beta \in [0,1]$ is a parameter (we adopt $\beta = 0.7$ in our implementation. See Section 4.4 for parameter study). The resulting peaks $\{\mathbf{p}_i^c\}_{i=1}^{m_c}$ and associated support masks $\{\mathbf{S}_i^c\}_{i=1}^{m_c}$ are then utilized to admit proposals belonging to the class c.

(2) Sort all the peaks $\{\mathbf{p}_i^c | c = 1, \ldots, C; i = 1, \ldots, m_c\}$ in the descending order of their values in the attention maps, i.e., $\{\mathbf{M}_c(\mathbf{p}_i^c)\}$. Then, for each ordered peak $\mathbf{p}_i^c$ and the associated $\mathbf{S}_i^c$, those proposals which overlap sufficiently with $\mathbf{S}_i^c$ are labeled as the class c, i.e., $z_n = c$ if

$$\text{IoU}(\mathbf{S}_i^c, \mathbf{R}_n) > 0.5, \tag{5}$$

where IoU stands for the Intersection-over-Union operation. Notice that one proposal is allowed to be exclusively assigned to one class only during the ordered labeling. For clarity, we summarize the AGPL algorithm above in Algorithm 1.

3.4. Entropic Open-Set Loss

Since our WS-RCNN is proposal-based, the proposals after pseudo labeling will unavoidably contain some background proposals, i.e., those labeled as none of the target classes. It is necessary to handle these background proposals in model training, otherwise a model trained only with samples from target classes will be distracted by the unseen background proposals in testing, degrading its robustness. A natural solution is to add a dummy class into the model to accommodate background proposals. However, since the background class is a class of "stuff", its variance is so large that it is hard to be modeled by any single class.

Algorithm 1 Attention-Guided Pseudo Labeling (AGPL)

Input: The label vector $\mathbf{y} = [y_1, \ldots, y_C] \in \{0,1\}^{C \times 1}$, the segment proposals $\{\mathbf{R}_n\}_{n=1}^N$ and the class-specific attention maps $\{\mathbf{M}_c\}_{c=1}^C$ (associated with the image $\mathbf{I}$); the parameter β.

Output: The one-hot pseudo class labels $\{\mathbf{z}_n\}_{n=1}^N$ for the proposals.

1: Initialize $\mathbf{z}_n = \mathbf{0}$, $\mathcal{Q} \leftarrow \emptyset$.
2: **for all** c with $y_c = 1$ (target class) **do**
3:     Find the local maxima $\{\mathbf{p}_i^c\}_{i=1}^{m_c}$ in $\mathbf{M}_c$;
4:     **for** $i = 1, \ldots, m_c$ **do**
5:         Calculate the support mask $\mathbf{S}_i^c$ using Equations (3) and (4);
6:         $\mathcal{Q} \leftarrow (\mathbf{p}_i^c, \mathbf{S}_i^c)$;
7:     **end for**
8: **end for**
9: Sort $\mathcal{Q}$ in the descending order of the values $\{\mathbf{M}_c(\mathbf{p}_i^c)\}$, denoted by $\tilde{\mathcal{Q}}$ the sorted set.
10: $\mathcal{L} \leftarrow \{1, \ldots, N\}$;
11: **for all** $(\mathbf{p}_i^c, \mathbf{S}_i^c) \in \tilde{\mathcal{Q}}$ **do**
12:     Find the proposals indexed by $\mathcal{I} \subseteq \mathcal{L}$ satisfying Equation (5);
13:     Set $(\mathbf{z}_n)_c = 1$, $\forall n \in \mathcal{I}$;
14:     $\mathcal{L} \leftarrow \mathcal{L} \setminus \mathcal{I}$;
15: **end for**

To address this issue, we observe that the task of background handling here is by nature an open-set recognition (OSR) problem, which has been well studied in robust pattern recognition. Hence, we propose to introduce the Entropic Open-Set Loss (OSEL), which is a representative method for OSR [25], to address our background handling problem. The basic idea of OSEL is to treat the samples from target and background classes separately in establishing the training loss. For target classes, the standard cross-entropy loss is used, while for the background class, an entropic loss is used to encourage predicting uniformly-distributed classification scores. Since the C scores sum up to 1 (output by the softmax layer), encouraging uniform distribution on background class will make these scores small and therefore suppressed during the Non-maximal Suppression (NMS) procedure. Formally, suppose the predicted score vector of a proposal is $\mathbf{x} = [x_1, \ldots, x_C] \in \mathbb{R}^{C \times 1}$ and the corresponding one-hot pseudo label vector is $\mathbf{z} = [z_1, \ldots, z_C] \in \{0,1\}^{C \times 1}$, the EOSL is defined by

$$
\ell_{\mathrm{EOSL}}(\mathbf{x}, \mathbf{z}) =
\begin{cases}
-z_c \log x_c, & \text{if } z_c \neq 0, \\
-\displaystyle\sum_{i=1}^{C} \log x_i, & \text{if } \mathbf{z} = \mathbf{0}.
\end{cases}
\tag{6}
$$

To our knowledge, this is the first work which addresses the background handling problem in object detection/instance segmentation from the perspective of open-set recognition.

3.5. Remarks

Despite the proposed WS-RCNN is very natural and simple, we will later experimentally demonstrate that it has overwhelming superiority in performance over the state-of-the-art methods, which is very likely due to two key insights of our approach. First, we adopt a learning-based approach to proposal scoring, which is advantageous in being able to learn to directly map feature representations of proposals to classification scores, rather than depending on hand-crafted scoring rules. Second, AGPL only utilizes very loose information to achieve pseudo labeling, i.e., some sparse points (peaks in attention maps) and their spatial relationship with proposal regions, which makes our approach much easier and thereby more robust to complex scenes in comparison with existing works that involve the consideration of spatial extents of objects. The proposed WSRCNN framework can be simply interpreted as "learning to score proposal under weak supervision", which may provide a promising new perspective for addressing WSIS.

4. Experiments

In this section, we perform extensive experiments to evaluate the effectiveness of the proposed WS-RCNN, mainly including: (1) comparison with the state-of-the-art WSIS methods, both quantitatively and qualitatively; (2) validation of the effectiveness of some key components by comparing with variant baselines; (3) parameter study; (4) analysis on failure patterns. Our method was implemented in PyTorch on a workstation with 2 Nvidia Titan XP GPUs, Intel Core(TM) i7-8700 3.70 GHz CPU, 32 GB RAM and Ubuntu 18.04 OS.

4.1. Experimental Setup

4.1.1. Datasets

Two well-known benchmarking datasets for instance segmentation, namely PASCAL VOC 2012 (termed as **VOC**) [53] and Microsoft COCO 2014 (termed as **COCO**) [54] are adopted for our experiments throughout this paper. **VOC** may be the most representative one for evaluating WSIS methods [8,13,33], which includes 10,582 images for training (*trainset*) and 1449 images for validation (*valset*) from 20 object classes. **COCO** is a much larger and more challenging dataset, including 82,783 images for training (*trainset*) and 40,775 images for validation (*valset*) from 80 object classes. This dataset has rarely been utilized for the task of WSIS before, and we consider it to enable more comprehensive evaluation. For both datasets, we take *trainset* for training and *valset* for testing, and no other annotations except for image-level class labels are used for training according to our problem definition of WSIS.

4.1.2. Performance Metrics

We adopt mAP_r, the most commonly-used performance metric for instance segmentation [8,13,33,53], for quantitative evaluation and comparison, where r is the IoU threshold utilized to calculate the metric [53,55]. In our experiments on WSIS, we use mAP_{50} ($r = 0.5$) as the major metric for comparison and analysis, but we also report mAP_{25} ($r = 0.25$) and mAP_{75} ($r = 0.75$) for more in-depth evaluation.

4.1.3. Implementation Details

We use VGG-16 [56] pre-trained on ImageNet as the backbone for the proposed WS-RCNN. The number of segment proposals per image is set to be $N = 200$ following [8,13]. For training WS-RCNN, the SGD optimizer is used with a initial learning rate of 5×10^{-4} for the first 3.5×10^4 iterations. In the following 10^4 iterations, the learning rate decreases to 5×10^{-5}. For data augmentation, we use five image scales $\{480, 576, 688, 864, 1200\}$ (for the shorter side) with horizontal flips for both training and testing.

4.2. Comparison with State-of-the-Art

The study on WSIS is still at its early stage and there have not been many works so far. We consider four state-of-the-art WSIS methods in the literature for our comparative study, termed as **PRM** [8], **IAM** [13], **Label-PEnet** [9] and **IRnet** [14] respectively. For PRM [8] and IRnet [14], we use the source codes as well as configurations provided by the authors themselves. For IAM [13] and Label-PEnet [9], since no source codes are publicly released, we directly cite the results in the original literature [9,13] wherever available (marked with "-" if the results are unavailable). Besides, we also construct a variant of WSDDN [44], which is a very impactful method for weakly supervised object detection (rather than instance segmentation), termed as **WSDDN-seg**. For this sake, WSDDN-seg can be easily adapted to our task utilizes segments as the proposals to replace the original bounding-boxes in WSDDN (while keeping everything else unchanged). We consider WSDDN-seg because this method is spiritually similar to ours in that it also deploys a deep network to learn to score proposals. For fair comparison, we use the same method [51,52] to generate segment proposals for all the compared methods except IRnet [14], because the method focuses on improving the component of proposal generation. In addition, some methods perform post-processing (typically taking the obtained instance masks as pseudo annotations to run Mask R-CNN like [14]) to further refine the results while others do not. For fair and comprehensive comparison, we consider the two settings separately, i.e., without and with (marked by "+p") using Mask R-CNN for refinement.

4.2.1. Results on VOC

The quantitative results obtained by various methods on VOC are reported in Table 1. As can be observed, our WS-RCNN outperforms the state-of-the-art (IRnet [14]) by 11.6% in terms of the major metric mAP_{50}, which also achieves the best performance in terms of other metrics, either without or with refinement. Relative to the pioneering work of PRM [8], all the other methods make an improvement to an extent by enhancing certain aspects, including localization maps (IAM [13] and IRnet [14]), proposal generation (IAM [13]) and combination with object detection (Label-PEnet [9]). Comparatively, our WS-RCNN is able to improve by a far larger margin, demonstrating its effectiveness. Interestingly, one can observe that WSDDN-seg [44] can manage to achieve comparable or even better results than PRM [8]. Notice that WSDDN-seg exploits a CNN-based model to simply learn to score proposals (like ours), even without using any localization map at all. This may suggest that proposal scoring is indeed critical to WSIS with a large room for improvement, and "learning to score" is a promising strategy worthy of further exploration. We also show some comparisons of instance segmentation performance under different supervision on VOC in Table 2, including SDI [10], Mask R-CNN [2] and our WS-RCNN. SDI [10] uses bounding box supervision and Mask R-CNN [2] uses full supervision.

Table 1. Quantitative results obtained by various WSIS methods on VOC.

Methods	mAP_{50}	mAP_{25}	mAP_{75}	$mAP_{50}(+p)$	$mAP_{25}(+p)$	$mAP_{75}(+p)$
PRM [8]	26.8	44.3	9.0	38.0	52.8	14.1
IAM [13]	28.8	45.9	11.9	-	-	-
Label-PEnet [9]	30.2	49.1	12.9	-	-	-
IRnet [14]	31.1	49.2	10.7	46.7	60.5	15.6
WSDDN-seg [44]	27.5	47.5	9.8	43.7	55.9	16.9
WS-RCNN (ours)	42.7	57.2	19.4	47.3	62.2	19.8

Table 2. Instance segmentation performance under different supervision on VOC.

Method	Supervision	mAP_{50}	mAP_{25}	mAP_{75}
SDI [10]	Box-Level	44.8	-	16.3
Mask R-CNN [2]	Fully-Supervised	69.0	76.7	52.5
WS-RCNN	Image-Level	47.3	62.2	19.8

To intuitively justify the motivation and merits of WS-RCNN, we comparatively visualize in Figure 5 two sets of representative intermediate results, namely, the scores of a same set of proposals acquired by PRM [8], WSDDN-seg [44] and our WS-RCNN (shown in the descending order of the proposal scores). One can observe that WSDDN-seg [44] tends to highlight one dominating instance while ignoring the others of the same class, and PRM [8] tends to highly score object parts or adjoining objects undesirably. Comparatively, our method can get more favorable proposal scores, which we argue is likely due to our advantageous learning mechanism. More representative results obtained by the various methods are further visualized in Figure 6.

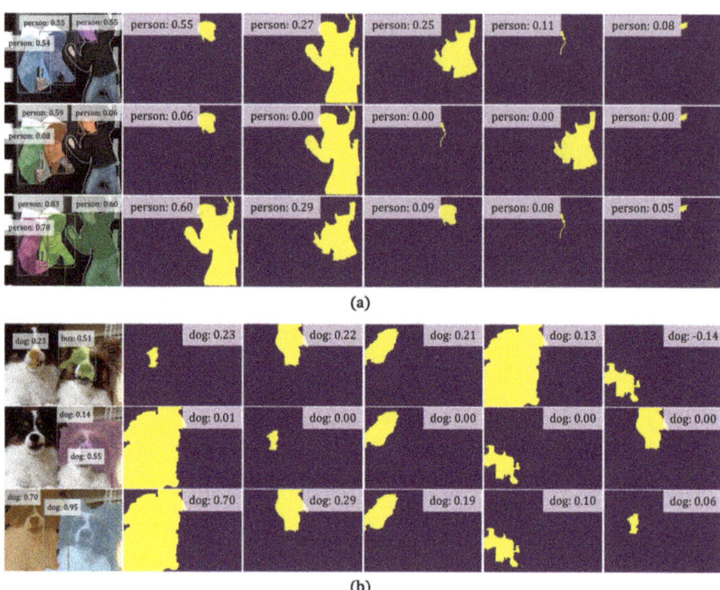

Figure 5. (**a**,**b**) are visualization of two sets of representative proposal scores obtained by various methods. In each set, from top to bottom are the results of PRM [8], WSDDN-seg [44] and our WS-RCNN respectively. In each row, the first column is the final WSIS result, and the second to sixth columns are the scores of the five proposals obtained by the method.

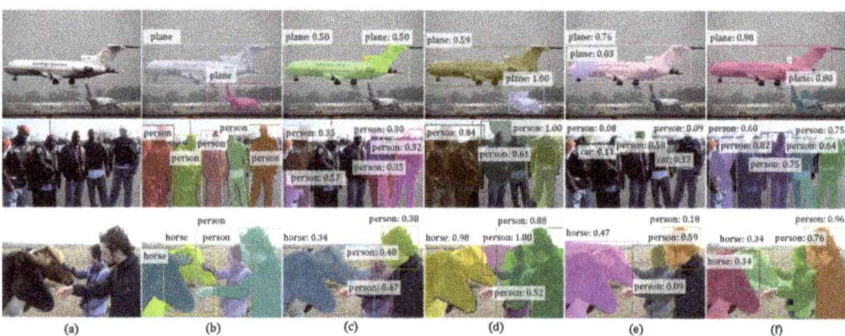

Figure 6. Representative results on VOC obtained by various methods. In each row, from left to right are (**a**) the input image, (**b**) the ground truth, and the results of (**c**) PRM [8], (**d**) IRnet [14], (**e**) WSDDN-seg [44] and (**f**) our WS-RCNN, respectively. Notice that we always output the same number of instances as that in the ground truth for all the methods to enable in-depth comparison.

4.2.2. Results on COCO

Table 3 shows the quantitative results on the COCO dataset. Notice that COCO has not been considered in previous works compared, and the source codes of IAM [13] and Label-PEnet [9] are not publicly released, so the results for these two methods are unavailable. This dataset is far more difficult than VOC, as can be seen by the much worse overall performance of all the methods. However, our WS-RCNN can still outperform the compared methods, with a remarkable margin of 10.7% in terms of mAP_{50} over the state-of-the-art (IRnet [14]), and 4.6% over WSDDN-seg constructed by ourselves. The relative performance of the compared methods in terms of other metric remains consistent with those on VOC, which further verifies the effectiveness of our approach. Some typical results are presented in Figure 7, which further demonstrate that our WS-RCNN can obtain more favorable instance masks, especially in case of multiple or adjoining instances.

Table 3. Quantitative results obtained by various WSIS methods on COCO.

Methods	mAP$_{50}$	mAP$_{25}$	mAP$_{75}$	mAP$_{50}$(+p)	mAP$_{25}$(+p)	mAP$_{75}$(+p)
PRM [8]	5.8	12.1	1.8	14.7	23.7	6.6
IRnet [14]	7.3	13.9	2.4	13.2	22.2	4.8
WSDDN-seg [44]	14.4	24.0	6.0	23.0	30.9	11.5
WS-RCNN (ours)	18.0	27.4	7.2	24.2	32.1	11.6

Figure 7. Representative results on COCO obtained by various methods. In each row, from left to right are (**a**) the input image, (**b**) the ground truth, and the results of (**c**) PRM [8], (**d**) IRnet [14], (**e**) WSDDN-seg [44] and (**f**) our WS-RCNN, respectively. Notice that we always output the same number of instances as that in the ground truth for all the methods to enable in-depth comparison.

4.3. Validation of Key Components

The superiority of the proposed WS-RCNN, to a large extent, should be attributed to the considerate design of its key components. We further carry out experiments on the VOC dataset to validate this point. For each component of concern, we construct some variants to replace the original one in WS-RCNN while keeping everything else unchanged, and compare the overall performance in terms of mAP_{50}.

4.3.1. Pseudo Labeling Strategy

We consider the following three variants of AGPL:

- PLS-1: For each $\mathbf{p}_i^c$ we assign those proposals which spatially include this point to the class c, i.e., assigning $z_n = c$ if $\mathbf{p}_i^c \in \mathbf{R}_n$, without considering the support mask $\mathbf{S}_i^c$ (removing Line 5-8 in Algorithm 1 and meanwhile changing the condition in Equation (5) to be "$\mathbf{p}_i^c \in \mathbf{R}_n$").
- PLS-2: For each $\mathbf{p}_i^c$ we thresold $\mathbf{M}_c$ (using the threshold value of 0.5) and take the connected component surrounding this peak as the support mask $\mathbf{S}_i^c$ (changing Line 6 in Algorithm 1).
- PLS-3: We simply adopt the method of PRM [8] for pseudo labeling, i.e., taking the proposal classification results obtained by the whole pipeline in [8] as the pseudo labels (changing Line 6 in Algorithm 1).

The results are reported in Table 4, which shows all these variants significantly underperform the proposed AGPL. The results validate the merit of AGPL, i.e., relying on peaks in attention maps and the spatial relationship between peak points and proposal regions to achieve pseudo labeling.

Table 4. Comparison with Various Pseudo Labeling Strategies on VOC.

PLS-1	PLS-2	PLS-3	Ours
24.0	25.2	36.2	**42.7**

4.3.2. Training Strategy

In our WS-RCNN, we adopt a particular two-phase strategy for network training (see Section 3.2. To verify the effectiveness of such a design, we construct two different training strategies for comparative study as follows:

- TS-coupled: When training the proposal scoring branch in the second phase, the backbone CNN is not reinitialized by the parameters pre-trained on ImageNet, but by those obtained during the training of the image-level classifier in the first phase.
- TS-joint: The image-level classifier is trained first, and the proposal scoring branch is trained jointly with the image-level classifier, i.e., combining the two training losses when training this branch.

The results in Table 5 demonstrate that the training strategy adopted by our WS-RCNN are more advantageous than the variants compared.

Table 5. Comparison with Various Network Structures and Training Strategies on VOC.

TS-Coupled	TS-Joint	Ours
36.5	37.6	**42.7**

4.3.3. Training Loss

To validate the effectiveness of our proposed EOSL loss, we compare it with a widely-used strategy for background handling as below

- LOSS-CE: We add a dummy class to accommodate background proposals and adopt the conventional binary cross-entropy to replace the EOSL loss in Equation (6).

As can be observed from Table 6 that our EOSL loss performs much better.

Table 6. Comparison with Different Training Loss for Background Handling on VOC.

LOSS-CE	Ours
34.5	**42.7**

4.4. Parameter Study

Our approach involves specifying two parameters, i.e., the number of proposals N and the threshold value β in AGPL. Here we further conduct experiments to analyze the impact of these parameters on performance.

4.4.1. The Number of Proposals N

We vary the number of proposals to be $N = \{50, 100, 150, 200, 300, 400, 500\}$, and report the results in Figure 8a. As can be seen, the performance will not increase significantly after $N > 200$, and we adopt $N = 200$ throughout our experiments for the best tradeoff between performance and efficiency.

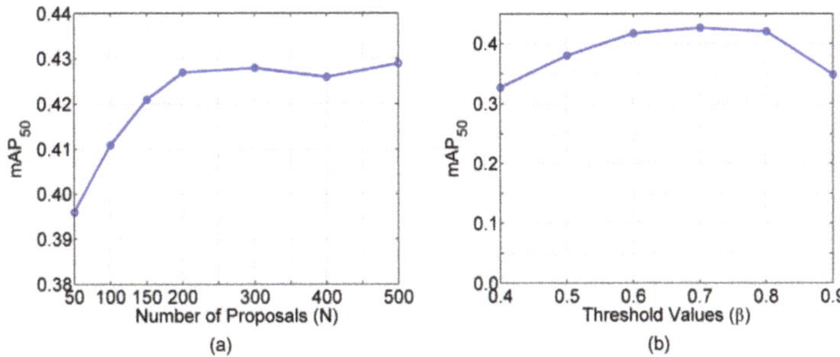

Figure 8. Impacts of varying the key parameters (**a**) the number of proposals N and (**b**) the threshold value β in AGPL.

4.4.2. The Threshold β in AGPL

We vary the threshold value to be $\beta = \{0.4, 0.5, 0.6, 0.7, 0.8, 0.9\}$, and the results are depicted in Figure 8b. One can observe that the performance is relatively stable when $\beta \in [0.5, 0.8]$ and we choose a fixed value of $\beta = 0.7$ throughout our experiments.

4.5. Failure Cases

Despite the effectiveness of WS-RCNN, WSIS is essentially a challenging task and the overall performance still has much room for improvement. Here we show several typical failure cases of WS-RCNN in Figure 9. Since our method is proposal-based, it relies much on the quality of proposals. If the proposal generator fails to cover the spatial extents of the true instance, our method will fail consequently (see Figure 9a). Another typical failure case is that if there are a number of instances of a class with overlap among or close to each other, our method may pick up those large proposals which cover multiple instances (see Figure 9b) or the small proposals which cover only a part of a instance (Figure 9c) (Notice that the desired proposals are present in (b) and (c).

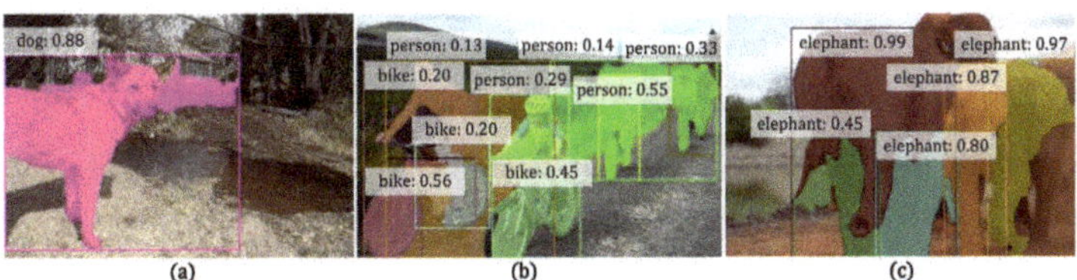

Figure 9. Typical failure cases of WS-RCNN. (**a**) Poor proposals. (**b**) Picking up large proposals covering multiple instances. (**c**) Picking up small proposals covering only a part of a instance.

5. Conclusions

In this paper, we have presented a simple, natural but surprisingly effective framework, termed as Weakly Supervised R-CNN (WS-RCNN), for weakly supervised instance segmentation (WSIS). The basic idea is to deploy a deep network to learn to score proposals under the particular setting of weak supervision. For this sake, a strategy called Attention-Guided Pseudo Labeling is proposed to address the key issue of proposal-level pseudo labeling. And a so-called Entropic Open-Set Loss is introduced for model training to further improve the robustness. Comprehensive experiments on two well-known datasets, i.e., PASCAL VOC 2012 and Microsoft COCO 2014, have demonstrated that the proposed WS-RCNN can significantly outperform the state-of-the-art. Experiments have also been carried out to validate effectiveness of the key components and to study the impacts of some key parameters. In our future work, we will consider integrating a learning-based proposal network to replace the current heuristic proposal generator in WS-RCNN in order to alleviate its dependency on the quality of proposals. We will also explore how to further improve our approach in dealing with complex scenes (e.g., complex background, multiple or adjoining instances of the same class) since there is generally much room for improvement.

Author Contributions: Methodology, J.-R.O.; writing—original draft preparation, J.-R.O., S.-L.D.; writing—review and editing, J.-R.O., S.-L.D., J.-G.Y.; project administration, J.-G.Y.; funding acquisition, J.-G.Y. All authors have read and agreed to the published version of the manuscript.

Funding: This work was supported by the Key R&D Program of Guangdong Province under Grant 2018B030339001 and National Natural Science Foundation of China under Grants 62076099, and also in part by the Guangzhou Science and Technology Program under Grant 201904010299.

Institutional Review Board Statement: The study did not involve humans or animals.

Informed Consent Statement: The study did not involve humans.

Data Availability Statement: The study did not report any data.

Conflicts of Interest: The authors declare no conflict of interest.

References

1. Hariharan, B.; Arbeláez, P.; Girshick, R.; Malik, J. Simultaneous detection and segmentation. In Proceedings of the European Conference on Computer Vision, Zurich, Switzerland, 6–12 September 2014; pp. 297–312.
2. He, K.; Gkioxari, G.; Dollár, P.; Girshick, R. Mask r-cnn. In Proceedings of the IEEE International Conference on Computer Vision, Venice, Italy, 22–29 Ocotber 2017; pp. 2961–2969.
3. Redmon, J.; Divvala, S.; Girshick, R.; Farhadi, A. You only look once: Unified, real-time object detection. In Proceedings of the IEEE Conference on Computer Vision and Pattern Recognition, Las Vegas, NV, USA, 27–30 June 2016; pp. 779–788.
4. Liu, S.; Qi, L.; Qin, H.; Shi, J.; Jia, J. Path aggregation network for instance segmentation. In Proceedings of the IEEE Conference on Computer Vision and Pattern Recognition, Salt Lake City, UT, USA, 18–23 June 2018; pp. 8759–8768.
5. Chen, L.C.; Hermans, A.; Papandreou, G.; Schroff, F.; Wang, P.; Adam, H. Masklab: Instance segmentation by refining object detection with semantic and direction features. In Proceedings of the IEEE Conference on Computer Vision and Pattern Recognition, Salt Lake City, UT, USA, 18–23 June 2018; pp. 4013–4022.
6. Chen, K.; Pang, J.; Wang, J.; Xiong, Y.; Li, X.; Sun, S.; Feng, W.; Liu, Z.; Shi, J.; Ouyang, W.; et al. Hybrid task cascade for instance segmentation. In Proceedings of the IEEE Conference on Computer Vision and Pattern Recognition, Long Beach, CA, USA, 16–20 June 2019; pp. 4974–4983.
7. Jeong, D.; Kim, B.G.; Dong, S.Y. Deep joint spatiotemporal network (DJSTN) for efficient facial expression recognition. *Sensors* **2020**, *20*, 1936. [CrossRef]
8. Zhou, Y.; Zhu, Y.; Ye, Q.; Qiu, Q.; Jiao, J. Weakly supervised instance segmentation using class peak response. In Proceedings of the IEEE Conference on Computer Vision and Pattern Recognition, Salt Lake City, UT, USA, 18–23 June 2018; pp. 3791–3800.
9. Ge, W.; Guo, S.; Huang, W.; Scott, M.R. Label-PEnet: Sequential label propagation and enhancement Networks for Weakly Supervised Instance Segmentation. In Proceedings of the IEEE International Conference on Computer Vision, Long Beach, CA, USA, 16–20 June 2019; pp. 3345–3354.
10. Khoreva, A.; Benenson, R.; Hosang, J.; Hein, M.; Schiele, B. Simple does it: Weakly supervised instance and semantic segmentation. In Proceedings of the IEEE Conference on Computer Vision and Pattern Recognition, Venice, Italy, 22–29 Ocotber 2017; pp. 876–885.
11. Remez, T.; Huang, J.; Brown, M. Learning to segment via cut-and-paste. In Proceedings of the European Conference on Computer Vision, Salt Lake City, UT, USA, 18–23 June 2018; pp. 37–52.
12. Kuo, W.; Angelova, A.; Malik, J.; Lin, T.Y. Shapemask: Learning to segment novel objects by refining shape priors. In Proceedings of the IEEE International Conference on Computer Vision, Long Beach, CA, USA, 16–20 June 2019; pp. 9207–9216.
13. Zhu, Y.; Zhou, Y.; Xu, H.; Ye, Q.; Doermann, D.; Jiao, J. Learning instance activation maps for weakly supervised instance segmentation. In Proceedings of the IEEE Conference on Computer Vision and Pattern Recognition, Long Beach, CA, USA, 16–20 June 2019; pp. 3116–3125.
14. Ahn, J.; Cho, S.; Kwak, S. Weakly supervised learning of instance segmentation with inter-pixel relations. In Proceedings of the IEEE Conference on Computer Vision and Pattern Recognition, Long Beach, CA, USA, 16–20 June 2019; pp. 2209–2218.
15. Arun, A.; Jawahar, C.; Kumar, M.P. Weakly supervised instance segmentation by learning annotation consistent instances. In Proceedings of the European Conference on Computer Vision, Glasgow, UK, 23–28 August 2020; pp. 254–270.
16. Zhou, B.; Khosla, A.; Lapedriza, A.; Oliva, A.; Torralba, A. Object detectors emerge in deep scene cnns. *arXiv* **2014**, arXiv:1412.6856.
17. Zhang, J.; Bargal, S.A.; Lin, Z.; Brandt, J.; Shen, X.; Sclaroff, S. Top-Down Neural Attention by Excitation Backprop. *Int. J. Comput. Vis.* **2018**, *126*, 1084–1102. [CrossRef]
18. Zhou, B.; Khosla, A.; Lapedriza, A.; Oliva, A.; Torralba, A. Learning deep features for discriminative localization. In Proceedings of the IEEE Conference on Computer Vision and Pattern Recognition, Las Vegas, NV, USA, 27–30 June 2016; pp. 2921–2929.

19. Zeiler, M.D.; Fergus, R. Visualizing and understanding convolutional networks. In Proceedings of the European Conference on Computer Vision, Zurich, Switzerland, 6–12 September 2014; pp. 818–833.

20. Simonyan, K.; Vedaldi, A.; Zisserman, A. Deep inside convolutional networks: Visualising image classification models and saliency maps. *arXiv* **2013**, arXiv:1312.6034.

21. Girshick, R. Fast r-cnn. In Proceedings of the IEEE International Conference on Computer Vision, Santiago, Chile, 13–16 December 2015; pp. 1440–1448.

22. Girshick, R.; Donahue, J.; Darrell, T.; Malik, J. Rich feature hierarchies for accurate object detection and semantic segmentation. In Proceedings of the IEEE Conference on Computer Vision and Pattern Recognition, Zurich, Switzerland, 6–12 September 2014; pp. 580–587.

23. Ren, S.; He, K.; Girshick, R.; Sun, J. Faster r-cnn: Towards real-time object detection with region proposal networks. In Proceedings of the Advances in Neural Information Processing Systems, Montreal, QC, Canada, 7–12 December 2015; pp. 91–99.

24. Geng, C.; Huang, S.j.; Chen, S. Recent advances in open set recognition: A survey. *IEEE Trans. Pattern Anal. Mach. Intell.* **2020**. [CrossRef] [PubMed]

25. Dhamija, A.R.; Günther, M.; Boult, T. Reducing network agnostophobia. In Proceedings of the Advances in Neural Information Processing Systems, Vancouver, BC, Canada, 3–8 December 2018; pp. 9157–9168.

26. Laradji, I.H.; Vazquez, D.; Schmidt, M. Where are the Masks: Instance Segmentation with Image-level Supervision. *arXiv* **2019**, arXiv:1907.01430.

27. Long, J.; Shelhamer, E.; Darrell, T. Fully convolutional networks for semantic segmentation. In Proceedings of the IEEE Conference on Computer Vision and Pattern Recognition, Santiago, Chile, 13–16 December 2015; pp. 3431–3440.

28. Kolesnikov, A.; Lampert, C.H. Seed, expand and constrain: Three principles for weakly-supervised image segmentation. In Proceedings of the European Conference on Computer Vision, Amsterdam, The Netherlands, 8–16 October 2016; pp. 695–711.

29. Roy, A.; Todorovic, S. Combining bottom-up, top-down, and smoothness cues for weakly supervised image segmentation. In Proceedings of the IEEE Conference on Computer Vision and Pattern Recognition, Honolulu, HI, USA, 21–26 July 2017; pp. 3529–3538.

30. Pinheiro, P.O.; Collobert, R. From image-level to pixel-level labeling with convolutional networks. In Proceedings of the IEEE Conference on Computer Vision and Pattern Recognition, Boston, MA, USA, 7–12 June 2015; pp. 1713–1721.

31. Papandreou, G.; Chen, L.C.; Murphy, K.; Yuille, A. Weakly-and semi-supervised learning of a DCNN for semantic image segmentation. *arXiv* **2015**, arXiv:1502.02734.

32. Huang, Z.; Wang, X.; Wang, J.; Liu, W.; Wang, J. Weakly-supervised semantic segmentation network with deep seeded region growing. In Proceedings of the IEEE Conference on Computer Vision and Pattern Recognition, Salt Lake City, UT, USA, 18–23 June 2018; pp. 7014–7023.

33. Ahn, J.; Kwak, S. Learning pixel-level semantic affinity with image-level supervision for weakly supervised semantic segmentation. In Proceedings of the IEEE Conference on Computer Vision and Pattern Recognition, Salt Lake City, UT, USA, 18–23 June 2018; pp. 4981–4990.

34. Wei, Y.; Xiao, H.; Shi, H.; Jie, Z.; Feng, J.; Huang, T.S. Revisiting dilated convolution: A simple approach for weakly-and semi-supervised semantic segmentation. In Proceedings of the IEEE Conference on Computer Vision and Pattern Recognition, Salt Lake City, UT, USA, 18–23 June 2018; pp. 7268–7277.

35. Zhang, X.; Wei, Y.; Feng, J.; Yang, Y.; Huang, T.S. Adversarial complementary learning for weakly supervised object localization. In Proceedings of the IEEE Conference on Computer Vision and Pattern Recognition, Salt Lake City, UT, USA, 18–23 June 2018; pp. 1325–1334.

36. Li, K.; Wu, Z.; Peng, K.C.; Ernst, J.; Fu, Y. Tell me where to look: Guided attention inference network. In Proceedings of the IEEE Conference on Computer Vision and Pattern Recognition, Salt Lake City, UT, USA, 18–23 June 2018; pp. 9215–9223.

37. Wang, X.; You, S.; Li, X.; Ma, H. Weakly-supervised semantic segmentation by iteratively mining common object features. In Proceedings of the IEEE Conference on Computer Vision and Pattern Recognition, Salt Lake City, UT, USA, 18–23 June 2018; pp. 1354–1362.

38. Hou, Q.; Jiang, P.; Wei, Y.; Cheng, M.M. Self-erasing network for integral object attention. In Proceedings of the Advances in Neural Information Processing Systems, Vancouver, BC, Canada, 3–8 December 2018; pp. 549–559.

39. Fan, R.; Hou, Q.; Cheng, M.M.; Yu, G.; Martin, R.R.; Hu, S.M. Associating inter-image salient instances for weakly supervised semantic segmentation. In Proceedings of the European Conference on Computer Vision, Salt Lake City, UT, USA, 18–23 June 2018; pp. 367–383.

40. Fan, R.; Cheng, M.M.; Hou, Q.; Mu, T.J.; Wang, J.; Hu, S.M. S4Net: Single stage salient-instance segmentation. In Proceedings of the IEEE Conference on Computer Vision and Pattern Recognition, Long Beach, CA, USA, 16–20 June 2019; pp. 6103–6112.

41. Zhang, C.; Platt, J.C.; Viola, P.A. Multiple instance boosting for object detection. In Proceedings of the Advances in Neural Information Processing Systems, Vancouver, BC, Canada, 4–7 December 2006; pp. 1417–1424.

42. Ren, W.; Huang, K.; Tao, D.; Tan, T. Weakly supervised large scale object localization with multiple instance learning and bag splitting. *IEEE Trans. Pattern Anal. Mach. Intell.* **2015**, *38*, 405–416. [CrossRef] [PubMed]

43. Wang, X.; Zhu, Z.; Yao, C.; Bai, X. Relaxed multiple-instance SVM with application to object discovery. In Proceedings of the IEEE International Conference on Computer Vision, Santiago, Chile, 13–26 December 2015; pp. 1224–1232.

44. Bilen, H.; Vedaldi, A. Weakly supervised deep detection networks. In Proceedings of the IEEE Conference on Computer Vision and Pattern Recognition, Las Vegas, NV, USA, 27–30 June 2016; pp. 2846–2854.

45. Diba, A.; Sharma, V.; Pazandeh, A.; Pirsiavash, H.; Van Gool, L. Weakly supervised cascaded convolutional networks. In Proceedings of the IEEE Conference on Computer Vision and Pattern Recognition, Honolulu, HI, USA, 21–26 July 2017; pp. 914–922.

46. Tang, P.; Wang, X.; Wang, A.; Yan, Y.; Liu, W.; Huang, J.; Yuille, A. Weakly supervised region proposal network and object detection. In Proceedings of the European Conference on Computer Vision, Munich, Germany, 8–14 September 2018; pp. 352–368.

47. Tang, P.; Wang, X.; Bai, S.; Shen, W.; Bai, X.; Liu, W.; Yuille, A. Pcl: Proposal cluster learning for weakly supervised object detection. *IEEE Trans. Pattern Anal. Mach. Intell.* **2018**, *42*, 176–191. [CrossRef] [PubMed]

48. Wan, F.; Wei, P.; Jiao, J.; Han, Z.; Ye, Q. Min-Entropy Latent Model for Weakly Supervised Object Detection. *IEEE Trans. Pattern Anal. Mach. Intell.* **2019**, *41*, 2395. [CrossRef] [PubMed]

49. Shen, Y.; Ji, R.; Yang, K.; Deng, C.; Wang, C. Category-Aware spatial constraint for weakly supervised detection. *IEEE Trans. Image Process.* **2019**, *29*, 843–858. [CrossRef] [PubMed]

50. Bendale, A.; Boult, T.E. Towards open set deep networks. In Proceedings of the IEEE Conference on Computer Vision and Pattern Recognition, Las Vegas, NV, USA, 27–30 July 2016; pp. 1563–1572.

51. Pont-Tuset, J.; Arbelaez, P.; Barron, J.T.; Marques, F.; Malik, J. Multiscale combinatorial grouping for image segmentation and object proposal generation. *IEEE Trans. Pattern Anal. Mach. Intell.* **2016**, *39*, 128–140. [CrossRef] [PubMed]

52. Maninis, K.K.; Pont-Tuset, J.; Arbeláez, P.; Van Gool, L. Convolutional oriented boundaries: From image segmentation to high-level tasks. *IEEE Trans. Pattern Anal. Mach. Intell.* **2017**, *40*, 819–833. [CrossRef] [PubMed]

53. Everingham, M.; Eslami, S.A.; Van Gool, L.; Williams, C.K.; Winn, J.; Zisserman, A. The pascal visual object classes challenge: A retrospective. *Int. J. Comput. Vis.* **2015**, *111*, 98–136. [CrossRef]

54. Lin, T.Y.; Maire, M.; Belongie, S.; Hays, J.; Perona, P.; Ramanan, D.; Dollár, P.; Zitnick, C.L. Microsoft coco: Common objects in context. In Proceedings of the European Conference on Computer Vision, Zurich, Switzerland, 6–12 September 2014; pp. 740–755.

55. Hariharan, B.; Arbeláez, P.; Bourdev, L.; Maji, S.; Malik, J. Semantic contours from inverse detectors. In Proceedings of the IEEE International Conference on Computer Vision, Barcelona, Spain, 6–13 November 2011; pp. 991–998.

56. Simonyan, K.; Zisserman, A. Very deep convolutional networks for large-scale image recognition. *arXiv* **2014**, arXiv:1409.1556.

Article

Superficial Characteristics and Functionalization Effectiveness of Non-Toxic Glutathione-Capped Magnetic, Fluorescent, Metallic and Hybrid Nanoparticles for Biomedical Applications

C. Fernández-Ponce [1,†], J. M. Mánuel [2,†], R. Fernández-Cisnal [1], E. Félix [2], J. Beato-López [3], J. P. Muñoz-Miranda [1], A. M. Beltrán [4], A. J. Santos [5], F. M. Morales [5], M. P. Yeste [5], O. Bomati-Miguel [2,*], R. Litrán [2,*] and F. García-Cózar [1]

[1] Department of Biomedicine, Biotechnology and Public Health, University of Cadiz and Institute of Biomedical Research Cadiz (INIBICA), 11003 Cadiz, Spain; ceciliamatilde.fernandez@uca.es (C.F.-P.); fecir80@gmail.com (R.F.-C.); juampmunoz@gmail.com (J.P.M.-M.); curro.garcia@uca.es (F.G.-C.)

[2] Department of Condensed Matter Physics, Faculty of Sciences, IMEYMAT: Institute of Research on Electron Microscopy and Materials, University of Cádiz, Puerto Real, 11510 Cadiz, Spain; jose.manuel@uca.es (J.M.M.); eduardo.felix@uca.es (E.F.)

[3] Departamento de Ciencias, Institute for Advanced Materials (INAMAT), Campus de Arrosadia, Universidad Pública de tgNavarra, 31006 Pamplona, Spain; juanjesus.beato@unavarra.es

[4] Departamento de Ingeniería y Ciencia de los Materiales y del Transporte, Sevilla 41011, Spain; abeltran3@us.es

[5] Department of Materials Science and Metallurgic Engineering, and Inorganic Chemistry, Faculty of Sciences, IMEYMAT: Institute of Research on Electron Microscopy and Materials University of Cadiz, Puerto Real, 11510 Cadiz, Spain; antonio.santos@gm.uca.es (A.J.S.); fmiguel.morales@uca.es (F.M.M.); pili.yeste@uca.es (M.P.Y.)

* Correspondence: oscar.bomati@uca.es (O.B.-M.); rocio.litran@uca.es (R.L.); Tel.: +34-667464426 (R.L.)

† Equal contribution.

Citation: Fernández-Ponce, C.; Mánuel, J.M.; Fernández-Cisnal, R.; Félix, E.; Beato-López, J.; Muñoz-Miranda, J.P.; Beltrán, A.M.; Santos, A.J.; Morales, F.M.; Yeste, M.P.; et al. Superficial Characteristics and Functionalization Effectiveness of Non-Toxic Glutathione-Capped Magnetic, Fluorescent, Metallic and Hybrid Nanoparticles for Biomedical Applications. *Metals* **2021**, *11*, 383. https://doi.org/10.3390/met11030383

Academic Editor: Houshang Alamdari

Received: 23 December 2020
Accepted: 22 February 2021
Published: 26 February 2021

Publisher's Note: MDPI stays neutral with regard to jurisdictional claims in published maps and institutional affiliations.

Abstract: An optimal design of nanoparticles suitable for biomedical applications requires proper functionalization, a key step in the synthesis of such nanoparticles, not only for subsequent crosslinking to biological targets and to avoid cytotoxicity, but also to endow these materials with colloidal stability. In this sense, a reliable characterization of the effectiveness of the functionalization process would, therefore, be crucial for subsequent bioconjugations. In this work, we have analyzed glutathione as a means to functionalize four of the most widely used nanoparticles in biomedicine, one of which is a hybrid gold-magnetic-iron-oxide nanoparticle synthetized by a simple and novel method that we propose in this article. We have analyzed the colloidal characteristics that the glutathione capping provides to the different nanoparticles and, using information on the Z-potential, we have deduced the chemical group used by glutathione to link to the nanoparticle core. We have used electron microscopy for further structural and chemical characterization of the nanoparticles. Finally, we have evaluated nanoparticle cytotoxicity, studying cell viability after incubation with different concentrations of nanoparticles, showing their suitability for biomedical applications.

Keywords: functionalization; glutathione; surface modifications; colloidal stability; cytotoxicity; electron microscopy; magnetic nanoparticle; gold nanoparticle; quantum dot

1. Introduction

Advances in biomedicine require new experimental tools that enable manipulation of biomolecules, and the study of biological processes at the molecular and cellular level. For some decades now, nanoparticles (NPs) have been broadly used in biomedicine [1–4]. A large variety of NPs have been utilized for the development of applications in diagnostics [5,6], therapy [7,8], biomarking [9,10], drug delivery [11], etc. [12]. In this context, three groups of NPs are traditionally considered of the highest interest for biomedical applications, due to their properties: fluorescent NPs (quantum dots, QDs), magnetic NPs and pure metallic NPs.

The biomedical interest in light-emitting NPs resides in the fact that many biological techniques are based on the use of fluorometric probes for the identification of specific biological species. Therefore, unless genetic expression is desired, NPs such as QDs represent a successful alternative for biological markers, because of their advantageous features when compared to conventional organic fluorophores or fluorescent proteins, especially due to their size-dependent fluorescence. Light emission of these QDs can be tuned by controlling their size at nanometric level, which is an interesting characteristic for their use as biological probes in multicolor experiments [13–15].

In the case of magnetic NPs, applications for protein immobilization [16], as hyperthermia agents for cancer therapy [17,18], for drug delivery [19], and as contrast agents for magnetic resonance imaging (MRI) [1,20] have been proposed [21]. Particularly, iron oxides and especially magnetite (Fe_3O_4) NPs are the most commonly used, due to their biocompatibility, low cytotoxicity and stability. MRI is a non-invasive clinical diagnostic technique with a high spatial resolution that is extensively used for anatomical imaging of soft body tissues [22]. However, this technique requires the use of contrast agents to enhance differences between damaged and normal tissues and is still lacking in technologies that allow for targeted identification of specific structures, such as tumors or infectious agents. Superparamagnetic iron oxide NPs are used in MRI as transverse relaxation time contrast agents, since they show excellent magnetic properties, in addition to their low toxicity and cytocompatibility [23,24].

Among the purely metallic NPs and over the last two decades, Au NPs have been widely studied and applied in the field of biomedicine, especially in cancer diagnostics [25], tumor therapy by optical hyperthermia [26], biosensing [27], virus detection and regulation of cell function [28] as well as X-ray computed tomography (CT) contrast agents [1,29,30]. The biomedical interest of these NPs resides in their good biocompatibility, low inherent toxicity [31], facile synthesis, high biofunctionalization capability and in their unique optical properties due to their associated localized surface plasmon resonance (LSPR) [32]. For X-ray CT, contrast agents that allow for the display of internal tissue structures, are usually necessary, in order to enhance the image contrast and to provide additional functional information. Gold NPs can potentially improve the contrast provided by iodine contrast agents currently in use for cardiovascular imaging at clinically relevant energies (80–140 kV) [33].

Moreover, synthesis of hybrid NPs, such as structures constituted by a magnetic core covered by a gold shell, allows for the combination of features from both constituents, which enhances NP possibilities for specific biomedical applications. There is great interest in these hybrid NPs for therapies such as tumor treatment by hyperthermia, while they also constitute highly attractive materials for image based medical diagnostics [34]. Moreover, the latter frequently require a combination of data obtained from different complementary techniques in order to achieve accurate diagnoses [35] and thus these core-shell NPs may be simultaneously imaged by several techniques offering a promising alternative to the fabrication of multimodal contrast agents [36]. In this sense, different synthesis for hybrid magnetic iron oxide/gold NPs have previously been proposed, either naked, functionalized with capping agents, or by preparing separately magnetic iron oxide (magnetite or maghemite) and gold NPs and promoting a subsequent conjugation. Many of these procedures involve temperatures higher than 190 °C as well as the use of organic or hydrophobic solvents such as 1,2—hexadecanediol, 1—octadecene, etc. Thus, ligand exchange steps are required to obtain hydrophilic functionalized NPs.

However, NPs used for biomedical applications must meet some highly important requirements: they must display low cytotoxicity, colloidal stability, and availability for subsequent bioconjugation with other biomolecules. Surface functionalization is a common strategy not only to promote subsequent crosslinking between NPs and specific biological species, but also to minimize their cytotoxicity and unspecific binding. Functionalization of the NP surface with species containing bio-active terminal groups, such as amino or carboxylic groups, allows for subsequent linking with relevant biomolecules for targeted applications [37]. NP functionalization can also contribute to avoid aggregation and unspecific cellular uptake, minimizing accumulation in organs and/or phagocyte activation, thus maintaining a prolonged circulation time [38–40]. In any case, obtaining all the information available on the effectiveness of the functionalization process, via a comprehensive characterization, is crucial for further bioconjugations and biomedical applications.

The tripeptide glutathione (GSH) constitutes an interesting asset for functionalization, as it is a low-cost reagent that harnesses in its small frame three highly used moieties for bioconjugation, such as a thiol terminal group and also an amino and two carboxylic groups. In this sense, the present work follows previous ones, in which we presented the functionalization of different NPs with GSH [14,41,42]. In this work, we have utilized GSH to functionalize four types of NP: fluorescent CdTe QDs, iron oxide magnetic NPs, gold NPs and hybrid iron oxide magnetic–gold NPs, labeled in the text, as QD-GSH, Mag-GSH, Au-GSH and Au-Mag-GSH, respectively. Glutathione has been chosen because it is a small molecule with three different types of functional groups that can easily be used in order to bind other proteins such as streptavidine or antibodies. Furthermore, the thiol moiety readily binds to gold which is of interest for gold containing NPs. In this work, we have designed a simple and novel hydrophilic route to prepare hybrid goldiron oxide magnetic NPs with GSH in situ functionalization. These NPs are labeled as Au-Mag-GSH NPs. We have compared the results for these NPs with those obtained for different types of GSH functionalized NPs: gold, magnetic iron oxide and CdTe QDs. We have applied chemical synthetic routes based on reduction reactions, using sodium borohydride and performing an in situ functionalization with GSH. In the literature there are other examples of GSH-capped NPs [43,44], but in all cases, the authors use different steps for the synthesis and perform an ex situ functionalization by a ligand exchange process, requiring, in some cases, the use of non-aqueous solvents. After a first characterization of the structural and physical properties of the NPs, we have evaluated the effectiveness of the functionalization process. With this aim, we have carried out a colloidal characterization, measuring the colloidal stability of the studied NPs at different pH values, different temperatures and different salinity, all fundamental features for biomedical applications. This study, combined with Fourier transform infrared (FTIR) spectroscopy, also provides information about the linking between NP and GSH. Depending on particle composition, GSH uses different terminal groups to link to the nanostructure. Moreover, we have combined electron microscopy-based structural and compositional nanoanalyses to obtain information about the composition and the thickness of the GSH layer surrounding the NPs. This characterization allows us to learn whether GSH has been properly linked to the NP surface thus functionalizing it. Finally, we have studied the cytotoxicity of the prepared NPs, analyzing cell viability after incubation of cells with different NP concentrations. The novelty of our work lies in the development of an easy and new synthetic method to prepare hybrid gold/iron oxide magnetic NPs, in situ functionalized with GSH. Our main objective is to deepen in the characterization of GSH functionalization and, in the knowledge of the colloidal characteristics of the resultant NP-based system, provide the basis for a material that could be successfully utilized in biomedical application.

In previous works [45], we showed an initial, mainly transmission electron microscopy (TEM)-focused, characterization of similarly prepared NPs but, due to their relevant biomedical properties, in this article we have deepened on their biomedical compatibility and surface capping both crucial for biomedical focused bioconjugation. Thus, here, we provide a new horizon for the design of GSH-capped gold/iron oxide magnetic NPs suited for biomedical applications and also expand in the mechanisms underlying GSH layer linkage, studying the effect that GSH capping causes to NP properties. For this purpose, different GSH NPs have been prepared and included in the study. Although GSH capped NPs have been previously studied, this work presents a deep characterization of their colloidal properties, the functionalization mechanism and its influence on cytotoxicity.

2. Materials and Methods

2.1. Chemicals

The following chemical reagents were utilized in this work: tetrachloroauric acid ($HAuCl_4$), cadmium chloride ($CdCl_2$), sodium borohidrure ($NaBH_4$), tellurium powder, sodium hydroxide (NaOH), iron (III) chloride ($FeCl_3$), sodium hydroxide (NaOH), diethylene glycol, cysteamine (CYS) and reduced glutathione (GSH). All these reagents were of analytic grade, purchased from Sigma-Aldrich (current "Millipore-Sigma", Burlington, MA, USA) and used as received. Milli-Q®(MQ) water was used for all experiments.

2.2. Synthesis of Glutathione-Capped Nanoparticles (GSH-NPs)

Based on our previous works, we have designed three rapid chemical routes to obtain small GSH-capped gold, magnetic iron oxide, and CdTe QDs [14,42]. In all these routes, we have promoted a reduction reaction, and we have used GSH to functionalize the NP surface. To synthetize the GSH-capped hybrid gold-iron oxide magnetic NPs, we have developed, in this work, a novel, original method. A diagram of the formation of these GSH-capped NP is shown in the graphical abstract.

2.2.1. GSH-Capped Gold NPs

Briefly, we started from 20 mL of an aqueous solution 0.01 M $HAuCl_4$ and 0.026 M GSH. The mixture was vigorously stirred for 30 min under a nitrogen atmosphere. A $NaBH_4$ aqueous solution was added dropwise, in a 1:5 $HAuCl_4/NaBH_4$ molar ratio. After 30 additional min, the red wine colored solution was stored at 8 °C protected from light. The code Au-GSH is used for this type of NPs. In this case, the Au^{3+} ions, introduced as $HAuCl_4$, are reduced by the $NaBH_4$. Although GSH could act as a reducing agent, the reaction would be slower than with a strong reducing agent, such as $NaBH_4$, that will promote a fast nucleation, leading to more homogeneous and smaller NPs.

2.2.2. GSH-Gapped Iron Oxide Magnetic NPs

In this case, we designed a preparation method based on the thermal decomposition of an iron salt, in the presence of GSH, to functionalize the NP, thus controlling its final size.

The synthesis of GSH-iron oxide magnetic NPs was carried out in two steps. The first consisted of the preparation of a NaOH solution that will be used as oxidizing agent for the co-precipitation of iron and formation of iron oxide. This solution was prepared by heating 20 g of NaOH and 20 mL of diethylene glycol at 120 °C for one hour. The resulting solution is stored at 70 °C for subsequent utilization.

The second step constituted the main thermal decomposition and co-precipitation reaction. 0.55 g of $FeCl_3$, used as iron precursor, have been solved into 15 mL of diethylene glycol, which is used as solvent. The solution obtained was placed in a 500 mL three-necked flask, where the GSH, used as capping agent, was added. The GSH:Fe molar ratio was 1:9. The solution was heated at 200 °C under reflux in a nitrogen atmosphere, in order to promote the thermal decomposition of iron salts. After heating for 30 min, a previously prepared 8 mL NaOH solution was added, to induce the formation of iron oxide. Once NaOH was added, the total solution was heated for 10 min, keeping the temperature at 200 °C. After cooling, the obtained solution was filtered three times (using a 0.1 μm Millipore membrane), after precipitation with ethanol, in order to eliminate an excess of precursor species, and/or, residues from the reaction. The Fe^{3+} cations, introduced as $FeCl_3$, were totally or partially reduced by the diethylene glycol, giving rise to the formation of either magnetite (Fe_3O_4) or maghemite ($\gamma-Fe_2O_3$), both iron-oxide magnetic phases equally appropriate for the objective of this work. The high GSH affinity for the NP surface led to GSH-functionalized magnetic NPs (Mag-GSH) as a black powder with magnetic character.

2.2.3. GSH-Capped Gold/Fe₃O₄ NPs

Starting from the Mag-GSH NPs previously prepared, we designed a method to easily prepare GSH capped gold/iron oxide magnetic NPs. In this novel synthetic route, we use the capability of GSH to link iron oxide and gold simultaneously. Briefly, 0.05 g Mag-GSH, prepared as descried above, were dispersed in 1 mL of MQ water while stirring for 1 h. Then, an aqueous solution containing 0.1 g HAuCl$_4$ was added. After 8 h of vigorous stirring, GSH was added in excess while stirring for one additional hour. The total synthesis was conformed in three simple steps. The first step was just the Mag-GSH preparation. In the second step, free terminal GSH moieties in Mag-GSH NPs bound to gold when we added the Au^{3+} precursor. In this case, a slow reduction of Au^{3+} by the same GSH used as capping agent was promoted, in order to avoid the formation of separated gold NPs, a process that could be induced by the addition of NaBH$_4$. During the last step, GSH was added in excess and linked to the final hybrid iron oxide-gold NPs. The colloidal solution was brought under magnetic decantation, in order to eliminate the iron oxide magnetic NPs which did not contain gold. As a result, a brown-pink and still magnetic colloidal solution was obtained. The solution could be filtered and washed, obtaining a magnetic, brown-black powder. As in the case of Mag-GSH, these NPs can easily be dissolved in an aqueous solution due to the hydrophilic character conferred by GSH. A schematic graphic of this particular synthesis is shown in Figure 1.

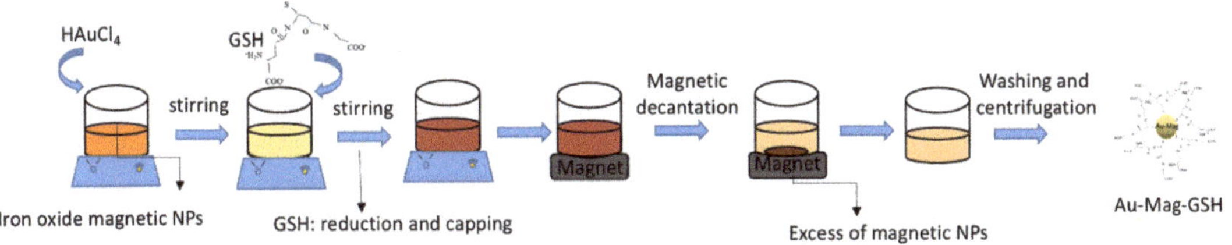

Figure 1. Diagram of the glutathione (GSH)-capped hybrid gold/iron oxide magnetic nanoparticles (NPs).

2.2.4. GSH-Capped QDs

Synthesis of GSH-CdTe QDs was carried out by a co-precipitation reaction, using CdCl$_2$ as cadmium precursor and NaHTe as tellurium precursor in the presence of GSH. For this purpose, 0.19 g GSH, used as a stabilizing agent, and 0.046 g CdCl$_2$, used as the cadmium precursor, were dissolved in 100 mL MQ water. In order to promote formation of Cd^{2+}-GSH complexes at the QD surface, the pH was adjusted to 8 by dropwise addition of a 1M NaOH solution, under vigorous stirring. The obtained solution was placed in a 500 mL three-necked flask, where 2 mL fresh NaHTe (previously prepared by reducing tellurium powder with NaBH$_4$ in a nitrogen atmosphere) was added. The Cd^{2+}:NaHTe:GSH molar ratio was 4:1:10. The solution was heated under reflux in a nitrogen atmosphere. The final QD size can be controlled by increasing temperature and/or reaction time, ranging from 90 to 145 °C and 90 to 360 min, respectively.

The different aqueous aliquots, containing the GSH-CdTe QDs, were cooled and filtered threefold (using a 0.1 μm Millipore membrane), after precipitation with a mixture of acetone and ethanol, in order to eliminate the excess of precursor species and/or, residues from the reaction. As a result, we obtained CdTe QDs capped with GSH molecules (QD-GSH) that can be easily dispersed in and aqueous solution due to the hydrophilic character conferred by GSH.

2.3. Characterization Techniques

2.3.1. Structural and Chemical Analyses

The identification of GSH-NPs has been carried out through structural and chemical techniques, associated with transmission and scanning-transmission electron microscopy, or (S)TEM. The general view, shape and atomic structure of the materials were observed using conventional bright field (BF-TEM), high-resolution (HRTEM) and high-angle annular dark field STEM (HAADF-STEM). Also, a corrected probe HAADF in high-resolution mode (HR-HAADF) was utilized for the sample Au-GSH, yielding images in which these crystals were easily distinguishable (due to the high difference of atomic numbers between the gold in the NP and the carbon in the supporting material, which leads to a high contrast in the image intensity for both structures) and simultaneously show atomic columns. Regarding chemical composition, energy-dispersive X-ray spectroscopy (EDX) was applied, so the structural information from imaging techniques is complemented, allowing an unequivocal identification of the NP and revealing the GSH shell. All these (S)TEM techniques were carried out using two STEM microscopes: a double-aberration corrected TITAN³ Themis and a Talos F200X ("ThermoFisher Scientific", Waltham, MA, USA), both operating at a 200 kV accelerating voltage. In order to observe the different materials under the electron beam, several samples were prepared by depositing 10 µL of an NP colloidal solution, drop-casted onto a holey-carbon coated Cu grid for TEM and dried for five hours. Alternatively, some samples, for Mag-GSH, were prepared by imbibing a grid over a dry amount of material, obtaining a sample with no significant difference in the observed NP state of aggregation when deposited on the sample grid. For all magnetic materials, a magnet was placed near to the TEM grid before inserting the sample in the electron microscope in order to remove the biggest NP clusters, which could be attracted by the magnetic lenses of the microscopes.

FTIR was utilized in order to obtain information about the linking between GSH and the NP surface. Experiments were recorded with Bruker Alpha System Spectrophotometer (KBr wafer technique), using the same quantity of sample in all measurements.

2.3.2. Colloidal Characterization

Dynamic light scattering (DLS) was utilized in order to measure NP hydrodynamic size. Measurements were carried out at 25, 37 and 40 °C, using a Malvern Zetasizer Nano-ZS ("Malvern Instruments", Worcestershire, UK), with a 1 cm path cell. This equipment also allows for the Z-potential to be measured which evaluate colloidal stability in solution, as well as NP surface charge.

2.3.3. Optical and Magnetic Properties

LSPR of the Au-GSH nanostructure was evaluated studying the ultraviolet–visible (UV-Vis) absorbance of each NP colloidal solution. Absorption spectra were acquired with a Lambda 19 Perkin Elmer spectrophotometer ("Perkin Elmer", Waltham, MA, USA). The position and profile of the LSPR band provide information about NP dispersion and allow for the estimation of NP average sizes.

Photoluminescence (PL) excitation and emission spectra were recorded in a PTI Quantamaster fluorometer using a Xenon arc Lamp at 150 W and a computer controller QuadraScopic monochromator.

Magnetic properties of magnetic-iron-oxide NPs were analyzed by measuring magnetization curves at room temperature in a high-sensitive Magnetic Faraday Balance ("Oxford Instruments", Abingdon, U.K.), applying magnetic fields up to 0.6 T.

2.3.4. Cytotoxicity Assay and Cell Culture

Jurkat cells ("American Type Culture Collection", Manassas, VA, USA) were cultured at 37 °C, in a 5% CO_2 atmosphere, in Dulbecco's Modified Eagle's Medium (DMEM) containing 2 mM L—glutamine, 10 mM Hepes, 10% (v/v) heat-inactivated fetal bovine serum (FBS), 1% (v/v) non-essential amino acids (NEAA), 1% (v/v) sodium pyruvate, 50 µM 2-mercaptoethanol, 100 U/mL penicillin and 100 µg/mL streptomycin (all from "Life Technologies", Carlsbad, CA, USA).

2.3.5. Cytotoxicity and Cell Viability.

We cultured 5×10^5 Jurkat cells cells in a 48-well plate, in the absence or presence of NPs at the indicated concentrations. Cell viability was analyzed 24 h after addition of NPs by a As previously commented, one key property (MTT)-based assay as previously described. Briefly, MTT reactant (thiazolyl blue tetrazolium bromide, TOX1-1KT, "Millipore-Sigma", Burlington, MA, USA) was added to the cells in a 1:10 ratio (MTT solution/culture medium) and incubated during 3 h at 37 °C. Then, formazan crystals, formed inside viable cells, were dissolved by adding MTT Solubilization Solution M-8910, ("Millipore-Sigma", Burlington, MA, USA) at 1:2 with vigorous pipetting. Optical density at 570 nm was evaluated to quantify the amount of formazan crystals, which is proportional to the number of viable cells (background absorbance was measured at 690 nm and subtracted from the 570 nm measurement). Viability of cell in the presence of NPs was compared to untreated controls (considered 100% viability). Media in the absence of cells was used as blank; while cells, cultured in the presence of 10% dimethyl sulfoxide (DMSO), was used as control for decreased cell viability, being always below 10%. Statistical analyses were performed using the Statgraphics software ("Statpoint Technologies", Warrenton, VA, USA). Significances were determined by using analyses of variance (ANOVA) and multiple range tests.

3. Results

As previously commented, in order to determine the potential of the presented NPs for biomedicine applications, it is required to previously study their structural properties, their optical behavior, as well as additional features such as colloidal stability at physiological values of salinity and pH, colloidal stability at temperatures ranging from 35 to 40 °C and cytotoxicity.

3.1. Structural Features and Physical Properties

Figure 2 shows results on the size and optical properties of Au-GSH samples, as well as an overview on the state of aggregation of the NPs when deposited on the holed carbon grid. A general view of these particles can be observed in Figure 2a, which shows a HAADF-STEM micrograph. This technique is ideal for imaging gold-based particles, due to the high atomic number of gold, which produces a high contrast with the carbon support in HAADF images. This micrograph reveals the formation of relatively small and well-dispersed NPs with nearly round shape. Figure 2b shows the NP size distribution obtained from more than 300 NPs in this and similar images. From this histogram, it can be affirmed that particle size is homogeneous, with an average diameter of 3.8 nm (see Table 1, in which particle diameter for all materials are collected), obtained from the fitting of this distribution to a Gaussian function. In any case, although this type of images shows, with a good contrast, the size and shape of the particles, hinting on the high atomic number of its constituent material, the exact chemical composition of the NPs is not determined with this technique. The presence of gold in this nanostructure can be confirmed by the UV-Visible absorption spectrum of the colloidal solution (Figure 2c). The inset in the figure is a photograph of the obtained red-wine colored colloidal solution. The spectrum shows the characteristic 520 nm LSPR absorption band, indicating the presence of small gold NPs. Position and width of LSPR bands give information on NP size and size homogeneity. The Au-GSH LSPR band appears as a shoulder, showing a wide profile which can be a consequence of the small average size of these NPs. The high number of surface atoms in small NPs increases the damping of oscillating electrons at conduction bands. This effect can be even higher for small thiol-capped NPs, due to the loss of itinerancy of electrons involved in Au-S bonding [46]. This ligand-metal charge transfer transition, produced due to the capping, can contribute to the relatively strong emission bands, in the blue range of the spectra, for the Au-GSH NPs, as shown in Figure 2d, where the PL emission spectra, recorded for the Au NPs colloidal solution (exciting at 317 nm), is shown. This emission could be an interesting feature for the use of this NPs for biomarker applications [42].

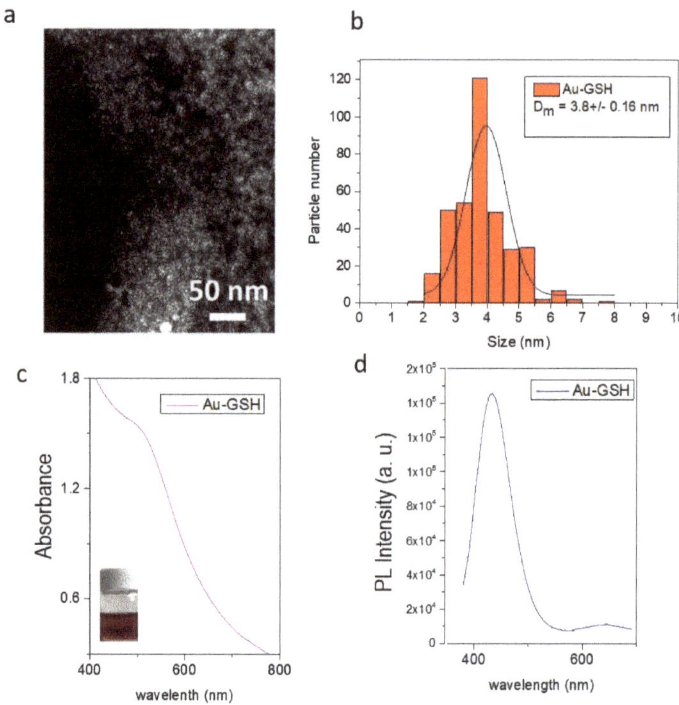

Figure 2. High-angle annular dark field (HAADF) image (**a**), as well as the histogram displaying particle size distribution (**b**) of Au-GSH NPs. (**c**) UV-Visible absorption spectrum and PL emission spectrum (**d**) of Au-GSH NPs. The inset in Figure 2c shows a digital photo image of an Au-GSH colloidal solution.

Table 1. Average diameter obtained via transmission electron microscopy (TEM), dynamic light scattering (DLS) hydrodynamic diameter and polydispersity index (PDI).

Sample	TEM Average Size (nm)	DLS Average Size (nm)	PDI (%)
Au-GSH	3.8 ± 0.2	6.2 ± 1.2	0.2
Mag-GSH	6.1 ± 0.4	12.7 ± 1.2	0.2
QD-GSH	2.9 ± 0.3	32.6 ± 1.3	0.1
Au-Mag-GSH	6.9 ± 0.4	42.3 ± 1.3	0.2

Figure 3 shows results obtained from the optical and structural characterization of Mag-GSH samples. A general view of the NPs is showed in Figure 3a. The BF-TEM micrograph reveals the obtained particle distribution for this preparation, less dispersed as in the case of gold NPs, most probably due to the small clustering promoted by magnetic interactions. Figure 3b displays and histogram showing the particle size distribution for this sample. Fitting of these results to a Gaussian function reveals a NP average size of 6.1 nm (Table 1). Figure 3c shows the magnetization curve, i.e., applied field versus magnetization, at room temperature. This curve shows the magnetic behavior of these NPs [47]. The inset in Figure 3c shows the response of Mag-GSH NPs (flask on the left) to a permanent magnetic field, applied using a permanent magnet adjacent to the flask that contains the NP colloidal dispersion.

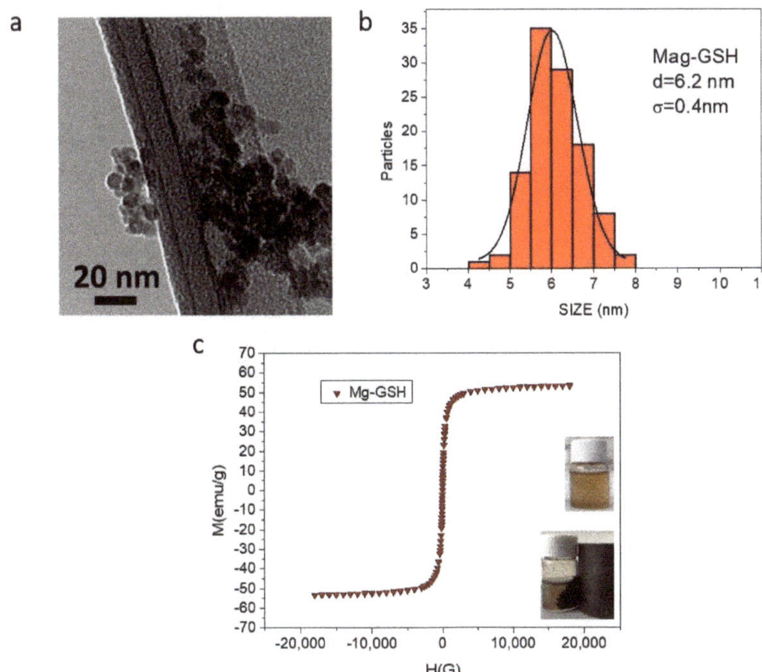

Figure 3. Bright field TEM (BF-TEM) micrograph (**a**), as well as the histogram displaying the particle size distribution (**b**) of Mag-GSH NPs. Magnetization curve of Mag-GSH NPs (**c**). The inset in this figure shows a digital photo image of Mag-GSH NPs placed in the absence and presence of a magnet located close to the vials.

Figure 4 shows a general view and optical features of QD-GSH NPs. Figure 4a corresponds to a HAADF-STEM image of an accumulation of these QD-GSH NPs. In this case, we can observe the presence of extremely small NPs immersed in the GSH capping. The histogram corresponding to these NPs is shown in Figure 4b. By fitting these results to a Gaussian function, an average NP diameter of 2.9 nm is obtained (Table 1). As we have previously commented, NP size is especially important for their optical properties, determining the wavelength of their emission, which is especially important in this case, since QD functionality in biomedical applications rely on their fluorescence. In this sense, the proposed synthesis allows control of NP final size and thus, its PL emission. Specifically, Figure 4c shows PL emission spectra obtained from the sample in Figure 4a (with average size reflected in Figure 4b), labelled as QD-GSH1 and prepared at a synthesis temperature of 90 °C for 30 min. Two additional QD-GSH samples of bigger average sizes (QD-GSH2, prepared at 90 °C for 3 h, and QD-GSH3, heating during 3 h at 90 °C and one more hour at 110 °C) are also included. As NP size increases (sizes for QD-GSH1, QD-GSH2 and QD-GSH3 are 2.9 nm, 3.6 nm and 3.9 nm, respectively), a red-shift in the emission peak maximum is observed, illustrating wavelength size-dependency of the emission. The photograph of colloidal solutions of these three samples, under UV radiation, is added to this figure, as an inset. As it can be observed from both, PL spectra and photograph of the colloidal solutions; by changing just a few nanometers in size, we can tune the emission color from blue to red.

Regarding hybrid Au-Mag-GSH NPs, one of the first and important results to be confirmed after their preparation was that each nanostructure is indeed formed by both iron oxide and gold. Taking into account that the synthesis has been initiated by heterogeneous nucleation of iron oxide seeds, the gold phase necessarily has to be located around the iron oxide phase. In this sense, a relevant measurement is shown in Figure 5a, namely the comparison of UV-Vis absorption spectra for the colloidal solutions of Au-GSH and Au-Mag-GSH samples. Both colloids, display the LSPR absorption band (associated to gold), however, the hybrid Au-Mag NP curve has a pronounced red-shift, from 560 nm to 520 nm. According to previous reported results, this red-shift can be attributed to the reduced electron deficiency of Au NPs caused by the interfacial communication between Au and magnetic-iron-oxide [48]. Therefore, these results reveal the formation of a hybrid NP, containing gold and magnetic-iron-oxide. The Au-Mag-GSH spectrum, shows a second and less intense near-infrared (NIR) absorption band that could be attributed to the presence of anisotropic NPs or even to iron oxide-gold interactions in the hybrid structure. This characteristic can be of interest in phototherapy applications [49,50]. A photograph of the Au-Mag-GSH colloidal solution is also shown in this figure (see inset image in Figure 5a). The spectra have not been normalized because the determinant parameters are the LSPR band position, the wavelength at which the LSPR is observed and the shape and width of the band. A BF-TEM micrograph of these particles is shown in Figure 5b, revealing the formation of homogeneously shaped structures, without the presence of clusters as dense as those observed for Mag-GSH. NP size distribution, and its fitting to a Gaussian function is shown in Figure 5c, showing an average particle size of 6.9 nm (Table 1).

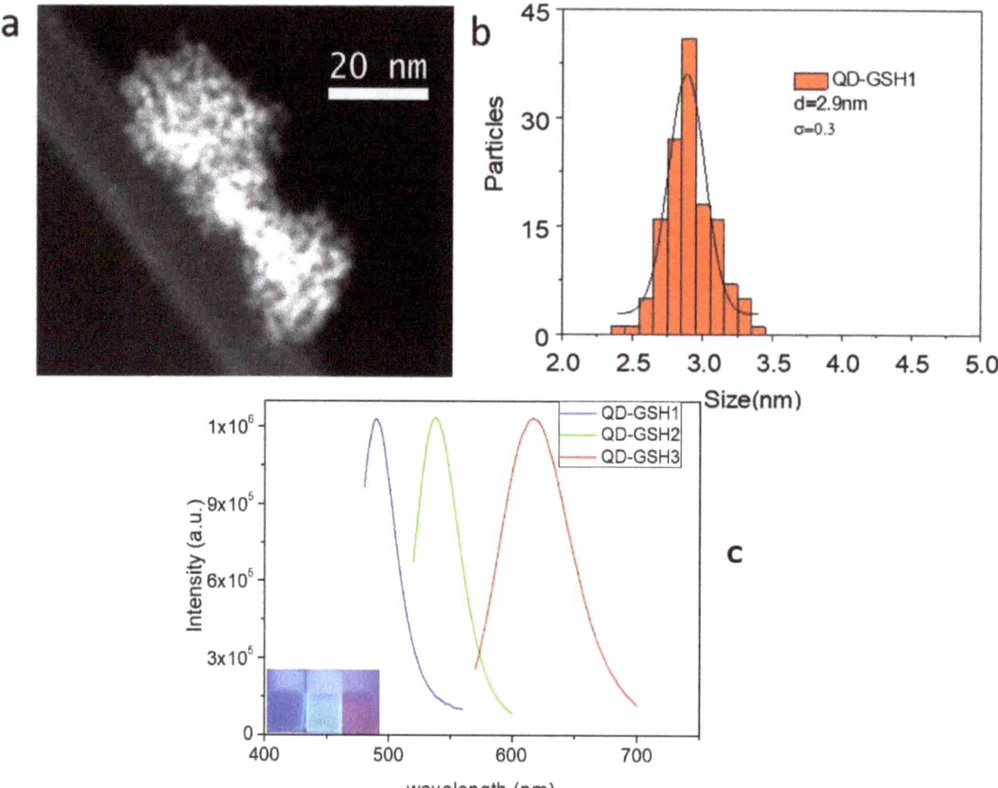

Figure 4. HAADF image (**a**), as well as the histogram displaying particle size distribution (**b**) of quantum dot (QD)-GSH1 NPs. (**c**) PL-emission spectra for QD-GSH1 (blue) as well as for QD-GSH2 (green) and QD-GSH3 (red), with higher average size. The inset in this figure shows a digital photo image of the three QD-GSH colloidal solutions used in this figure, under UV irradiation.

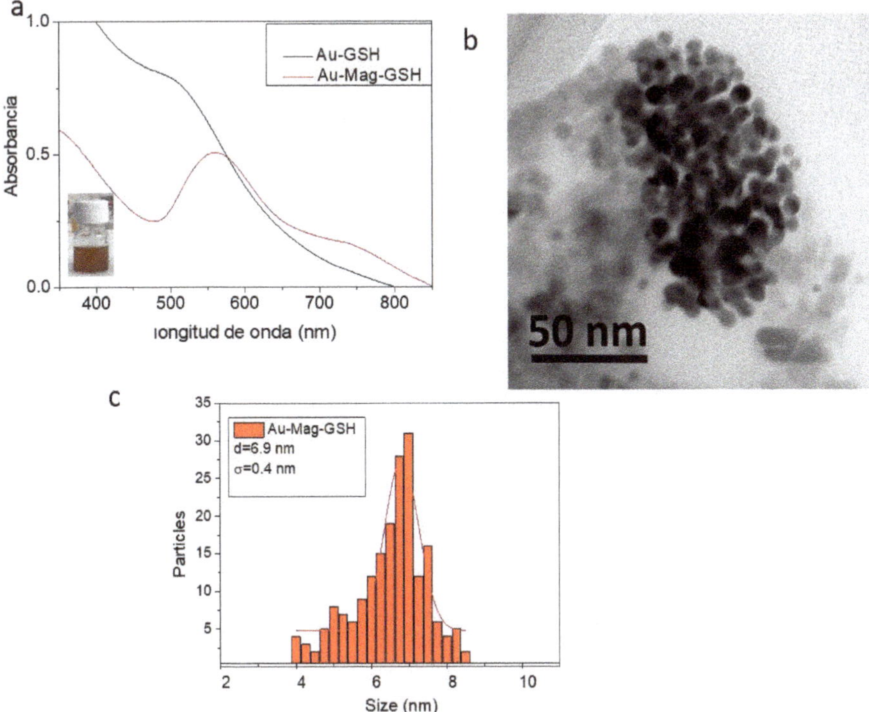

Figure 5. Ultraviolet (UV)–visible spectra for Au-GSH and Au-Mag-GSH colloidal solutions (**a**); TEM micrograph of Au-Mag-GSH (**b**) and its corresponding histogram displaying NP size distribution (**c**).

Figure 6 summarizes the main (S)TEM results on the structure and composition of the different GSH-capped NP samples. At this point, it is worth mentioning that the study via electron beam-based techniques of these materials produced a large amount of data and information. Therefore, more details on the (S)TEM characterization can be found elsewhere [45]. Figure 6a,b show respectively HRTEM and HR-HAADF micrographs, for Au-GSH NPs taken from the same region. The comparison of these modes, for the same NP, not only allows visualization of the crystalline structure, but also indicates the chemical homogeneity of such structure. Also, it is possible to visualize (Figure 6a) a thin amorphous layer (GSH) covering the NP core. Images such as the ones in Figure 6a,b allow measuring inter-planar distances of 2.17 and 2.38 Å, which respectively agree well with the spacing between {002} and {111} families of atomic planes for cubic Au. These values of inter-planar atomic distances have been obtained using the fast Fourier transform, FFT, (Figure 6a, inset) of a large enough region of the NP that show a crystalline structure with the same orientation. Using a generic diffraction software [51] to compare the experimental FFT with a simulated diffraction diagram, the zone axis can be identified and the diffraction peaks indexed. Note that FFT data are equivalent to electron diffraction diagrams when in Fourier condition, as is the case for the microscope optics. Similar analyses (not shown here) also revealed that crystalline structures were formed, in all samples and for each type of NP, corresponding to the expected elements and compounds.

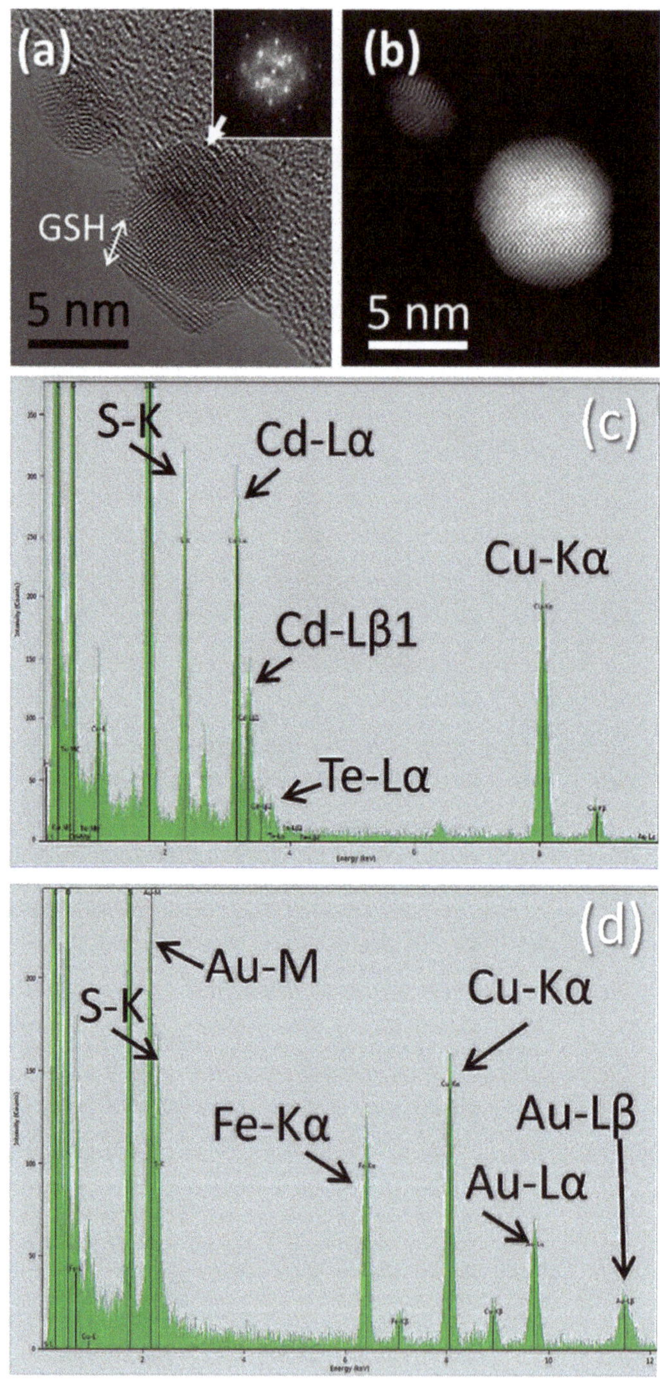

Figure 6. High-resolution TEM (HRTEM) (**a**) and HR-HAADF (**b**) images of NPs in sample Au-GSH. Punctual energy-dispersive X-ray (EDX) spectrum from QD-GSH (**c**) and from Au-Mag-GSH (**d**) NPs.

Compositional sensitive STEM techniques, such as EDX, can complement these results, allowing analyze the uniformity of the GSH layer. In this sense, we have utilized the sulfur (S) peak associated to the K-level transition in order to detect GSH. Figure 6c presents an EDX-punctual spectrum for a NP of the same sample. Such a spectrum is taken from the area defined by the scanning STEM probe, which is estimated to correspond to 5 nm × 5 nm. Thus, the spectrum corresponds, approximately, to a region spanning, the size of one NP. Note, that other expected peaks, such as the ones associated to carbon and oxygen, also presented in the spectrum, are mostly due to both the carbon support and organic residual particles in the electron microscopy grid. In this sense, the copper signal, visible in all spectra is produced by the copper in such grid. In a previous work [45], EDX-maps for the different elements were presented, so a more comprehensive view of the STEM study for these samples can be obtained from this communication. Note that EDX results, such as those shown in Figure 5f from the study presented in reference 45, only indicate the iron-containing composition of the NP. The affirmation that the NP contains iron oxide would need further evidence. On the other hand, Figure 6d presents the EDX punctual spectrum of isolated NPs from a sample of Au-Mag-GSH. All the main studied elements were found, revealing a hybrid structure in a large amount of NPs.

Base on all these results, it can be affirmed that the synthetic routes utilized to prepare the different NPs lead to the formation of small NPs, with homogeneous size distribution, a crystalline structure and a homogeneous composition, both for the core and for the polymer GSH covering layer. The in situ functionalization with GSH probably contributes to this morphological feature. These NPs show suitable characteristics for different biomedical applications, depending on the core composition. However, the use of NPs as potential tools in biomedicine, requires potential capability to be bioconjugated with biological species. The effectiveness of the functionalization process is, in any case, fundamental for subsequent bioconjugations.

3.2. Colloidal Stability and Superficial Properties

3.2.1. Hydrodynamic Size

To evaluate the colloidal stability of the different NPs, DLS experiments have been carried out. Figure 7 shows the DLS diameter distribution (represented in number) obtained for the different GSH NPs. Each distribution has been fitted to a log-normal function, also shown in the graphics. Colloidal average diameters obtained from these fittings, as well as the PDI (polydispersity index) are shown in Table 1.

All NPs show a relatively narrow size distribution with colloidal average sizes smaller than 50 nm, and with a relatively low polydispersity index (PDI in Table 1). As expected, all NPs show DLS hydrodynamic sizes bigger than those obtained from TEM, since this last technique measures mostly the crystalline volume of the particles (as explained before, in most particles, the polymer layer is veiled by the carbon support in the images). This effect is not only due to the GSH layer, but also to the presence of extra hydrate layers in aqueous medium. However, the similar sizes found from both TEM and DLS, in the case of Au-GSH, reveals the high colloidal stability of these NPs. Also, in the case of QD-GSH, Mag-GSH and Mag-Au-GSH, the formation of clusters and particle associations in solution could influence the DSL measurements. In any case, the four types of synthetized GSH-NPs show DLS diameters in the appropriate range for biomedical applications. DLS measurements over time show no important size differences after 48 h post-redispersion for any type of NP, being the size increase lower than 1.2% in all cases.

3.2.2. Z-Potential and Surface Characteristics

To find out the electric charge surrounding NP surface we calculated the Z-potential. Figure 8a displays Z-potential versus pH for different NP colloidal solutions.

This figure indicates that all NP systems in this work show suitable features for biomedical applications, since the values in the Z-potential graphic reveals a relatively high colloidal stability in the studied pH range, and particularly at physiological pH.

These measurements shed information about the sign of the charge on the NP surface (as can be seen, all GSH samples present a negative Z-potential, which indicate an excess electric negative charge at the surface), that is also relevant for potential crosslinking applications. In general, the GSH molecule has thiol, amine and two carboxylic terminal groups available for linking to the NP surface. In the case of CdTe, due to the affinity between cadmium and sulfur, the GSH probably links the NP by its thiol group, keeping free the carboxylic terminal groups, responsible for the negative net charge. In order to study the linking mechanism for gold and iron oxide surfaces, we have prepared iron oxide and gold NPs, using a synthetic method similar to that used for Au-GSH and Mag-GSH NPs, but replacing the GSH with cysteamine (CYS), a molecule containing only two terminal groups: thiol and a primary amine. Table 2 shows Z-potential values at pH 7.2 for Au-GSH and Mag-GSH NPs, as well as for Mag-CYS and Au-CYS NPs. In a previous work [42], we demonstrated that the CYS molecule, links to the gold NP surface by the thiol group, due to the high affinity of thiols for the gold core, thus these CYS-capped NPs show positive Z-potential values due to the free amine terminal groups. However, in this work, we have obtained negative Z-potential for a Mag-CYS colloidal solution. This result points out that the affinity of amine terminal groups for the iron oxide core is higher than for the thiol group. Consequently, the GSH amine terminal group could have higher affinity for the iron oxide core than the GSH thiol group. We deduce that there are two different mechanisms for the linking between CYS and gold versus iron oxide NPs (Figure 8b,c). Interestingly, Mag-GSH NPs show the highest negative Z-potential values in all pH range (Figure 8a). This maximum negative charge could be better explained if we propose a different linking mechanism for this NP, in which GSH binds the NP surface by the amine group, with the thiol and the two carboxylic groups remaining free and, thus, contributing to the negative superficial charge. These results, which indicate that the iron oxide surface shows a higher affinity for amine than for thiol groups, can explain the higher negative Z-potential for the Mag-GSH NPs, due to the GSH linking to the iron oxide surface by its amine group instead of by the thiol group, contrary to what happens for gold NPs (Figure 8d,e).

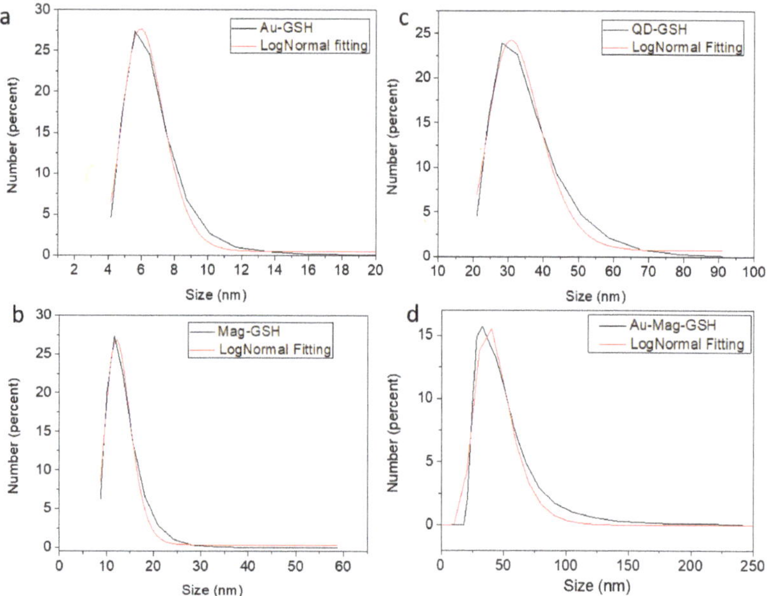

Figure 7. DLS size distributions of Au-GSH (**a**), Mag-GSH (**b**), CdTe-GSH1 (**c**) and Au-Mag-GSH (**d**) in PBS (phosphate-buffered saline) colloidal solutions.

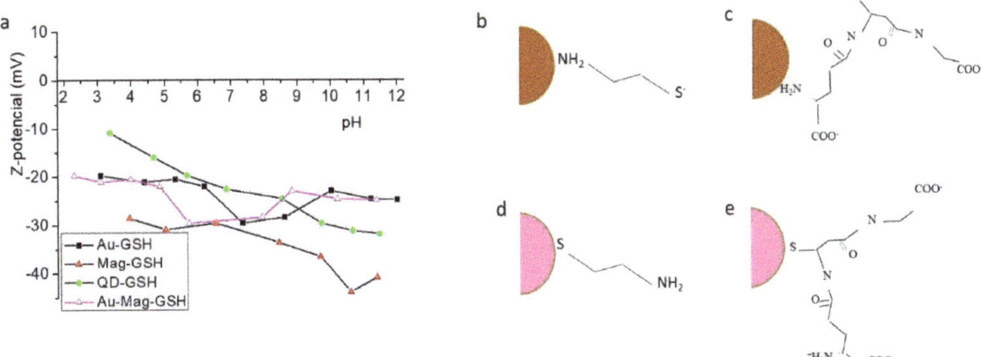

Figure 8. Z-potential versus pH for the different GSH capped NPs (**a**). Diagram of the cysteamine (CYS)-capped (**b**) and GSH-capped (**c**) magnetic-iron-oxide NPs. Diagram of the CYS-capped (**d**) and GSH-capped (**e**) gold NPs.

Table 2. Z-potential values obtained for gold and magnetic-iron-oxide NPs capped with CYS and with GSH at 25 °C and in PBS solution.

Sample	Z-Potential (mV)
Au-GSH	−32.8
Mag-GSH	−33.5
Au-CYS	+88.0
Mag-CYS	−27.5

In order to deepen knowledge of the differences in linking mechanisms between gold and iron oxide surfaces, we performed FTIR spectroscopy. Figure 9 shows the FTIR spectra for the free GSH ligand, as well as for Au-GSH, Mag-GSH and Au-Mag-GSH. The GSH spectrum present bands characteristics of amino (3350 cm^{-1}), carboxylic (1600 cm^{-1}) and thiol (2530 cm^{-1}) groups. The peak at 2530 cm^{-1} that corresponds to the S–H stretching vibration of GSH clearly disappeared in the Au-GSH and Au-Mag-GSH spectra, indicating that GSH anchors on the surface of these NPs through Au-S bonding. Conversely, a band at this position can be appreciated for Mag-GSH. The presence of the S–H stretching vibration band in this case, indicates a different GSH anchorage for this type of NP [52].

These FTIR spectra not only show a different GSH linking mechanism for gold and iron oxide NPs, but also suggests that gold is covering the iron oxide core in the case of hybrid iron oxide/gold NPs, as GSH linking occurs as it does for gold, with disappearance of the S–H stretching vibration of GSH.

These differences in the linking mechanism could also introduce differences in the shell thickness, as we have previously commented. A schematic representation of this linking mechanism is shown in Figure 8b through Figure 8e.

3.2.3. Temperature Effect

NPs used for biological applications require appropriate colloidal stability at human body temperature, ranging from 37 to 40 °C, as well as at different saline concentrations. We studied the stability of the NP dispersions under different environmental conditions (temperature and salinity). DLS experiments were also performed at 37 and 40 °C; and also subjecting NPs to an environment of a higher ionic strength, after dispersion of NPs in a 0.18 M solution of NaCl. Table 3 shows NP diameters obtained by DLS for temperatures of 25, 37 and 40 °C for all NPs in PBS dispersions, as well as at 25 °C in a 0.18 M NaCl dispersion. Results reveal that the only NPs that present a colloidal stability dependence with temperature is Au-GSH, which has an extremely low hydrodynamic size at 25 °C.

The rest of the NPs do not show a strong variation in the dynamic size when temperature increases. However, when the ionic strength changes, a significant increase in the dynamic diameter is observed for all GSH-NPs. In all cases NPs remain in the appropriate range for biomedical applications.

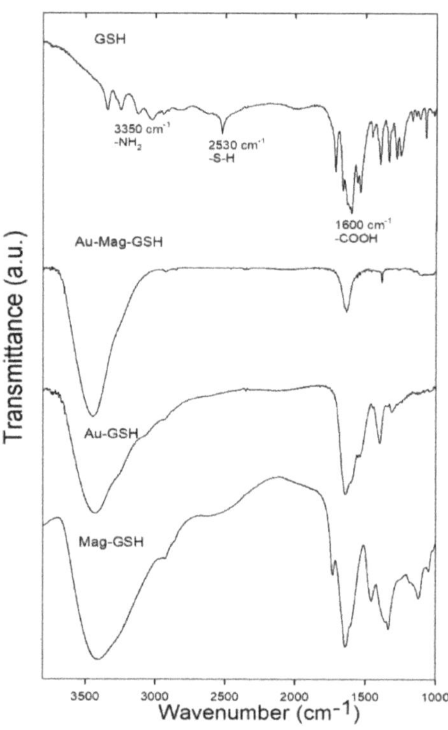

Figure 9. Fourier transform infrared (FTIR) spectra for free GSH ligand as well as for Au-Mag-GSH, Au-GSH and Mag-GSH NPs.

Table 3. NPs and their corresponding dynamic light scattering (DLS) hydrodynamic sizes at different temperatures.

Sample	Temperature (°C)	Dispersion Medium	DLS Average Size (nm)
Au-GSH	25	PBS	6.2 ± 1.2
	37	PBS	24.9 ± 2.2
	40	PBS	13.5 ± 1.5
	25	0.18 NaCl	28.2 ± 2.5
QD-GSH	25	PBS	32.6 ± 1.3
	37	PBS	39.3 ± 2.7
	40	PBS	36.3 ± 2.7
	25	0.18 NaCl	58.1 ± 2.5
Mag-GSH	25	PBS	12.7 ± 1.2
	37	PBS	11.3 ± 1.6
	40	PBS	12.9 ± 1.7
	25	0.18 NaCl	29.2 ± 2.6

All these colloidal characterization results confirm that the NP core is capped by GSH molecules that properly functionalize the NP, allowing for subsequent biofunctionalization and thus corroborating previously presented (S)TEM results. Functionalized NPs show colloidal stability at physiological pH and temperature values. Indeed, GSH-functionalized NPs show a higher colloidal stability at physiological pH and temperatures than citrate-capped gold NPs [42].

3.3. Cell Viability and Cytotoxicity

As previously commented, one key property of NPs to consider for medical applications is cytotoxicity, and thus we performed MTT cell viability assays for Jurkat cells incubated with the synthesized NPs. In Figure 10, viability for cells cultured in the presence of NPs at 1.5 µg/mL (a) or 15 µg/mL (10.b) is shown. These graphics show mean values and standard errors for the percentage of viable cells when cultured in the presence of Au-GSH NPs (A), Mag-GSH (B), Au-Mag-GSH NPs (C) or QD-GSH NPs (D) considering 100% the viability of cells cultured in the absence of NPs (NP($-$)) as a control. Each experiment was performed independently in triplicate.

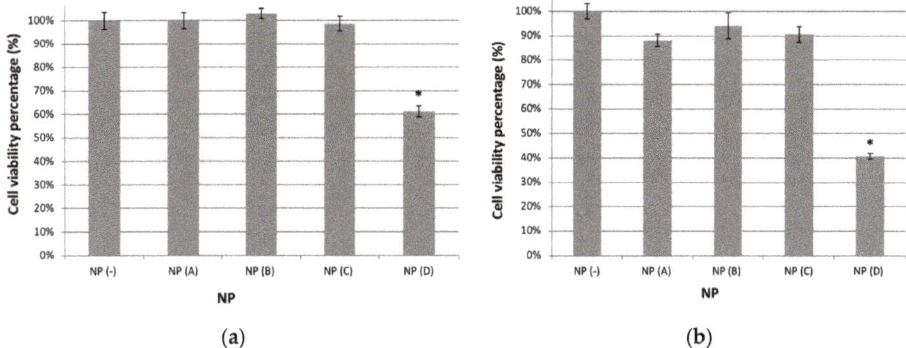

(a)   (b)

Figure 10. MTT cell viability assay for Jurkat tumor cells incubated with the indicated type of NP at 1.5 µg/mL (**a**) or 15 µg/mL (**b**). Percentage of viable cells as compared to cells cultured in the absence of NP (considered as 100% viability) is shown. Mean and standard error are shown for each sample. A representative experiment out of three is shown. Statistically significant differences ($p < 0.05$) are marked with a star (*).

As shown in Figure 10a, all NPs, except for QD-GSH, are not toxic at concentrations usually utilized in the literature (1.5 µg/mL) with viabilities similar to those observed in control cultures in the absence of NPs, indicating the extremely low cytotoxicity of these NPs. Even at concentrations 10 times higher (15 µg/mL), these NPs have only a marginal effect on cell viability (90% viability) (Figure 10b). At this concentration (15 µg/mL), cell viability is 88% for Au-GSH, 94% for Mag-GSH and between 88% and 94% for the hybrid Au-Mag-GSH, indicating that their toxicity profile is very favorable for in vivo applications, with Mag-GSH being the NP with the best toxicity profile (Figure 10b).

On the other hand QD-GSH NPs, have an effect on cell viability with percentages of 61% and 41% for NP concentrations of 1.5 and 15 µg/mL, respectively.

To better ascertaining the toxicity profile, we performed a dose respond experiment with NP concentrations up to 960 µg/mL. As shown in Figure 11, Mag-GSH and Au-Mag-GSH have a very favorable toxicity profile, while QD-GSH are more toxic, with Au-GSH having an intermediate toxicity.

The lower viability of cells that were cultured in the presence of QD-GSH is due to the inherent toxicity of the CdTe material, that prevails even when capped with GSH. Therefore, until an alternative capping, able to seal off the CdTe material, is engineered, this type of NP should be utilized as a biomarker only for in vitro experiments, but not for in vivo applications. Interestingly, hybrid Au-Mag-GSH NPs show better toxicity profile than Au-GSH NPs, which can be due to the small size of this gold NPs [53].

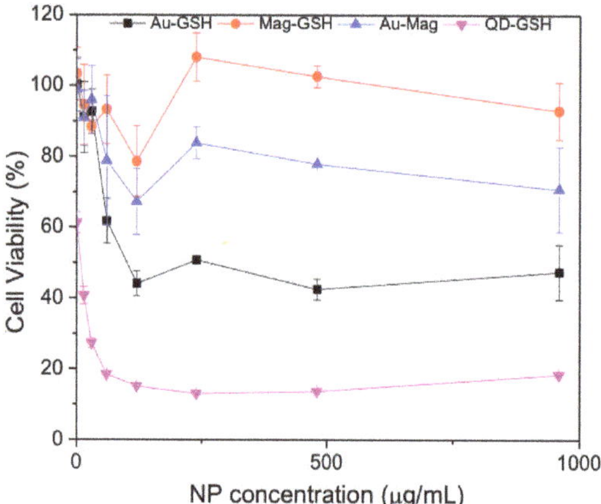

Figure 11. MTT cell viability assay for Jurkat tumor cells incubated with the indicated NP at 1.5; 15; 30; 60; 120; 240; 480 and 960 µg/mL. Percentage of viable cells, as compared to cells cultured in the absence of NP (considered as 100% viability), is shown. Mean values and standard error for three experiments are shown. Viability of cells cultured in the presence of 10% dimethyl sulfoxide (DMSO) is 8.79%.

4. Discussion

In this work we used glutathione as a capping agent to functionalize gold/iron oxide NPs for biomedical applications, prepared by a novel and simple route. We also compared the results obtained for this GSH-gold/magnetic iron oxide NPs with those obtained from other GSH capped NPs. First, we confirmed that the functionalization preserves the structure and physical properties that confer each of these NPs with their usefulness for biomedical applications. The work was focused on determining GSH-capping effectiveness, linkage characteristics between GSH and NP surfaces and the influence of GSH capping on the physical properties and cytotoxicity of NPs.

We described a route to obtain Au-Mag-GSH, a hybrid NP encompassing the biomedical performances of both, magnetic iron oxide and gold. Although different authors have proposed different routes to prepare hydrophilic synthesizes [54,55], we have designed a simple route to obtain, in two simple steps, hydrophilic functionalized magnetic iron oxide-gold NPs. The formation of hybrid NPs in which both phases, iron oxide and gold, are present in the same NP was confirmed by results obtained from TEM and UV–visible absorption. This last technique allowed us to identify the characteristic gold LRSP band, but with a clear red shift when compared to the Au-GSH LSPR band, which according to Bhatia et al. [56] can be attributed to the interfacial charge transfer between gold and magnetic-iron-oxide. TEM results revealed formation of spherical and homogeneous GSH-NPs of 7 nm average size with a core formed by gold and either iron or an iron compound. EDX results also revealed the existence of both, iron and gold in the same NP. Considering that our first synthesis step was the formation of an iron oxide core, it is reasonable to suggest formation of a gold NP shell surrounding the magnetic-iron-oxide. In any case, since no region of pure gold was found via EDX maps for iron-gold NPs (not shown here), but areas with very close diameters for the spectral signals of Au and Fe, if the gold was forming a shell, it would be assumed to be thin, although at least a few monolayers-thick since the gold structure was detected by the beam. Thus, the nature of this NP, whether it is formed by an iron compound or magnetite, or even by an alloy of gold and iron could not have been concluded by the applied (S)TEM techniques. In any case, both the red-shift in the LSPR absorption band of these NPs and the sequence followed during the synthesis method employed in the study (iron oxide nucleation followed by gold addition) seemed to indicate that a core–shell structure should have been formed. Moreover the fact that the

FTIR results showed a linkage mechanism for GSH to gold/iron oxide akin to that seen for gold NPs, further strengthened the hypothesis that a core iron oxide, that confers magnetic properties, is surrounded by gold that justifies the linking mechanism.

We have compared results obtained for these hybrid NPs with those obtained for pure gold, magnetic-iron-oxide and CdTe, also capped with GSH. In the case of QDs (CdTe) and according to recent work [57], extremely small NPs immersed in GSH capping appeared. This was probably due to the excess of GSH added during the synthesis and to the high affinity of the GSH thiol group for the CdTe surface. However, the process allowed for size control, providing a tunable photoluminescence [14].

GSH capping contributes to the formation of spherical and homogeneous NPs in the case of iron oxide and gold NPs. Magnetic NPs (Mag-GSH) of 6 nm average size are obtained, while smaller, and less aggregated NPs are formed in the case of pure gold NPs (Au-GSH). For the latter, the presence of gold was confirmed by UV–visible absorption and (S)TEM. These NPs show a PL emission in the blue range that, in agreement with our previous studies, can be explained considering the small size of gold cores and the ligand-metal charge transfer transition due to the thiol group of the GSH capping. This characteristic could be an interesting feature, making these NPs suitable not only for diagnostics, therapy or further bioconjugations but also for direct biomarker applications [42].

We used the HRTEM technique to prove the homogeneous composition, spherical shape and size (under 10 nm) of NPs, and also to visualize the GSH layer. In order to prove that this layer was composed of GSH, the sulfur present in the glutathione was followed by energy-dispersive X-ray spectroscopy. In this way, the presence of such element, detected at the NP surface, confirms that the amorphous layer at the NP surface is the GSH capping. Additionally, results obtained from Z-potential measurements allowed us to deepen the GSH linkage mechanism. The GSH molecule contains different terminal functional groups: thiol, amine and two carboxylic groups. Knowing the group that molecules use to link to the NP surface and consequently the remaining free moieties, is crucial for the successful design of subsequent crosslinking processes. With this goal, we used a simple molecule, cysteamine, to functionalize gold and iron oxide NPs. Our results, which yielded negative Z-potential values for gold NPs functionalized with cysteamine, confirmed the high affinity between thiol and gold surface. On the other hand, the positive Z-potential obtained for cysteamine capped iron oxide, revealed a higher affinity between the iron oxide surface and amine than between this iron oxide and the thiol group. This suggests that GSH does not use the thiol group to link to iron oxide GSH-NPs as it does for Au containing NPs. Whereas Au-GSH and Au-Mag-GSH lose the characteristic S–H stretching vibration peak, this band was observed for Mag-GSH, indicating that in this last case the thiol group was not used to link the NP surface. This is in agreement with recent works that propose a conjugation of GSH capped iron oxide NPs by the free thiol group [58]. Several authors propose the iron oxide functionalization by carboxylic groups [59]. However, the relatively high negative Z-potential values obtained for Mag-GSH in all the studied pH range together with our cysteamine-iron oxide results led us to propose the linkage by the amine group for Mag-GSH. NPs using gold (Au-GSH and Au-Mag-GSH) were clearly linked by the thiol group.

Concerning colloidal stability, DLS experiments reveal hydrodynamic sizes bigger than those obtained by HRTEM (NP core plus GSH layer). In the case of Au-GSH, this difference could be explained by the presence of extra hydrate layers in aqueous medium. However, for all other NPs differences between DLS and TEM diameters were significantly bigger. For Mag-GSH and Au-Mag-GSH the magnetic character of NPs originated in magnetic interactions between particles, especially for Mag-GSH, in which the observed aggregation was higher. It is interesting to remark that DLS average sizes were inversely proportional to the estimated GSH layer thickness. A wider GSH layer provided less aggregation in aqueous colloidal solution. In the case of QD-GSH, as we observed by TEM, NPs were not well dispersed, and the cores were immersed in the GSH capping. This could be the reason for the relatively high DLS size (32.6 nm) for non-magnetic NPs with a core size under 3 nm.

Nevertheless, at the studied conditions, all NPs showed DLS diameters smaller than 50 nm, and colloidal stability in an appropriate range, although further work is in process to check colloidal stability under different conditions to perform additional steps of multimodal crosslinking.

Finally, we showed how GSH functionalization was also able to confer biocompatibility to most of the studied NPs. Au-GSH, Mag-GSH and Au-Mag-GSH show cell viability percentages similar to controls at concentrations up to 30µg/mL; with Au-GSH and Mag-GSH, maintaining the low toxicity profile up to concentrations of 960 µg/mL. Interestingly enough, hybrid Au-Mag-GSH NPs showed a better toxicity profile than Au-GSH NPs.

However, this extremely low cytotoxicity was not observed for QD-GSH particles, in which the GSH capping cannot passivate cadmium inherent toxicity. Different authors propose different mechanisms for CdTe toxicity, such as photo oxidation and the consequent formation of degradation products that generate superoxide anions, which might lead to rust and corrosion of NP surfaces [60]. Concomitantly, in the case of QDs made of cadmium, cytotoxicity is a consequence of the release of free cadmium ions (Cd^{2+}) that are highly toxic. However, QD-GSH NPs can still be used as biomarkers for in vitro experiments.

5. Conclusions

We have designed a new and simple synthetic method to prepare hydrophilic metallic NPs capped with GSH (including, gold/iron oxide hybrid NPs). We have studied the structural, morphological and colloidal properties, as well as the cytotoxicity, for these new hybrid NPs and for iron-containing NPs, gold NPs and fluorescent NPs, previously prepared for us and also covered with GSH. We have obtained, in all cases, spherical highly crystalline and homogeneous NPs with sizes under 10 nm. The combination of TEM and STEM techniques also allowed visualization of the amorphous GSH layer, confirming an adequate functionalization. All GSH-capped NPs showed colloidal stability, after 48-h post-redispersion, especially at the physiological pH range. This colloidal stability did not significantly decrease by increasing temperature up to 40 °C. Colloidal characterization allowed us to determine the moiety in GSH that binds to the NP surface. We have deduced that GSH links iron-containing NP surface by its amine group and not by the thiol groups, as is the case for the rest of studied NPs. This knowledge is crucial to determine the free terminal groups in the GSH capping, which is pivotal for the design of subsequent crosslinking strategies. The developed NPs show high biocompatibility with low cytotoxicity even at high concentrations, with the exception of QD-GSH, that could be only suitable for in vitro biomedical applications.

Author Contributions: Conceptualization: R.L., O.B.-M. and F.G.-C.; Nanoparticle Synthesis: C.F.-P., J.B.-L. and R.L.; Magnetic experiments: J.B.-L.; Optical measurements: R.L. and R.F.-C.; DLS experiments: E.F. and O.B.-M.; FTIR experiments: M.P.Y. Electron microscopy: J.M.M., A.M.B., F.M.M. and A.J.S., Cell viability and cytotoxicity: C.F.-P., J.P.M.-M., R.F.-C. and F.G.-C.; Data curation: All author; Writing—original draft preparation: R.L., J.M.M., C.F.-P. and F.G.-C.; writing—review and editing: All authors; Project administration: R.L.; Funding acquisition: F.G.-C. and O.B.-M. All authors have read and agreed to the published version of the manuscript.

Funding: This work was supported by Iniciativa Territorial Integrada, Junta de Andalucia (PI-0030-2017) and by the Ministry of Education and Science, Instituto Salud Carlos III (PI16-00784) fondo covid-19 COV20/00173) for F.G-C and by the Spanish Ministry of Economy and Competitiveness Program "Plan I+D+i", subprogram "Retos" (MAT2015-67354-R) for O.B-M.

Institutional Review Board Statement: Not applicable.

Informed Consent Statement: Not applicable.

Data Availability Statement: Not applicable.

Acknowledgments: The authors would like to thank the Core Biomedical Research (sc-IBM) Facility and the Core Science and Technology Research Facility (sc-ICyT) of University Cadiz for the use of core infrastructure, as well as to the microscopy facilities of CITIUS-Universidad de Sevilla (Spain).

Conflicts of Interest: The authors declare no conflict of interest.

References

1. Bakhtiari-Asl, F.; Divband, B.; Mesbahi, A.; Gharehaghaji, N. Bimodal magnetic resonance imaging-computed tomography nanoprobes: A Review. *Nanomed. J.* **2020**, *7*, 1–12. [CrossRef]
2. Gharpure, S.; Akash, A.; Ankamwar, B. A Review on Antimicrobial Properties of Metal Nanoparticles. *J. Nanosci. Nanotechnol.* **2020**, *20*, 3303–3339. [CrossRef] [PubMed]
3. Giljohann, D.A.; Mirkin, C.A. Drivers of biodiagnostic development. *Nature* **2009**, *462*, 461–464. [CrossRef] [PubMed]
4. Wang, J.; Li, P.; Yu, Y.; Fu, Y.; Jiang, H.; Lu, M.; Sun, Z.; Jiang, S.; Lu, L.; Wu, M.X. Pulmonary surfactant-biomimetic nanoparticles potentiate heterosubtypic influenza immunity. *Science* **2020**, *367*, 869. [CrossRef]
5. Lu, L.; Qi, S.; Chen, Y.; Luo, H.; Huang, S.; Yu, X.; Luo, Q.; Zhang, Z. Targeted immunomodulation of inflammatory monocytes across the blood-brain barrier by curcumin-loaded nanoparticles delays the progression of experimental autoimmune encephalomyelitis. *Biomaterials* **2020**, *245*, 119987. [CrossRef]
6. Mirzaei, M.; Akbari, M.E.; Mohagheghi, M.A.; Ziaee, S.A.M.; Mohseni, M. A novel biocompatible nanoprobe based on lipoproteins for breast cancer cell imaging. *Nanomed. J.* **2020**, *7*, 73–79. [CrossRef]
7. Shi, S.; Vissapragada, R.; Abi Jaoude, J.; Huang, C.; Mittal, A.; Liu, E.; Zhong, J.; Kumar, V. Evolving role of biomaterials in diagnostic and therapeutic radiation oncology. *Bioact. Mater.* **2020**, *5*, 233–240. [CrossRef]
8. Van Elk, M.; Murphy, B.P.; Eufrasio-da-Silva, T.; O'Reilly, D.P.; Vermonden, T.; Hennink, W.E.; Duffy, G.P.; Ruiz-Hernandez, E. Nanomedicines for advanced cancer treatments: Transitioning towards responsive systems. *Int. J. Pharm.* **2016**, *515*, 132–164. [CrossRef]
9. Chen, D.; Zhao, C.; Ye, J.; Li, Q.; Liu, X.; Su, M.; Jiang, H.; Amatore, C.; Selke, M.; Wang, X. In Situ Biosynthesis of Fluorescent Platinum Nanoclusters: Toward Self-Bioimaging-Guided Cancer Theranostics. *Acs Appl. Mater. Interfaces* **2015**, *7*, 18163–18169. [CrossRef] [PubMed]
10. Rehman, F.U.; Du, T.; Shaikh, S.; Jiang, X.; Chen, Y.; Li, X.; Yi, H.; Hui, J.; Chen, B.; Selke, M.; et al. Nano in nano: Biosynthesized gold and iron nanoclusters cargo neoplastic exosomes for cancer status biomarking. *Nanomed. Nanotechnol. Biol. Med.* **2018**, *14*, 2619–2631. [CrossRef] [PubMed]
11. Meng, W.; He, C.; Hao, Y.; Wang, L.; Li, L.; Zhu, G. Prospects and challenges of extracellular vesicle-based drug delivery system: Considering cell source. *Drug Deliv.* **2020**, *27*, 585–598. [CrossRef]
12. He, Z.; Jiang, R.; Long, W.; Huang, H.; Liu, M.; Feng, Y.; Zhou, N.; Ouyang, H.; Zhang, X.; Wei, Y. Red aggregation-induced emission luminogen and Gd3+ codoped mesoporous silica nanoparticles as dual-mode probes for fluorescent and magnetic resonance imaging. *J. Colloid Interface Sci.* **2020**, *567*, 136–144. [CrossRef]
13. Larson, D.R.; Zipfel, W.R.; Williams, R.M.; Clark, S.W.; Bruchez, M.P.; Wise, F.W.; Webb, W.W. Water-soluble quantum dots for multiphoton fluorescence imaging in vivo. *Science* **2003**, *300*, 1434–1436. [CrossRef]
14. Beato-Lopez, J.J.; Espinazo, M.L.; Fernandez-Ponce, C.; Blanco, E.; Ramirez-del-Solar, M.; Dominguez, M.; Garcia-Cozar, E.; Litran, R. CdTe quantum dots linked to Glutathione as a bridge for protein crosslinking. *J. Lumin.* **2017**, *187*, 193–200. [CrossRef]
15. Qu, Z.; Liu, L.; Sun, T.; Hou, J.; Sun, Y.; Yu, M.; Diao, Y.; Lu, S.; Zhao, W.; Wang, L. Synthesis of bifunctional carbon quantum dots for bioimaging and anti-inflammation. *Nanotechnology* **2020**, *31*, 175102. [CrossRef] [PubMed]
16. Kayili, H.M.; Salih, B. Fast and efficient proteolysis by reusable pepsin-encapsulated magnetic sol-gel material for mass spectrometry-based proteomics applications. *Talanta* **2016**, *155*, 78–86. [CrossRef]
17. Jose, J.; Kumar, R.; Harilal, S.; Mathew, G.E.; Parambi, D.G.T.; Prabhu, A.; Uddin, M.S.; Aleya, L.; Kim, H.; Mathew, B. Magnetic nanoparticles for hyperthermia in cancer treatment: An emerging tool. *Environ. Sci. Pollut. Res.* **2019**. [CrossRef]
18. Sharma, S.K.; Shrivastava, N.; Rossi, F.; Le Duc, T.; Nguyen Thi Kim, T. Nanoparticles-based magnetic and photo induced hyperthermia for cancer treatment. *Nano Today* **2019**, *29*. [CrossRef]
19. Karimzadeh, I.; Dizaji, H.R.; Aghazadeh, M. Development of a facile and effective electrochemical strategy for preparation of iron oxides (Fe3O4 and gamma-Fe2O3) nanoparticles from aqueous and ethanol mediums and in situ PVC coating of Fe3O4 superparamagnetic nanoparticles for biomedical applications. *J. Magn. Magn. Mater.* **2016**, *416*, 81–88. [CrossRef]
20. Bomati-Miguel, O.; Lahoz, R.; Lennikov, V.; Naghilou, A.; Subotic, A.; Angel Rodriguez, M.; Rentenberger, C.; Kautek, W.; IEEE. Liquid-Assisted Pulsed Laser Ablation: A Novel Route to Produce Multifunctional Contrast Agents for Multimodal Imaging Diagnosis. In Proceedings of the European Conference on Lasers and Electro-Optics and European Quantum Electronic Conference, Munich, Germany, 23–27 June 2017.
21. Du, Y.; Liu, X.; Liang, Q.; Liang, X.-J.; Tian, J. Optimization and Design of Magnetic Ferrite Nanoparticles with Uniform Tumor Distribution for Highly Sensitive MRI/MPI Performance and Improved Magnetic Hyperthermia Therapy. *Nano Lett.* **2019**, *19*, 3618–3626. [CrossRef]
22. Soares, P.I.P.; Laia, C.A.T.; Carvalho, A.; Pereira, L.C.J.; Coutinho, J.T.; Ferreira, I.M.M.; Novo, C.M.M.; Borges, J.P. Iron oxide nanoparticles stabilized with a bilayer of oleic acid for magnetic hyperthermia and MRI applications. *Appl. Surf. Sci.* **2016**, *383*, 240–247. [CrossRef]

23. Scialabba, C.; Puleio, R.; Peddis, D.; Varvaro, G.; Calandra, P.; Cassata, G.; Cicero, L.; Licciardi, M.; Giammona, G. Folate targeted coated SPIONs as efficient tool for MRI. *Nano Res.* **2017**, *10*, 3212–3227. [CrossRef]

24. Neuwelt, E.A.; Varallyay, C.G.; Manninger, S.; Solymosi, D.; Haluska, M.; Hunt, M.A.; Nesbit, G.; Stevens, A.; Jerosch-Herold, M.; Jacobs, P.M.; et al. The potential of ferumoxytol nanoparticle magnetic resonance imaging, perfusion, and angiograpgy in central nervous system malignancy: A pilot study. *Neurosurgery* **2007**, *60*, 601–611. [CrossRef]

25. Bai, X.; Wang, Y.; Song, Z.; Feng, Y.; Chen, Y.; Zhang, D.; Feng, L. The Basic Properties of Gold Nanoparticles and their Applications in Tumor Diagnosis and Treatment. *Int. J. Mol. Sci.* **2020**, *21*, 2480. [CrossRef]

26. Sugumaran, S.; Jamlos, M.F.; Ahmad, M.N.; Bellan, C.S.; Schreurs, D. Nanostructured materials with plasmonic nanobiosensors for early cancer detection: A past and future prospect. *Biosens. Bioelectron.* **2018**, *100*, 361–373. [CrossRef] [PubMed]

27. Ru, F.; Du, P.; Lu, X. Efficient ratiometric fluorescence probe utilizing silicon particles/gold nanoclusters nanohybrid for "on-off-on" bifunctional detection and cellular imaging of mercury (II) ions and cysteine. *Anal. Chim. Acta* **2020**, *1105*, 139–146. [CrossRef] [PubMed]

28. Patel, P.C.; Giljohann, D.A.; Daniel, W.L.; Zheng, D.; Prigodich, A.E.; Mirkin, C.A. Scavenger Receptors Mediate Cellular Uptake of Polyvalent Oligonucleotide-Functionalized Gold Nanoparticles. *Bioconjug. Chem.* **2010**, *21*, 2250–2256. [CrossRef]

29. Rawson, S.D.; Maksimcuka, J.; Withers, P.J.; Cartmell, S.H. X-ray computed tomography in life sciences. *BMC Biol.* **2020**, *18*, 21. [CrossRef]

30. Chen, M.; Tang, S.; Guo, Z.; Wang, X.; Mo, S.; Huang, X.; Liu, G.; Zheng, N. Core-Shell Pd@Au Nanoplates as Theranostic Agents for In-Vivo Photoacoustic Imaging, CT Imaging, and Photothermal Therapy. *Adv. Mater.* **2014**, *26*, 8210–8216. [CrossRef] [PubMed]

31. Lewinski, N.; Colvin, V.; Drezek, R. Cytotoxicity of nanoparticles. *Small* **2008**, *4*, 26–49. [CrossRef]

32. Hu, M.; Petrova, H.; Sekkinen, A.R.; Chen, J.; McLellan, J.M.; Li, Z.-Y.; Marquez, M.; Li, X.; Xia, Y.; Hartland, G.V. Optical properties of Au-Ag nanoboxes studied by single nanoparticle spectroscopy. *J. Phys. Chem. B* **2006**, *110*, 19923–19928. [CrossRef] [PubMed]

33. Piergies, N.; Ocwieja, M.; Paluszkiewicz, C.; Kwiatek, W.M. Nanoparticle stabilizer as a determining factor of the drug/gold surface interaction: SERS and AFM-SEIRA studies. *Appl. Surf. Sci.* **2021**, *537*, 147897. [CrossRef]

34. Takemura, K.; Lee, J.; Suzuki, T.; Hara, T.; Abe, F.; Park, E.Y. Ultrasensitive detection of norovirus using a magnetofluoroimmunoassay based on synergic properties of gold/magnetic nanoparticle hybrid nanocomposites and quantum dots. *Sens. Actuators B-Chem.* **2019**, *296*, 147897. [CrossRef]

35. Carril, M.; Fernandez, I.; Rodriguez, J.; Garcia, I.; Penades, S. Gold-Coated Iron Oxide Glyconanoparticles for MRI, CT, and US Multimodal Imaging. *Part. Part. Syst. Charact.* **2014**, *31*, 81–87. [CrossRef]

36. Lv, W.; Shen, Y.; Yang, H.; Yang, R.; Cai, W.; Zhang, J.; Yuan, L.; Duan, Y.; Zhang, L. A Novel Bimodal Imaging Agent Targeting HER2 Molecule of Breast Cancer. *J. Immunol. Res.* **2018**, *2018*, 6202876. [CrossRef]

37. Albanese, A.; Tang, P.S.; Chan, W.C.W. The Effect of Nanoparticle Size, Shape, and Surface Chemistry on Biological Systems. In *Annual Review of Biomedical Engineering*; Yarmush, M.L., Ed.; University of Toronto: Toronto, ON, Canada, 2012; Volume 14, pp. 1–16.

38. Uzun, O.; Hu, Y.; Verma, A.; Chen, S.; Centrone, A.; Stellacci, F. Water-soluble amphiphilic gold nanoparticles with structured ligand shells. *Chem. Commun.* **2008**, *2*, 196–198. [CrossRef]

39. Verma, A.; Stellacci, F. Effect of Surface Properties on Nanoparticle-Cell Interactions. *Small* **2010**, *6*, 12–21. [CrossRef]

40. Verma, A.; Uzun, O.; Hu, Y.; Hu, Y.; Han, H.-S.; Watson, N.; Chen, S.; Irvine, D.J.; Stellacci, F. Surface-structure-regulated cell-membrane penetration by monolayer-protected nanoparticles (vol 7, pg 588, 2008). *Nat. Mater.* **2013**, *12*, 376. [CrossRef]

41. Beato-Lopez, J.J.; Fernandez-Ponce, C.; Blanco, E.; Barrera-Solano, C.; Ramirez-del-Solar, M.; Dominguez, M.; Garcia-Cozar, F.; Litran, R. Preparation and Characterization of Fluorescent CdS Quantum Dots used for the Direct Detection of GST Fusion Proteins Regular Paper. *Nanomater. Nanotechnol.* **2012**, *2*, 2012. [CrossRef]

42. Fernandez-Ponce, C.; Munoz-Miranda, J.P.; de los Santos, D.M.; Aguado, E.; Garcia-Cozar, F.; Litran, R. Influence of size and surface capping on photoluminescence and cytotoxicity of gold nanoparticles. *J. Nanopart. Res.* **2018**, *20*, 305. [CrossRef]

43. Simpson, C.A.; Salleng, K.J.; Cliffel, D.E.; Feldheim, D.L. In vivo toxicity, biodistribution, and clearance of glutathione-coated gold nanoparticles. *Nanomed. Nanotechnol. Biol. Med.* **2013**, *9*, 257–263. [CrossRef]

44. Chen, G.; Zhang, Y.; Peng, Z.; Huang, D.; Li, C.; Wang, Q. Glutathione-capped quantum dots for plasma membrane labeling and membrane potential imaging. *Nano Res.* **2019**, *12*, 1321–1326. [CrossRef]

45. Beltran, A.M.; Manuel, J.M.; Litran, R.; Felix, E.; Santos, A.J.; Morales, F.M.; Bomati-Miguel, O. (S)TEM structural and compositional nanoanalyses of chemically synthesized glutathione-shelled nanoparticles. *Appl. Nanosci.* **2020**, *10*, 2295–2301. [CrossRef]

46. Crespo, P.; Litran, R.; Rojas, T.C.; Multigner, M.; de la Fuente, J.M.; Sanchez-Lopez, J.C.; Garcia, M.A.; Hernando, A.; Penades, S.; Fernandez, A. Permanent magnetism, magnetic anisotropy, and hysteresis of thiol-capped gold nanoparticles. *Phys. Rev. Lett.* **2004**, *93*, 087204. [CrossRef] [PubMed]

47. Beato-Lopez, J.J.; Dominguez, M.; Ramirez-del-Solar, M.; Litran, R. Glutathione-magnetite nanoparticles: Synthesis and physical characterization for application as MRI contrast agent. *Sn Appl. Sci.* **2020**, *2*, 1202. [CrossRef]

48. Umut, E.; Pineider, F.; Arosio, P.; Sangregorio, C.; Corti, M.; Tabak, F.; Lascialfari, A.; Ghigna, P. Magnetic, optical and relaxometric properties of organically coated gold-magnetite (Au-Fe3O4) hybrid nanoparticles for potential use in biomedical applications. *J. Magn. Magn. Mater.* **2012**, *324*, 2373–2379. [CrossRef]

49. Himstedt, R.; Rusch, P.; Hinrichs, D.; Kodanek, T.; Lauth, J.; Kinge, S.; Siebbeles, L.D.A.; Dorfs, D. Localized Surface Plasmon Resonances of Various Nickel Sulfide Nanostructures and Au-Ni3S2 Core-Shell Nanoparticles. *Chem. Mater.* **2017**, *29*, 7371–7377. [CrossRef]

50. Paramasivam, G.; Kayambu, N.; Rabel, A.M.; Sundramoorthy, A.K.; Sundaramurthy, A. Anisotropic noble metal nanoparticles: Synthesis, surface functionalization and applications in biosensing, bioimaging, drug delivery and theranostics. *Acta Biomater.* **2017**, *49*, 45–65. [CrossRef]
51. Bernal, S.; Botana, F.J.; Calvino, J.J.; Lopez-Cartes, C.; Perez-Omil, J.A.; Rodriguez-Izquierdo, J.M. The interpretation of HREM images of supported metal catalysts using image simulation: Profile view images. *Ultramicroscopy* **1998**, *72*, 135–164. [CrossRef]
52. Gao, N.; Zhou, H.; Tan, H.; Qi, R.; Chen, J.; Zhang, S.; Xu, J. Turn-on fluorescence detection of cysteine with glutathione protected silver nanoclusters. *Methods Appl. Fluoresc.* **2019**, *7*, 034004. [CrossRef]
53. Enea, M.; Pereira, E.; Costa, J.; Soares, M.E.; da Silva, D.D.; Bastos, M.d.L.; Carmo, H.F. Cellular uptake and toxicity of gold nanoparticles on two distinct hepatic cell models. *Toxicol. Vitr.* **2021**, *70*, 105046. [CrossRef] [PubMed]
54. Maniglio, D.; Benetti, F.; Minati, L.; Jovicich, J.; Valentini, A.; Speranza, G.; Migliaresi, C. Theranostic gold-magnetite hybrid nanoparticles for MRI-guided radiosensitization. *Nanotechnology* **2018**, *29*, 315101. [CrossRef]
55. Guzman, F.V.; Mercadal, P.A.; Coronado, E.A.; Encina, E.R. Near-Field Enhancement Contribution to the Photoactivity in Magnetite-Gold Hybrid Nanostructures. *J. Phys. Chem. C* **2019**, *123*, 29891–29899. [CrossRef]
56. Bhatia, P.; Verma, S.S.; Sinha, M.M. Optical Properties Simulation of Magneto-Plasmonic Alloys Nanostructures. *Plasmonics* **2019**, *14*, 611–622. [CrossRef]
57. Parani, S.; Oluwafemi, O.S. Selective and sensitive fluorescent nanoprobe based on AgInS2-ZnS quantum dots for the rapid detection of Cr (III) ions in the midst of interfering ions. *Nanotechnology* **2020**, *31*, 395501. [CrossRef] [PubMed]
58. Santos, M.C.; Seabra, A.B.; Pelegrino, M.T.; Haddad, P.S. Synthesis, characterization and cytotoxicity of glutathione- and PEG-glutathione-superparamagnetic iron oxide nanoparticles for nitric oxide delivery. *Appl. Surf. Sci.* **2016**, *367*, 26–35. [CrossRef]
59. Dheyab, M.A.; Aziz, A.A.; Jameel, M.S.; Noqta, O.A.; Khaniabadi, P.M.; Mehrdel, B. Simple rapid stabilization method through citric acid modification for magnetite nanoparticles. *Sci. Rep.* **2020**, *10*, 10793. [CrossRef]
60. Gomes, S.A.O.; Vieira, C.S.; Almeida, D.B.; Santos-Mallet, J.R.; Menna-Barreto, R.F.S.; Cesar, C.L.; Feder, D. CdTe and CdSe Quantum Dots Cytotoxicity: A Comparative Study on Microorganisms. *Sensors* **2011**, *11*, 11664–11678. [CrossRef]

Article

Synergistic Effect of rhBMP-2 Protein and Nanotextured Titanium Alloy Surface to Improve Osteogenic Implant Properties

Andrea Mesa-Restrepo [1,2,*], Ana Civantos [2,3], Jean Paul Allain [2,3], Edwin Patiño [4], Juan Fernando Alzate [5], Norman Balcázar [6], Robinson Montes [7], Juan José Pavón [7], José Antonio Rodríguez-Ortiz [8] and Yadir Torres [8]

[1] Department of Biomedical Engineering, Pennsylvania State University, State College, PA 16802, USA
[2] The Ken and Mary Alice Lindquist Department of Nuclear Engineering, Pennsylvania State University, State College, PA 16802, USA; ancife@illinois.edu (A.C.); allain@psu.edu (J.P.A.)
[3] Department of Nuclear, Plasma and Radiological Engineering, University of Illinois at Urbana-Champaign, Urbana, IL 61801, USA
[4] Structural Biochemistry of Macromolecules, Chemistry Institute, University of Antioquia, Medellin 50022, Colombia; edwin.patiño@udea.edu.co
[5] National Center for Genomic Sequencing-CNSG, School of Medicine, University of Antioquia, Medellin 50022, Colombia; jfernando.alzate@udea.edu.co
[6] Department of Physiology and Biochemistry, School of Medicine, University of Antioquia, Medellin 50022, Colombia; norman.balcazar@udea.edu.co
[7] Grupo de Biomateriales Avanzados y Medicina Regenerativa BAMR, Facultad de Ingeniería, Universidad de Antioquia, Medellín 50022, Colombia; robinson.montes@udea.edu.co (R.M.); juan.pavon@udea.edu.co (J.J.P.)
[8] Departamento de Ingeniería y Ciencia de los Materiales y del Transporte, Escuela Politécnica Superior, Universidad de Sevilla, 41003 Sevilla, Spain; jarortiz@us.es (J.A.R.-O.); ytorres@us.es (Y.T.)
* Correspondence: aqm6463@psu.edu; Tel.: +1-2177213197

Citation: Mesa-Restrepo, A.; Civantos, A.; Allain, J.P.; Patiño, E.; Alzate, J.F.; Balcázar, N.; Montes, R.; Pavón, J.J.; Rodríguez-Ortiz, J.A.; Torres, Y. Synergistic Effect of rhBMP-2 Protein and Nanotextured Titanium Alloy Surface to Improve Osteogenic Implant Properties. *Metals* **2021**, *11*, 464. https://doi.org/10.3390/met11030464

Academic Editors: Asit Kumar Gain and Leszek Adam Dobrzanski

Received: 29 December 2020
Accepted: 9 March 2021
Published: 11 March 2021

Publisher's Note: MDPI stays neutral with regard to jurisdictional claims in published maps and institutional affiliations.

Abstract: One of the major limitations during titanium (Ti) implant osseointegration is the poor cellular interactions at the biointerface. In the present study, the combined effect of recombinant human Bone Morphogenetic Protein-2 (rhBMP-2) and nanopatterned Ti6Al4V fabricated with Directed irradiation synthesis (DIS) is investigated in vitro. This environmentally-friendly plasma uses ions to create self-organized nanostructures on the surfaces. Nanocones ($\approx$36.7 nm in DIS 80°) and thinner nanowalls ($\approx$16.5 nm in DIS 60°) were fabricated depending on DIS incidence angle and observed via scanning electron microscopy. All samples have a similar crystalline structure and wettability, except for sandblasted/acid-etched (SLA) and acid-etched/anodized (Anodized) samples which are more hydrophilic. Biological results revealed that the viability and adhesion properties (vinculin expression and cell spreading) of DIS 80° with BMP-2 were similar to those polished with BMP-2, yet we observed more filopodia on DIS 80° ($\approx$39 filopodia/cell) compared to the other samples (<30 filopodia/cell). BMP-2 increased alkaline phosphatase activity in all samples, tending to be higher in DIS 80°. Moreover, in the mineralization studies, DIS 80° with BMP-2 and Anodized with BMP-2 increased the formation of calcium deposits (>3.3 fold) compared to polished with BMP-2. Hence, this study shows there is a synergistic effect of BMP-2 and DIS surface modification in improving Ti biological properties which could be applied to Ti bone implants to treat bone disease.

Keywords: surface modification; bone morphogenetic proteins; directed irradiation synthesis; nanopatterning; advanced biointerfaces; osseointegration

1. Introduction

Bone loss affects more than half a million patients in the United States and represents over \$2.5 billion in health costs. Indeed, trauma, tumor recessions or developmental defects limit bone's ability to self-repair after an injury, creating large non-healing fragments that require the necessary use of implants and other medical devices [1]. Current treatments use bone metal implants based on titanium (Ti) and Ti-based alloys to replace a diversity of bone tissue, including dental implants, repair large bone defects, spinal bone implants, and a host of reconstructive and regenerative solutions for millions of patients. Ti and its alloys are widely used in the dentistry and orthopedic industry due to their excellent biological and chemico-mechanical properties. When Ti is exposed to air, it forms a passive oxide layer (TiO_2) that is responsible for its chemical inertness, corrosion resistance, and biocompatibility. Unfortunately, conventional Ti surfaces (polished) show a limited cellular interaction at the biointerface, limiting the formation of a strong direct chemical bond between implant and tissue, also known as osseointegration. During the implantation process, tissues are damaged, and an acute inflammation process occurs—if this becomes chronic, the implant is encapsulated in fibrotic tissue and finally is rejected [2–6]. Thus, there is an urgent need to improve titanium bone contact to accelerate healing times, avoid implant failure, and minimize secondary surgeries to remove it, reducing the associated costs and increasing the patient's quality of life [3,7,8].

Framed into this special issue, researchers from a multidisciplinary field (e.g., engineers, biologists, physicists, pharmacists) have joined their knowledge and expertise to develop smart biomaterials that address essential health concerns. In this sense, this study seeks to instill in students the value of multidisciplinary research in the development of medical devices that regenerate living tissues.

One key strategy to improve Ti bioactivity is the modification of surface properties such as roughness and topography [9]. These surface properties have shown a pivotal role in governing cellular interaction at the material-bone interface. Once the titanium implant is in contact with living tissues, blood proteins adsorb to the implant surface depending upon its physicochemical properties. These proteins, in turn, will interact with cells and guide their cellular processes, resulting in implant osseointegration [10–13].

Conventional surface treatments have been shown to enhance cellular processes by increasing roughness, changing the topography, or adding bioactive molecules, leading to a stronger bone to implant contact (BIC) than polished surfaces [14–17]. Some commercially available surface modifications are sandblasting/acid-etching (SLA) and anodization. SLA consists of the surface bombardment with ceramics and later submerged in strong acids (such as hydrofluoric, nitric and sulfuric acid), while anodization is an electrochemical method where Ti is used as an anode in an electrochemical cell with sodium hydroxide or hydrofluoric or sulfuric acid as electrolytes. The structures created in SLA are random and in the range of micro and nanometers scale, which have shown to promote implant osseointegration [16,18,19]. On the other hand, the anodization process produces ordered and high-aspect-ratio nanotubes, depending on anodization conditions, which has also enhanced the osteoblast response [9,20]. Additionally, multiple studies have used nanotubes as drug reservoirs to deliver growth factors or antibacterial agents to enhance the osteogenic properties of Ti [21–24]. For example, the usage of members of the Bone Morphogenetic Protein family (BMPs) has improved Ti osseointegration in in vitro and in vivo conditions, highlighting BMP-2 protein as one of the most osteogenic factors used in clinical applications [25–27]. Nevertheless, using SLA or anodized surfaces have encountered several limiting factors regarding the reproducibility of geometrical features, the use of highly toxic reagents or additional steps, which might increase the cost of production and reduce the industry scalability [9,16,28,29]. Therefore, new methods to nanopattern surfaces with high tunability on nanofeatures morphology, size and surface chemistry-independent modification are being intensively investigated. Under this premise, ion irradiation is a scalable industrial method that tailors ion-surface interactions to form reproducible and geometrically interesting nanostructures on the surface in a controlled manner [30,31]. Directed irradiation synthesis (DIS) is an advanced ion irradiation method that uses gradient energetic ion beams to create hierarchical micro/nanostructures in a bottom-up fashion [30,32,33]. When energetic ions collide and interact with the solid, ions transfer energy to the surface atoms. If these atoms have sufficient kinetic energy, they will leave their position in the atomic lattice (sputtering) and create point defects (vacancies and interstitials). The movement of these defects in the lattice may form complex structures or result in phase changes due to species accumulation in an area. Thus, ion irradiation may result in substantial chemical and morphological changes on the surface [30,34].

DIS shows multiple advantages compared to conventional nanopatterning techniques or new sintering methods, it is a fast process, reliable, with strong capacity to tune small nanofeatures (10 to 100's nm) without the use of masks [29], high temperature [35] or toxic reagents [32,33]. In addition, this bottom-up technique has previously shown the capability to tailor nanofeatures on Ti surfaces keeping the bulk properties stable when using low fluences. This bioactive nanotopography has been shown to modulate cytoskeleton orientation and cell adhesion, as well as cell viability, guiding the tissue regeneration process [32,33].

Based on previous results in our group, in this study, we have characterized the surface properties of DIS-treated titanium samples irradiated using high fluences and different incidence angles and evaluated the effect of combining these active nanotopographies with effective biologics, such as BMP-2, to help elicit a cellular biological response, which may have a synergistic effect in promoting osteoblast differentiation in a BMP-2 responsive cellular model such as C2C12 cells [36]. To the best of our knowledge, this is the first study that does a systematic study using ion-induced surface patterning techniques in synergy with biologics compared with industry-leading anodized and sandblast surface treatment technologies.

2. Materials and Methods

For this purpose, we will first characterize the surface properties (topography, chemistry and wettability) (Figure 1a) and evaluate the effect of the different samples on cell adhesion and spreading (number of filopodia, cell and nucleus shape, vinculin expression, total cells attached), cell viability by measuring cell metabolic activity, osteogenic differentiation and surface mineralization) (Figure 1b).

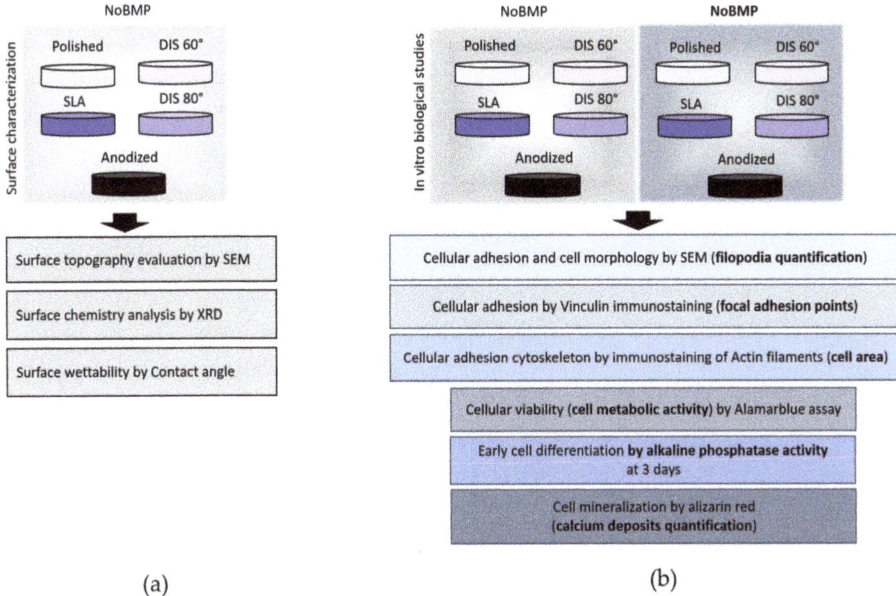

Figure 1. Schematic design of materials and methods of this research study. (**a**) surface characterization and (**b**) biological characterization. Abbreviations: Scanning electron microscopy (SEM), X-ray diffraction (XRD), Alkaline Phosphatase (ALP), NoBMP (without BMP-2 protein), BMP (with BMP-2 protein), DIS 60° (Ti6Al4V irradiated using argon ions and 60 incidence angle), DIS 80° (Ti6Al4V irradiated using argon ions and 80 incidence angle), Sandblasted and acid etched Ti6Al4V (SLA), Acid etched and anodized Ti6Al4V (Anodized).

2.1. Titanium Sample Preparation

Titanium alloy samples (Ti4Al6V, area 0.25 cm^2, Stryker, Kalamazoo, MI, USA) were polished to mirror finished and cleaned before DIS irradiation. The samples were grinded using 320, 1200 and 2400 sandpaper in an Ecomet III grinder (Buehler, Lake Bluff, IL, USA) then were mirror-polished using a ChemoMet cloth (Buehler, Lake Bluff, IL, USA) with a 0.05 µm silica solution (MasterMet, Buehler, Lake Bluff, IL, USA). Afterwards, the samples were ultrasonically cleaned in acetone, isopropanol, ethanol, and water for 15 min each. As controls sandblasted/acid-etched (SLA, commercially pure Ti grade IV, area 0.21 cm^2) and acid-etched/anodized (Anodized), area 0.25 cm^2) Ti6Al4V surfaces were used. SLA surfaces were provided by Tissue Engineering Group (TEG) of the Complutense University of Madrid, Spain. DIS samples were irradiated with argon ions at 1000 eV of energy and 1×10^{18} ions/cm^2 using two different incidence angles: 60° and 80° degrees naming the samples DIS 60° and DIS 80°, respectively. All samples were ultrasonically cleaned and autoclaved before using them for the in vitro studies.

2.2. Surface Characterization of Titanium Samples

The surface topography was analyzed by scanning electron microscopy (SEM, S-4800, Hitachi, Tokyo, Japan) at two ranges of magnification: 10 k–22 k to detect microstructures and 70 k–100 k to detect submicron and nanostructures. The surface chemistry was examined by X-ray diffraction (XRD, PANalytical Phillips X′pert MRD system #2, Malvern Panalytical, Malvern, United Kingdom) with Cu Kα radiation wavelength (λ = 0.15418 nm), generated at a voltage of 45 kV and a filament emission of 40 mA. Theta angle (2θ) was collected from 2θ = 30–80°, with a step size of 0.02°, and the analysis was performed with Origin and Jade software. The surface hydrophilicity was determined by contact angle (CA, Ramé-hart 250 Contact Angle Goniometer, Ramé-hart, Succasunna, NJ, USA) with DROPimage Advanced Software. The sessile method of CA analysis was employed, using 3 µL of deionized water drops to measure the CA of each sample; 4 samples per condition were used for the measurements.

2.3. In Vitro Cell Culture

2.3.1. C2C12 Cell Line

C2C12 cells are multipotent cells that can differentiate, in addition to myotubes, into osteoblasts and adipocytes under specific culture conditions [37]. They have been widely used to study BMP-2 bioactivity since they display low basal BMP signaling activity and show good BMP-2 responsiveness. e.g., expressing bone differentiation markers (early phase such as alkaline phosphatase (ALP) or late-stage such as osteocalcin (OCN)), and inducing the formation of calcium nodules in ascorbic acid, β-glycerol phosphate and dexamethasone rich media [36,38,39]. Murine C2C12 myoblasts were purchased from the American Type Culture Collection (ATCC® CRL-1772). This cell line was maintained in Dulbecco's Modified Eagle Medium (DMEM) (ATCC® 30-2002™) supplemented with 10% Fetal Bovine Serum (FBS) (Invitrogen, Waltham, MA, USA) in a 37 °C, 5% CO$_2$ incubator. At 80% cellular confluence, cells were trypsinized in 0.25% Trypsin/EDTA (Invitrogen, Waltham, MA, USA). C2C12 passages <20 were cultured in Ti samples at densities of 60,000 cell/cm^2 (adhesion assay) or 100,000 cell/cm^2 (viability, osteogenic differentiation and mineralization assay) in 48 well-plates (Corning Costar) with DMEM media, 10% FBS and 3 µg/mL of BMP-2. The cells were initially cultured in 10 µL drops with BMP-2 on Ti surfaces for 4 h to promote cell attachment and then the rest of the media was added (490 µL).

Cellular Adhesion

To determine the combined effect of BMP2 and DIS surface treatment during initial cellular attachment of C2C12, cells were cultured as mentioned above on the different Ti surfaces (DIS 60°, DIS 80°, and polished, SLA, and Anodized as controls). After 4 h, Ti samples were prepared for SEM and confocal laser scanning microscopy. For SEM, the samples were washed twice with Phosphate-buffered saline (PBS) and fix with 2.5% glutaraldehyde (Sigma, St. Louis, MO, USA) overnight at 4 °C. Then, samples were dehydrated using an ethanol gradient in PBS (30%, 50%, 70%, 80%, 90% and 100%), each step for 15 min. Afterwards, they were critically point dried, sputter with gold-palladium and observed using a scanning electron microscope (JSM-6490LV, Jeol, Tokyo, Japan) at 2 k–3.5 k magnification. For confocal microscopy, the Ti samples were washed twice with PBS and fixed with 5% Formalin (Sigma, MO, USA) for 10 min, permeabilized with 0.1% Triton X-100 (Sigma, St. Louis, MO, USA) for 5 min, washed with 0.01% BSA/PBS and incubated for 30 min at room temperature with Texas red phalloidin (Invitrogen, Waltham, MA, USA) (1:75), DAPI (Thermofisher, Waltham, MA, USA) (1:1000) and Alexa 647 vinculin (Invitrogen, Waltham, MA, USA) (1:125) to stain actin filaments, nuclei and vinculin, respectively. Finally, the samples were examined via confocal microscopy (Leica SP8 Laser Confocal Microscope Microsystems, Wetzlar, Germany). The number of filopodia, vinculin intensity, total cells, cell area and nucleus area were quantified using FIJI software.

Cell Viability

C2C12 cells were seeded on titanium discs as mentioned above. After 3 days, the Ti samples were incubated for 3 h with Alamarblue (Invitrogen, Waltham, MA, USA) following the manufacturer's recommendations. Briefly, media was removed to avoid counting unattached cells and fresh media with Alamarblue in a 1:10 ratio was added to the wells. Alamarblue is a resazurin-based solution, a cell-permeable compound that upon entering living cells is reduced to resorufin, a fluorescent compound. After incubating the samples in the dark for 3 h, the fluorescence signal was measured using a microplate reader (Synergy HT, BioTek, Winooski, VT, USA) at 530 nm/590 nm Ex/Em.

Evaluation of Cell Differentiation: Alkaline Phosphatase Activity

C2C12 cells were seeded on titanium discs as mentioned above. After 3 days, cells were washed twice with PBS, and lysed by the addition of 100 µL of buffer lysate and subjecting the samples to 3 cycles at $-80\ °C/37\ °C$ for 30 min. ALP activity was measured according to the manufacturer's instructions using a commercial ALP kit (ab83369, Abcam, Cambridge, MA, USA). ALP hydrolyses phosphate esters in alkaline conditions, generating an organic radical and an inorganic phosphate. This ALP kit uses p-nitrophenyl phosphate (pNPP) as a phosphatase substrate which is dephosphorylated by ALP, generating a yellow compound (p-nitrophenol). The absorbance was measured at 405 nm on a microplate reader (Synergy HT, BioTek, Winooski, VT, USA). ALP enzyme activity was expressed U/L.

Evaluation of Cell Mineralization: Calcium Deposits Production

C2C12 cells were seeded on Ti discs in a 48 well-plate at 37 °C in a 5% CO_2 incubator at a density of 100,000 cell/cm^2 in DMEM with 10% FBS, 3 µg/mL of BMP-2, 100 nM of dexamethasone (Sigma, St. Louis, MO, USA, USA), 1 mM of β-Glycerophosphate (Sigma, St. Louis, MO, USA) and 50 µg/mL of L-ascorbic acid (Sigma, St. Louis, MO, USA). The media was changed every 2–3 days. After 21 days, the samples were washed with PBS, fixed with 4% formaldehyde for 15 min, stained with 40 mM of alizarin red solution (Sigma, St. Louis, MO, USA) for 30 min and washed 5 times with deionized water to remove the excess of the alizarin red solution. The red deposits were quantified in FIJI. The mineralization percentage (%) was calculated by dividing the area covered by the red deposits to cell culture area times 100.

2.4. Statistical Analysis

All experiments were done in triplicates using two to three samples per condition, except for the confocal and SEM experiments where we examined 3–5 different fields of a sample. Analysis of variance (ANOVA) and Bonferroni's Multiple Comparison Test were used to determine statistically significant differences at 0.05 level of significance by using Origin lab and Graph Prism 5 software.

3. Results and Discussion

3.1. Characterization of Irradiated Titanium Samples

3.1.1. Evaluation of Surface Topography of Irradiated Titanium Samples

Ion irradiation transfers energy and momentum via ion-atom collisions. This results in erosive and diffusive regimes, which drive the surfaces to self-nanopatterned, generating surfaces with attractive topographies [40]. Previous work conducted on Ti alloy and pure titanium (porous scaffolds) revealed that changing the incidence angle, the nanopatterning process was governed by two regimes: diffusion and erosion. At normal incidence angle (0°) and low fluences (ion/cm^2), ion diffusive process predominates, generating short nanoripples and nanorod-like structures. Moving from 0 to 60°, in small or low off-normal angles, there is a combination of diffusive and erosive regimes where ripples and nanorods are combined on the surface, increasing the uniformity of the surface treatment and length. At highly oblique off-normal angles ($\geq$80°), an erosion regime predominates, in which ions from the source crash and sputter the atoms of the outmost layer of the surface. This complex process showed more elongated nanoripples and no nanorods [32,33]. In this study, we have observed that at higher fluences and off-normal incidence angles, nanoripples have grown in height, turning into nanowalls/nanocones. As observed in Figure 2, scanning electron microscopy (SEM) images of the studied specimens surface topography, DIS 60° generated nanowalls homogeneously distributed on the surface of 16.5 ± 1 nm of thickness, and 41.2 ± 2.9 nm of spacing distance between nanowalls (white arrow); whereas DIS 80° samples presented nanocone-like structures of a width of 36.7 ± 9.9 nm (white arrow) spaced throughout the surface.

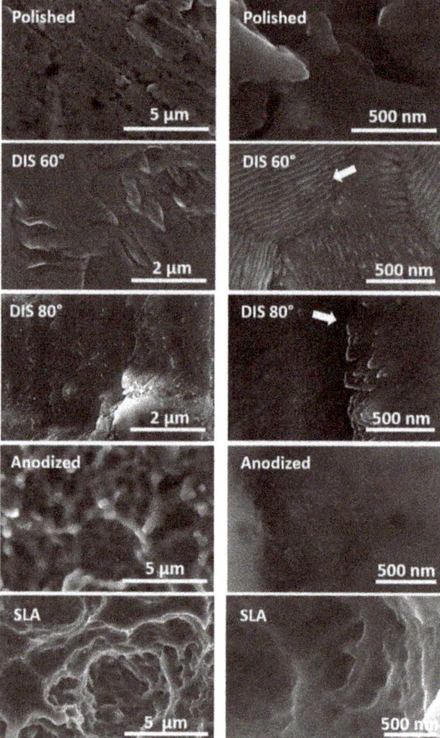

Figure 2. Scanning electron micrographs of modified Ti samples, white arrows indicate nanostructures.

3.1.2. Evaluation of Surface Chemistry and Wettability of Irradiated Titanium Alloy Specimens

Our samples are (α + β) titanium alloy that contains α stabilizer element aluminum and β stabilizer element vanadium. α phase could be observed using (100), (002) and (101) α peaks. β phase could be observed at (110) β peaks (black arrow). We did not observe any major changes in the α and β phase of the modified DIS Ti samples, which confirms the similar surface crystalline structure of Ti6Al4V modified samples as polished Ti (Figure 3) [41,42]. These surface modifications do not change the bulk crystallographic orientation of Ti (microstructure) due to the low ion penetration around 3–4 nm depth [32,33].

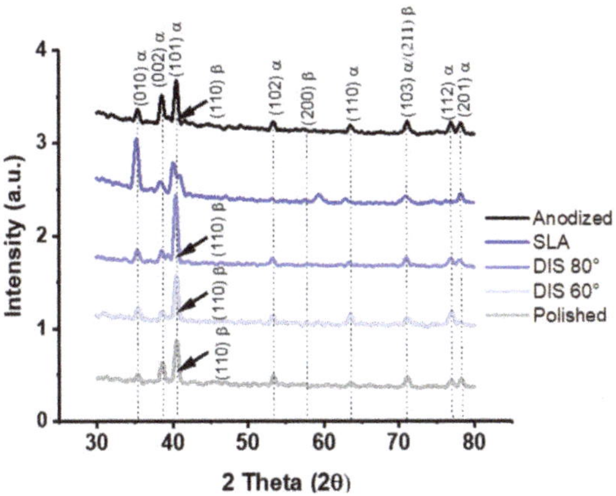

Figure 3. XRD pattern of modified Ti samples. Black arrows indicate (110) β peak.

On the other hand, we did observe a reduction of CA values on DIS samples compared to polished surfaces. DIS surfaces were slightly more hydrophilic than polished Ti even though no statistical differences were detected ($p > 0.05$). It should be noticed that SLA ($40.83 \pm 3.42°$) and Anodized ($20.57 \pm 0.9°$) showed statistical differences with polished and DIS-treated samples. Therefore, SLA and Anodized samples were more hydrophilic surfaces compared to polished or DIS-treated samples which had similar wettability, polished ($79.43 \pm 3.44°$), DIS 60° ($65.33 \pm 3.89°$) and DIS 80° ($71.33 \pm 1.27°$) (Figure 4).

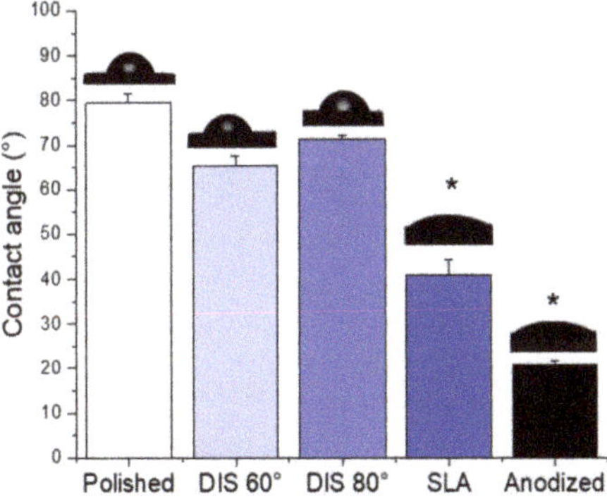

Figure 4. Determination of hydrophilicity/hydrophobicity of Ti samples by contact angle measurement. Mean + SE, N = 3 * p-values < 0.05 compared to polished, DIS 60° and DIS 80°.

3.2. In Vitro Biological Characterization of Titanium Samples

3.2.1. Evaluation of the Cellular Attachment on Irradiated Titanium Samples

After the implantation-derived immune response, mesenchymal stem cells (MSC) migrate to the implant site, where growth factors are released, and start their differentiation process into bone-forming cells [6,43]. In light of this, we evaluated cellular attachment of undifferentiated bone-forming cells 4 h after seeding on the BMP-2 conjugated modified Ti alloy surfaces through the analysis of SEM microphotographs (compiled in Figure 5). We observed the presence of attached cells in all samples. However, SEM images suggested that with the addition of BMP-2, the filopodia number per cell increases, particularly for DIS 80° and Anodized samples and slightly for SLA samples. We observed a 2.25-fold increase on DIS 80° and 0.6-fold increase on Anodized samples compared to their counterpart without BMP-2. However, DIS 80° samples seemed to have more filopodia (39 ± 20 filopodia) compared to the other samples averaging less than 30 filopodia per cell (see Figure 5a,b). Although these results are promising, it will be interesting in future studies to measure the adhesion strength of cells cultured on these surfaces to corroborate this data. Filopodia protrusions are composed of bundles of actin filaments which play a role in the initial cell adhesion, spreading and migration. Cells use filopodia to sense and tether to their surroundings, which require the development of strong tensile forces. As time progresses, these tensile forces will further stabilize the cells, recruiting cell adhesion receptors (integrins) and force-regulating proteins (vinculin, tailin and zyxin), which participate in forming mature adhesions [44–46].

Due to the values observed on titanium surfaces regarding cellular adhesion structures (filopodia prolongations), we decided to evaluate the adhesion process by the immunostaining and quantification of vinculin protein expression. Vinculin protein participates in focal adhesion complexes and plays a fundamental role in cell-cell and cell-matrix interactions and regulates adhesion through binding, polymerizing and remodeling actin fibers [47,48]. Figure 6a shows the confocal images of C2C12 cells, in which vinculin proteins appeared in green, actin in red, and cell nuclei in blue color. Cell number and vinculin intensity from these images were quantified using FIJI software and the results were compiled in Figure 6b. We observed that the addition of BMP-2 increased vinculin expression on cells growing on polished (465%), DIS 80° (200%) and Anodized samples (110%) compared to the samples without BMP-2, although no statistical differences between DIS 80° with BMP-2 and polished BMP-2 were found.

In addition, cell spreading was evaluated through the analysis of the confocal microphotographs (compiled in Figure 7). The cell cytoskeleton, which is composed of actin filaments, was measured to determine the area of the cell (Figure 7a). We observed that the presence of BMP-2 in cells growing on DIS 80° samples showed higher surface area, increasing cellular spreading by 120% compared to its counterpart without the protein, but no statistical differences were found compared to polished samples with BMP-2 (Figure 7b). Although we did not observe significant differences in nucleus area among the different samples, it seemed that the addition of BMP-2 increased the nucleus area of cells cultured on DIS 80° samples (Figure 7c).

Figure 8 shows cell viability results after 3 days of cell culture of all studied surfaces measured by the metabolic activity of C2C12 cells. Percentages of cell viability were calculated using polished without BMP-2 as reference (100%), observing that all surfaces achieved cell viability percentages from 80 to 120%.

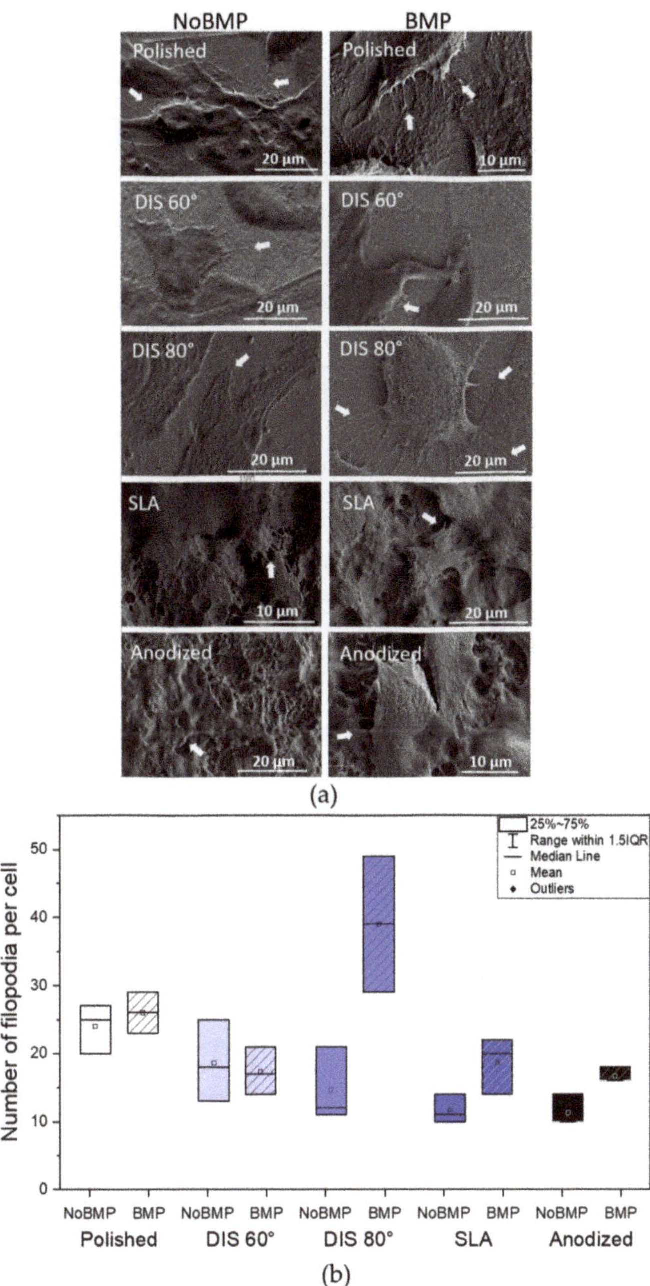

Figure 5. Evaluation of cell morphology on Ti alloy surfaces with and without the incorporation of BMP-2. (**a**) SEM micrographs of C2C12 cultured on modified Ti samples for 4 h, white arrows indicate filopodia prolongations (**b**) Filopodia number quantification. Median and IQR, N = 3 fields.

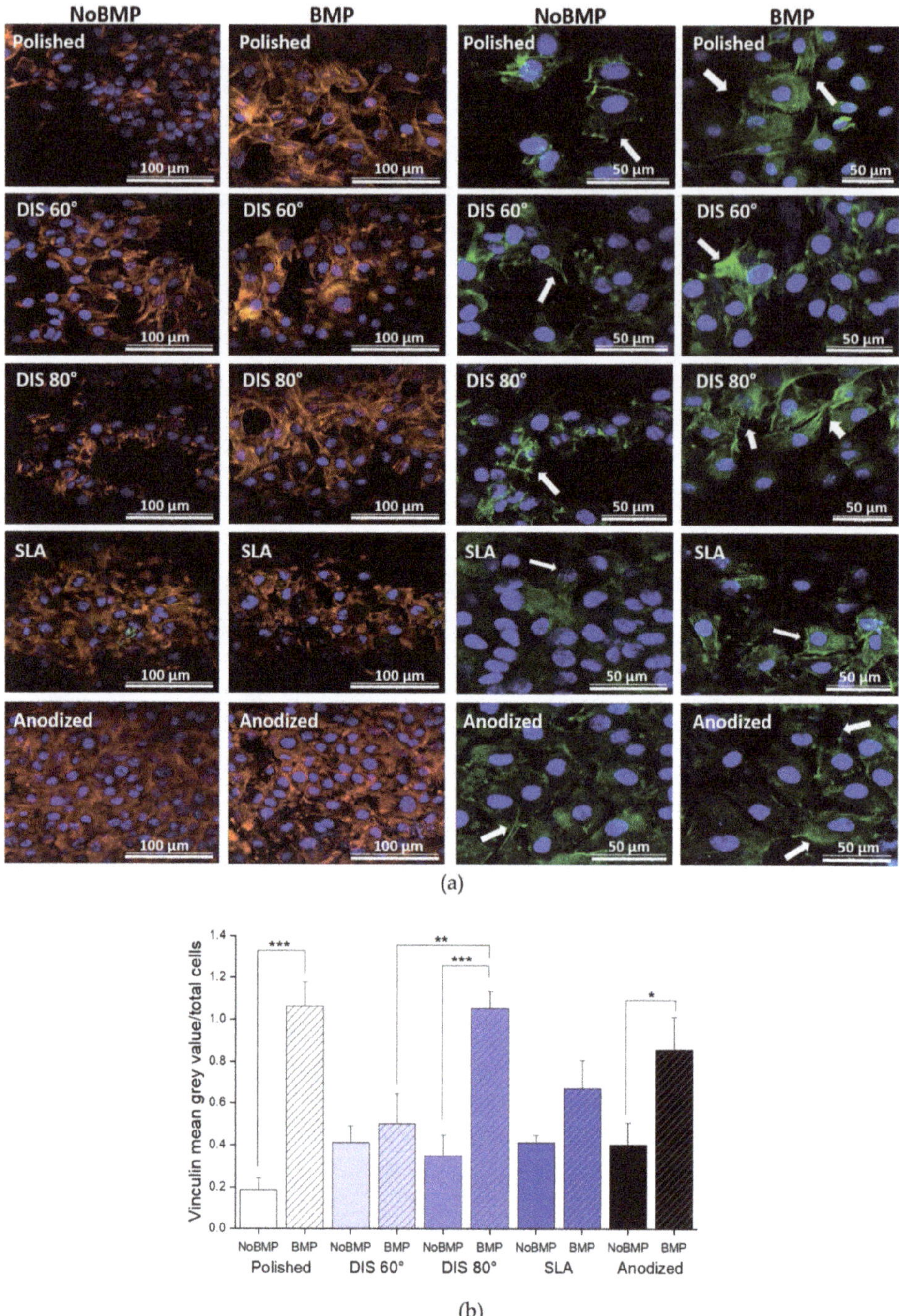

Figure 6. Determination of cellular attachment. (**a**) Confocal microscopy images of C2C12 cells cultured on Ti samples with and without the incorporation of BMP-2. Vinculin protein was stained green (white arrows), actin filaments in red and cell nuclei in blue; (**b**) Vinculin mean grey value of samples with and without the incorporation of BMP-2 normalized to total cells. Mean + SE N = 5 fields, *** $p < 0.001$, ** $p < 0.001$, * $p < 0.05$.

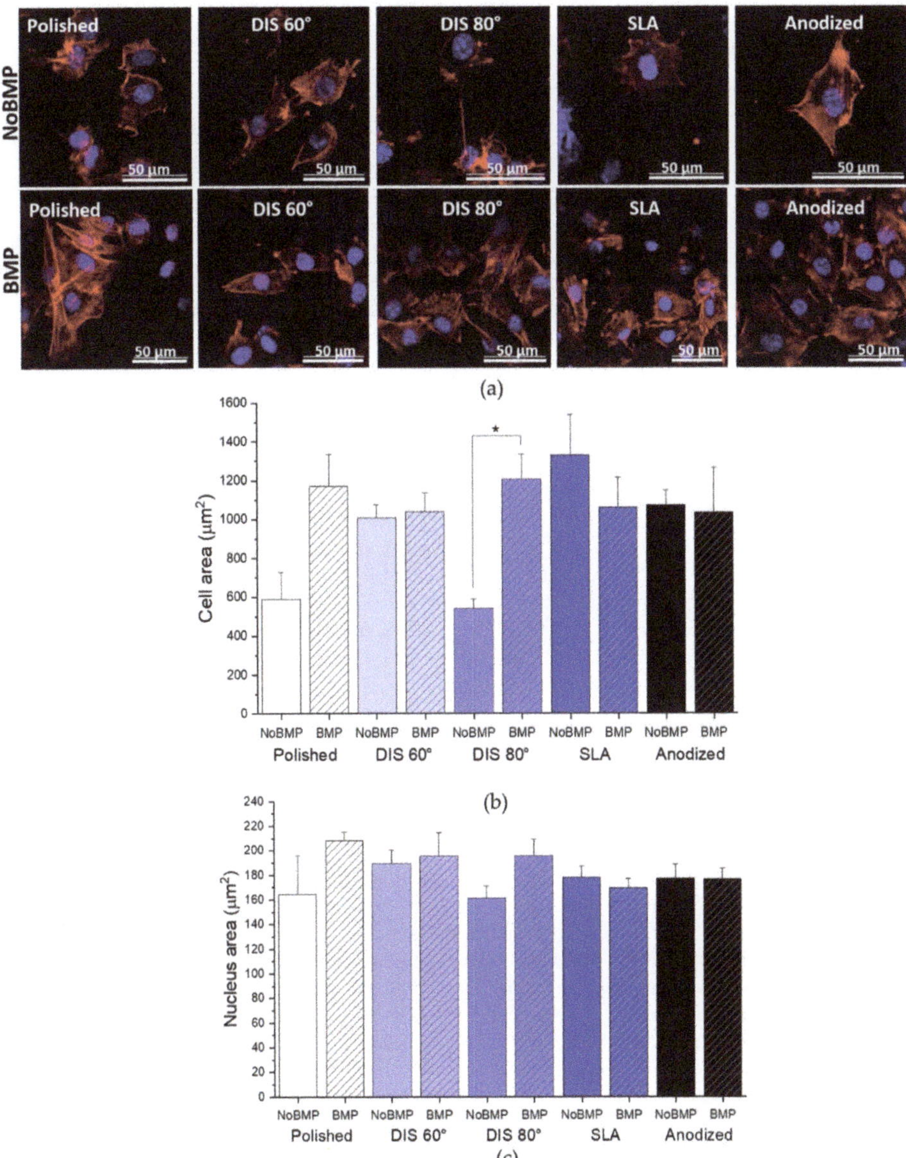

Figure 7. Cell spreading. (**a**) Confocal microscopy of C2C12 cultured on Ti samples with and without the incorporation of BMP-2. The cytoskeleton actin filaments were stained red and nuclei blue; (**b**) Cell area measured by FIJI software. Mean + SE, N = 3 fields, * $p < 0.05$; (**c**) Nucleus area. Mean + SE, N < 3 fields.

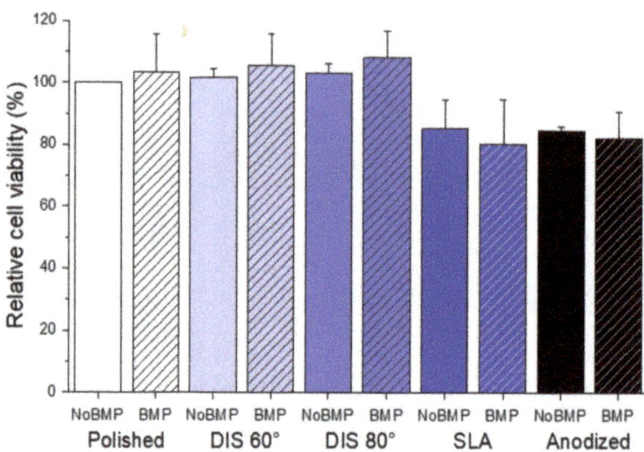

Figure 8. Relative C2C12 viability on Ti samples with and without the incorporation of BMP-2 normalized to polished without BMP-2 after 3 days in culture. Mean + SE, N = 3.

Integrins α and β-subunit are linked to the actin cytoskeleton via adaptor proteins (talin and vinculin). Integrins agglomerate and form focal adhesions, which translate mechanical stimuli into biochemical signals that will start gene expression either by activating signaling pathways, such as extracellular-signal-regulated kinase (ERK)/mitogen-activated protein kinase (MAPK) or through actomyosin contractility, which will distort the shape of the nucleus and allow the translocation of transcription factors [47–49]. Fourel and collaborators observed an increase in $\beta3$ integrin expression in C2C12 cultured on soft polymeric films with matrix-bound BMP-2, which promoted cell spreading and adhesion [50]. BMP-2 enhancement of C2C12 cellular adhesion has also been reported in hydroxyapatite and magnesium surfaces by Huang and collaborators [51]. In this study, we used nanopatterned metallic substrates, which might serve as BMP-2 nanoreservoirs, promoting localized BMP-2 signaling. We observed an increase in vinculin intensity and changes in cell area in surfaces with BMP-2, particularly in DIS 80° treated surfaces. Although we did not measure integrin expression, the vinculin and cell spreading results suggest that integrin-mediated signaling is involved in this process and could explain the changes we observed in cellular attachment and cell fate.

Moreover, topographical features have been found to influence cell attachment [13,49]. For example, Pan and collaborators evaluated the behavior of MSC on a macropore/nanowire structure fabricated via vacuum plasma spraying and etching with sodium hydroxide (NaOH), which mimics bone hierarchical microenvironment. These cells were elongated and spread in multiple directions with well-developed focal adhesions. This led to an increase in cytoskeletal tension and yes-associated protein (YAP) activity and nuclear translocation. YAP is a member of the Hippo Signaling pathways that shuttles between the cytoplasm and the nucleus under specific physical cues, e.g., stiffness and topography, acting as a promoter for osteogenic transcription factors; thus, inducing MSC differentiation [52]. Other structures such as nanotubes (diameters of 10 nm and 30 nm), which have spacings of less than 50 nm provide an effective length promoting integrin clustering/focal contact formation and increasing detachment forces [53,54]. It is hypothesized that this is caused by the structures matching the integrin diameters (10 nm) [55]. Alternatively, this phenomenon might be caused by an increase in surface contact area to which more integrins can bind to. For example, Babchenko and collaborators have found that nanopatterned surfaces with nanocones, which were densely packed and distributed homogeneously on the surface, increased vinculin signaling and promoted focal adhesion kinase (FAK) activation in osteoblastic-like cells (Saos-2) [56,57]. Considering this, the dimensions of our nanostructures could further improve integrin-mediated adhesion and traction forces; and thus, their involvement in mechanotransduction pathways. Integrin expression and its role on cell morphology and behavior is the focus of a future study which will include cells at different stages of osteogenic differentiation and different stages of cellular adhesion.

Other factors that influence osteoblast cell spreading and proliferation are surface texturing/patterning [8,58]. Although we did not measure roughness (Sa) in this study, similar Ar^+ irradiation conditions (with lower fluence 2.5×10^{17} ions/cm^2) were performed on titanium alloy and showed Sa of 10 to 50 nm reaching higher values on SLA (49 nm) [32]. However, it will be interesting in further studies to correlate these measurements with the DIS conditions used in this study.

3.2.2. Evaluation of Osteogenic Differentiation and Mineralization on Irradiated Titanium Samples

Generally, there are three major stages during MSC osteogenic differentiation: differentiation, matrix maturation, and mineralization. First, attached MSC will differentiate into osteoprogenitor cells expressing Runt-related transcription factor 2 (Runx2), Distal-less homeobox 5 (Dlx5) and Osterix (Osx). Once committed to an osteogenic lineage, these cells (preosteoblasts) will express ALP and collagen I (Col I) and eventually will differentiate into mature osteoblasts, expressing and secreting Col I, OCN, and osteopontin (OPN), which form the matrix. Afterwards, ALP will aid in the mineralization process by releasing phosphate ions, which will be combined with calcium to produce hydroxyapatite. In mouse models, the mineralization phase peeks at 14–21 days [59–62]. In C2C12, ALP is expressed by stimulating the cells with 300 ng/mL of BMP-2 after 2–3 days in culture, which induces their osteogenic differentiation and inhibits myotube formation [38,63]. From previous studies and what has been established in the literature, BMP-2 UdeA optimal concentration was 3 µg/mL to detect significant ALP production after 3 days of C2C12 cell culture [64].

For this purpose, we cultured BMP-2 and C2C12 cells directly on modified Ti alloy surfaces and evaluated the osteogenic differentiation and mineralization. We observed statistical differences in ALP activity produced by cells on Ti surfaces with BMP-2 *versus* without BMP-2 due to the intrinsic osteoinductive effect of BMP-2. We observed an increase in ALP production for DIS 80° (265%), Anodized (195%) and SLA samples (190%), yet we did not observe statistical differences among the different samples with BMP-2. This fact suggests that BMP-2 plays a more dominant role in cell differentiation than the surface nanotopography. Nevertheless, it should be noticed the slightly higher ALP activity on cells growing on DIS 80° surfaces with BMP-2 compared to the other samples with BMP-2, which suggests a possible synergistic effect between the nanopatterned irradiated surface at 80° and BMP-2s (Figure 9a). This synergistic effect might be similar to the one MSCs face in their native environment, where they have physical and biological cues, leading to their differentiation into bone cells. On the other hand, the late phase of osteoblast differentiation or cell mineralization phase is characterized by the presence of calcium and phosphate deposits, which form hydroxyapatite and collagen fiber production. Through Alizarin red staining of free calcium nodules, we could determine the extent of the C2C12 cell mineralization process on Ti surfaces in the presence of mineralization media [39]. DIS 80° and Anodized surfaces with BMP-2 showed a greater mineralization area after 21 days in cell culture compared to other samples, particularly DIS 80° with BMP-2 increased the percentage of calcium nodules by 329% compared to polished samples with BMP-2 (Figure 9b). The differences between DIS-based modified surfaces and BMP-2 mineralization results could, in fact, still be within the variance as the statistical sample was limited.

Surface treatment can influence the adsorption of bone extracellular matrix proteins such as BMP-2 and modulate cell behavior. For instance. Xiao and collaborators evaluated BMP-2 adsorption on polished, etched and anodized Ti surfaces. They found that etched samples had the highest BMP-2 absorption, yet the protein changed its conformation and reduced its bioactivity in MSC in the long term, in contrast to BMP-2 adsorbed on anodized samples [65]. They also studied the synergistic effect of BMP-2/fibronectin adsorption on these surfaces and observed that it promoted MSC spreading and osteoblast differentiation [66]. These results agree with those observed in our study in which DIS 80° in the presence of BMP-2 showed the highest levels of vinculin expression (FAK components) and cell spreading (cell area). Nevertheless, future studies are needed to evaluate how our surfaces modulate protein conformation, adsorption, bioactivity at different concentrations and timepoints and their effects on cellular behavior.

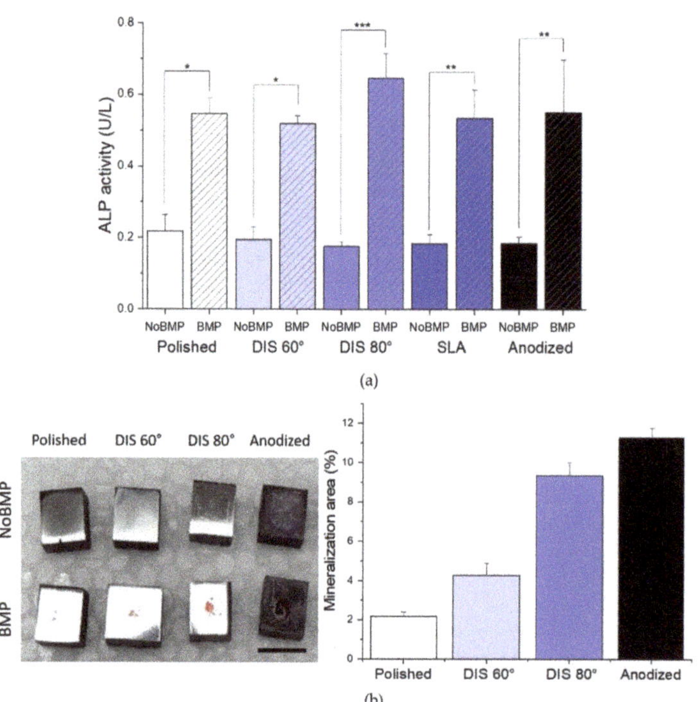

Figure 9. C2C12 osteoblast differentiation and mineralization. (**a**) ALP production after 3 days on Ti samples with and without the incorporation of BMP-2. Mean + SE, N = 3, *** $p < 0.001$, ** $p < 0.001$, * $p < 0.05$; (**b**) Cell mineralization. Cell culture with mineralization media on Ti sample with and without the incorporation of BMP-2, black bar indicates 0.5 cm (left panel), quantification of mineralization area compared to the culture area of samples with BMP-2 (right panel). Mean + SD, N = 3 fields.

Researchers have observed that cells can respond to nanopatterns of less than 13 nm in height or orthogonal or hexagonal patterns of nanopits with diameters from 35 to 200 nm [67]. Nanopits of 22 nm promote osteogenic gene expression via integrin signaling [68]. Our results agree with other studies using different nanopatterning techniques, which stimulate bone formation by targeting integrin signaling [69–71] yet DIS presents many advantages over these methods. By modulating DIS irradiation parameters, we can generate reproducible and scalable nanopatterns directly on the material in a short amount of time (usually a few minutes) without using toxic chemicals that pose health risks or require extra steps to discard them. Future studies will focus on evaluating integrin-mediated bone formation on these surfaces.

BMP-2 controlled delivery is one of the main challenges in bone tissue engineering as uncontrolled drug release can cause serious side effects such as ectopic bone formation, bone reabsorption and hematomas in soft tissues [72]. Currently, these and other growth factors can be tethered to the implant surface via physisorption or covalent binding to control their release. Physical adsorption relies on electrostatic interactions, hydrogen bonding, or hydrophobic interactions. In covalent binding, the substrate is treated with plasmas, chemical etching and surface coatings to functionalize the surface. Also, bioconjugation reactions such as amidation, esterification and click reactions through carbodiimides, silanes, mussel-inspired bioconjugation have been used [73,74]. Another strategy is to encapsulate growth factors in scaffolds, e.g., titania nanotubes, to control the release of BMP-2 and induce bone formation [21,73]. Therefore, it will be interesting in the future to evaluate the functionalization of these surfaces via ion irradiation with non-inert gases (nitrogen and oxygen) to generate functional groups in which BMP-2 can be tether to and control its release.

4. Conclusions

The combination of growth factors (BMP-2) and active nanotopographies (nanocones and nanowalls-like structures) have demonstrated a synergy in cell adhesion, differentiation and mineralization of a non-osteoblastic cell lineage (C2C12 cells). The following findings can be drawn:

1. DIS allowed the design of nanostructures on titanium alloys surfaces, resulting in nanocones and nanowalls of size below 50 nm by changing the incidence angle from 60 to 80 degrees with high fluences.
2. The crystalline structure of DIS samples was unmodified; and although no statistical differences were observed in terms of wettability, DIS samples seemed more hydrophilic than polished samples.
3. The presence of BMP-2 plays an important role in cellular adhesion and spreading. In this study, BMP-2 addition seemed to increase filopodia number per cell and vinculin expression in most surfaces and cell spreading on DIS 80° and polished surfaces compared to surfaces without BMP-2. However, surface topography or nanopatterning by itself does not contribute significantly to these processes, except by increasing slightly vinculin expression in DIS 80° nanocone-patterned surfaces.
4. DIS 80° nanocone-patterned surfaces in conjunction with BMP-2 increase almost 1.2-fold cell spreading and 2-fold vinculin expression, reaching values similar to polished samples with BMP-2. Moreover, we observed a 2.25-fold increase in the number of filopodia per cells (39 ± 20) in these surfaces compared to all surfaces with or without BMP-2, suggesting a synergistic effect in cell adhesion when we combine DIS 80° treatment with BMP-2.
5. Cell differentiation and mineralization, determined by ALP activity and calcium nodules formation, respectively, were enhanced in the presence of BMP-2 for all samples. In particular, we observed that this effect was more pronounced on DIS 80° and Anodized samples with BMP-2 (>2-fold increase in ALP activity compared to their counterparts without BMP-2 and >3.3-fold increase in cell mineralization compared to polished samples with BMP-2).
6. Finally, the nanocone-like structures generated at an incidence angle of 80° in combination with BMP-2 have shown a stronger synergistic effect in modulating cellular processes when compared to DIS 60° and polished observing this nanocone topography more suitable to improve cellular interactions. Thus, DIS treatment in conjunction with BMP-2 may improve Ti implants osseointegration by guiding cell differentiation toward bone formation.

Author Contributions: Conceptualization, project administration, supervision, methodology, J.P.A., J.J.P., R.M., E.P., J.F.A., N.B., Y.T., J.A.R.-O., A.C. and A.M.-R., Investigation, formal analysis, validation, A.M-R. and A.C. Discussion and writing—original draft preparation, all the authors. All authors have read and agreed to the published version of the manuscript.

Funding: This work was supported by the Ministry of Science and Innovation of Spain under the grant PID2019-109371GB-I00, by the Junta de Andalucía–FEDER (Spain) through the Project Ref. US-1259771, Colciencias (Departamento Administrativo de Ciencia, Tecnología e Innovación) under the grant COL-13-2-16 and University of Antioquia Master's Scholarship fund. The BMP-2 production work was supported by AM LTDA and Transferencia Tecnológica from University of Antioquia under the grant CODI-UdeA # 2017-18192.

Data Availability Statement: Not applicable.

Acknowledgments: The authors would like to thank University of Antioquia and Colciencias for the funding, Sandra Arias and Camilo Jaramillo for the DIS treatment and Viviana Posada for taking some of the SEM micrographs as well as professors Luz Marina Restrepo, Junes Villarraga and Gabriel Bedoya for their guidance. Surface characterization was performed in Frederick Seitz Materials Research Lab and all the in vitro experiments were conducted in Micro and Nanotechnology Lab in UIUC.

Conflicts of Interest: The authors declare no conflict of interest.

References

1. de Witte, T.M.; Fratila-Apachitei, L.E.; Zadpoor, A.A.; Peppas, N.A. Bone tissue engineering via growth factor delivery: From scaffolds to complex matrices. *Regen. Biomater.* **2018**, *5*, 197–211. [CrossRef]
2. Jin, W.; Chu, P.K. Orthopedic implants. In *Encyclopedia of Biomedical Engineering*; Elsevier: Amsterdam, The Netherlands, 2019; pp. 425–439. [CrossRef]
3. Kirmanidou, Y.; Sidira, M.; Drosou, M.E.; Bennani, V.; Bakopoulou, A.; Tsouknidas, A.; Michailidis, N.; Michalakis, K. New Ti-alloys and surface modifications to improve the mechanical properties and the biological response to orthopedic and dental implants: A review. *BioMed Res. Int.* **2016**, *2016*, 1–21. [CrossRef] [PubMed]
4. Shah, F.A.; Thomsen, P.; Palmquist, A. Osseointegration and current interpretations of the bone-implant interface. *Acta Biomater.* **2019**, *84*, 1–15. [CrossRef] [PubMed]
5. Spriano, S.; Yamaguchi, S.; Baino, F.; Ferraris, S. A critical review of multifunctional titanium surfaces: New frontiers for improving osseointegration and host response, avoiding bacteria contamination. *Acta Biomater.* **2018**, *79*, 1–22. [CrossRef] [PubMed]
6. Ramazanoglu, M.; Oshi, Y. Osseointegration and bioscience of implant surfaces—Current concepts at bone-implant interface. In *Implant Dentistry—A Rapidly Evolving Practice*; Turkyilmaz, I., Ed.; InTech: Rijeka, Croatia, 2011. Available online: http://www.intechopen.com/books/implant-dentistry-a-rapidly-evolving-practice/osseointegration-and-bioscience-of-implant-surfaces-current-concepts-at-bone-implant-interface (accessed on 4 October 2014).
7. Landgraeber, S.; Jäger, M.; Jacobs, J.J.; Hallab, N.J. The pathology of orthopedic implant failure is mediated by innate immune system cytokines. *Mediat. Inflamm.* **2014**, *2014*, 1–9. [CrossRef] [PubMed]
8. Durmus, N.G.; Webster, T.J. Nanostructured titanium: The ideal material for improving orthopedic implant efficacy? *Nanomedicine* **2012**, *7*, 791–793. [CrossRef] [PubMed]
9. Mansoorianfar, M.; Khataee, A.; Riahi, Z.; Shahin, K.; Asadnia, M.; Razmjou, A.; Hojjati-Najafabadi, A.; Mei, C.; Orooji, Y.; Li, D. Scalable fabrication of tunable titanium nanotubes via sonoelectrochemical process for biomedical applications. *Ultrason. Sonochem.* **2020**, *64*, 104783. [CrossRef]
10. Beutner, R.; Michael, J.; Schwenzer, B.; Scharnweber, D. Biological nano-functionalization of titanium-based biomaterial surfaces: A flexible toolbox. *J. R. Soc. Interface* **2009**, *7*, S93–S105. [CrossRef]
11. Feller, L.; Jadwat, Y.; Khammissa, R.A.G.; Meyerov, R.; Schechter, I.; Lemmer, J. Cellular responses evoked by different surface characteristics of intraosseous titanium implants. *BioMed Res. Int.* **2015**, *2015*, 1–8. [CrossRef] [PubMed]
12. Lavenus, S.; Louarn, G.; Layrolle, P. Nanotechnology and dental implants. *Int. J. Biomater.* **2010**, *2010*, 1–9. [CrossRef]
13. Aminuddin, N.I.; Ahmad, R.; Akbar, S.A.; Pingguan-Murphy, B. Osteoblast and stem cell response to nanoscale topographies: A review. *Sci. Technol. Adv. Mater.* **2016**, *17*, 698–714. [CrossRef] [PubMed]
14. Bosshardt, D.D.; Chappuis, V.; Buser, D. Osseointegration of titanium, titanium alloy and zirconia dental implants: Current knowledge and open questions. *Periodontol. 2000* **2017**, *73*, 22–40. [CrossRef] [PubMed]
15. Gittens, R.A.; Scheideler, L.; Rupp, F.; Hyzy, S.L.; Geis-Gerstorfer, J.; Schwartz, Z.; Boyan, B.D. A review on the wettability of dental implant surfaces II: Biological and clinical aspects. *Acta Biomater.* **2014**, *10*, 2907–2918. [CrossRef] [PubMed]
16. Jemat, A.; Ghazali, M.J.; Razali, M.; Otsuka, Y. Surface modifications and their effects on titanium dental implants. *BioMed Res. Int.* **2015**, *2015*, 1–11. [CrossRef] [PubMed]
17. Civantos, A.; Martínez-Campos, E.; Ramos, V.; Elvira, C.; Gallardo, A.; Abarrategi, A. Titanium coatings and surface modifications: Toward clinically useful bioactive implants. *ACS Biomater. Sci. Eng.* **2017**, *3*, 1245–1261. [CrossRef]
18. Souza, J.C.; Sordi, M.B.; Kanazawa, M.; Ravindran, S.; Henriques, B.; Silva, F.S.; Aparicio, C.; Cooper, L.F. Nano-scale modification of titanium implant surfaces to enhance osseointegration. *Acta Biomater.* **2019**, *94*, 112–131. [CrossRef] [PubMed]
19. İzmir, M.; Ercan, B. Anodization of titanium alloys for orthopedic applications. *Front. Chem. Sci. Eng.* **2019**, *13*, 28–45. [CrossRef]
20. Gao, A.; Hang, R.; Bai, L.; Tang, B.; Chu, P.K. Electrochemical surface engineering of titanium-based alloys for biomedical application. *Electrochim. Acta* **2018**, *271*, 699–718. [CrossRef]
21. Tao, B.; Deng, Y.; Song, L.; Ma, W.; Qian, Y.; Lin, C.; Yuan, Z.; Lu, L.; Chen, M.; Yang, X.; et al. BMP2-loaded titania nanotubes coating with pH-responsive multilayers for bacterial infections inhibition and osteogenic activity improvement. *Colloids Surfaces B Biointerfaces* **2019**, *177*, 242–252. [CrossRef]
22. Li, Y.; Song, Y.; Ma, A.; Li, C. Surface immobilization of TiO2 nanotubes with bone morphogenetic protein-2 synergistically enhances initial Preosteoblast adhesion and osseointegration. *BioMed Res. Int.* **2019**, *2019*, 1–12. [CrossRef] [PubMed]

23. Wigmosta, T.B.; Popat, K.C.; Kipper, M.J. BMP -2 delivery from polyelectrolyte multilayers enhancesosteogenic activityon nanostructured titania. *J. Biomed. Mater. Res. Part A* **2020**. [CrossRef] [PubMed]

24. Oliveira, W.F.; Arruda, I.R.; Silva, G.M.; Machado, G.; Coelho, L.C.; Correia, M.T. Functionalization of titanium dioxide nanotubes with biomolecules for biomedical applications. *Mater. Sci. Eng. C* **2017**, *81*, 597–606. [CrossRef] [PubMed]

25. Teng, F.Y.; Tai, I.C.; Ho, M.L.; Wang, J.W.; Weng, L.W.; Wang, Y.J.; Wang, M.W.; Tseng, C.C. Controlled release of BMP-2 from titanium with electrodeposition modification enhancing critical size bone formation. *Mater. Sci. Eng. C* **2019**, *105*, 109879. [CrossRef]

26. Wu, C.; Lu, H. Smad signal pathway in BMP-2-induced osteogenesis a mini review. *J. Dent. Sci.* **2008**, *3*, 13–21.

27. Teng, F.Y.; Chen, W.C.; Wang, Y.L.; Hung, C.C.; Tseng, C.C. Effects of osseointegration by bone morphogenetic protein-2 on titanium implants in vitro and in vivo. *Bioinorg. Chem. Appl.* **2016**, *2016*, 1–9. [CrossRef]

28. Kang, Y.; Ren, X.; Yuan, X.; Ma, L.; Xie, Y.; Bian, Z.; Zuo, J.; Wang, X.; Yu, Z.; Zhou, K.; et al. The effects of combined micron-scale surface and different nanoscale features on cell response. *Adv. Mater. Sci. Eng.* **2018**, *2018*, 1–9. [CrossRef]

29. Greer, A.I.; Goriainov, V.; Kanczler, J.; Black, C.R.; Turner, L.A.; Meek, R.M.; Burgess, K.; MacLaren, I.; Dalby, M.J.; Oreffo, R.O.; et al. Nanopatterned titanium implants accelerate bone formation in vivo. *ACS Appl. Mater. Interfaces* **2020**, *12*, 33541–33549. [CrossRef]

30. Allain, J.P.; Shetty, A. Unraveling atomic-level self-organization at the plasma-material interface. *J. Phys. D Appl. Phys.* **2017**, *50*, 283002. [CrossRef]

31. Averback, R. Ion-irradiation studies of cascade damage in metals. *J. Nucl. Mater.* **1982**, *108–109*, 33–45. [CrossRef]

32. Civantos, A.; Barnwell, A.; Shetty, A.R.; Pavón, J.J.; El-Atwani, O.; Arias, S.L.; Lang, E.; Reece, L.M.; Chen, M.; Allain, J.P. Designing nanostructured Ti6Al4V bioactive interfaces with directed irradiation synthesis toward cell stimulation to promote host–tissue-implant integration. *ACS Biomater. Sci. Eng.* **2019**, *5*, 3325–3339. [CrossRef]

33. Civantos, A.; Allain, J.P.; Pavón, J.J.; Shetty, A.; El-Atwani, O.; Walker, E.; Arias, S.L.; Gordon, E.; Rodríguez-Ortiz, J.A.; Chen, M.; et al. Directed irradiation synthesis as an advanced plasma technology for surface modification to activate porous and "as-received" titanium surfaces. *Metals* **2019**, *9*, 1349. [CrossRef]

34. Krasheninnikov, A.V.; Nordlund, K. Ion and electron irradiation-induced effects in nanostructured materials. *J. Appl. Phys.* **2010**, *107*, 071301. [CrossRef]

35. Ghasali, E.; Baghchesaraee, K.; Orooji, Y. Study of the potential effect of spark plasma sintering on the preparation of complex FGM/laminated WC-based cermet. *Int. J. Refract. Met. Hard Mater.* **2020**, *92*, 105328. [CrossRef]

36. de Gorter, D.J.; van Dinther, M.; Dijke, P.T. Measurement of constitutive activity of BMP type I receptors. *Methods Enzymol.* **2010**. [CrossRef]

37. Mancini, A.; El Bounkari, O.; Norrenbrock, A.F.; Scherr, M.; Schaefer, D.; Eder, M.; Banham, A.H.; Pulford, K.; Lyne, L.; Whetton, A.D.; et al. FMIP controls the adipocyte lineage commitment of C2C12 cells by downmodulation of C/EBPalpha. *Oncogene* **2006**, *26*, 1020–1027. [CrossRef]

38. Katagiri, T.; Yamaguchi, A.; Komaki, M.; Abe, E.; Takahashi, N.; Ikeda, T.; Rosen, V.; Wozney, J.M.; Fujisawa-Sehara, A.; Suda, T. Bone morphogenetic protein-2 converts the differentiation pathway of C2C12 myoblasts into the osteoblast lineage. *J. Cell Biol.* **1994**, *127*, 1755–1766. [CrossRef]

39. Hidaka, Y.; Chiba-Ohkuma, R.; Karakida, T.; Onuma, K.; Yamamoto, R.; Fujii-Abe, K.; Saito, M.M.; Yamakoshi, Y.; Kawahara, H. Combined effect of midazolam and bone morphogenetic protein-2 for differentiation induction from C2C12 myoblast cells to osteoblasts. *Pharmaceutics* **2020**, *12*, 218. [CrossRef]

40. Mishra, I.; Joshi, S.R.; Majumder, S.; Manna, A.K.; Varma, S. Low energy ion irradiation of TiO 2 (110)—Understanding evolution of surface morphology and scaling studies. *Radiat. Eff. Defects Solids* **2016**, *171*, 594–605. [CrossRef]

41. Xue, L. Laser Consolidation—A rapid manufacturing process for making net-shape functional components. In *Advances in Laser Materials Processing*; Elsevier: Amsterdam, The Netherlands, 2018; pp. 461–505. [CrossRef]

42. Pederson, R.; Babushkin, O.; Skystedt, F.; Warren, R. Use of high temperature X-ray diffractometry to study phase transitions and thermal expansion properties in Ti-6Al-4V. *Mater. Sci. Technol.* **2003**, *19*, 1533–1538. [CrossRef]

43. Albertini, M.; Yagüe, M.-A.F.; Lázaro, P.; Herrero-Climent, M.; Rios-Santos, J.-V.; Bullon, P.; Gil Mur, F.J. Advances in surfaces and osseointegration in implantology. Biomimetic surfaces. *Med. Oral Patol. Oral Cir. Bucal* **2015**, *20*, e316–e325. [CrossRef] [PubMed]

44. Hu, W.; Wehrle-Haller, B.; Vogel, V. Maturation of filopodia shaft adhesions is upregulated by local cycles of lamellipodia advancements and retractions. *PLoS ONE* **2014**, *9*, e107097. [CrossRef]

45. Jacquemet, G.; Hamidi, H.; Ivaska, J. Filopodia in cell adhesion, 3D migration and cancer cell invasion. *Curr. Opin. Cell Biol.* **2015**, *36*, 23–31. [CrossRef]

46. Mattila, P.K.; Lappalainen, P. Filopodia: Molecular architecture and cellular functions. *Nat. Rev. Mol. Cell Biol.* **2008**, *9*, 446–454. [CrossRef]

47. Bays, J.L.; DeMali, K.A. Vinculin in cell–cell and cell–matrix adhesions. *Cell. Mol. Life Sci.* **2017**, *74*, 2999–3009. [CrossRef]

48. Kechagia, J.Z.; Ivaska, J.; Roca-Cusachs, P. Integrins as biomechanical sensors of the microenvironment. *Nat. Rev. Mol. Cell Biol.* **2019**, *20*, 457–473. [CrossRef]

49. Bertrand, A.A.; Malapati, S.H.; Yamaguchi, D.T.; Lee, J.C. The intersection of mechanotransduction and regenerative osteogenic materials. *Adv. Healthc. Mater.* **2020**, *9*, 2000709. [CrossRef]

50. Fourel, L.; Valat, A.; Faurobert, E.; Guillot, R.; Bourrin-Reynard, I.; Ren, K.; Lafanechère, L.; Planus, E.; Picart, C.; Albiges-Rizo, C. β3 integrin–mediated spreading induced by matrix-bound BMP-2 controls Smad signaling in a stiffness-independent manner. *J. Cell Biol.* **2016**, *212*, 693–706. [CrossRef] [PubMed]

51. Huang, B.; Yuan, Y.; Li, T.; Ding, S.; Zhang, W.; Gu, Y.; Liu, C. Facilitated receptor-recognition and enhanced bioactivity of bone morphogenetic protein-2 on magnesium-substituted hydroxyapatite surface. *Sci. Rep.* **2016**, *6*, 24323. [CrossRef]

52. Pan, H.; Xie, Y.; Zhang, Z.; Li, K.; Hu, D.; Zheng, X.; Fan, Q.; Tang, T. YAP-mediated mechanotransduction regulates osteogenic and adipogenic differentiation of BMSCs on hierarchical structure. *Colloids Surf. B Biointerfaces* **2017**, *152*, 344–353. [CrossRef]

53. Park, J.; Bauer, S.; Von Der Mark, K.; Schmuki, P. Nanosize and vitality: TiO2Nanotube diameter directs cell fate. *Nano Lett.* **2007**, *7*, 1686–1691. [CrossRef]

54. Selhuber-Unkel, C.; Erdmann, T.; López-García, M.; Kessler, H.; Schwarz, U.; Spatz, J. Cell adhesion strength is controlled by intermolecular spacing of adhesion receptors. *Biophys. J.* **2010**, *98*, 543–551. [CrossRef]

55. Jäger, M.; Jennissen, H.P.; Dittrich, F.; Fischer, A.; Köhling, H.L. Antimicrobial and osseointegration properties of nanostructured titanium orthopaedic implants. *Materials* **2017**, *10*, 1302. [CrossRef]

56. Babchenko, O.; Kromka, A.; Hruska, K.; Kalbacova, M.H.; Broz, A.; Vanecek, M. Fabrication of nano-structured diamond films for SAOS-2 cell cultivation. *Phys. Status Solidi A* **2009**, *206*, 2033–2037. [CrossRef]

57. Kalbacova, M.; Broz, A.; Babchenko, O.; Kromka, A. Study on cellular adhesion of human osteoblasts on nano-structured diamond films. *Phys. Status Solidi B* **2009**, *246*, 2774–2777. [CrossRef]

58. Ferreira, M.R.W.; Fernandes, R.R.; Assis, A.F.; Dernowsek, J.A.; Passos, G.A.; Variola, F.; Bombonato-Prado, K.F. Oxidative nanopatterning of titanium surface influences mRNA and MicroRNA expression in human alveolar bone osteoblastic cells. *Int. J. Biomater.* **2016**, *2016*, 1–15. [CrossRef]

59. Capulli, M.; Paone, R.; Rucci, N. Osteoblast and osteocyte: Games without frontiers. *Arch. Biochem. Biophys.* **2014**, *561*, 3–12. [CrossRef]

60. James, A.W. Review of signaling pathways governing MSC osteogenic and adipogenic differentiation. *Scientifica* **2013**, *2013*, 1–17. [CrossRef]

61. Karperien, M.; Roelen, B.A.; Poelmann, R.E.; Groot, A.C.G.-D.; Hierck, B.P.; DeRuiter, M.C.; Meijer, D.; Gibbs, S. Tissue formation during embryogenesis. *Tissue Eng.* **2015**. [CrossRef]

62. Blair, H.C.; Larrouture, Q.C.; Li, Y.; Lin, H.; Beer-Stoltz, D.; Liu, L.; Tuan, R.S.; Robinson, L.J.; Schlesinger, P.H.; Nelson, D.J. Osteoblast differentiation and bone matrix formation in vivo and in vitro. *Tissue Eng. Part B Rev.* **2017**, *23*, 268–280. [CrossRef]

63. Vaes, B.L.T.; Dechering, K.J.; Feijen, A.; Hendriks, J.M.A.; Lefévre, C.; Mummery, C.L.; Olijve, W.; Van Zoelen, E.J.J.; Steegenga, W.T. Comprehensive microarray analysis of bone morphogenetic protein 2-induced osteoblast differentiation resulting in the identification of novel markers for bone development. *J. Bone Miner. Res.* **2002**, *17*, 2106–2118. [CrossRef]

64. Mesa-Restrepo, A.; Alzate, J.F.; Patiño-Gonzalez, E. Bone morphogenetic protein 2: Heterologous expression and potential in bone regeneration. *Actual. Biól.* **2021**, *43*, 1–10, in press. [CrossRef]

65. Xiao, M.; Biao, M.; Chen, Y.; Xie, M.; Yang, B.; Xiao, M.; Biao, M.; Chen, Y.; Xie, M.; Yang, B. Regulating the osteogenic function of rhBMP 2 by different titanium surface properties. *J. Biomed. Mater. Res. Part A* **2016**, *104*, 1882–1893. [CrossRef]

66. Biao, M.N.; Chen, Y.M.; Xiong, S.B.; Wu, B.Y.; Yang, B.C. Synergistic effects of fibronectin and bone morphogenetic protein on the bioactivity of titanium metal. *J. Biomed. Mater. Res. Part A* **2017**, *105*, 2485–2498. [CrossRef] [PubMed]

67. Cavalcanti-Adam, E.A.; Aydin, D.; Hirschfeld-Warneken, V.C.; Spatz, J.P. Cell adhesion and response to synthetic nanopatterned environments by steering receptor clustering and spatial location. *HFSP J.* **2008**, *2*, 276–285. [CrossRef] [PubMed]

68. Rosa, A.; Kato, R.; Raucci, L.C.; Teixeira, L.; De Oliveira, F.; Bellesini, L.; De Oliveira, P.; Hassan, M.; Beloti, M. Nanotopography drives stem cell fate toward osteoblast differentiation through α1β1 integrin signaling pathway. *J. Cell. Biochem.* **2014**, *115*, 540–548. [CrossRef]

69. Lotz, E.M.; Olivares-Navarrete, R.; Berner, S.; Boyan, B.D.; Schwartz, Z. Osteogenic response of human MSCs and osteoblasts to hydrophilic and hydrophobic nanostructured titanium implant surfaces. *J. Biomed. Mater. Res. Part A* **2016**, *104*, 3137–3148. [CrossRef] [PubMed]

70. Fernández-Yagüe, M.; Antoñanzas, R.P.; Roa, J.J.; Biggs, M.; Gil, F.J.; Pegueroles, M. Enhanced osteoconductivity on electrically charged titanium implants treated by physicochemical surface modifications methods. *Nanomed. Nanotechnol. Biol. Med.* **2019**, *18*, 1–10. [CrossRef] [PubMed]

71. Lopes, H.B.; Freitas, G.P.; Elias, C.N.; Tye, C.; Stein, J.L.; Stein, G.S.; Lian, J.B.; Rosa, A.L.; Beloti, M.M. Participation of integrin β3 in osteoblast differentiation induced by titanium with nano or microtopography. *J. Biomed. Mater. Res. Part A* **2019**, *107*, 1303–1313. [CrossRef]

72. El Bialy, I.; Jiskoot, W.; Nejadnik, M.R. Formulation, delivery and stability of bone morphogenetic proteins for effective bone regeneration. *Pharm. Res.* **2017**, *34*, 1152–1170. [CrossRef]

73. Dang, M.; Saunders, L.; Niu, X.; Fan, Y.; Ma, P.X. Biomimetic delivery of signals for bone tissue engineering. *Bone Res.* **2018**, *6*, 1–12. [CrossRef] [PubMed]

74. Wang, Z.; Wang, Z.; Lu, W.W.; Zhen, W.; Yang, D.; Peng, S. Novel biomaterial strategies for controlled growth factor delivery for biomedical applications. *NPG Asia Mater.* **2017**, *9*, e435. [CrossRef]

Article

Analysis of the Main Aspects Affecting Bonding in Stainless Steel Rebars Embedded in a Hydraulic Medium

Fernando Ancio [1], Esperanza Rodriguez-Mayorga [2],* and Beatriz Hortigon [1]

[1] Departamento de Mecánica de Medios Continuos y Teoría de Estructuras, Escuela Politécnica Superior, Universidad de Sevilla, 41011 Seville, Spain; ancio@us.es (F.A.); bhortigon@us.es (B.H.)

[2] Departamento de Estructuras de Edificación e Ingeniería del Terreno, Escuela Técnica Superior de Arquitectura, Universidad de Sevilla, 41012 Seville, Spain

* Correspondence: espe@us.es; Tel.: +34-954-556-602

Abstract: The use of stainless steel rebars to reinforce masonry structures has become established as an eminently efficient methodology. From among the numerous techniques available, bed-joint structural repointing and superficial reinforcement with rebars or meshes attached to surfaces have become widespread, thanks to the excellent results they have produced in recent decades. Both techniques imply the use of diameters less than 6 mm and thin coverings. This article deals with the characterization of the bonding behavior of the rebar under these special circumstances. To this end, several finite element analyses have been carried out to identify the possible relationships between pull-out forces in various situations. These models allow certain conclusions to be drawn regarding the influence of the thickness of covering, boundary conditions, and geometrical aspects of the rebars in bonding. Certain mathematical expressions that relate the various conclusions from this research are finally laid out.

Keywords: masonry reinforcement; stainless rebar; finite element analysis; bonding behavior; boundary conditions

Citation: Ancio, F.; Rodriguez-Mayorga, E.; Hortigon, B. Analysis of the Main Aspects Affecting Bonding in Stainless Steel Rebars Embedded in a Hydraulic Medium. *Metals* **2021**, *11*, 786. https://doi.org/10.3390/met11050786

Academic Editor: Fernando Castro

Received: 24 March 2021
Accepted: 10 May 2021
Published: 12 May 2021

Publisher's Note: MDPI stays neutral with regard to jurisdictional claims in published maps and institutional affiliations.

1. Introduction

The repair of masonry structures concerns society. Historical towns not only form part of the historical and cultural heritage but also constitute a major source of income for the economy of many countries worldwide. Historical towns are usually composed of mainly humble dwellings with a few impressive and magnificent buildings, such as cathedrals and palaces. The origin of all these buildings commonly dates back to the XVIIIth century or even earlier. This fact leads to the conclusion that the constructive system in most of these buildings is that of masonry. On the other hand, the suburbs generally accommodate the majority of the population in large cities, a number of which are ancient neighborhoods without any historical or artistic value, but have also been built with masonry. In this last case, the repair of masonry becomes a social problem for administrations since the repair must be carried out on a reduced budget while preventing the reallocation of residents as far as possible. The development of repair and consolidation techniques for all types of masonry has therefore awakened the interest of many administrations.

Historical masonry is usually composed of three layers: two external layers of stone units infilled with rubble masonry, all pointed with poor lime mortar. In more recent masonries, brick pieces are commonly joined with cement mortars. Many of the repair and consolidation techniques usually applied to masonry structures include the introduction of connectors and/or reinforcements. These have traditionally been steel pieces located either in parallel or perpendicular to the wall faces [1,2]. This reinforcement is frequently introduced into the bed-joints, thereby preventing damage to the masonry units [3–5]. This repair and consolidation methodology, particularly known as bed-joint structural repointing, is especially suitable for historic masonries since it almost totally respects the original materials that compose the structural elements. This technique is currently applied by substituting the steel rebars with fibers, thus attaining a more respectful way to repair walls since smaller sections of reinforcements are required than when steel pieces are used [6]. The main disadvantage of this technique is, obviously, its cost. The cost of this technique makes it unaffordable in many cases, not only due to the cost of the fibers but also to the execution of the work itself. The use of steel rebars instead of fiber plates contributes significantly towards reducing the cost of the application of the technique. Even if stainless steel is chosen, the cost can be reduced by approximately 500% when compared with fibers. The use of stainless steel is almost compulsory when repairing historical masonries since traditional dwellings are usually affected by rising damp [7]. Furthermore, if masonry without any historical or artistic value is being repaired, then there is no issue with introducing the reinforcement into drills and grooves or placing them superficially inside renders.

These reinforcements, whose effectiveness is widely demonstrated, constitute one of the most financially feasible techniques of repair and consolidation available today. Steel rebars with diameters up to 6 mm normally provide the reinforcements. A particular feature shared by these reinforcements is the mortar thicknesses that cover the rebars, which are usually thinner than those existing in reinforced concrete. The reason is obviously linked to the small cavity in which the rebar is usually introduced, as well as the thin layer of mortar that embeds the rebars when they are superficially placed [8,9]. This fact has traditionally been disregarded, however, since there are no special standardized tests to determine the bonding behavior of rebars placed as described. Furthermore, the fact that only bars with small diameters are used in these repairs also modifies their bonding behavior since no linear relationship has been identified between the diameter and load transfer of bars [10]. Lastly, the shape of the bars exerts a strong influence on bonding, but this is poorly documented regarding bars with diameters of up to 10 mm embedded in a medium different from concrete [11,12].

The beam test, as the standard bonding test collected in codes [13] aims to determine the force needed to extract a rebar with diameter of up to 16 mm from a prism of concrete that has effective coverings of 50 mm in three of the faces. The lack of codes for the particular situation of bars reinforcing masonries leads to the necessity for the adaptation of them from similar fields. The bonding of anchors can be regulated by the British Standard BS EN 1881:2006 [14], while BS EN 846-2:2001 [15] can be applicable to bed-joint structural repointing. The codes establish the measurement of the axial force to pull out the rebar from 30-mm grouted drills or from its position in the brick joint, respectively. On the other hand, RILEM-TC RC6 [16] opens the door to a simpler test from the point of view of the development of the samples. The aim of this test is to determine the pull-out force necessary to pull out a bar embedded in a cube whose edge measures 10 times the diameter of the bar and is fixed in the face in which the bar is inserted. A similar test is proposed for fibers when used to reinforce masonry structures [17]. The variety of codes available drives undoubtedly to the fact that results from various studies are incomparable, since they depend on the criteria followed when obtained [18–20].

The bonding of bars to reinforce masonry is barely documented. The aim of this paper is to research the influence that the different requirements of codes have on the final bond behavior of the bar. Since this research is oriented towards the reinforcement of masonry, stainless steel rebars embedded in prisms of hydraulic materials are analyzed under a variety conditions: (i) effective covering; (ii) boundary conditions; (iii) Young's modulus value of covering; and (iv) ribs of bar geometry.

The finite element (FE) method has been chosen for the analysis of the bonding behavior in the aforementioned specific circumstances. In this research, a complete 3D analysis of the ribbed bars was carried out in order to attain a more precise reproduction of the behavior of the rebars in terms of bonding than those achieved using macro models [21,22]. Finally, several conclusions are drawn regarding the influence of these different parameters on bonding.

2. Materials and Methods

Rebars were modelled by means of the software CAD Rhinoceros 3D. This software, together with its plug-in Grasshopper 3D, eases the parameterization and model generation for their later finite element analysis. In all cases, rebars with 30 mm length and 5 mm diameter have been analyzed. The bar shape adopted for these analyses is based on the standard stainless-steel bar, of which the transverse section is composed of the filleted intersection of three arcs (Figures 1 and 2a).

Figure 1. Standard stainless-steel rebar.

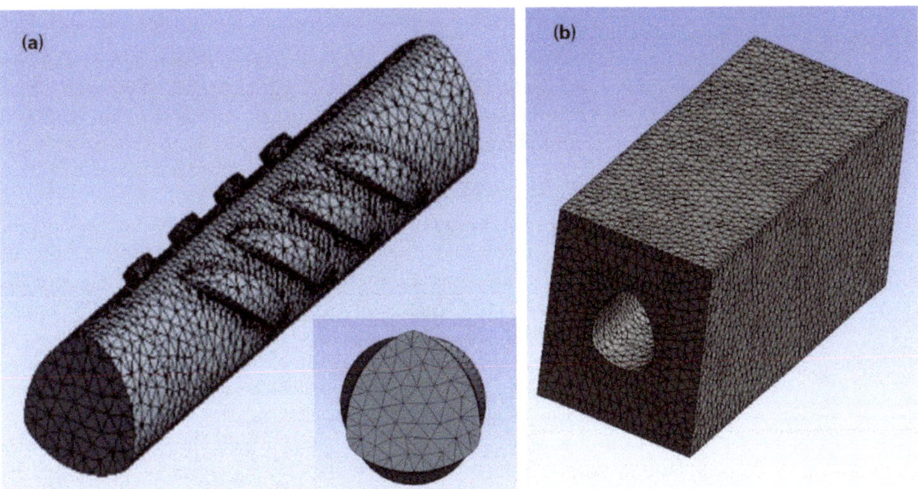

Figure 2. Finite element model for the rebar embedded in a prism of mortar whose base measures 12 mm × 12 mm: (**a**) model of the rebar; (**b**) model of the prism of mortar that surrounds the rebar.

Altogether, 280 3D solid finite element analyses have been carried out by means of the software ANSYS R19.3 for the development of this research. All the models consist of a rebar embedded in a prism of hydraulic material, reproducing the media in which rebar are inserted. Since rebars are usually surrounded by mortar or grout made with hydraulic binders (mainly lime, cement or mixtures of both), the properties assigned to the prisms are those of a generic repairing mortar. The contact between rebars and the mortar joints was defined through a cohesive model zone. In all these samples, geometrical and mechanical parameters were changed iteratively, measuring, this way, the influence of each of them in the final results in terms of bonding. Thus, (i) 16 samples were analyzed in order to calibrate the influence of the resistant properties in the cohesive model zone; (ii) in 96 samples, effective covering values ranging from 6 to 25 mm were combined with four different values of Young's modulus of the mortar joints that embedded the bars and four different boundary conditions; and finally, (iii) 168 analyses were carried out based on 42 different rib shapes, which were analyzed effectively, covered with 6 and 25 mm, as well as under two different boundary conditions.

In all the analyses, the failure of masonry by pull-out forces has been dismissed since it falls outside the scope of this research. The failure of the samples can be therefore deduced as being due to (i) excessive tensile stresses in reinforcements, which is highly improbable; (ii) cone failure of the mortar embedment, which occurs when a small part of the covering around the bar fails; (iii) sliding of the set rebar–medium along the interface between the masonry and the mortar; and (iv) sliding along the steel–mortar interface. Failure mode (ii) is taken into account when modelling the mortar joints through the microplane model by finite element analysis (Section 2.2). Regarding sliding failures, this research only considers (iv), since (iii) is considered negligible in comparison with (iv), especially regarding rebars with thin coverings [10,23]. Figure 2 depicts one of the finite element samples. As a final result, the number of nodes and elements of models ranges from 22,159 nodes to 29,223 nodes, and 98,010 elements to 124,890 elements. The descriptions of the numerical models of both the failure modes considered are addressed in the following subsections.

2.1. Modelling Rebars by Finite Elements

When rebars are employed to reinforce masonry structural elements, they are always expected to be generously less than the proportional limit of steel. For this reason, steel has been defined as an isotropic elastic linear material with E = 210 GPa and Poisson modulus ν = 0.3. Elements type 3D Solid have been employed to model the rebars and the mortar joints that surrounds them. Specifically, a tetrahedral 8-node solid element with three degrees of freedom at each node, i.e., Solid 185, was used [24]. This element has properties such as plasticity, stress-stiffening, large deflection, and large strain capabilities, which make it suitable for modelling rebars (Figure 2a).

2.2. Modelling Mortar Joints by Finite Elements

Mortars present non-linear behavior as well as differences in their responses under tensile and compressive stresses. Regarding tensile strength (f_t), sudden softening occurs accompanied by a reduction in stiffness. On the other hand, in compressive strength (f_c), stress–strain behavior will first involve ductile hardening followed by softening and reduction in the stiffness. This behavior, defined as quasi-brittle, cannot be represented by the slip theory of plasticity, since the inelastic behavior in the microscale does not physically represent a slip. The microplane model is suitable for quasi-brittle materials, since it has provided good results in similar previous experiences [12,25].

The microplane model was first enunciated by Taylor [26,27] and later developed and settled by the authors [28–33]. Microplane theory discretizes materials into 42 symmetrically located planes that define a 42-faced polyhedron (tetracontadihedron). By applying a simple constitutive law to each plane, the theory extrapolates the initial plane model to a consistent three-dimensional model. Microplane theory analyzes the physical phenomena occurring in the microstructure of the material by analyzing the physical phenomena occurring on each plane, thus reproducing the anisotropic damage.

The microplane model establishes for each of the 42 microplanes the value of equivalent strain energy (η^{mic}) as obtained by [34]:

$$\eta^{mic} = k_0 I_1 + \sqrt{k_1^2 I_1^2 + k_2 J_2}$$

(1)

where I_1 and J_2 are the first invariant of the strain tensor and the second invariant of the deviatoric part of the strain tensor ε, respectively:

$$I_1 = \varepsilon_1 + \varepsilon_2 + \varepsilon_3$$

(2)

$$J_2 = \frac{1}{6}\left[(\varepsilon_1 + \varepsilon_2)^2 + (\varepsilon_2 + \varepsilon_3)^2 + (\varepsilon_1 + \varepsilon_3)^2\right]$$

(3)

and k_0, k_1, and k_2 are constants whose values are set as follows [35–37]:

$$k_0 = k_1 = (k-1)/(2k(1-2v))$$

(4)

$$k_2 = 3/(k(1+v)^2)$$

(5)

where v is Poisson's ratio and k is the ratio between the compressive and tensile strength of the material $k = f_c/f_t$.

When the material is working under its proportional limit, the strain tensor ε can be easily obtained through Young's modulus and Poison's ratio. This elastic behavior ceases when in any of the 42 microplanes η^{mic} reaches a certain value γ_0^{mic}. Then, the material damage is considered to be initiated, and the material response is affected by the damage parameter d^{mic}:

$$d^{mic} = 1 - \frac{\gamma_0^{mic}}{\eta^{mic}}\left[1 - \alpha^{mic} + \alpha^{mic}\cdot exp\left(\beta^{mic}\left(\gamma_0^{mic} - \eta^{mic}\right)\right)\right]$$

(6)

where α^{mic} is the maximum degradation coefficient; β^{mic} is the rate of damage evolution (shape of the softening curve), and γ_0^{mic} is the equivalent strain energy when the material damage starts. When the damage starts, the material response is reduced by the factor $(1 - d^{mic})$, with d^{mic} ranging from 0 to 1, where 0 is an undamaged material and 1 a totally damaged material where the stiffness is lost.

The microplane model with damage is implemented in ANSYS R19.3 through the element Solid 185, described in Section 2.1. The microplane model in ANSYS R19.3 is defined through Young's modulus and Poisson's ratio as well as via the six constants C_1, C_2, C_3, C_4, C_5, and C_6. The values of these constants were set as k_0, k_1, k_2, γ_0^{mic}, α^{mic}, and β^{mic}, respectively. In this research, in accordance with the mechanical properties of a hydraulic mortar, previous experiences [34,38] and by comparison with results obtained from laboratory tests [39], the C constants were set to 0.729, 0.729, 0. 26, 6×10^{-5}, 0.75, and 100. These values provide quality results since the behavior of the sample predicted by finite element analysis is compatible with that published in literature for mortars specially designed to be used in repairing and retrofitting tasks (Figure 3). Young's modulus iteratively takes the values 5.6, 10, 20, and 50 GPa, thereby considering several possibilities of hydraulic mortars usually employed in repairing masonries, from poor-quality lime mortars to the high-performance grouts based on micro fine hydraulic binders. Poisson's ratio was set to 0.2 [40–44].

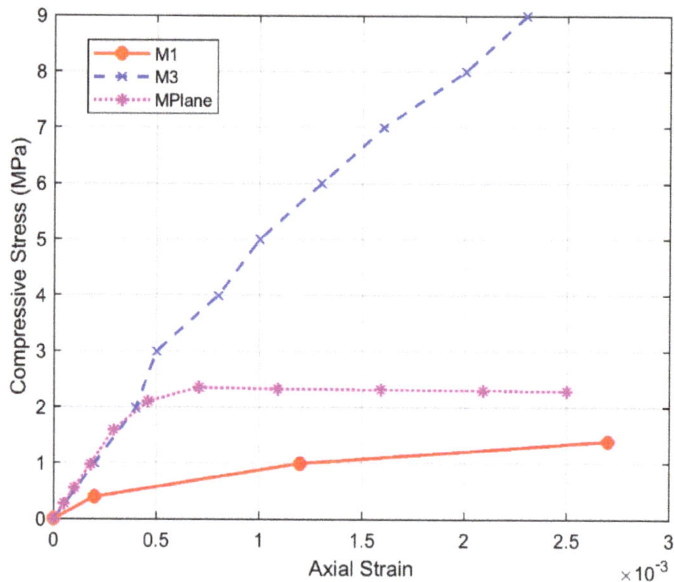

Figure 3. Comparison of the stress-strain behavior of samples of hydraulic mortars obtained in the laboratory (M1 and M3) from literature [39] and analyzed by finite elements for this research (MPlane).

2.3. Modelling Contacts

The correct reproduction of the bonding behavior of the samples by the FE models depends on the definition of contact parameters. Three phenomena occur in the interface [45–48]: (i) friction, due to the joint action of dilation slip and shear failure of the grout; (ii) mechanical interlock, mainly dependent on rib shape and rib separation; and (iii) chemical adhesion between the rebar and the media in which it is inserted. The cohesive zone model with mixed debonding interface (henceforth, CZM) reflects the behavior of the conjunct in terms of points (i) and (ii), since (iii) is usually disregarded due to its low influence in bonding [46]. CZM is modelled through elements CONTA174 and TARGE170. Elements type CONTA are used to represent contact and sliding between surfaces. The contact is pair based, with the target surface defined by the 3D target element type [24].

A CZM is defined though maximum values of bonding stress that, when exceeded, lead to the failure of the contact by slippage and/or separation between surfaces. This failure can be due to normal stress or shear stress, both being limited by the values of the maximum bond tensile stress and maximum bond shear stress. Many studies use macro models or simplified axisymmetric models for ribbed bars [21,49,50]. These works involve modeling plain bars where bond tensile stress does not influence the final results, and fictitious values of bond shear stress supply the absence of the rib. Since this research involves the consideration of real rib shapes to evaluate their influence on bonding, a sensitivity analysis was previously carried out to calibrate the values for maximum bond shear stress and maximum bond tensile stress given to the CZM. This analysis consisted of the finite element analysis of a rebar embedded in a cube with 100 mm edge, simulating a pull-out test. A longitudinal displacement of 5×10^{-5} mm was imposed on the rebar while the face of the cube in which the rebar was inserted was maintained as fixed. In this study, maximum bond shear and bond tensile stresses were fixed at 0.10, 0.25, 0.50, and 1.00 MPa. The results were analyzed in terms of the reaction force (F) transferred from the rebar to the cube of the mortar. The results are depicted in Figure 4 and can be found in greater detail in Appendix A.

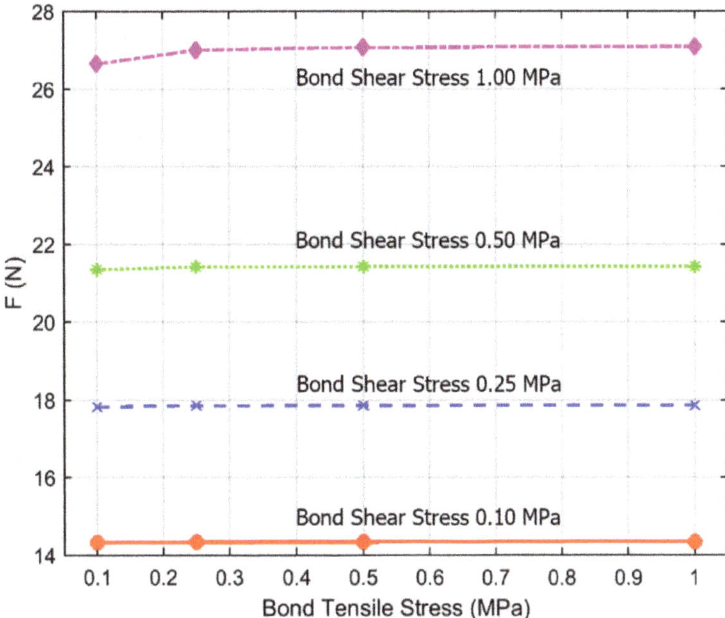

Figure 4. Values of the reaction force transferred from the bar to the surrounding media (100-mm-edge cube) for a displacement equal to 5×10^{-5} mm.

These 16 calculations confirm that maximum bond tensile stress has effects on the bonding behavior of the rebar. This thesis was previously enunciated for plain bars [23]. It is hereby confirmed that changing the limit of the bond tensile stress, which appears only in one of the rib faces, fails to significantly modify the bonding behavior of rebars. Thus, bonding behavior depends on the value of the maximum bond shear stress that is set, as well as on the compression strength of the mortar joints opposed to the compressed face of the ribs. When high values of bond shear stress are set, a homogeneous distribution of frictional stresses is obtained: this distribution is obviously unfeasible. In contrast, setting the maximum bond shear stress at lower values involves a stress distribution that is perfectly feasible (Figure 5). Therefore, maximum values were set to 0.10 MPa for bond shear stress and 1 MPa for bond tensile stress.

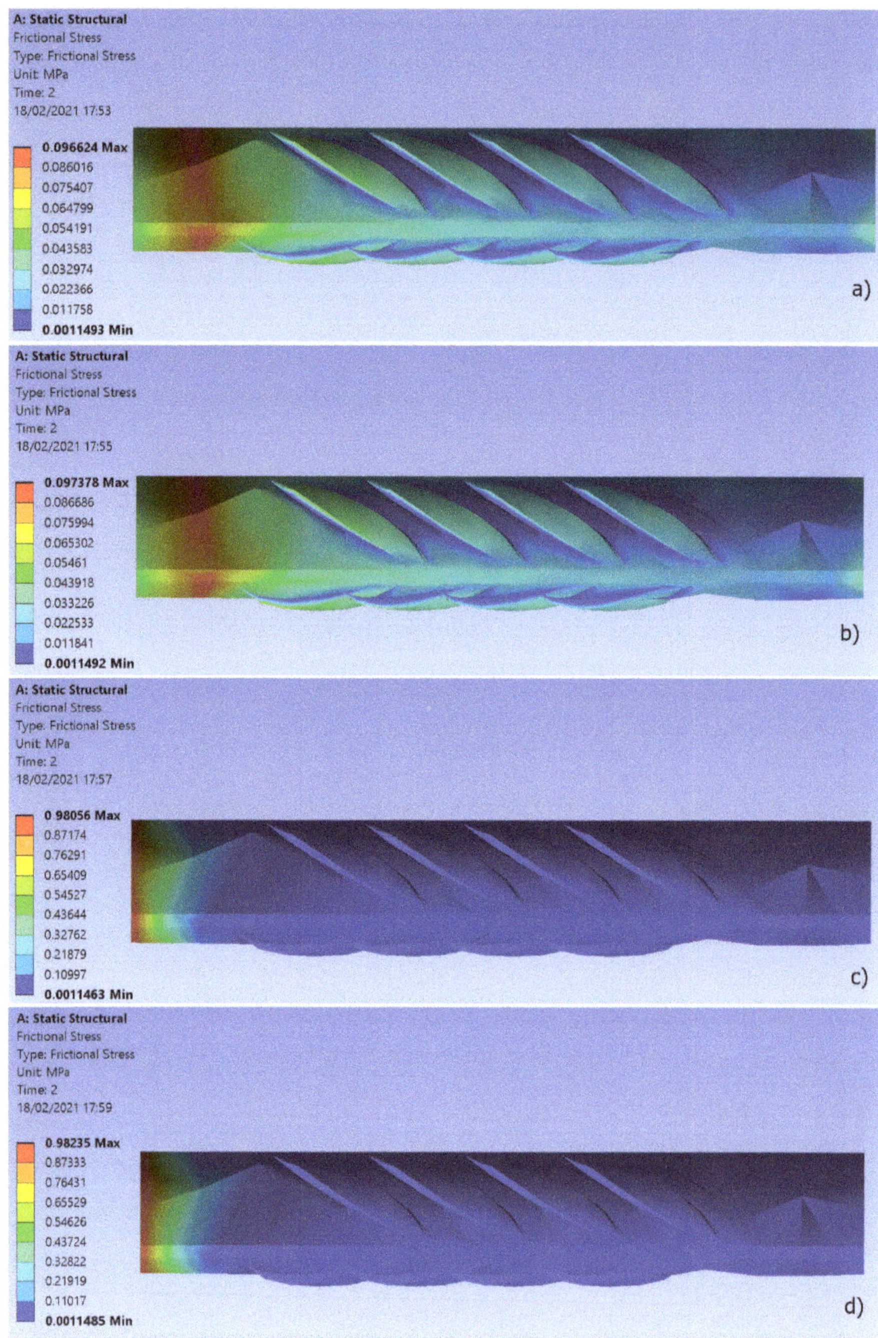

Figure 5. Bond shear stress (MPa) measured at the rebar–mortar interface for a displacement equal to 5×10^{-5} mm from a 100-mm-edge cube, where maximum bond shear stress and bond tensile stress (both in MPa) were set to (**a**) 0.10/0.10; (**b**) 0.10/1.00; (**c**) 1.00/0.10; (**d**) 1.00/1.00.

3. Results

The results obtained from all the analyses are presented in this section and are subsequently discussed in Section 4. As in the previous sections, the results are presented in terms of the reaction force (F) induced in the medium when the rebar was longitudinally displaced 5×10^{-5} mm from its initial position. Ninety-six samples were analyzed taking into account several values of the thickness of the covering and Young's modulus of the mortar in which the rebar is embedded as well as different boundary conditions (Section 3.1). On the other hand, 42 samples were obtained for different rib shapes and were analyzed under different boundary conditions and with two values of effective covering (Section 3.2).

3.1. Thickness of Covering, Boundary Conditions, and Material

The results of the effect of the thickness of covering, boundary conditions, and Young's modulus of material in which the rebar is embedded are shown in Table 1. In these analyses, different values for the parameters were combined while the fixed values of 30 mm and 5 mm were taken for the rebar length and rebar diameter, respectively.

Table 1. Reaction force F (N) needed to longitudinally displace a rebar 5×10^{-5} mm from a prism of mortar under various geometrical and mechanical conditions.

Boundary Conditions	Effective Covering (mm)	F (N)			
		* E_1	* E_2	* E_3	* E_4
Fixed base (1)	6	4.846	8.07	13.769	26.074
	7.5	6.479	10.779	18.287	34.375
	10	8.636	14.384	24.432	44.732
	12.5	10.251	17.075	28.89	49.997
	15	10.539	17.648	30.021	50.605
	25	12.633	20.941	34.734	49.328
Two fixed lateral faces (2, 3)	6	12.91	22.275	35.564	50.379
	7.5	12.67	21.987	37.243	50.698
	10	10.33	18.066	33.671	50.172
	12.5	8.91	15.637	29.947	50.406
	15	6.82	12.021	23.319	47.906
	25	5.512	9.741	19.029	42.943
Three fixed lateral faces (2, 3, 4)	6	17.899	30.457	43.622	49.251
	7.5	17.358	29.775	44.931	50.366
	10	13.916	24.158	42.513	50.126
	12.5	11.894	20.745	38.599	50.094
	15	9.024	15.834	30.348	48.786
	25	7.446	13.112	25.371	49.535
Four fixed lateral faces (2, 3, 4, 5)	6	20.502	34.492	43.613	50.869
	7.5	19.552	33.238	44.414	51.268
	10	15.441	26.706	44.889	50.568
	12.5	13.089	22.768	41.415	50.317
	15	9.858	17.262	32.917	48.821
	25	8.124	14.287	27.546	48.663

* Values of Young's modulus (GPa): $E_1 = 5.6$, $E_2 = 10$, $E_3 = 20$, $E_4 = 50$.

Regarding the thickness of covering, rebars were embedded in mortar prisms with base A × A, where A was changed iteratively: 12, 15, 20, 25, 30, and 50 mm. This assumes effective coverings of 6, 7.5, 12.5, 15, and 25 mm, respectively, which always remain below the minimum values established by codes, as was presented in Section 1. On the other hand, Young's modulus of hydraulic material that composes prisms surrounding the rebars had values of 5.6, 10, 20, and 50 GPa. Finally, prisms that surround rebars were fixed in their base (face number 1), two lateral faces (face numbers 2 and 3), three lateral faces (face numbers 2, 3, and 4), and four lateral faces (face numbers 2, 3, 4, and 5) iteratively in each analysis (Figure 6).

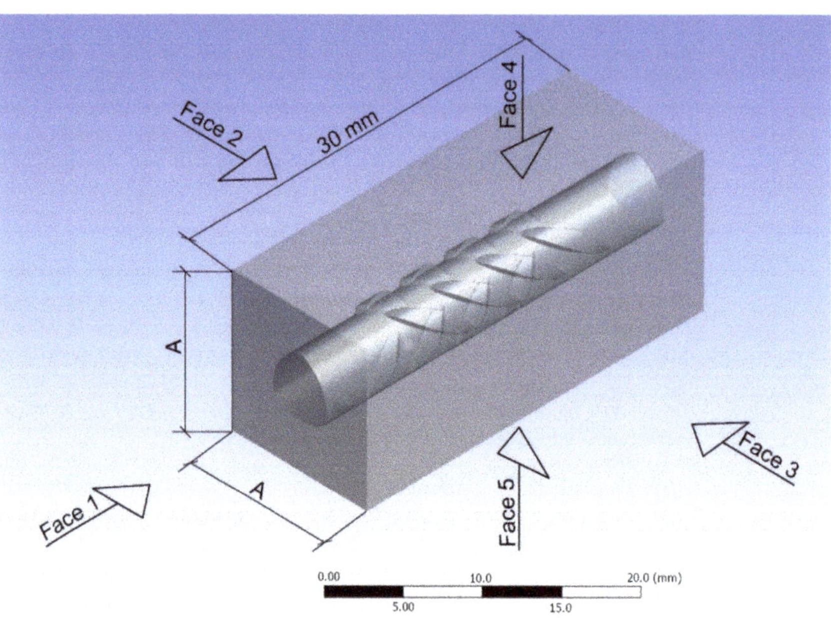

Figure 6. Diagram depicting the parameterization of samples to set the different mechanical and geometrical characteristics of each of the finite element analyses carried out.

To better understand load transference from rebar to mortar joints, as well as the failure mechanisms of rebars with small diameters thinly covered, an analysis of the slippage failure was carried out. This analysis consisted of augmenting the value of the displacement imposed up to failure of the interface. It was carried out twice. In both analyses, mortar with Young's modulus of 5.6 MPa and rebar effectively covered with 7.5 mm were set. These samples were analysed when (i) the base (face number 1) was fixed and when (ii) three lateral faces were fixed (face numbers 2, 3, 4). Figure 7 represents the force transferred to the mortar in the function of the longitudinal displacement of the rebar for every load step-up to failure in cases (i) and (ii). The maximum force in (ii) assumes a 90.2% increment with respect to (i). Stress distribution for (i) for 40%, 60%, and 100% of the force transferred from the rebar to the mortar when failure occurs is presented in Figure 8, while Figure 9 depicts similar load steps for (ii).

Figure 8a–c show bonding shear stress distribution for 40%, 60%, and 100% of the failure load, respectively, when the rebar is embedded in a prism fixed in the base (face number 1). In the first steps of the simulation (Figure 8a), the load is mainly transferred to the area of the rebar close to the application of the displacement. Despite the fact that only 40% of the maximum force was transferred, the maximum value of bonding shear stress (0.1 MPa) was reached in some points. Later in the simulation, the area of the rebar close to the application of the displacement (Figure 8b,c) fails due to slippage and does not transfer any force. Figure 8d–f show bonding normal stresses for 40%, 60%, and 100% of the failure load, respectively. These three pictures show clearly how only the compressed area of the ribs closest to the area where the displacement was applied contributed to the force transference.

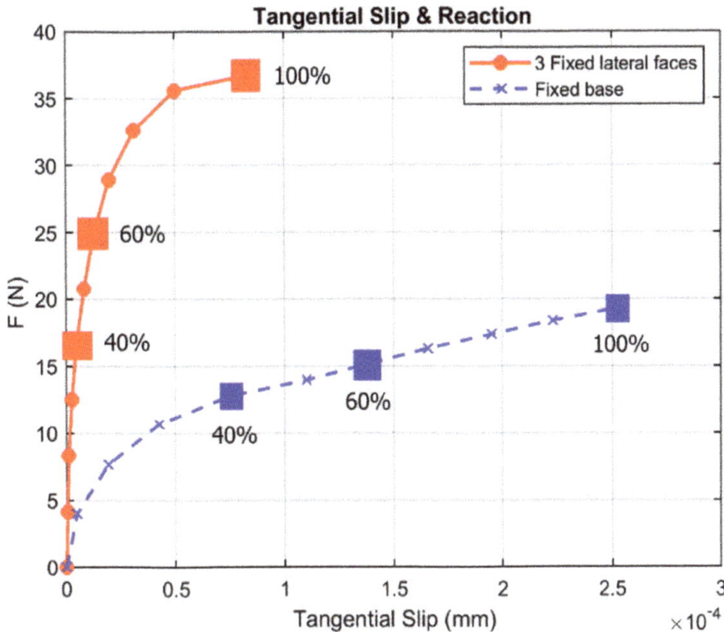

Figure 7. Displacement (mm) vs. force transferred to the mortar joints by rebars embedded in mortar with 5.6 GPa Young's modulus and 7.5 mm of effective covering up to failure under two different boundary conditions.

Figure 9a–c depict bonding shear stress distribution when the rebar transferred 40%, 60%, and 100% of the force transferred in failure, respectively, and the rebar was embedded in a mortar prism fixed in three of the four longitudinal faces (faces number 2, 3, and 4). From the beginning of the simulation, these pictures show that the whole rebar contributes to the load transfer mechanism. Figure 9d–f represent bonding normal stress distribution for the load steps previously described. Although only in the last load steps did the ribs work under compression, it is remarkable that all ribs contributed to the force transference mechanism. As a consequence, compressive values were significantly lower than those attained when only the base face (face number 1) was fixed: 1.6459 MPa (Figure 8f) compared to 0.61233 MPa (Figure 9f).

3.2. Rib Shape

Rib shape is one of the key values regarding bonding in rebars. Various rebar shapes were analyzed in order to check their performance in terms of bonding behavior. Rebar geometry was parameterized as depicted in Figure 10, and the values of the parameters ranged as follows between those recommended in the codes [51–53]: (i) central width, Wc, from 1.0 to 3.5 mm; (ii) extreme width, We, from 1.0 to 3.5 mm; (iii) angle between the rib and rebar axes, B, from 35° to 75°; (iv) rib-face angles, Bf, from 45° to 90°; (v) rib height in the central transversal section of the rib, hr, from 0.15 to 0.75 mm; (vi) rib spacing; s, from 3 to 6 mm.

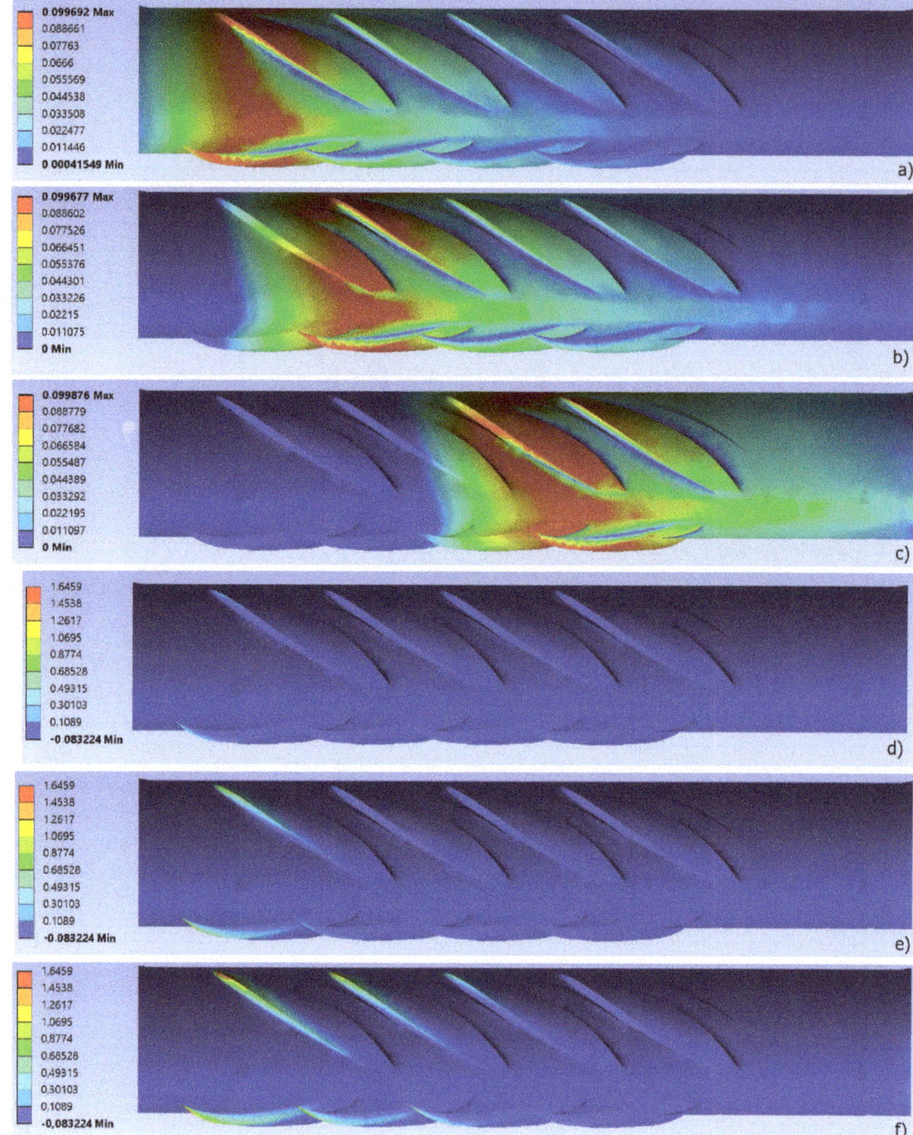

Figure 8. Bonding stresses (MPa) in the interface of a rebar embedded in a prism of mortar with Young's modulus of 5.6 GPa and 7.5 mm of effective covering when the base face (face number 1) was fixed. Bonding shear stress distribution when the force transferred from the rebar to the mortar was (**a**) 40% of that transferred under failure; (**b**) 60% of that transferred under failure; and (**c**) 100% of that transferred under failure. Bonding normal stress distribution when the force transferred from the rebar to the mortar was (**d**) 40% of that transferred under failure; (**e**) 60% of that transferred under failure; and (**f**) 100% of that transferred under failure.

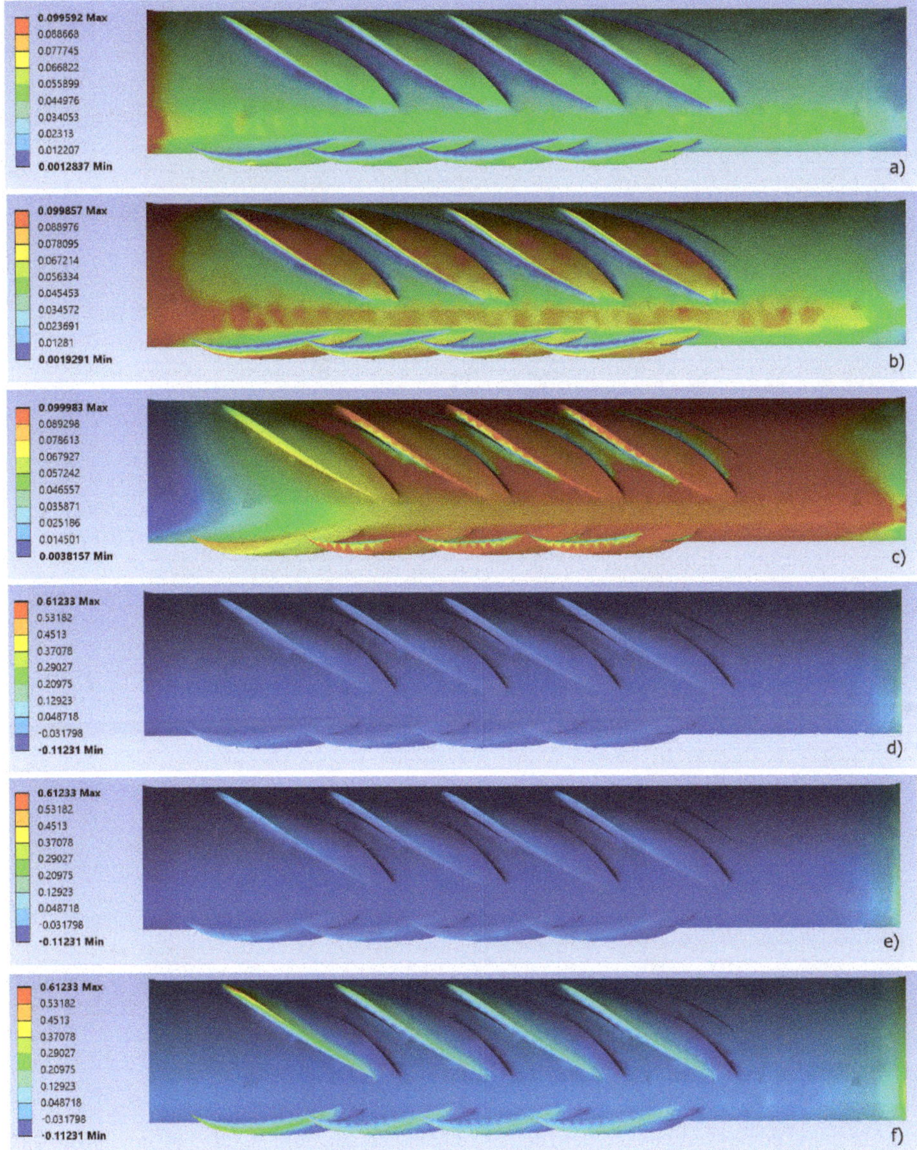

Figure 9. Bonding stresses (MPa) in the interface for a rebar embedded in a prism of mortar with Young's modulus of 5.6 GPa and 7.5 mm of effective covering when three lateral faces (faces number 2, 3, and 4) were fixed. Bonding shear stress distribution when the force transferred from the rebar to the mortar was (**a**) 40% of that transferred under failure; (**b**) 60% of that transferred under failure; and (**c**) 100% of that transferred under failure. Bonding normal stress distribution when the force transferred from the rebar to the mortar was (**d**) 40% of that transferred under failure; (**e**) 60% of that transferred under failure; and (**f**) 100% of that transferred under failure.

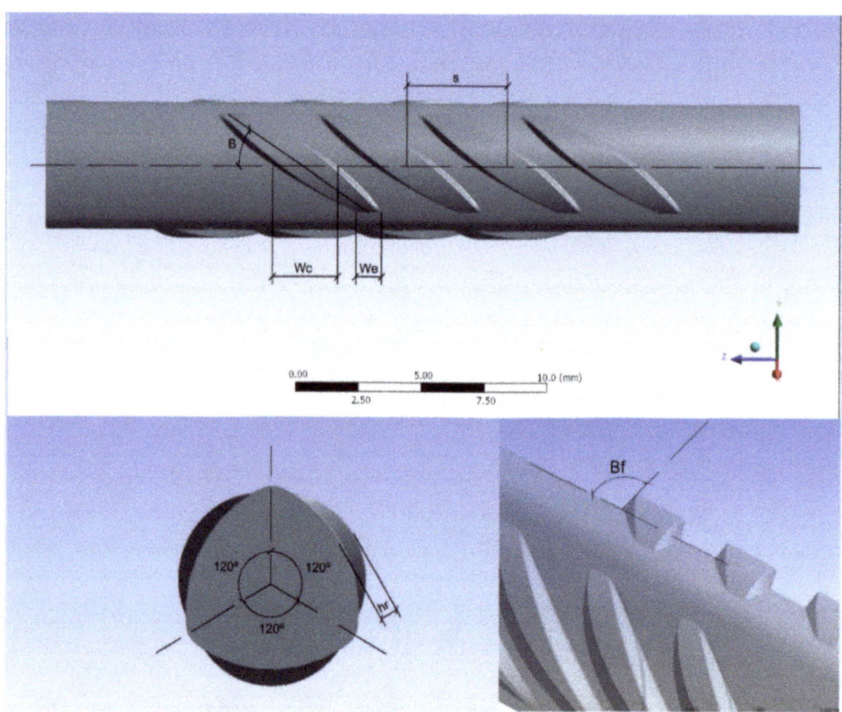

Figure 10. Geometrical parameters of the rebar that ranged in iterative analyses.

In all these analyses, Young's modulus of mortar was set to 5.6 GPa. Boundary conditions of fixation along three and four of the longitudinal faces (faces 3, 4, and 5) of the prism were considered, as well as effective covering for rebars of 6 mm and 25 mm. Results are laid out in Table 2. In order to avoid noise in subsequent calculations, the rebar geometrical parameters were not changed simultaneously.

Table 2. Reaction force F (N) needed to displace a 5×10^{-5} mm rebar with different shapes, with effective covering (eff.cov.) 6 mm/ 25 mm and Young´s modulus 5.6 GPa; as boundary conditions, fixation in three or four of the longitudinal faces of the prism.

W_c (mm)	W_e (mm)	B (°)	B_f (°)	h_r (mm)	s (mm)	* F_{3LF}	** F_{4LF}	* F_{3LF}	** F_{4LF}
						eff.cov. = 6 mm		eff.cov. = 25 mm	
1						17.685	20.23	6.494	7.046
1.5						17.77	20.339	6.506	7.059
2	2.5	55	67.5	0.45	4	17.861	20.454	6.518	7.074
2.5						17.949	20.567	6.517	7.073
3						18.029	20.66	6.527	7.084
3.5						18.085	20.737	6.536	7.094
	1					17.9	21.122	6.573	7.136
	1.5					17.911	20.72	6.53	7.088
2.5	2	55	67.5	0.45	4	17.935	20.724	6.534	7.092
	2.5					17.949	20.723	6.53	7.088
	3					17.948	20.726	6.532	7.092
	3.5					17.972	20.742	6.532	7.09

Table 2. *Cont.*

W_c (mm)	W_e (mm)	B (°)	B_f (°)	h_r (mm)	s (mm)	* F_{3LF}	** F_{4LF}	* F_{3LF}	** F_{4LF}
						eff.cov. = 6 mm		eff.cov. = 25 mm	
2.5	2.5	15	67.5	0.45	4	18.393	20.756	6.528	7.086
		20				18.071	20.546	6.518	7.073
		25				18.075	20.567	6.517	7.073
		30				18.074	19.958	6.457	7.002
		35				18.078	20.13	6.474	7.022
		40				18.093	20.327	6.494	7.045
		45				18.1	20.567	6.517	7.072
		50				17.934	20.817	6.545	7.105
		55				17.949	21.085	6.574	7.14
2.5	2.5	55	45	0.45	4	17.981	21.363	6.606	7.177
			50			17.907	20.604	6.569	7.132
			55			17.931	20.513	6.519	7.074
			60			17.935	20.544	6.52	7.075
			65			17.95	20.546	6.518	7.074
			70			17.951	20.568	6.519	7.074
			75			17.95	20.567	6.516	7.071
			80			17.943	20.564	6.522	7.078
			85			17.952	20.557	6.526	7.083
			90			17.954	20.567	6.525	7.063
2.5	2.5	55	67.5	0.15	4	17.464	20.571	6.525	7.082
				0.25		17.604	20.503	6.514	7.069
				0.35		17.761	20.517	6.514	7.069
				0.45		17.949	20.548	6.517	7.072
				0.55		18.15	20.567	6.517	7.073
				0.65		18.358	20.563	6.525	7.082
				0.75		18.58	20.597	6.52	7.076
2.5	2.5	55	67.5	0.45	3	18.203	20.888	6.551	7.112
					4	17.949	20.567	6.517	7.073
					5	17.793	20.367	6.503	7.056
					6	17.632	20.166	6.489	7.039

* Reaction force F (N) with three fixed longitudinal faces ** Reaction force F (N) with four fixed longitudinal faces.

4. Discussion

In this section, the results presented in Section 3 are analyzed and discussed. To better evaluate the influence that the aspects taken into account exert on the final results, several graphical depictions of the results are provided. Thus, Figure 11, which depicts data in Table 1, shows that the Young's modulus of the material that surrounds the rebar clearly influences the final value of F.

Regarding the relationships between values of F, the improvement rate of this parameter using mortars with Young's modulus of either 5.6 or 10 GPa is almost linear, and this improvement is also independent of boundary conditions and the thickness of effective coverings. That is, while Young's modulus increases by 178.6%, the median of the values of F assumes 173.2% with a standard deviation of 0.03. With higher values of the Young's modulus, the correlation becomes less clear, since it is much more dependent on the effective covering and boundary conditions. When this increases by 297% (that is, it takes a value of 20 GPa), there is almost a linear relationship between this value and F only when Face 1 is fixed, since this value assumes 283% with a standard deviation of 0.01, while the respective values for two, three and four fixed lateral faces (Faces 2, 3, 4, and 5) are 331% and 0.40; 315% and 0.84; 304% and 1.48, respectively. This undoubtedly leads to the fact that the results from the pull-out test cannot be extrapolated to the usual situation of rebars in which the boundary conditions are different to those of the test. When Young's modulus of mortars is 50 GPa, this correlation is impossible since the standard deviation reaches inadmissible values. On the other hand, a certain proportionality between values is found in the results for 20 GPa. When effective coverings are either 6 mm or 7.5 mm, no relevant difference can be found between values obtained from prisms with three or four fixed faces. In contrast, when effective coverings are 10, 12.5, 15 or 25 mm, differences of 2.38%, 2.82%, 2.57% and 2.18%, respectively, are found. Regarding global results in terms of Young's modulus and effective coverings, the higher these results, the more homogeneous F becomes. Figure 11d depicts this fact clearly, where fixation of Face 1 and 6 mm of effective covering assumes approximate values of F less than 50% of those attained with an effective covering of 12.5 mm or higher.

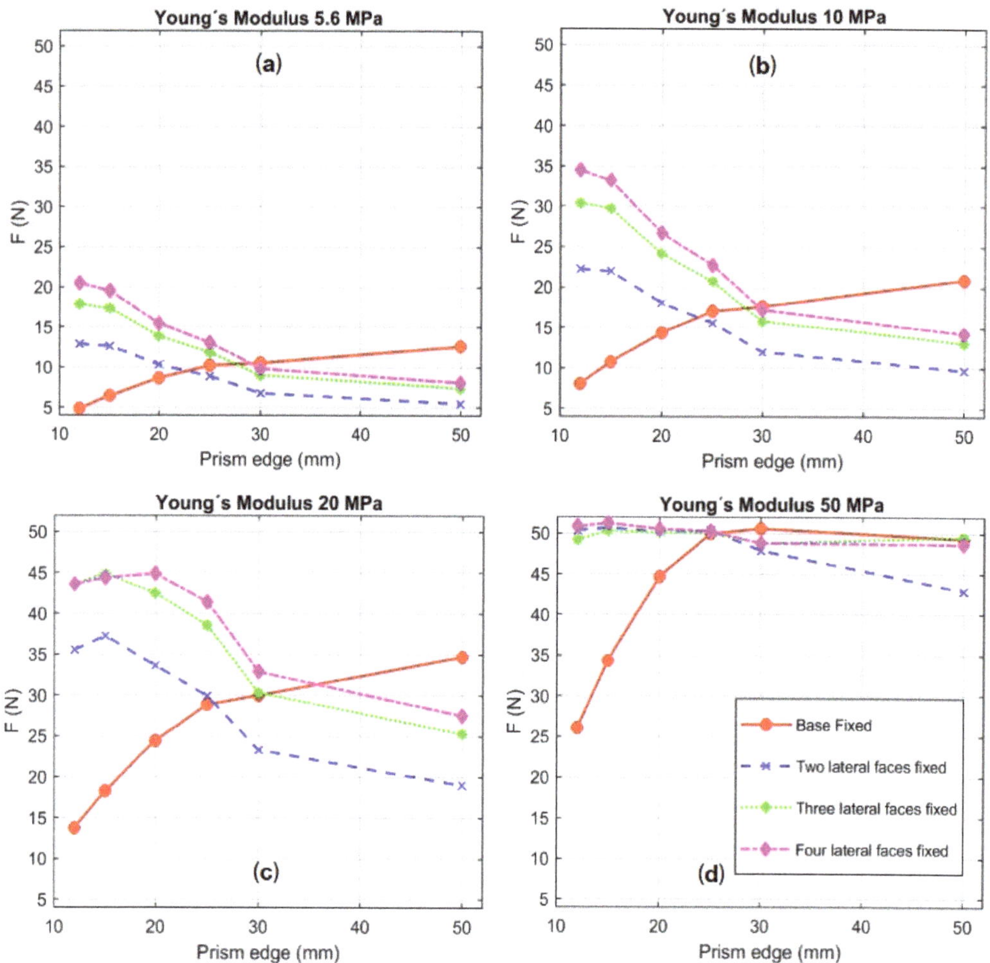

Figure 11. Chart depicting the reaction force F (N) produced by a 5×10^{-5} mm displacement of a rebar embedded in mortar prisms with different edges, different Young's modulus and different boundary conditions: (**a**) Mortar with Young's modulus of 5.6 GPa; (**b**) mortar with Young's modulus of 10 GPa; (**c**) mortar with Young's modulus of 20 GPa; (**d**) mortar with Young's modulus of 50 GPa.

Figure 11 also demonstrates that boundary conditions constitute key data for the definition of the value of F. It is easy to observe that, while lateral fixations assume a decrease when the effective covering is thicker, fixation in Face 1 produces the opposite effect. This fact has a clear implication in the quantification of bonding by pull-out tests in rebars subjected to these conditions, since the bonding behavior is not correctly reproduced in this test. The number of faces that are fixed also affects the value of F: the more fixed faces there are, the higher F becomes. This fact is linked to the equivalent strain energy (η^{mic}) in the microplane model (Equation (1)).

When only Face 1 is fixed and the rebar is embedded in a 12×12 mm^2 section prism, the highest values of equivalent strain energy are concentrated close to the bounded face (Figure 12 right). Ribs transfer load to the medium, but, under these circumstances, the ribs placed in positions that are far from the fixation hardly make any contribution to this transfer. As the effective covering increases, a higher number of ribs contribute to this mechanism and, consequently, the value of F increases (Figure 12 left).

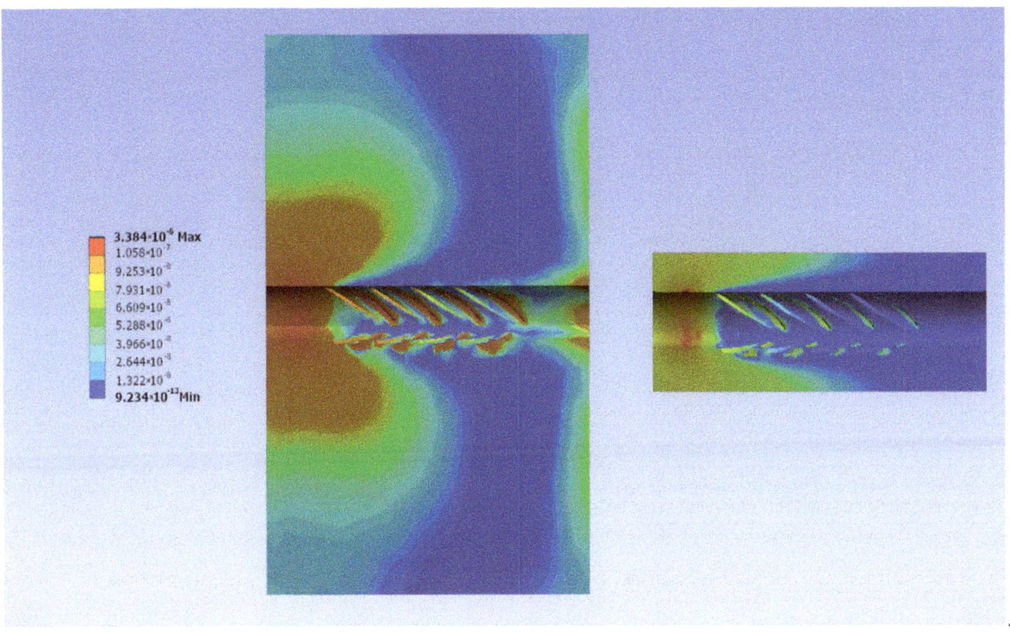

Figure 12. Equivalent strain energy distribution in the mortar joints (E = 5.6 GPa) when face 1 of the prism is fixed and 5×10^{-5} mm displacement is applied to the base of the rebar: (**left**) base section of 50 x 50 mm^2; (**right**) base section of 12×12 mm^2.

When lateral faces of the prism are fixed, bonding shear stress decreases while effective covering increases. The value of F, as the integral of the stresses, obviously also decreases. Fixation in lateral faces produces a confinement effect in the bar (especially in the case of reduced coverings) that notably improves its bonding behavior. Since this effect does not exist when only the base face of the prism (Face 1) is fixed, the bonding behavior of rebars is completely different (Figure 13). Bond shear stress is higher in the case of reduced covering (Figure 13a). This justifies the reduction of F observed in Figure 11 when the covering increases in the cases of lateral restraint (Figure 13b).

Regarding the rib shape, the Pearson product moment correlation matrix based on data presented in Table 2 is obtained, thus obtaining the influence for each one of the six geometrical parameters of the rib (Wc, We, B, Bf, hr, s). This matrix is non-dependent on the number of faces that are fixed. In contrast, it is strongly dependent on the thickness of the effective covering (Table 3).

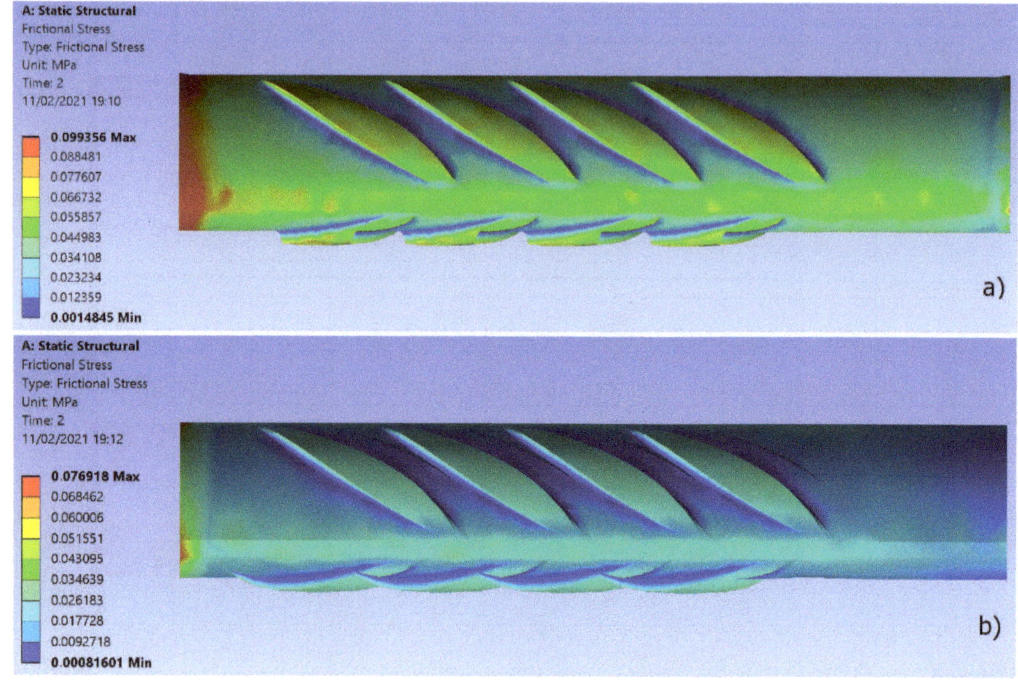

Figure 13. Distribution of bond shear stress (MPa) in the interface in a prism of mortar (E = 5.6 MPa) for a displacement equal to 5×10^{-5} mm when Faces 2, 3, and 4 are fixed, for a prism with dimensions: (**a**) 12×12 mm^2; (**b**) 50×50 mm^2.

Table 3. Pearson correlation coefficient of each geometrical parameter of the rib in bonding behavior.

Effective Covering	W_c	W_e	B	B_f	h_r	s
6 mm	0.30	0.06	0.38	0.01	0.78	−0.35
25 mm	0.22	0.07	0.28	−0.11	0.80	−0.29

The most influential geometrical parameter is, undoubtedly, h_r, while both B and s exert only a medium influence. Rib height in the center (h_r) has also medium influence in bonding, but this influence is reduced to 27% as the effective covering increases. The effect of B_f and W_e in bonding is irrelevant. Regarding s, although it is demonstrated that high values of this parameter positively influence bonding [45,54], a negative value in the coefficient (Table 3) implies the opposite. Spacing between ribs is also linked to the number of ribs that fit into a fixed length of bar: the greater the rib spacing, the fewer ribs in the 30-mm-length bar. In this research, the lowest number of ribs that involves high rib spacing, and the subsequent negative influence on bonding, carry more weight than the positive effect of higher values of this parameter.

A regression analysis with the values obtained in this research (Table 2) reveals the relationship between the geometrical parameter of the ribs and the value of F for the case of the rebar embedded in the prism with three and four lateral faces (F_{3LF}-F_{4LF}) fixed in prisms of 12×12 and 50×50 mm^2.

$$F_{3LF(12X12)} = 17.0659 + 0.1693W_c + 0.0316W_e + 0.0067B + 0.0002B_f + 1.8732h_r - 0.176s \tag{7}$$

$$F_{4LF(12X12)} = 19.4548 + 0.2138W_c + 0.0405W_e + 0.0083B + 0.0002B_f + 2.3625h_r - 0.2224s \tag{8}$$

$$F_{3LF(50X50)} = 6.4397 + 0.0162W_c + 0.0048W_e + 0.0067B + 0.0004B_f + 0.2493h_r - 0.0191s \tag{9}$$

$$F_{4LF(50X50)} = 6.9887 + 0.0186W_c + 0.0054W_e + 0.0007B + 0.0006B_f + 0.2932h_r - 0.0224s \tag{10}$$

Equations (7)–(10) are highly reliable since (i) R-square coefficients are 94.64%, 94.64%, 86.83%, and 87.50%; (ii) the residual standard deviations are 0.0497, 0.0627, 0.0101, and 0.0115; (iii) the mean absolute errors (MAEs) are 0.0298, 0.0377, 0.0064, and 0.0073, and (iv) the Durbin Watson (DB) statistic is 2.1857, 2.1717, 1.9809, and 2.1285. These equations, together with Figure 10, allow us to relate the results from this research to different shapes of ribs.

5. Conclusions

This paper deals with the bonding behavior of rebars under the special circumstances that occur when masonry is reinforced. Thicknesses of coverings that are lower than usual, together with variable boundary conditions, involve different behavior of rebars in terms of bonding.

This research covers the cases of stainless steel rebars with 5 mm diameter, embedded in mortar joints with Young's modulus of 5.6, 10, 20, and 50 GPa, and effective coverings of 6, 7.5, 10, 12.5, 15, and 25 mm. Furthermore, the variability of the boundary conditions is taken into account by the fixation of two, three or four longitudinal faces of the prisms into which the bars are embedded, as well as their bases. In this way, several of the most frequent performances of this reinforcement are reproduced: bed joint structural repointing, transversal anchors in walls, meshes attached to wall surfaces, and the conditions of the standard pull-out test. By changing Young's modulus, the use of standard poor mortars to high-strength binders is encompassed.

A pull-out test with no embracement of the samples does not reproduce the behavior of the bars under these conditions. When the prisms that surround the rebars are not embraced, maximum reaction force increases with effective covering, decreasing in the opposite case.

Regarding values of this, when the Young's modulus of the mortar reaches 10 GPa, the relationship between both parameters is linear. In this way, the results can be extrapolated for various materials. This fact only occurs in high-performance mortars when only the base of the prism is fixed. These facts lead to the conclusion that rebars must be tested under the boundary conditions in which they will work.

Regarding the shape of the rebars, the most influential geometrical aspects of the ribs are identified. Although rib height is obviously the key value in bonding, the contribution of the other aspects, such as central width, angle between rib and rebar axes, and rib spacing, depends on boundary conditions and effective coverings. The rib central width is of major importance when effective covering is low and the bar is highly confined, but this importance decreases when effective covering increases. For a fixed length of bar, as used in this research, the spacing between ribs has a negative influence on bonding. As a result, several relationships between the rebar shape and the results obtained in terms of bonding are attained.

Author Contributions: Conceptualization, F.A., E.R.-M. and B.H.; investigation, F.A. and E.R.-M.; Methodology, F.A., E.R.-M. and B.H.; project administration, E.R.-M. and B.H.; resources, F.A.; supervision, B.H.; writing–original draft, E.R.-M.; writing–review & editing, F.A. and B.H. All authors have read and agreed to the published version of the manuscript.

Funding: This research has been carried out under the project PGC2018-098185-A-I100, funded by: FEDER/Ministerio de Ciencia e Innovación-Agencia Estatal de Investigación of Spain.

Institutional Review Board Statement: Not applicable.

Informed Consent Statement: Not applicable.

Acknowledgments: The authors wish to thank the undergraduate students David Perejon and Marcos García for their contribution.

Conflicts of Interest: The authors declare no conflict of interest.

Appendix A

Table summarizing the values of reaction force F (N) needed to pull a 5-mm rebar out a distance of 5×10^{-5} mm from its initial position embedded in a 100-mm edge cube of mortar, when different combinations of bond shear stress and bond tensile stress in the CZM are set and the frontal face of the cube is fixed.

Maximum Bond Shear Stress (MPa)	Maximum Bond Tensile Stress (MPa)	Reaction Force F (N)
0.10	0.10	14.333
	0.25	14.346
	0.50	14.348
	1.00	14.349
0.25	0.10	17.825
	0.25	17.859
	0.50	17.864
	1.00	17.866
0.50	0.10	21.362
	0.25	21.432
	0.50	21.444
	1.00	21.446
1.00	0.10	26.654
	0.25	27.003
	0.50	27.074
	1.00	27.092

References

1. Candela, M.; Borri, A.; Corradi, M.; Righetti, L. Effect of transversal steel connectors on the behavior of rubble stone-masonry walls: Two case studies in Italy. In Proceedings of the Brick Block Mason. Trends, Innov. Challenges—Proc. International Brick and Block Masonry Conference IBMAC 2016, Padova, Italy, 26–30 June 2016; pp. 2029–2038. [CrossRef]
2. Corradi, M.; Castori, G.; Borri, A. Repairing brickwork panels using titanium rods embedded in the mortar joints. *Eng. Struct.* **2020**, *221*, 111099. [CrossRef]
3. Valluzzi, M.R.; Binda, L.; Modena, C. Mechanical behaviour of historic masonry structures strengthened by bed joints structural repointing. *Constr. Build. Mater.* **2005**, *19*, 63–73. [CrossRef]
4. Binda, L.; Modena, C.; Saisi, A.; Tongini-Folli, R.; Valluzzi, M.R. Bedjoint structural repointing of historic masonry structures. In *Proceedings of the 9th Canadian Masonry Symposium, Fredericton, NB, Canada, 4–6 June 2001*; Bischoff, P.H., Dawe, J.L.D., Eds.; Department of Civil Engineering, University of New Brunswick: Frederickton, NB, Canada, 2001.
5. Anzani, A.; Cardani, G.; Condoleo, P.; Garavaglia, E.; Saisi, A.; Tedeschi, C.; Tiraboschi, C.; Valluzzi, M.R. Understanding of historical masonry for conservation approaches: The contribution of Prof. Luigia Binda to research advancement. *Mater. Struct. Constr.* **2018**, *51*, 1–27. [CrossRef]
6. Valluzzi, M.R. On the vulnerability of historical masonry structures: Analysis and mitigation. *Mater. Struct. Constr.* **2007**, *40*, 723–743. [CrossRef]
7. Corradi, M.; Di Schino, A.; Borri, A.; Rufini, R. A review of the use of stainless steel for masonry repair and reinforcement. *Constr. Build. Mater.* **2018**, *181*, 335–346. [CrossRef]
8. Meriggi, P.; de Felice, G.; De Santis, S. Design of the out-of-plane strengthening of masonry walls with fabric reinforced cementitious matrix composites. *Constr. Build. Mater.* **2020**, *240*, 122452. [CrossRef]

9. Ascione, L.; Carozzi, F.G.; D'Antino, T.; Poggi, C. New Italian guidelines for design of externally bonded Fabric-Reinforced Cementitious Matrix (FRCM) systems for repair and strengthening of masonry and concrete structures. *Procedia Struct. Integr.* **2018**, *11*, 202–209. [CrossRef]

10. Ceroni, F.; Darban, H.; Luciano, R. Analysis of bond behavior of injected anchors in masonry elements by means of Finite Element Modeling. *Compos. Struct.* **2020**, *241*, 112099. [CrossRef]

11. Gentilini, C.; Finelli, F.; Girelli, V.A.; Franzoni, E. Pull-out behavior of twisted steel connectors employed in masonry: The influence of the substrate. *Constr. Build. Mater.* **2021**, *274*, 122115. [CrossRef]

12. Mayorga, E.R.; Ancio, F.; Hortigon, B. Optimisation of stainless steel rebars to repair masonry structures. In Proceedings of the REHABEND Construction Pathology, Rehabilitation Technology and Heritage Management, Granada, Spain, 28–30 September 2020.

13. Siderurgia, A. 36 UNE 36740: Adherence Test for Steel for Reinforcement of Concrete. Beam Test. *UNE*. 1998. Available online: https://www.techstreet.com/ashrae/standards/une-36740-1998?product_id=1300670 (accessed on 10 May 2021).

14. BS EN 1881:2006. *Products and Systems for the Protection and Repair of Concrete Structures—Test Methods—Testing of Anchoring Products by Pull-Out Method*; British Standard Institute: London, UK, 2006; ISBN 0580 49827 1.

15. BS EN 846-2:2000. *Methods of Test for Ancillary Components for Masonry—Part 2: Determination of Bond Strength of Prefabricated Bed Joint Reinforcement in Mortar Joints*; British Standard Institute: London, UK, 2000; ISBN 0580 34889 X.

16. RILEM (Ed.) RILEM-TC RC 6 Bond test for reinforcement steel. 2. Pull-out test, 1983. In *RILEM Recommendations for the Testing and Use of Constructions Materials*; E & FN SPON: New York, NY, USA, 1994; pp. 218–220. ISBN 2351580117.

17. ACI Committee 440 Guide Test Methods for Fiber-Reinforced Polymers (FRPs) for Reinforcing or Strengthening Concrete Structures, ACI 440.3R-12. 2012. Available online: https://www.concrete.org/store/productdetail.aspx?ItemID=440312&Language=English&Units=US_AND_METRIC (accessed on 10 May 2021).

18. Chu, S.H.; Kwan, A.K.H. A new method for pull out test of reinforcing bars in plain and fibre reinforced concrete. *Eng. Struct.* **2018**, *164*, 82–91. [CrossRef]

19. Ceroni, F.; Di Ludovico, M. Traditional and innovative systems for injected anchors in masonry elements: Experimental behavior and theoretical formulations. *Constr. Build. Mater.* **2020**, *254*, 119178. [CrossRef]

20. Paganoni, S.; D'Ayala, D. Testing and design procedure for corner connections of masonry heritage buildings strengthened by metallic grouted anchors. *Eng. Struct.* **2014**, *70*, 278–293. [CrossRef]

21. Miranda, M.P.; Morsch, I.B.; Brisotto, D.D.S.; Bittencourt, E.; Carvalho, E.P. Steel-concrete bond behavior: An experimental and numerical study. *Constr. Build. Mater.* **2021**, *271*, 121918. [CrossRef]

22. Ceroni, F.; Darban, H.; Caterino, N.; Luciano, R. Efficiency of injected anchors in masonry elements: Evaluation of pull-out strength. *Constr. Build. Mater.* **2021**, *267*, 121707. [CrossRef]

23. Hu, X.; Peng, G.; Niu, D.; Wu, X.; Zhang, L. Bond behavior between deformed steel bars and cementitious grout. *Constr. Build. Mater.* **2020**, *262*, 120810. [CrossRef]

24. ANSYSInc Element Library. *Ansys® Academic Research Mechanical, Release 20.2 Help System*; ANSYS Inc.: Canonsburg, PA, USA, 2020.

25. Hortigón, B.; Ancio, F.; Rodriguez-Mayorga, E. Parameterization of stainless steel rebars to improve bonding strength in masonry repairing. In Proceedings of the PROHITECH 2020 4th International Conference on Protection of Historical Constructions, Athens, Greece, 25–27 October 2021.

26. Taylor, G.I. Plastic strain in metals. *J. Inst. Met.* **1938**, *62*, 307–324.

27. Batdorf, S.B.; Budiansky, B. *A Mathematical Theory of Plasticity Based on the Concept of Slip*; National Advisory Committee for Aeronautics, Ed.; NACA Technology: Langley, VA, USA, 1949.

28. Bažant, Z.P.; Prat, P.C. Microplane model for brittle-plastic material: I. Theory. *J. Eng. Mech.* **1988**, *114*, 1672–1688. [CrossRef]

29. Bažant, Z.P.; Prat, P.C. Microplane model for brittle-plastic material: II. verification. *J. Eng. Mech.* **1988**, *114*, 1689–1702. [CrossRef]

30. Bazant, Z.P.; Gambarova, P.G. Crack Shear in Concrete: Crack Band Microplane Model. *J. Struct. Eng.* **1984**, *110*, 2015–2036. [CrossRef]

31. Carol, I.; Bažant, Z.P. Damage and plasticity in microplane theory. *Int. J. Solids Struct.* **1997**, *34*, 3807–3835. [CrossRef]

32. Bazant, Z.P.; Adley, M.D.; Carol, I.; Jirásek, M.; Akers, S.A.; Rohani, B.; Cargile, J.D.; Caner, F.C. Large-strain generalization of microplane model for concrete and application. *J. Eng. Mech.* **2000**, *126*, 971–980. [CrossRef]

33. Caner, F.C.; Bazant, Z.P. Microplane model M4 for concrete. II: Algorithm and calibration. *J. Eng. Mech.* **2000**, *126*, 954–961. [CrossRef]

34. de Vree, J.H.P.; Brekelmans, W.A.M.; van Gils, M.A.J. Comparison of nonlocal approaches in continuum damage mechanics. *Comput. Struct.* **1995**, *55*, 581–588. [CrossRef]

35. Mazars, J.; Pyaudier-Cabot, G. Continuum damage theory—Application to concrete. *J. Eng. Mech.* **1989**, *115*, 345–365. [CrossRef]

36. Peerlings, R.H.J.; de Borst, R.; Brekelmans, W.A.M.; Geers, M.G.D. Gradient-enhanced damage modelling of concrete fracture. *Mech. Cohesive-Frictional Mater.* **1998**, *3*, 323–342. [CrossRef]

37. Geers, M.G.D.; De Borst, R.; Brekelmans, W.A.M.; Peerlings, R.H.J. Strain-based transient-gradient damage model for failure analyses. *Comput. Methods Appl. Mech. Eng.* **1998**, *160*, 133–153. [CrossRef]

38. Zreid, I.; Kaliske, M. A gradient enhanced plasticity–damage microplane model for concrete. *Comput. Mech.* **2018**, *62*, 1239–1257. [CrossRef]

39. Binda, L.; Fontana, A.; Frigerio, G. *Mechanical Behaviour of Brick Masonries Derived from Unit and Mortar Characteristics. Proceedings of the 8th International Brick and Block Masonry Conference IB2MAC, Dublin, Ireland, 19–21 September 1988*; Courcy, J.W.T., Ed.; Elsevier Applied Science: Amsterdam, The Netherlands; Dublin, Ireland, 1988; pp. 205–216.

40. Kaklis, K.N.; Maurigiannakis, S.P.; Agioutantis, Z.G.; Maravelaki-Kalaitzaki, P. Characterization of pozzolanic lime mortars used as filling material in shaped grooves for restoring member connections in ancient monuments. *Int. J. Archit. Herit.* **2018**, *12*, 75–90. [CrossRef]

41. Baltazar, L.G.; Henriques, F.M.A.; Cidade, M.T. Grouts with improved durability for masonry consolidation: An experimental study with non-standard specimens. In *Key Engineering Materials*; Trans Tech Publications Ltd.: Bach, Switzerland, 2017; Volume 747.

42. Luso, E.; Lourenço, P.B. Experimental laboratory design of lime based grouts for masonry consolidation. *Int. J. Archit. Herit.* **2017**, *11*, 1143–1152. [CrossRef]

43. Arandigoyen, M.; Alvarez, J.I. Pore structure and mechanical properties of cement-lime mortars. *Cem. Concr. Res.* **2007**, *37*, 767–775. [CrossRef]

44. Thamboo, J.; Dhanasekar, M. Assessment of the characteristics of lime mortar bonded brickwork wallettes under monotonic and cyclic compression. *Constr. Build. Mater.* **2020**, *261*, 120003. [CrossRef]

45. Tao, W.; Chen, C.; Jun, H.; Ting, R. Effect of bolt rib spacing on load transfer mechanism. *Int. J. Min. Sci. Technol.* **2017**, *27*, 431–434. [CrossRef]

46. Barbosa, M.T.; Sánchez Filho, E.d.S.; de Oliveira, T.M.; Santos, W.J. dos Analysis of the relative rib area of reinforcing bars pull out tests. *Mater. Res.* **2008**, *11*, 453–457. [CrossRef]

47. Chiriatti, L.; Mercado-Mendoza, H.; Apedo, K.L.; Fond, C.; Feugeas, F. A study of bond between steel rebar and concrete under a friction-based approach. *Cem. Concr. Res.* **2019**, *120*, 132–141. [CrossRef]

48. Araújo, A.S.; Oliveira, D.V.; Lourenço, P.B. Numerical study on the performance of improved masonry-to-timber connections in traditional masonry buildings. *Eng. Struct.* **2014**, *80*, 501–513. [CrossRef]

49. Kabir, R.; Islam, M. Bond stress behavior between concrete and steel rebar: Critical investigation of pull-out test via Finite Element Modeling. *Int. J. Civ. Struct. Eng.* **2014**, *5*, 80–90. [CrossRef]

50. Zhao, W.; Zhu, B. Basic parameters test and 3D modeling of bond between high-strength concrete and ribbed steel bar after elevated temperatures. *Struct. Concr.* **2017**, *18*, 653–667. [CrossRef]

51. Allen, J.H.; Felder, A.L.; McDermott, J.F.; Frosch, R.J.; Mitchell, D.; Eligehausen, R.; Leon, R.T.; Pantazopoulou, S.J.; Azizinamini, A.; Fagundo, F.E.; et al. ACI Committee 408 Bond and Development of Straight Reinforcing Bars in Tension Reported by ACI Committee 408. *ACI* **2003**, *408–03*, 1–49.

52. International Organization for Standardization. *Standardization, I.O. for ISO 6935-2:2019 Preview Steel for the Reinforcement of Concrete. Part 2: Ribbed Bars*; ISO: Geneva, Switzerland, 2019.

53. Astm. *A 615/A 615M—09B Standard Specification for Deformed and Plain Carbon-Steel Bars for Concrete Reinforcement*; Astm: West Conshohocken, PA, USA, 2012. [CrossRef]

54. Lorrain, M.S.; Caetano, L.F.; Silva, B.V.; Gomes, L.E.S.; Barbosa, M.P.; Silva Filho, L.C.P. Bond strength and rib geometry: A comparative study of the influence of deformation patterns on anchorage bond strength. In Proceedings of the PCI Annual Convention & 3rd International FIB Congress FIB, Washington, DC, USA, 29 May–2 June 2010.

Article

Development of Elastoplastic-Damage Model of AlFeSi Phase for Aluminum Alloy 6061

Hailong Wang [1], Wenping Deng [1], Tao Zhang [1], Jianhua Yao [1] and Sujuan Wang [1,2,*]

[1] State Key Laboratory of Precision Electronic Manufacturing Technology and Equipment, Guangdong University of Technology, Guangzhou 510006, China; wanghl@gdut.edu.cn (H.W.); 2112001273@mail2.gdut.edu.cn (W.D.); 2111901201@mail2.gdut.edu.cn (T.Z.); 1111701015@mail2.gdut.edu.cn (J.Y.)

[2] Guangdong Provincial Key Laboratory of Micro-Nano Manufacturing Technology and Equipment, Guangdong University of Technology, Guangzhou 510006, China

* Correspondence: grace.wangsj@gdut.edu.cn

Abstract: Material properties affect the surface finishing in ultra-precision diamond cutting (UPDC), especially for aluminum alloy 6061 (Al6061) in which the cutting-induced temperature rise generates different types of precipitates on the machined surface. The precipitates generation not only changes the material properties but also induces imperfections on the generated surface, therefore increasing surface roughness for Al6061 in UPDC. To investigate precipitate effect so as to make a more precise control for the surface quality of the diamond turned Al6061, it is necessary to confirm the compositions and material properties of the precipitates. Previous studies have indicated that the major precipitate that induces scratch marks on the diamond turned Al6061 is an AlFeSi phase with the composition of $Al_{86.1}Fe_{8.3}Si_{5.6}$. Therefore, in this paper, to study the material properties of the AlFeSi phase and its influences on ultra-precision machining of Al6061, an elastoplastic-damage model is proposed to build an elastoplastic constitutive model and a damage failure constitutive model of $Al_{86.1}Fe_{8.3}Si_{5.6}$. By integrating finite element (FE) simulation and JMatPro, an efficient method is proposed to confirm the physical and thermophysical properties, temperature-phase transition characteristics, as well as the stress–strain curves of $Al_{86.1}Fe_{8.3}Si_{5.6}$. Based on the developed elastoplastic-damage parameters of $Al_{86.1}Fe_{8.3}Si_{5.6}$, FE simulations of the scratch test for $Al_{86.1}Fe_{8.3}Si_{5.6}$ are conducted to verify the developed elastoplastic-damage model. $Al_{86.1}Fe_{8.3}Si_{5.6}$ is prepared and scratch test experiments are carried out to compare with the simulation results, which indicated that, the simulation results agree well with those from scratch tests and the deviation of the scratch force in X-axis direction is less than 6.5%.

Keywords: AlFeSi phase; elastoplastic-damage; aluminum alloy 6061 (Al6061); FEM

1. Introduction

Aluminum alloy 6061 (Al6061) belongs to the age-hardenable 6000 series aluminum alloys for which the chemical composition by wt % is: Mg0.92, Si0.76, Fe0.28, Cu0.22, Ti0.10, Cr0.07, Zn0.06, Mn0.04 and Al balance, in which Mg and Si contribute to strengthening by precipitating intermetallic phases during heat treatment. Due to its favorable combination of medium strength, good machinability, corrosion resistance and good strength, weldability as well as heat treatability, Al6061 has become one of the most extensively used extruded products in different areas [1]. Especially in the optical industry, Al6061 is one of the preferred materials for mirrors or optical lenses in spaceborne applications [2]. However, the precipitation in Al6061 not only contributes to changing its mechanical properties, like hardness and elastic module, but also results in some detrimental effects on the surface finish in the machining process [3].

Ultra-precision diamond cutting (UPDC) is one of the popular and feasible manufacturing technologies for the fabrication of optical functional components with sophisticated

Metals **2021**, *11*, 954. https://doi.org/10.3390/met11060954   https://www.mdpi.com/journal/metals

geometrical features and high-quality requirements since this technology can directly achieve sub-micrometric form accuracy and nanometric surface finishing. The achieved surface finish is an important factor by affecting optical functions, like the reflectance. Lots of studies have been conducted to investigate the influencing factors for surface finishing in UPDC. Zhang et al. [4] summarized the influencing factors including machine tool systems, cutting parameters, cutting tool geometry, environmental conditions and material properties. Among them, the material properties affecting surface roughness in UPDC include anisotropy [5], plastic side flow and elastic recovery of materials [6,7] and the crystallographic properties [8]. Some extra factors induce imperfection on surface finishing and increase the surface roughness by generating micro-defects, pits and cracks. Cheung et al. [9] found that the hard SiC of Al6061/15SiCp generated pits and cracks on the diamond machined surface. Simoneau et al. [10] reported that surface micro-defects, such as dimples occurring at hard-soft grain boundaries, influenced the surface roughness during micro-scale cutting. Harlow et al. [11] studied the effects of particles in Al7075-T6 on fatigue damage evolution based on fatigue cycling experiments subjected to constant amplitude loading of a 7075-T651 aluminum alloy, and reported that the particles obviously play a major role during the evolution of 7075-T6 fatigue damage, and about 87% of the observed particles in the high stress area are Fe-bearing. Wang et al. [3,12–14] found that the cutting-induced heat in ultra-precision raster milling (UPRM) and single point diamond turning (SPDT) generated two types of hard precipitates on the machined Al6061 and created scratch marks, pits and cracks on the raster milled surface, which increased the surface roughness.

Meanwhile, in our previous studies, it was indicated that an AlFeSi phase induced cracks, scratch marks and pits on diamond machined Al6061 and also affected the cutting forces [15]. The variation trend of the friction coefficient of Al6061 under different heating conditions agrees especially well with that of the number of AlFeSi particles. However, all of these published studies are experimental investigations, since an analytical or mechanical model for chip formation, surface generation and cutting force in diamond cutting of Al6061 with precipitation effect is difficult. For example, limited by current measurement technologies, some important parameters related to analytical models cannot accurately be measured by experimental methods, such as shear angle, strain, working temperature and the size of dead metal zone (DMZ). In this case, the alternative approaches are numerical methods in which the finite element (FE) methods are the most frequently used.

The material properties needed in FE simulations for metal cutting processes can be divided into two main parts: (1) the physical and thermophysical properties including density, melting point, thermal conductivity, Young's modulus, Poisson's ratio, specific heat and thermal expansion coefficient, and (2) the mechanical properties including the flow stress and the yield strength. In general, the confirmation of these material property parameters needs a lot of test experiments, which is a time-consuming and expensive process. Therefore, this paper presents a method to determine the material properties of AlFeSi phases in Al6061 and proposes an elastoplastic-damage model to build an elastoplastic constitutive model and a damage failure constitutive model of AlFeSi phase for FE simulating diamond cutting of Al6061.

2. Determination of Material Properties of AlFeSi Phase

As mentioned, the material properties for FE simulations include two parts: the physical and thermophysical properties and the mechanical properties. In this paper, JMatPro (Sente Software Company, United Kingdom, The period of validity: 15 April 2021), a phase diagram calculation and performance simulation software, is used to confirm all of the material properties for the AlFeSi phases in Al6061.

Referring to the previous study [15], the white block-like particles (Area I in Figure 1a) and needle-like particles (Area II in Figure 1a) are α-AlFeSi and β-AlFeSi particles, respectively [16], as presented in Figure 1. According to the EDX results, $Al_{86.1}Fe_{8.3}Si_{5.6}$ is used in this paper to represent the equivalent compositions of AlFeSi phase in Al6061, and its mean grain size is about 2 μm under the analysis of a large number of SEM images.

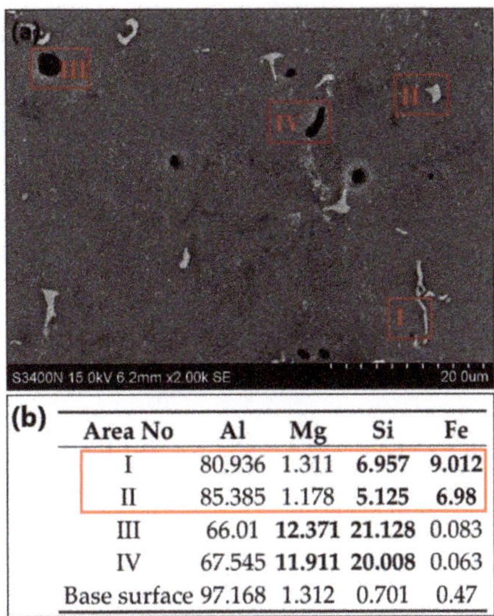

Area No	Al	Mg	Si	Fe
I	80.936	1.311	6.957	9.012
II	85.385	1.178	5.125	6.98
III	66.01	12.371	21.128	0.083
IV	67.545	11.911	20.008	0.063
Base surface	97.168	1.312	0.701	0.47

Figure 1. Chemical composition of AlFeSi in Al6061 from the previous study [15]: (**a**) SEM of the Al6061 samples; (**b**) EDX results of the Al6061 samples (Atomic %).

According to the types and compositions of AlFeSi in Al6061, the temperature-phase transition characteristics of AlFeSi are calculated by JMatPro, as presented in Figure 2. It shows that both the α-AlFeSi and β-AlFeSi begin to dissolve at 620 °C and dissolve completely at 690 °C. Meanwhile, the physical and thermophysical properties of $Al_{86.1}Fe_{8.3}Si_{5.6}$ are calculated in JMatPro, including density, thermal conductivity, Young's modulus, Poisson's ratio, specific heat and thermal expansion coefficient, as shown in Figure 3. From this figure, each curve shows a sudden change when the temperature reaches to 620 °C, which agrees well with the start of dissolution points of α-AlFeSi and β-AlFeSi in Figure 2.

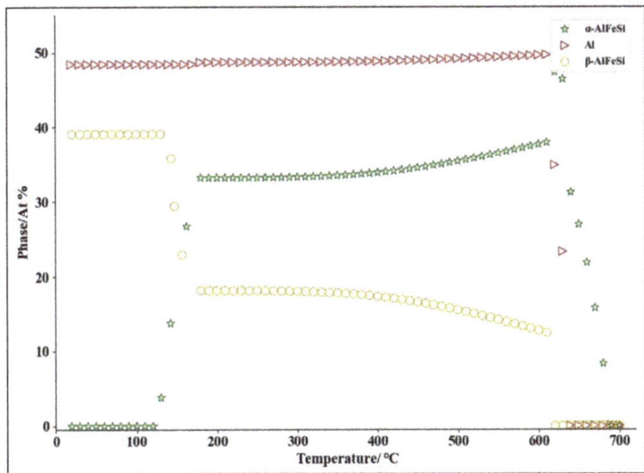

Figure 2. The temperature-phase transition characteristics of α-AlFeSi and β-AlFeSi from JMatPro.

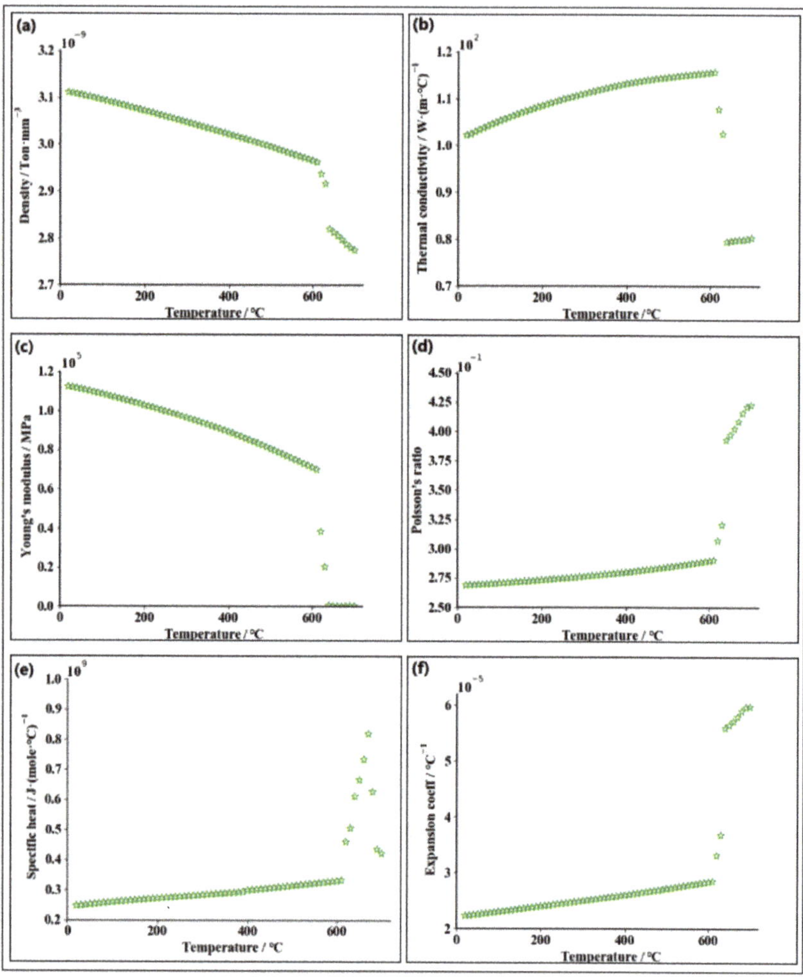

Figure 3. The characterization results of (**a**) density; (**b**) thermal conductivity; (**c**) Young's modulus; (**d**) Poisson's ratio; (**e**) specific heat; (**f**) thermal expansion coefficient.

Figure 4 shows the yield strength of the AlFeSi ($Al_{86.1}Fe_{8.3}Si_{5.6}$) phase under quasi-static conditions (the solid solution temperature is 520 °C and the holding time is 2 h, the aging temperature is 200 °C and aging time is 2 h) from JMatPro. It indicates that the yield strength decreases with the increase of grain size, and it can be found that, the yield strength is 662.33 MPa when the grain size is 2 μm.

During the metal cutting process, under different cutting speeds, the workpiece materials undergo different plastic deformations and present different strain rates. The plastic deformation can be divided into static plastic deformation (strain rates less than $0.00001\ s^{-1}$), quasi-static plastic deformation (strain rates between 0.00001 and $0.1\ s^{-1}$) and dynamic plastic deformation (strain rates greater than $0.1\ s^{-1}$). Figure 5 lists the stress–strain curves of AlFeSi in Al6061 under the condition of quasi-static (strain rates as $0.001\ s^{-1}$, $0.01\ s^{-1}$, $0.1\ s^{-1}$), and dynamic plastic strain (strain rates as $10\ s^{-1}$, $100\ s^{-1}$, $1000\ s^{-1}$) are obtained from JMatPro. It shows that, when the temperature is lower than 400 °C, the stress of AlFeSi increases with the increase of strain. When the temperature is higher than 400 °C, a damage phenomenon of AlFeSi is present even for small strain rates (strain rates less than $0.1\ s^{-1}$), as shown in Figure 5a,b, while the stress of the AlFeSi phase increases with the increase of strain under a larger strain rate (strain rates greater than 0.1 s^{-1}), as shown in Figure 5c–f.

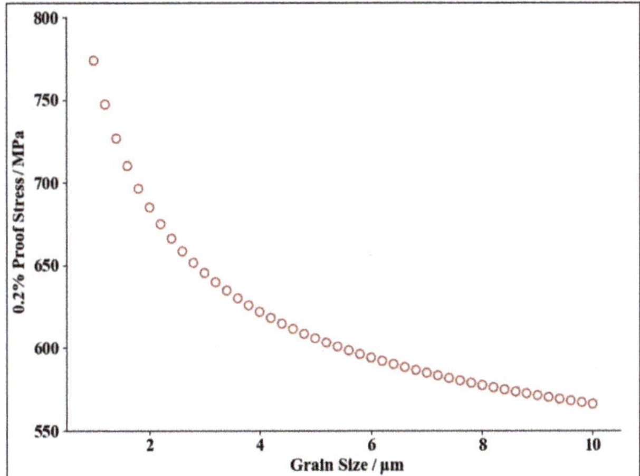

Figure 4. The yield strength of AlFeSi phase under quasi-static conditions.

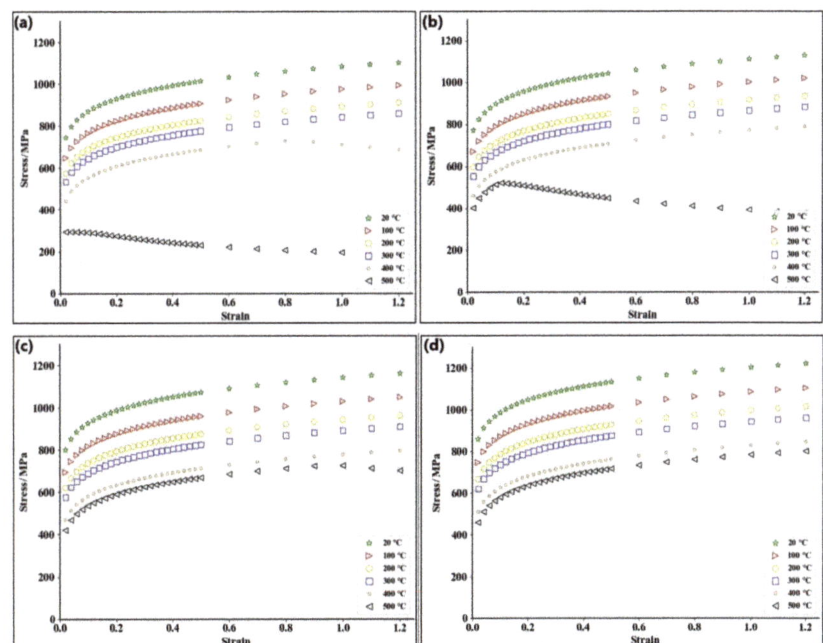

Figure 5. *Cont.*

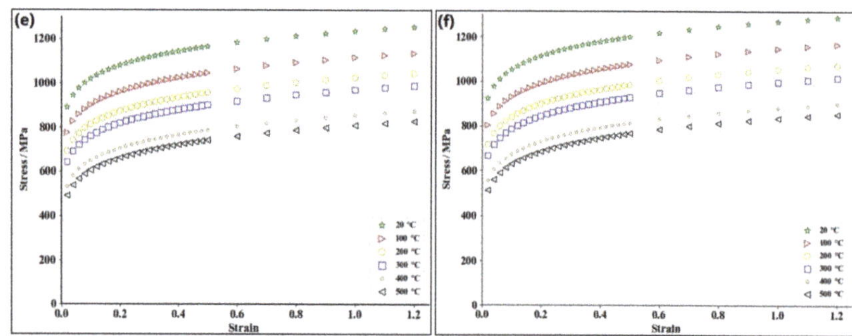

Figure 5. The stress–strain curve of AlFeSi phase under different strain rates: (**a**) strain rate 0.001 s^{-1}; (**b**) strain rate 0.01 s^{-1}; (**c**) strain rate 0.1 s^{-1}; (**d**) strain rate 10 s^{-1}; (**e**) strain rate 100 s^{-1}; (**f**) strain rate 1000 s^{-1}.

3. Elastoplastic-Damage Model of AlFeSi Phase

3.1. Solution of Elastoplastic Constitutive Equation

Johnson–Cook elastoplastic constitutive equation (J–C) takes the strain hardening effect, the strain rate effect and the temperature effect of the flow stress into consideration and can be represented as [17,18]

$$\sigma = \left(A + B\varepsilon_p{}^n\right)\left(1 + C\ln\frac{\dot{\varepsilon}}{\dot{\varepsilon}_0}\right)\left[1 - \left(\frac{T - T_0}{T_{melt} - T_0}\right)^m\right] \tag{1}$$

where, σ is the stress of material, A is the yield strength, B is the hardening parameter of strain and C is the strengthening parameter of strain rate. ε_p is the equivalent plastic strain of the material, $\dot{\varepsilon}$ is the equivalent plastic strain rate and $\dot{\varepsilon}_0$ is the reference strain rate. T_0 is the room temperature and T_{melt} is the melting point. n is the hardening index of strain. m refers to the thermal softening parameter.

Based on the stress–strain curves at different temperatures and strain rates of AlFeSi (Figure 5), the unknown parameters (A, B, C, n and m) of J–C elastoplastic constitutive equation can be confirmed by the following steps.

Step 1: Assume at quasi-static room temperature (the strain rates as 0.001 s^{-1}, 0.01 s^{-1}, 0.1 s^{-1} and the temperature at 20 °C), Equation (1) can be simplified as: $\sigma = A + B\varepsilon_p{}^n$, therefore it can be rewritten as: $\ln(\sigma - A) = n \cdot \ln\varepsilon_p + \ln B$. From Figure 4, when grain size is 2 μm: A = 662.33 MPa. Additionally, the values of factors (A, B, n) can be confirmed, as shown in Figure 6.

Step 2: Assume at dynamic normal temperature (the strain rates as 10 s^{-1}, 100 s^{-1}, 100 s^{-1} and the temperature at 20 °C), Equation (1) can be simplified as: $\sigma = \left(A + B\varepsilon_p{}^n\right)\left[1 + C \cdot \ln(\dot{\varepsilon}/\dot{\varepsilon}_0)\right]$, therefore can be rewritten as: $\sigma / \left(A + B\varepsilon_p{}^n\right) - 1 = C \cdot \ln(\dot{\varepsilon}/\dot{\varepsilon}_0)$. Referring to [19]: $\dot{\varepsilon}_0 = 1$, the value of C can be confirmed, see Figure 7.

Step 3: Assume at quasi-static high temperature (the strain rates as 0.001 s^{-1}, 0.01 s^{-1}, 0.1 s^{-1} and the temperature higher than 20 °C), Equation (1) can be simplified as $\sigma = \left(A + B\varepsilon_p{}^n\right)\left\{1 - \left[(T - T_0)/(T_{melt} - T_0)\right]^m\right\}$ and be changed as: $\ln\left[1 - \sigma/\left(A + B\varepsilon_p{}^n\right)\right] = m \cdot \ln\left[(T - T_0)/(T_{melt} - T_0)\right]$. T_0 = 20 °C and T_{melt} = 690 °C. Due to the small cutting depth and low cutting speed in Ultra-precision machining (UPM), the cutting-induced heat generation is less and the temperature T can be set as: 50 °C, 100 °C, 150 °C and 200 °C, and m can be confirmed, as shown in Figure 8.

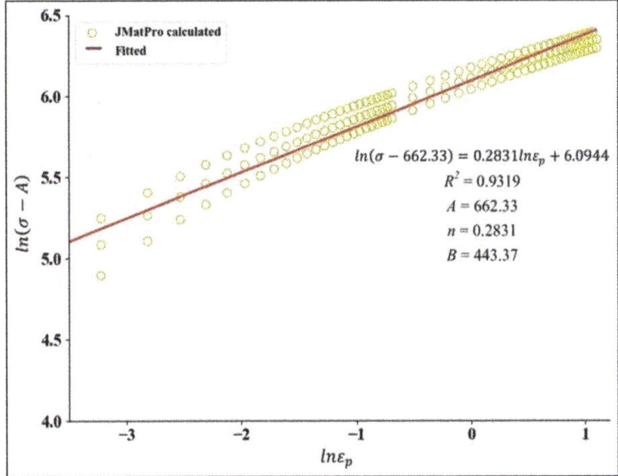

Figure 6. Solutions of three parameters (A, B and n).

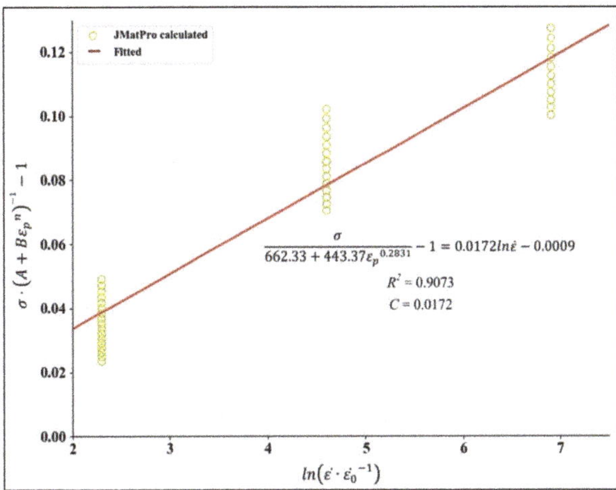

Figure 7. Solution of parameter C.

3.2. Solution of Damage Constitutive Equation

Johnson et al. [17,18] proposed a fracture criterion of the material with the consideration of the effects of stress triaxiality, strain, strain rate and temperature on the material failure:

$$D = \sum \frac{\Delta \varepsilon_f}{\varepsilon_f} \tag{2}$$

Meanwhile, the material breaks when the equivalent effect increment ($\Delta \varepsilon_f$) is equal to the failure strain ($\overline{\varepsilon}_f$) ($D = 1$), where $\Delta \varepsilon_f$ can be calculated by Equation (3).

$$\Delta \varepsilon_f = \varepsilon_f = [D_1 + D_2 \exp(-\eta D_3)](1 + D_4 \ln \frac{\dot{\varepsilon}}{\varepsilon_0})(1 + D_5 \frac{T - T_0}{T_{melt} - T_0}) \tag{3}$$

where, D_1–D_5 are fitting coefficients, η is the stress triaxiality.

The damage constitutive equation of the AlFeSi phase is based on the J–C damage constitutive equation. To obtain the fracture stress (σ_k), the strain and the strain rate of AlFeSi, multi-group dynamic rotating disk impact tensile experiments at room temperature and high-temperature environment are needed. In this paper, an FEM simulation approach is proposed to calculate the damage stress–strain curves of the AlFeSi phase under each group of D_1–D_5 parameters, which will be compared with those from JMatPro to determine the convergence of the calculation.

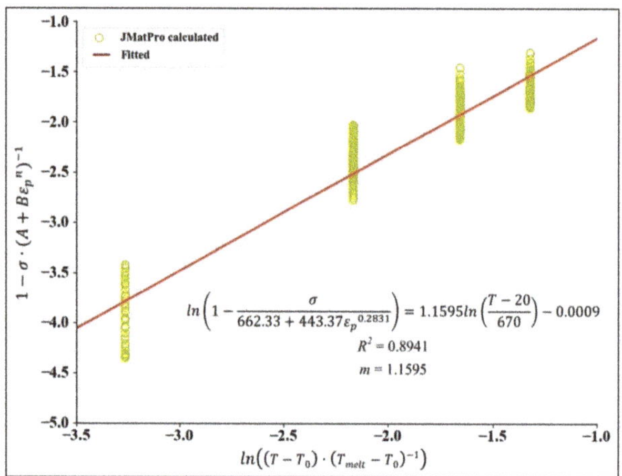

Figure 8. Solution of parameter m.

In general, to obtain the five coefficients (D_1–D_5) and to guarantee the validity of experimental results, several groups of quasi-static stretching experiments and dynamic rotary disk impact stretching experiments need to be conducted at different temperatures to evaluate the fracture stress, strain and strain rate of the tested materials. The experimental process is not only time-consuming but also requires a large number of material samples with consistent material properties, especially for the unknown materials ($Al_{86.1}Fe_{8.3}Si_{5.6}$). In this paper, a method based on FE simulation and JMatPro is proposed to solve the coefficients (D_1–D_5) of the material damage equation, which is scheduled as follows:

Step 1: Conduct FE simulations to achieve the stress–strain curve of AlFeSi at different temperatures and different strain rates by Abaqus software 2019 (Dassault SIMULIA, France, The period of validity: 20 June 2021) and Python (Version 3.5, accessed on 15 February 2021). Figure 9 shows the FE simulation results for the tensile fracture damage of AlFeSi, in which the mesh grid cell type is C3D8RT and the total element grid of the model is 12,870.

Step 2: Set the ambient temperature as 20 °C, 50 °C, 100 °C, 200 °C, 300 °C, 400 °C and 500 °C, the default value of D_1–D_5 as 1 and the step size (S_k) as 0.5, 0.1, 0.05 and 0.01. Table 1 lists the boundary conditions and the total step size of the model at 20 °C.

Table 1. The parameters of tensile fracture damage FEM for AlFeSi phase at 20 °C.

Strain Rate/s^{-1}	Force/N	The Total Time/s
0.001	60,000	0.25
0.01	90,000	0.15
0.1	120,000	0.1
10	650,000	0.025
100	6,000,000	0.005
1000	200,000,000	0.001

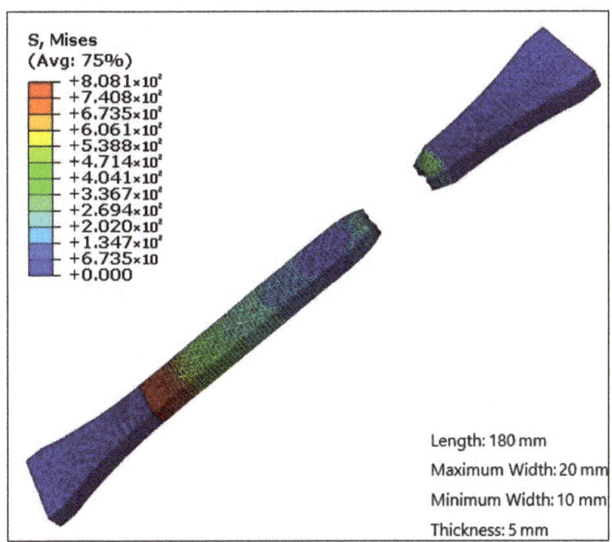

Figure 9. FE simulated tensile fracture damage of AlFeSi phase.

Step 3: Calculate the five parameters (D_1–D_5) as: $D_i = D_i \pm S_k \cdot j$, and $k = 1, 2, 3, 4$, $i = 1, 2, 3, 4, 5$, $j = 0, 1, 2, 3, 4$. The value of the step size (S_k) ranges from the largest value (0.5) to the smaller one (0.1), and final to the smallest one (0.01).

Step 4: Achieve an optimized combination of all five parameters (D_1–D_5) with the smallest mean square error (δ):

$$\min(\delta) = \min\left(\frac{\sqrt{\frac{1}{n}\sum_{i=1}^{i=n}(x_i - y_i)^2}}{\frac{1}{n}\sum_{i=1}^{i=n} y_i} \right) \tag{4}$$

where, $x_i = f(i)$, $y_i = g(i)$ represent the damage stress–strain curves of AlFeSi from JMatPro and FE simulation, respectively.

Figure 10a,b presents the calculation results of D_1–D_5 under two different conditions and the mean square errors in Figure 10a,b are 0.023 and 0.029, respectively.

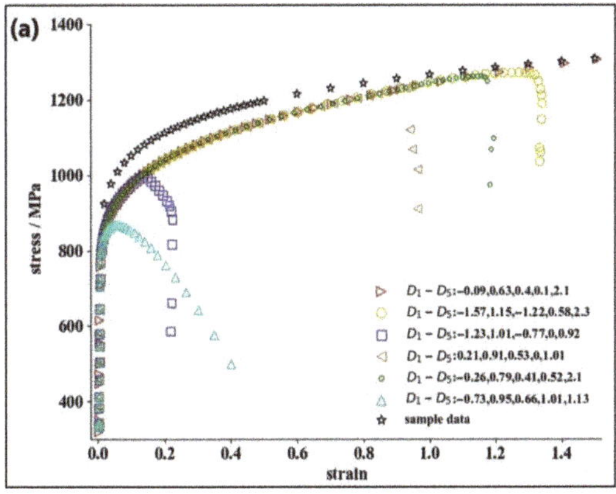

Figure 10. *Cont.*

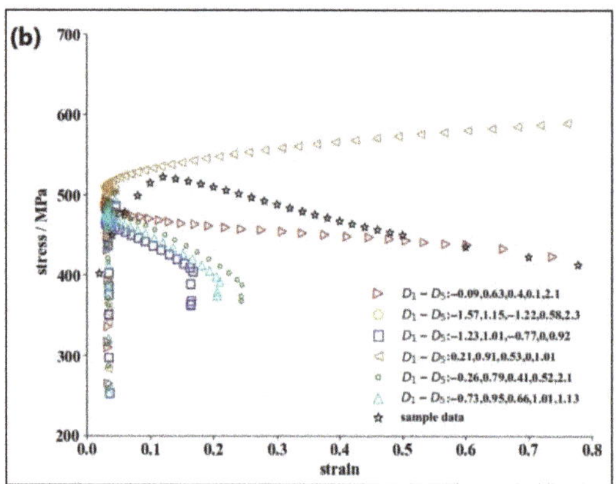

Figure 10. The stress–strain curve of AlFeSi phase with D_1–D_5: (**a**) strain rate of 1000 s^{-1}, temperature of 20 °C; (**b**) strain rate of 0.01 s^{-1}, temperature of 500 °C.

4. Experimental Verification

4.1. Material Preparation for AlFeSi

To verify the proposed method for the material properties, AlFeSi ($Al_{86.1}Fe_{8.3}Si_{5.6}$) is prepared and the material properties are examined to compare with the calculation results. The material preparation of AlFeSi is scheduled as (Figure 11): (1) Dissolving the pure aluminum at 750 °C; (2) increasing the proportion of alloy composition to meet the national standard limit of AlFeSi; (3) casting the solution into ingots; (4) hot press molding with heating temperature at 400 °C and pressure at 200 MPa; and (5) T6 treatment.

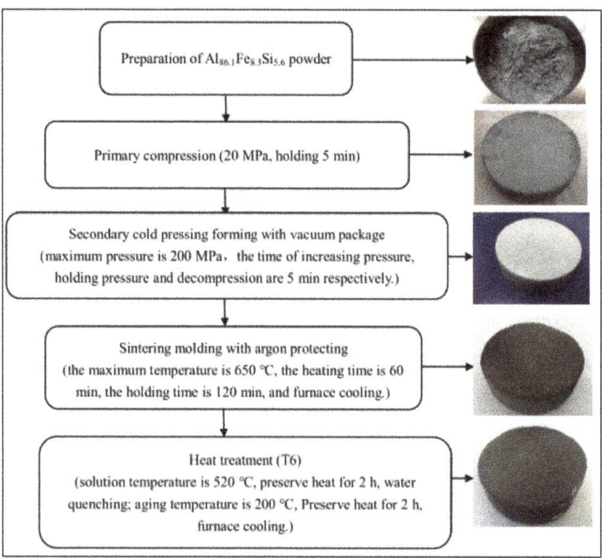

Figure 11. Preparation of the AlFeSi phase.

4.2. Scratch Experiment

Scratch experiments are carried out to examine the prepared AlFeSi at room temperature (20 °C) on Bruker UMT Tribolab (from Berlin, Germany, Figure 12). The X-axis speed and the scratch length are set as 5 mm/s and 10 mm, respectively. The Z-axis loads are changed as 1 N, 2 N, 3 N and 4 N. Figure 13 shows the geometric parameters of the tungsten carbide (WC) cemented carbide tool (Rockwell hardness 93). After the scratch test, the polished surface and the scratch depth are evaluated by an optical profiler (BRUKER Contour GT-X, from Berlin, Germany), as shown in Figure 14.

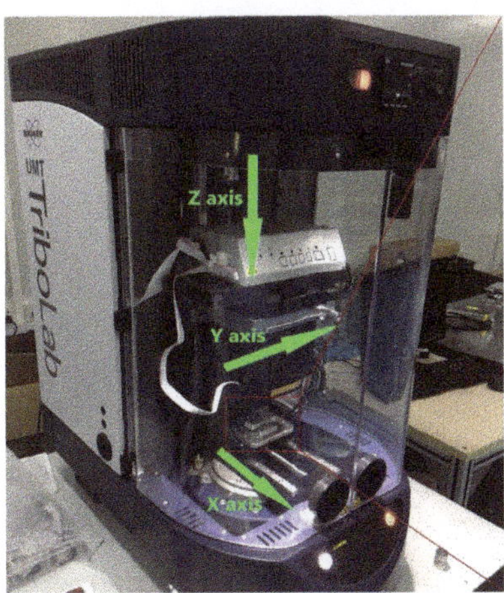

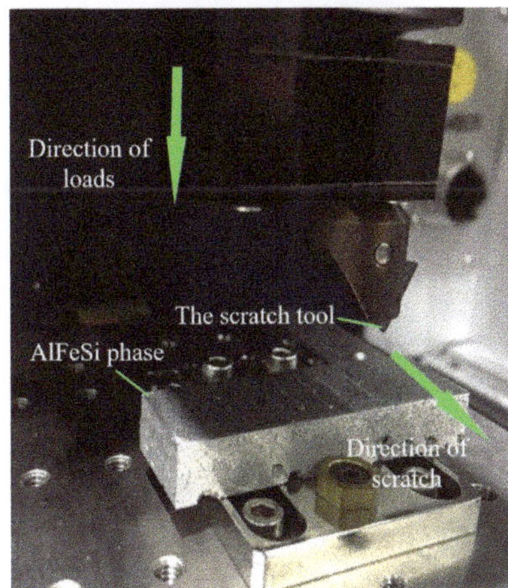

Figure 12. The scratch experiment of AlFeSi phase.

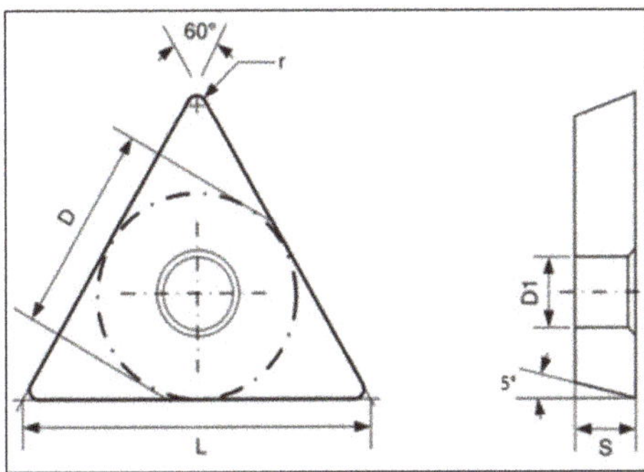

Figure 13. The parameters of the scratch tool (L = 10.5 mm, D = 6.01 mm, D1 = 2.56 mm, S = 2.74 mm, r = 0.4 mm).

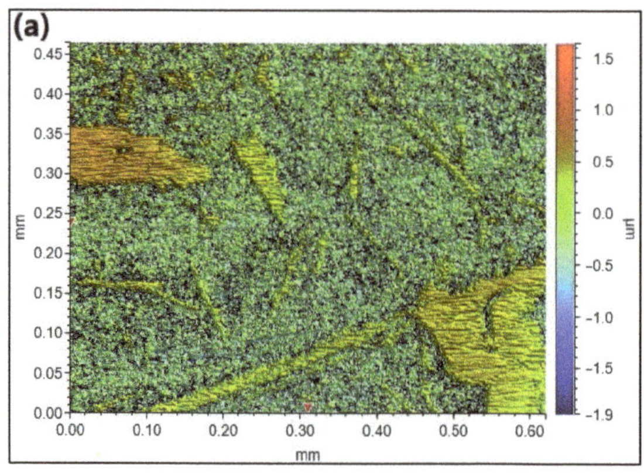

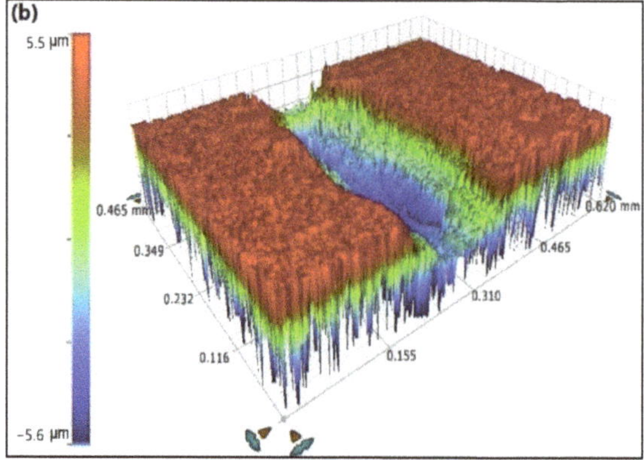

Figure 14. (**a**) The measured surface and (**b**) scratch of AlFeSi sample from the optical profiler.

4.3. Results and Discussions

Based on the calculation results of the AlFeSi material properties from the proposed method in this paper and a well-established elastoplastic-damage model, the FE model of the scratch test for AlFeSi is built by Abaqus software (as shown in Figure 15). The mesh element type is C3D8RT and the total grid quantity is 129,510. Table 2 lists all of the material properties for the WC cemented carbide tool used in FE simulation [20]. The boundary conditions of the FE model are set to the same parameters of the scratch experiment.

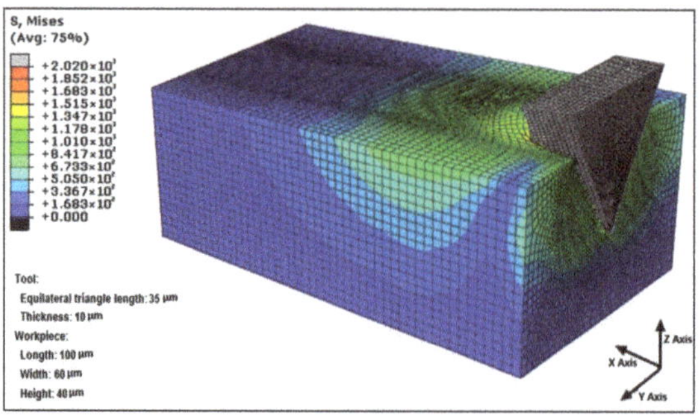

Figure 15. FE simulation model for the scratch experiment of AlFeSi phase.

Table 2. The parameters of WC cemented carbide tool (20 °C).

Density. (ton/mm^3)	Young's Modulus (MPa)	Poisson's Ratio	Specific Heat (mJ/(ton·°C))	Coefficient of Thermal Expansion (1/°C)	Thermal Conductivity (W/(m·°C))
11.9×10^{-9}	534,000	0.22	0.4×10^{-9}	4.7×10^{-6}	50

Figure 16 shows the comparison if the X-axis scratching force between the experiment and the simulation results from FE model. From this figure, as the scratch depth increases from 6.5 to 20 µm, both the simulated scratch forces and the experimental increase, and the simulated forces are close to the experimental ones. The probability of the X-axis scratch forces errors being less than 6.5% is 98% between FE simulation results and the scratch experiment. Moreover, the fluctuation ranges of the simulated forces and the experimental results increase with the increase of scratch depth, while the former is smaller than the latter. All of these infer that the method proposed in this study can be applied to confirm that all the parameters for AlFeSi and the developed elastoplastic-damage model of AlFeSi phase are feasible and effective.

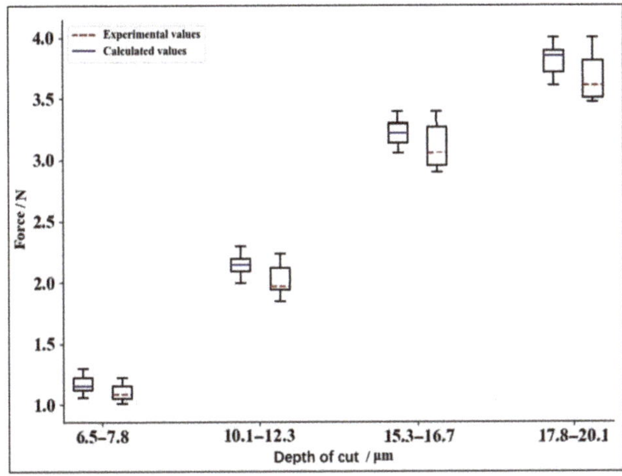

Figure 16. Comparison between FE simulation results and the scratch experiment.

5. Conclusions

This paper performs an investigation into the development of an elastoplastic-damage model for AlFeSi phase in Al6061. Based on the FE method and JMatPro, a new method is proposed in this study to efficiently calculate all of the parameters of J–C elastoplastic constitutive equation and J–C constitutive damage failure equation for AlFeSi. $Al_{86.1}Fe_{8.3}Si_{5.6}$ is prepared and the scratch experiments are conducted to study the scratch force error between FE simulation results and the scratch experiment. It shows that the errors of scratch forces between the simulation and experimental results are less than 6.5%.

Therefore, this study not only helps to propose a new method for studying the properties of a new material, but also contributes to providing a better understanding for the mechanisms of surface generation and material removal in ultra-precision machining of Al-Mg-Si alloys.

Author Contributions: Conceptualization, H.W. and S.W.; methodology, H.W. and S.W.; validation, H.W., T.Z. and S.W.; formal analysis, H.W., T.Z., W.D. and S.W.; investigation, H.W. and S.W.; data curation, H.W., T.Z., W.D., J.Y. and S.W.; writing—original draft preparation, H.W.; writing—review and editing, S.W.; supervision, S.W. All authors have read and agreed to the published version of the manuscript.

Funding: This research was funded by the National Natural Science Foundation of China, grant number 51975128.

Data Availability Statement: All data, models, and code generated or used during the study appear in the submitted article.

References

1. Khan, M.; Ud Din, R.; Wadood, A.; Syed, W.H.; Akhtar, S.; Aune, R.E. Effect of Graphene Nanoplatelets on the Physical and Mechanical Properties of Al6061 in Fabricated and T6 Thermal Conditions. *J. Alloys Compd.* **2019**, *790*, 1076–1091. [CrossRef]
2. Sekhar, A.P.; Mandal, A.B.; Das, D. Mechanical Properties and Corrosion Behavior of Artificially Aged Al-Mg-Si Alloy. *J. Mater. Res. Technol.* **2020**, *9*, 1005–1024. [CrossRef]
3. Wang, S.; To, S.; Chen, X.; Chen, X. An Investigation on Surface Finishing in Ultra-Precision Raster Milling of Aluminum Alloy 6061. *Proc. Inst. Mech. Eng. Part B J. Eng. Manuf.* **2015**, *229*, 1289–1301. [CrossRef]
4. Zhang, S.J.; To, S.; Wang, S.J.; Zhu, Z.W. A Review of Surface Roughness Generation in Ultra-Precision Machining. *Int. J. Mach. Tools Manuf.* **2015**, *91*, 76–95. [CrossRef]
5. Zhang, W.; Cao, Y.; Wang, J.; Gao, Y.; Zhou, M. Anisotropy Research of Ultra-Precision Machining on KDP Crystal. In Proceedings of the SPIE 7655, 5th International Symposium on Advanced Optical Manufacturing and Testing Technologies: Advanced Optical Manufacturing Technologies, Dalian, China, 6 October 2010; Volume 7655. [CrossRef]
6. Zong, W.J.; Huang, Y.H.; Zhang, Y.L.; Sun, T. Conservation Law of Surface Roughness in Single Point Diamond Turning. *Int. J. Mach. Tools Manuf.* **2014**, *84*, 58–63. [CrossRef]
7. He, C.L.; Zong, W.J.; Sun, T. Origins for the Size Effect of Surface Roughness in Diamond Turning. *Int. J. Mach. Tools Manuf.* **2016**, *106*, 22–42. [CrossRef]
8. Wang, M.; Wang, W.; Lu, Z. Anisotropy of Machined Surfaces Involved in the Ultra-Precision Turning of Single-Crystal Silicon—A Simulation and Experimental Study. *Int. J. Adv. Manuf. Technol.* **2012**, *60*, 473–485. [CrossRef]
9. Cheung, C.F.; Chan, K.C.; To, S.; Lee, W.B. Effect of Reinforcement in Ultra-Precision Machining of Al6061/SiC Metal Matrix Composites. *Scr. Mater.* **2002**, *47*, 77–82. [CrossRef]
10. Simoneau, A.; Ng, E.; Elbestawi, M.A. Surface Defects during Microcutting. *Int. J. Mach. Tools Manuf.* **2006**, *46*, 1378–1387. [CrossRef]
11. Harlow, D.G.; Nardiello, J.; Payne, J. The Effect of Constituent Particles in Aluminum Alloys on Fatigue Damage Evolution: Statistical Observations. *Int. J. Fatigue* **2010**, *32*, 505–511. [CrossRef]
12. Wang, S.J.; To, S.; Cheung, C.F. Effect of Workpiece Material on Surface Roughness in Ultraprecision Raster Milling. *Mater. Manuf. Process.* **2012**, *27*, 1022–1028. [CrossRef]
13. Wang, S.J.; Chen, X.; To, S.; Ouyang, X.B.; Liu, Q.; Liu, J.W.; Lee, W.B. Effect of Cutting Parameters on Heat Generation in Ultra-Precision Milling of Aluminum Alloy 6061. *Int. J. Adv. Manuf. Technol.* **2015**, *80*, 1265–1275. [CrossRef]
14. Wang, S.; Xia, S.; Wang, H.; Yin, Z.; Sun, Z. Prediction of Surface Roughness in Diamond Turning of Al6061 with Precipitation Effect. *J. Manuf. Process.* **2020**, *60*, 292–298. [CrossRef]
15. Wang, H.; Zhang, T.; Wang, S.; To, S. Characterization of the Friction Coefficient of Aluminum Alloy 6061 in Ultra-Precision Machining. *Metals* **2020**, *10*, 336. [CrossRef]
16. Lan, X.; Li, K.; Wang, F.; Su, Y.; Yang, M.; Liu, S.; Wang, J.; Du, Y. Preparation of millimeter scale second phase particles in aluminum alloys and determination of their mechanical properties. *J. Alloys Compd.* **2019**, *784*, 68–75. [CrossRef]
17. Johnson, G.R.; Cook, W.H. A constitutive model and data for metals subjected to large strains, high strain rates and high temperatures. In Proceedings of the 7th International Symposium on Ballistics, The Hague, The Netherlands, 19–21 April 1983; pp. 541–547.
18. Johnson, G.; Cook, W. Fracture Characteristics of Three Metals Subjected to Various Strains, Strain Rates, Temperatures and Pressures. *Eng. Fract. Mech.* **1985**, *21*, 31–48. [CrossRef]
19. Akram, S.; Jaffery, S.H.I.; Khan, M.; Fahad, M.; Mubashar, A.; Ali, L. Numerical and Experimental Investigation of Johnson–Cook Material Models for Aluminum (Al6061-T6) Alloy Using Orthogonal Machining Approach. *Adv. Mech. Eng.* **2018**, *10*, 1687814018797794. [CrossRef]
20. Asad, M.; Ijaz, H.; Khan, M.A.; Mabrouki, T.; Saleem, W. Turning Modeling and Simulation of an Aerospace Grade Aluminum Alloy Using Two-Dimensional and Three-Dimensional Finite Element Method. *Proc. Inst. Mech. Eng. Part B J. Eng. Manuf.* **2014**, *228*, 367–375. [CrossRef]

Article

Effects of Mould Temperature on Rice Bran-Based Bioplastics Obtained by Injection Moulding

María Alonso-González [1],*, Manuel Felix [2], Antonio Guerrero [2] and Alberto Romero [1]

[1] Departamento de Ingeniería Química, Facultad de Química, Universidad de Sevilla, 41012 Sevilla, Spain; alromero@us.es
[2] Departamento de Ingeniería Química, Escuela Politécnica Superior, Universidad de Sevilla, 41011 Sevilla, Spain; mfelix@us.es (M.F.); aguerrero@us.es (A.G.)
* Correspondence: maralonso@us.es; Tel.: +34-635-313-411

Abstract: The high production rate of conventional plastics and their low degradability result in severe environmental problems, such as plastic accumulation and some other related consequences. One alternative to these materials is the production of oil-free bioplastics, based on wastes from the agro-food industry, which are biodegradable. Not only is rice bran an abundant and non-expensive waste, but it is also attractive due to its high protein and starch content, which can be used as macromolecules for bioplastic production. The objective of this work was to develop rice-bran-based bioplastics by injection moulding. For this purpose, this raw material was mixed with a plasticizer (glycerol), analysing the effect of three mould temperatures (100, 130 and 150 °C) on the mechanical and microstructural properties and water absorption capacity of the final matrices. The obtained results show that rice bran is a suitable raw material for the development of bioplastics whose properties are strongly influenced by the processing conditions. Thus, higher temperatures produce stiffer and more resistant materials (Young's modulus improves from 12 ± 7 MPa to 23 ± 6 and 33 ± 6 MPa when the temperature increases from 100 to 130 and 150 °C, respectively); however, these materials are highly compact and, consequently, their water absorption capacity diminishes. On the other hand, although lower mould temperatures lead to materials with lower mechanical properties, they exhibit a less compact structure, resulting in enhanced water absorption capacity.

Keywords: bioplastics; rice bran; injection moulding; water absorption capacity

check for updates

Citation: Alonso-González, M.; Felix, M.; Guerrero, A.; Romero, A. Effects of Mould Temperature on Rice Bran-Based Bioplastics Obtained by Injection Moulding. *Polymers* **2021**, *13*, 398. https://doi.org/10.3390/polym13030398

Academic Editor: Alonso-González M.
Received: 13 January 2021
Accepted: 25 January 2021
Published: 27 January 2021

Publisher's Note: MDPI stays neutral with regard to jurisdictional claims in published maps and institutional affiliations.

1. Introduction

Nowadays, research and development of biopolymers has grown due to the increasing interest in using renewable and natural sources in the polymer processing industry. This is a consequence of the depletion of oil reserves and the serious environmental issues caused by plastic accumulation. Biopolymers can be divided into two categories: biodegradable polyesters, which are petroleum-based yet biodegradable, and polymers from renewable sources, such as protein and starch-based biopolymers, which are produced from natural sources and are readily biodegradable [1]. Proteins and starches are widely found in wastes and byproducts from the food and agricultural industries [2,3]. In the best scenario, these materials are used for animal feeding, as they are a low-added-value byproduct, even though this is a better end-of-life option than incineration or composting, which in turn are better than landfill disposal or leakage, which is by far the worst-case scenario. Therefore, biodegradable polymers can make significant contributions to material recovery (producing high added value products), reduction of landfill and utilization of renewable resources [4].

Starch-based biopolymers have gained attention and importance since they show thermoplastic-like processability after the application of suitable temperature and shear. In fact, a well-known type of starch-based polymer is the so-called thermoplastic starch (TPS) [5]. TPS is similar to other synthetic polymers (in terms of structure, molar mass, glass transition temperature, crystallinity, melting temperature, etc.). However, it melts and fluidizes in the presence of a low molecular plasticizer such as water or glycerol, high temperatures and shearing, being suitable for injection moulding, extrusion and blowing facilities, in the same way as synthetic plastics [6]. Similar to starches, proteins are macromolecules with continuous and cohesive matrices, receiving enormous attention for the production of biodegradable plastic, edible films and sheets [7]. Plant proteins that can be used for bio-based plastic include soy protein, corn zein, wheat protein, etc. [7–9]. Animal proteins such as blood meal, gelatin, collagen, keratin, egg protein, . . . etc., can also be used as feedstocks of such bio-based plastics [10–12]. For this purpose, plasticizers are generally added to the protein matrix during thermoplastic processes such as extrusion and injection moulding to improve its processability, reduce brittleness and modify the properties of the final structure [13]. Typically, the manufacture of protein-based bioplastics involves chemically, thermally, or pressure-induced protein denaturation as a first stage. Due to the diversity in the assembling of protein networks and their unique structures, a large variety of biodegradable materials can be produced, offering a wide range of functional properties [14]. Although the thermoplastic processing of starch and protein-based plastics is being widely studied, they still have weaknesses in their characteristics, such as weak mechanical properties, poor long-term stability and sensitivity to water; thus, there is scientific interest for studying new bio-based materials and optimizing processing conditions to find well-established alternatives to conventional plastics [15].

In this way, rice bran (RB) is a by-product of brown rice production. It contains up to 20% protein, 45% carbohydrates (mainly starch), 10% fibre and some varying percentages of moisture, lipids and ashes [16]. Globally, rice cultivation covers 145 million ha. In the European Union alone, the area dedicated to this crop is approximately 410,000 ha. Furthermore, it is estimated that 100 million tons of rice residues and byproducts are generated each year [17]. Currently, rice residues are treated as waste products, so they are incinerated for energy purposes or used as animal feed. For these reasons, the composition of rice bran suggests its possible successful employment for bioplastic production, which would benefit from reliable, renewable sources, creating high-value-added products that are also biodegradable, i.e., beneficial for the environment. Bioplastics have already been obtained from rice husk protein concentrate [17]. However, more research is required to produce competitive bio-based plastics from food industry by-products such as RB.

This research work aims to develop RB-based bioplastics via injection moulding, using water and glycerol as plasticizers. Furthermore, the effects of mould temperature on the properties of the final bioplastics were also studied. To this end, rheological measurements and mechanical, functional and microstructural characterization were carried out on the final samples.

2. Materials and Methods

2.1. Materials

RB was provided by Herba Ingredients (San José de la Rinconada, Seville, Spain). The RB supplied by this company is a byproduct obtained from the polishing process that produces the variety "vaporized indica white rice". Water (w) and glycerin (gly) were employed as plasticizers; the former was deionized-grade water, whereas the latter was supplied by PANREAC S.A. (Seville, Spain). All other reagents were supplied by Sigma Aldrich (St. Louis, MO, USA).

2.2. Chemical Composition

The chemical composition of RB was characterized by following the A.O.A.C. methods [18]. The protein content was determined as % N $\times$ 6.25 using a LECO TRUSPEC CHNS-932 nitrogen micro analyser (Leco Corporation, Saint Joseph, MI, USA) [19]. Ash content was determined by heating a small amount (5 g) of RB at 550 °C in a muffle furnace (Hobersal HD-230, Barcelona, Spain) for 5 h in air atmosphere. The sample was then cooled to room temperature in a desiccator before being weighted again to calculate the mass difference. The lipid content was quantified by the Soxhlet extraction method [20]. For this purpose, hexane was used as the solvent in a Soxhlet device. The hexane in contact with the sample dragged the lipids in subsequent cycles until the whole lipid content was removed. The exact lipid amount was also calculated by mass difference. Moisture was determined by placing 3 g of sample in a conventional oven (Memmert B216.1126, Schwabach, Germany) at 105 °C for 24 h.

2.3. Sample Preparation

Blends containing sieved rice bran (<500 μm), water and glycerin (RB/w/gly) were prepared by a two-stage thermomechanical procedure. Water was used as a plasticizer for starch, since it can break hydrogen bonds, allowing starch gelatinization. However, it cannot be used alone, as the products obtained are brittle, and consequently, it should be used with some other plasticizer, such as glycerol or sorbitol [21]. For this reason, selected blends containing 55% RB and 45% total plasticizer (2:1 w/Gly) were mixed in a two-blade counter-rotating batch mixer HAAKE POLYLAB QC (Thermo Scientific, Waltham, MA, USA) at 200 rpm and 80 °C, which was also selected as the intermediate temperature used by other authors (the process is usually carried out between 70 and 90 °C) [22]. This stage was carried out during 1 h based on a previous research work that employed a variety of rice bran plasticized by extrusion [16]. Secondly, the obtained doughs were processed by injection moulding using a MiniJet Piston Moulding System II (Thermo Scientific, Waltham, MA, USA). The temperature of the cylinder was fixed to 50 °C, and the pressure applied to force the material into the mould cavity was 500 bar; both conditions were applied for 15 s injection time. Three mould temperatures (100, 130 and 150 °C) were employed to study the influence of this parameter on the final properties of the bioplastics. These temperatures were selected based on previous works, and the results obtained from the temperature sweep tests performed on the processed doughs [23,24]. Thus, three systems were obtained, all of them processed at 500 bar injection pressure for 200 s. The fixed processing conditions were selected based on previous works carried out with similar samples. Rectangular bioplastic samples measuring $60 \times 10 \times 1$ mm were obtained for further mechanical and microstructural characterization.

2.4. Ageing of Doughs

The obtained doughs were not suitable for injection moulding right after the mixing stage; however, when kept in opened containers with <45% humidity, they evolved, acquiring a firmer appearance to the naked eye and losing some of their moisture content. In this way, after a certain amount of time, they were successfully processed by injection moulding. In order to quantify this ageing process, the homogeneous blends were characterized every 24 h through rheological measurements, while the water content was determined until a steady state was reached. A Dynamic Mechanical Analyser (RSA3, TA Instruments, New Castle, DE, USA) was employed in compression mode using a plate–plate geometry (8 mm diameter). Strain sweep tests were performed in order to establish the linear viscoelastic range. Subsequently, the lower critical strain was selected to carry out frequency sweep tests between 0.01 and 20 Hz at room temperature. Finally, once the dough remained unchanged, a temperature sweep test between 30 and 160 °C at 5 °C/min was also carried out to study the rheological behaviour of the dough to be injected with temperature. In these measurements, the storage modulus (E'), loss modulus (E'') and loss tangent (tan δ) were determined for the whole studied ranges. The water content was characterized following the same protocol mentioned in Section 2.2 for moisture content.

2.5. Bioplastics Characterization

2.5.1. Dynamic Mechanical Thermal Analysis (DMTA)

These tests were performed with an RSA3 dynamic mechanical analyser (TA Instruments, New Castle, DE, USA) on the obtained bioplastics using two grips, one at the bottom and one at the top, to study their rheological behaviour in tension mode. Again, the linear viscoelastic range was first determined, and the lower critical strain was used in the subsequent experiments. Frequency sweep tests at room temperature were performed between 0.01 and 20 Hz for the three manufactured systems (100, 130 and 150 °C mould temperature). Finally, temperature sweep tests at 1 Hz between 30 and 140 °C were carried out for all systems using a heating rate of 5 °C/min.

2.5.2. Tensile Tests

Tensile tests were performed using the RSA3 device in continuous deformation mode following a modified protocol (using rectangular specimens) based on ISO 570-2:2012, regarding the assessment of mechanical properties of polymeric samples [25]. At least three strain–stress curves were obtained for each evaluated system (100, 130 and 150 °C) and the Young's modulus (E), maximum tensile strength (σ_{max}) and elongation at break (ε_{max}) were evaluated with a strain rate of 1 mm/min at room temperature (25 ± 1 °C).

2.5.3. Water Uptake Capacity and Soluble Matter Loss

Water uptake capacity was measured according to the ASTM D570 method [26], using a third of the rectangular bioplastic specimens, that is, a rectangular sample measuring $20 \times 10 \times 1$ mm. The specimens were dried in a conventional oven (UN 55, Memmert, Schwabach, Germany) at 50 °C for 24 h to determine the dry weight. Immediately after, they were immersed in distilled water for 24 h. Finally, they were submitted to a freeze-drying step at -80 °C under vacuum in a LyoQuest freeze-dryer with a Flask M8 head (Telstar, Seville, Spain) and their weights after these last two steps were employed to determine their water uptake capacity and soluble matter loss according to the following equations:

$$\text{Water Uptake Capacity (\%)} = 100 \cdot (w_2 - w_3)/w_3, \tag{1}$$

$$\text{Soluble Matter Loss (\%)} = 100 \cdot (w_1 - w_3)/w_1, \tag{2}$$

where w_1, w_2 and w_3 are the weight of the bioplastic after dehydrothermal treatment, the immersion step and the freeze-drying stage, respectively.

2.5.4. Scanning Electron Microscopy (SEM)

The microstructure of the final bioplastics after the freeze-drying step was studied by scanning electron microscopy using a Zeiss EVO device (Carl Zeiss Microscopy, White Plains, NY, USA) in order to establish a relationship between the processing conditions and the structure of the final specimens. The samples were first coated with Pd/Au (10 nm) by sputtering using a Leica AC600 Metallizer, and then they were observed at 10 kV acceleration voltage and 500× magnification. Finally, the mean crack diameter was calculated using the software ImageJ using at least 5 measurements.

2.6. Statistical Analyses

At least three replicates of each measurement were carried out. Statistical analyses were performed using t-test and one-way analysis of variance (ANOVA) (significance value (p) < 0.05) using the STATGRAPHICS 18 software (Statgraphics Technologies, Inc., The Plains, VA, USA). Standard deviations from some selected parameters were calculated. Significant differences are indicated by different letters; that is, all mean values labelled with the same letter did not show significant differences.

3. Results

3.1. Chemical Composition

The RB variety analysed (vaporized Indica), shows a composition of $7.06 \pm 0.09\%$ moisture, $10.50 \pm 0.16\%$ ashes, $13.22 \pm 0.52\%$ proteins and $22.77 \pm 1.33\%$ lipids. There is a remaining percentage, accounting for 46.45% of the sample, which can be attributed to carbohydrates.

3.2. Doughs Characterization (Ageing)

Figure 1a shows the ageing process of the doughs obtained after mixing. Two parameters are considered: the elastic modulus (E′) and moisture content (%). $E_{1'}$ are the elastic modulus values at 1 Hz obtained from different frequency sweep tests performed each day for the corresponding doughs. As can be seen, E′ increased with time, reaching its maximum value after 6 days ageing time, with no further increases for longer times. Moreover, the water content followed the same tendency, stabilizing its value after 6 days. However, its values decreased with time from almost 40% moisture right after mixing to less than 30% at the end of the characterization (10 days ageing).

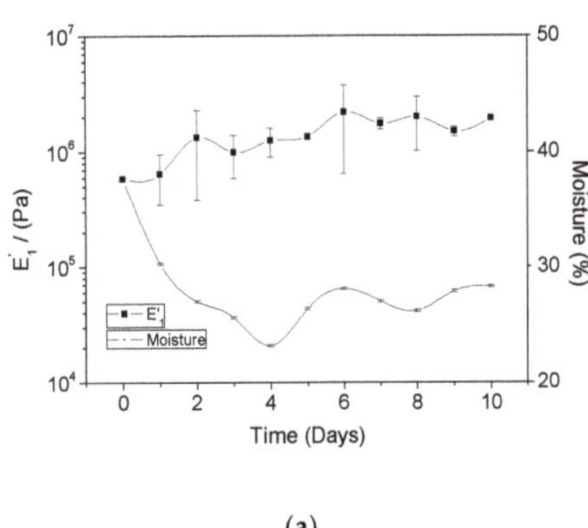

(a)

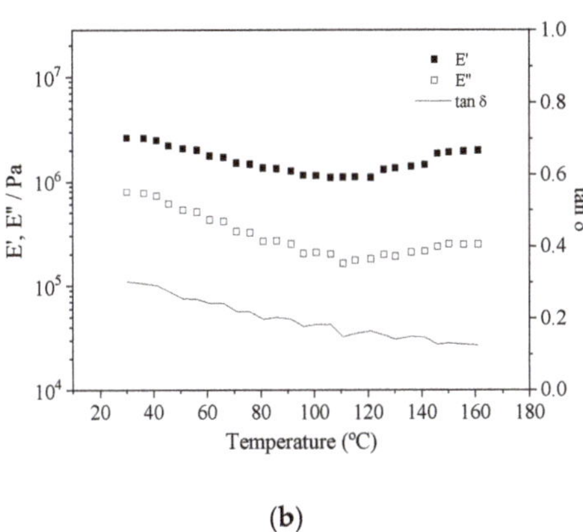

(b)

Figure 1. Viscoelastic properties at a constant frequency (1 Hz) of the doughs obtained after mixing. (**a**) Evolution of the elastic modulus (E′) and moisture content (%) over time. (**b**) Temperature sweep test of the stabilized dough between 30 and 160 °C.

Once the doughs were stabilized (after 6 days) and their parameters remained constant, they were subjected to temperature sweep tests between 30 and 160 °C. Figure 1b shows the elastic (E′) and viscous (E″) moduli with increasing temperature. Both viscoelastic moduli followed the same behaviour, i.e., a decrease with temperature up to 100–120 °C (E′ decreased from 2.63×10^6 to 1.10×10^6 Pa) and, subsequently, an increase for temperatures above 120 °C (E′ up to 1.99×10^6 Pa), indicating the thermosetting potential of the doughs before their processing by injection moulding [12]. The loss tangent (tan δ) decreased for the whole studied range; thus, no phase transitions were recorded [27].

3.3. Bioplastics Characterization
3.3.1. Dynamic Mechanical Thermal Analysis (DMTA)

Figure 2a shows the results of the frequency sweep tests carried out on the systems obtained with different mould temperatures (100, 130 and 150 °C) between 0.01 and 20 Hz. As can be seen, E′ was always higher than E″, indicating the predominantly elastic character of the obtained bioplastics, where both moduli increased with frequency, showing a noticeable dependence. The systems were also subjected to temperature sweep tests between 30 and 140 °C (Figure 2b). Two different behaviours were observed in this figure. First, the systems processed at 130 and 150 °C showed a similar response, with E′ and E″ decreasing with increasing temperature up to 80 °C. From this temperature to 140 °C, none of the two moduli changed significantly ($p < 0.05$). On the other hand, the system processed at 100 °C, although with significantly lower E′ ad E″ values, exhibited a similar behaviour up to 80 °C, with both moduli decreasing with increasing temperature. However, from 80 °C to 110 °C, E′ and E″ increased to values closer to those obtained for the other evaluated systems, being the only samples with certain thermosetting potential [28].

Finally, both viscoelastic moduli exhibited an abrupt decrease for the higher temperatures considered.

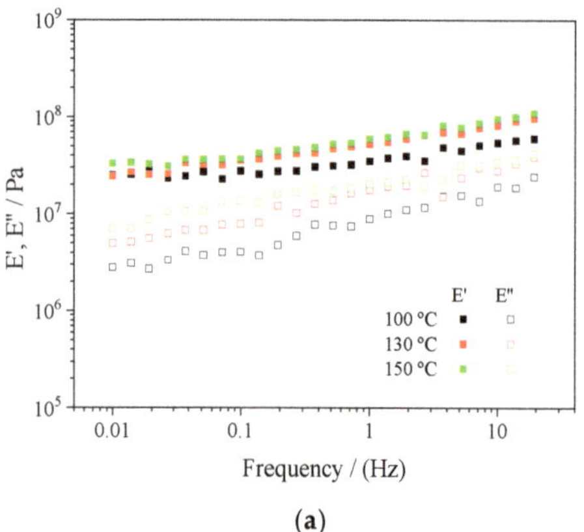

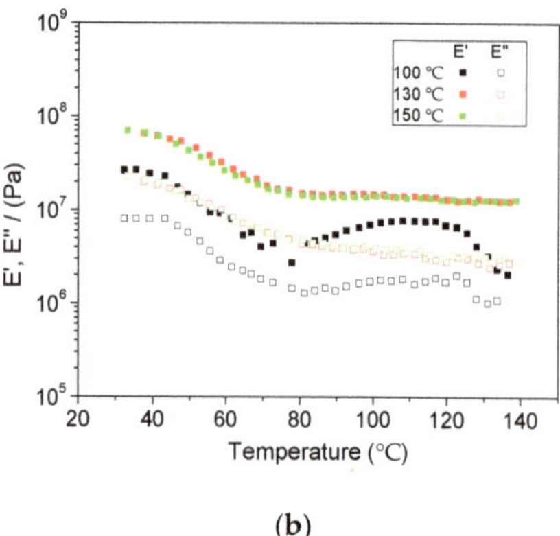

Figure 2. DMTA tests performed on the final bioplastics. (**a**) Frequency sweep tests from 0.01 to 20 Hz at room temperature and (**b**) temperature sweep tests from 30 to 140 °C at 1 Hz.

3.3.2. Tensile Tests

The stress–strain curves obtained from tensile tests are shown in Figure 3. The mechanical properties of the processed bioplastics improved for higher injection moulding temperatures, increasing the maximum tensile strength and strain at break. Thus, the bioplastics obtained at 130 and 150 °C are stiffer and exhibit greater toughness, especially the system obtained at 150 °C. For a more accurate comparison, Young's modulus (E), tensile strength (σ_{max}) and strain at break (ε_{max}) are shown in Table 1 for all studied systems.

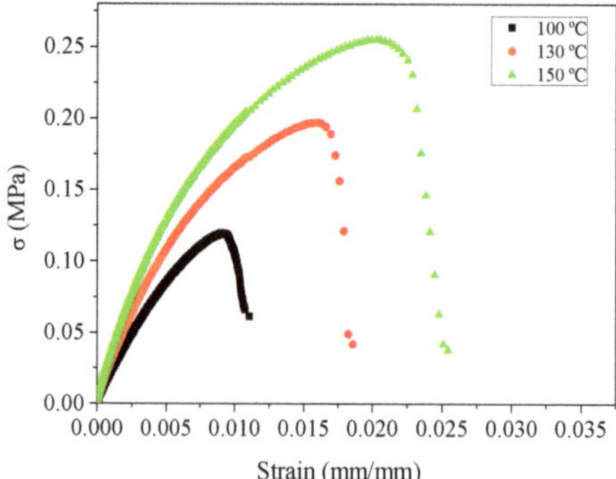

Figure 3. Stress–strain curves obtained from tensile tests for the bioplastics obtained for different mould temperatures (100, 130 and 150 °C).

Table 1 shows that all parameters are improved when higher processing temperatures are used. Young's modulus increased from 12 ± 7 to 23 ± 6 MPa when the temperature increased from 100 to 130 °C, reaching 33 ± 6 MPa for the higher temperature employed (150 °C). Tensile strength also increased from the 0.08 ± 0.04 MPa, exhibited by the sample obtained at 100 °C to 0.27 ± 0.01 MPa for the 150 °C sample, with the 130 °C sample having an intermediate value of 0.18 ± 0.03 MPa. Finally, the elongation at break also improved with temperature, being $1.1 \pm 0.3\%$ for the 100 °C sample, $1.6 \pm 0.4\%$ for the

130 °C sample and 2.4 ± 0.3% for the sample processed at 150 °C. It is worth mentioning that the elongations exhibited by the samples led to no significant changes in their cross-sections, being below 1% in all cases. Similar results were found by Felix et al. [28] for crayfish-based bioplastics. However, these results are not usually found in protein-based systems, where the Young's modulus also increases with temperature but the elongation at break decreases [29].

Table 1. Young's modulus, maximum tensile strength and elongation at break of the different processed systems. Different superscript letters within a column indicate significant differences ($p < 0.05$).

System	Young's Modulus (MPa)	Maximum Tensile Strength (MPa)	Elongation at Break (%)
100 °C	12 ± 7 [a]	0.12 ± 0.04 [a]	1.1 ± 0.3 [a]
130 °C	23 ± 6 [a,b]	0.18 ± 0.03 [b]	1.6 ± 0.4 [a]
150 °C	33 ± 6 [b]	0.26 ± 0.01 [c]	2.4 ± 0.3 [b]

3.3.3. Water Uptake Capacity and Soluble Matter Loss

The results obtained from the water absorption tests (Figure 4) show that an increase in the injection moulding temperature leads to poorer water uptake capacities. In this way, the water uptake capacity of the systems obtained at 100 °C was 254 ± 14%, while for the systems processed at 130 and 150 °C, these values decreased to 179 ± 33 and 137 ± 5%, respectively. Soluble matter loss also decreased with increasing temperature, although the differences were slighter, changing from 32.2 ± 0.8 to 28.0 ± 0.4 and 27.3 ± 0.9% for the systems obtained at 100, 130 and 150 °C, respectively.

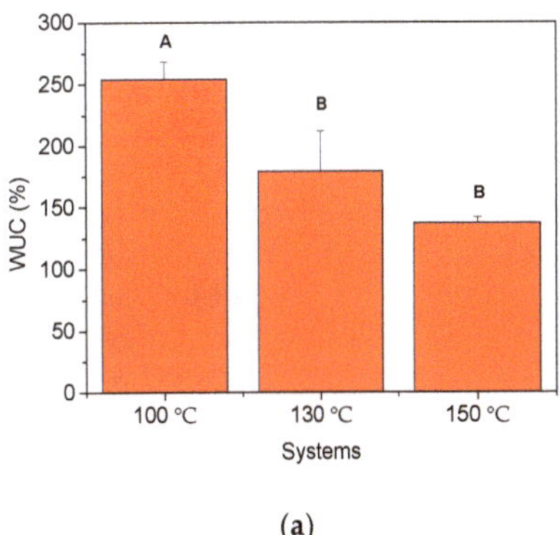

(a)

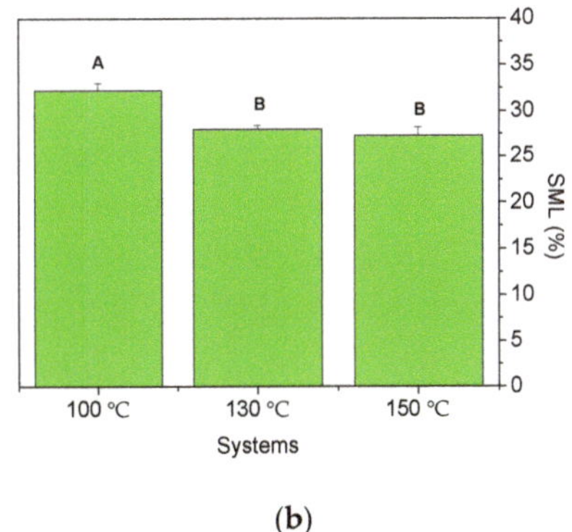

(b)

Figure 4. (a) Water uptake capacity (WUC) and (b) Soluble Matter Loss (SML) for the bioplastics obtained for different mould temperatures (100, 130 and 150 °C). Different letters above each bar indicate significant differences ($p < 0.05$).

3.3.4. Scanning Electron Microscopy (SEM)

Figure 5 shows SEM micrographs after 24 h of water immersion and subsequent freeze-drying of the obtained bioplastics. The SEM images indicate that a continuous structure was obtained, where the homogeneity of the system depends on the processing temperature. Thus, this procedure did not lead to porous matrices but compact bioplastics, where the effect of increasing processing temperature could be analysed. The higher the injection moulding temperature, the more compact the morphology of the bioplastic is, showing a more uniform surface with fewer cracks and with the mean crack diameter decreasing from 121 ± 24 μm for the 100 °C system to 54 ± 16 and 29 ± 16 μm for the

130 and 150 °C systems. The effect was most noticeable when the highest temperature was applied.

(a) (b) (c)

Figure 5. SEM micrographs after lyophilization of the different samples at (**a**) 100 °C; (**b**) 130 °C; (**c**) 150 °C.

4. Discussion

4.1. Chemical Composition

The chemical composition of the sample was successfully evaluated, leading to similar results to those obtained by Siswanti et al. [30] and Thiranan et al. [31] for red and defatted rice bran. According to these research works, the majority of the sample accounts for carbohydrates; thus, the non-identified percentage of the rice bran under evaluation can be attributed to this basic food group. Furthermore, the characterization carried out by Klanwan et al. [16] revealed a variety of this byproduct with a very similar composition, which was employed for the same purpose (bioplastic production), suggesting the suitability of rice bran to produce bioplastics obtained by injection moulding. Besides, these authors highlighted the presence of fibre and starch in the carbohydrate fraction, as two of the main components, along with protein, for this valorisation process. This fact supports the importance of using strategies based on these biopolymers for the development of bioplastics.

4.2. Doughs Characterization (Ageing)

Under the selected conditions, the obtained doughs achieved a steady state after ageing for six days, and after this time, they were suitable for injection moulding. As can be seen in Figure 1a, the elastic modulus increased when the moisture content decreased; thus, it can be assumed that both parameters are related. In order to determine whether there is some structuration of the doughs during ageing when no water is lost, the doughs were kept in closed containers (the moisture content remained unchanged) and their rheological parameters were measured with time. In this experiment, the obtained doughs did not change, with the elastic modulus remaining constant for one week (results shown in Supplementary Material, Figure S1). Therefore, the moisture content was selected as the main parameter to be considered when the doughs are evaluated before injection moulding. Thus, excess water leads to incomplete filling of the mould cavity, since the water evaporates in the mould, causing voids and cracks. However, water is required during dough formation since it promotes the breakage of the hydrogen bonds, leading to the subsequent drop in glass transition and melting temperatures, improving processability [13].

Regarding the temperature sweep tests shown in Figure 1b, the thermosetting potential exhibited above 120 °C indicates that the mechanical properties of the final bioplastics will probably improve when processing at these higher temperatures [32]. This response can be attributed to the protein fraction that undergoes some heat-induced strengthening phenomena (e.g., aggregation and gelation). In this way, the best results are expected for the systems obtained at 130 and 150 °C, especially for the latter. Further characterization of obtained bioplastics by injection moulding will elucidate this thermomechanical response.

4.3. Bioplastics Characterization

4.3.1. Dynamic Mechanical Thermal Analysis (DMTA)

The frequency sweep tests obtained for the final bioplastics indicate a predominantly elastic character with certain frequency dependence, which has been obtained in previous studies for different successful protein-based biodegradable bioplastics [7,33]. Moreover, the rheological parameters (viscoelastic moduli) improved for higher temperatures, as expected from the results obtained in Section 3.2 regarding the rheological behaviour of the doughs with temperature.

On the other hand, the only system that followed a different tendency when evaluating the rheological properties with temperature was the one processed at 100 °C, whose moduli began to diminish with temperature until an important increase was observed between 80 and 120 °C, indicating the remaining thermosetting potential. It seems very clear that better rheological properties were obtained for higher processing temperatures, with no significant differences between the systems processed at 130 and 150 °C, which followed the same behaviour with also very similar values. In this case, the increase in processing temperature, with the corresponding higher economic and energetic costs, would not be justified, since no further improvements in the rheological parameters were obtained. Similar results for temperature sweep tests were obtained by Felix et al. [23] for bioplastics obtained from albumen, soy and pea protein isolate, as well as rice husk protein concentrate.

4.3.2. Tensile Tests

The results obtained from the tensile tests shown in Figure 3 and Table 1 confirm the above-mentioned behaviours, where higher processing temperatures produced bioplastics with improved mechanical properties. In this case, the three studied parameters (i.e., σ, ε and E) improved not only when the temperature increased from 100 to 130 °C, but also for the system obtained at 150 °C, which justifies the further increase in temperature that led to better mechanical properties in tension, besides the fact that these differences were not observed in the previous section. Finally, it is worth mentioning that similar Young's modulus values, although with slightly higher tensile strengths, have been obtained in two previous studies for crayfish, albumen and rice husk materials [23,28], where the authors stated that the protein-based bioplastics exhibited suitable characteristics to replace conventional plastics in different applications. However, the samples obtained in the present research work exhibited higher elongation at break. In addition, the obtained results are very similar to those obtained by Klanwan et al. [16] and Zárate-Ramírez et al. [24] for protein/carbohydrate-based bioplastics, in the first case obtained by extrusion using kraft lignin for plastification purposes, especially to those obtained for the higher kraft lignin content. This research shows the synergetic effect of the combination of proteins and carbohydrates in the biopolymer matrix since the increase in the Young's modulus also brings an increase in the elongation at break, which is not typically observed in protein-based materials [34]. In any case, all the bioplastics obtained at different temperatures exhibit tensile properties lower than synthetic polymers such as LDPE. Thus, the tensile strength, elongation at break and Young's modulus show values that reach as much as 2%, 1% and 10%, respectively, of the values for ASTM-normalized LDPE [35].

4.3.3. Water Uptake Capacity and Soluble Matter Loss

The water uptake capacity values decreased from 254 to 137% when temperature increased from 100 to 150 °C, indicating that the mechanical properties improved with higher processing temperatures. Thus, the systems with stronger structuration and better mechanical behaviour retain less water within their structure [36]. In this way, the systems with higher physical integrity, that is, those with stronger structure during processing, will undergo minor matter loss during immersion. As can be seen, this can be observed in Figure 4b, which shows the lower soluble matter loss values exhibited for higher temperatures. Finally, it should be highlighted that this behaviour has been observed before in previous studies, where higher mechanical properties led to poorer water uptake capacities and lower soluble matter loss results in pea protein-based bioplastic systems [37].

There is a remarkable difference with the results found for bioplastic matrices containing protein as the only biopolymer, which generally release the whole plasticizer content at the immersion stage [29]. These results suggest that the starch fraction contributes to retaining an important proportion of the plasticizer (either water, glycerol or both).

4.3.4. Scanning Electron Microscopy (SEM)

The micrographs shown in Figure 5 match all results obtained before. Firstly, the most homogeneous structures correspond to those systems that exhibited the highest mechanical properties, which then retained less water and underwent minor matter losses during immersion. Thus, the water uptake capacities and mechanical properties exhibited can be related to the microstructure generated, with greater homogeneity producing higher mechanical properties and lower water absorption. In this sense, the systems with poorer mechanical behaviour and better water absorption properties correspond to the less compact structures, showing more voids and cracks. From these results, it can be assumed that higher processing temperatures might induce more protein–protein interactions, producing more compact structures with higher physical integrity, resulting in improved mechanical properties, as was observed before for wheat gluten-based bioplastics obtained using a very similar process [38].

5. Conclusions

The results obtained in this work support the suitability of RB to develop bioplastics obtained by injection moulding. The initial high content of carbohydrates (~46%) indicates that the processing conditions selected must be compatible with both proteins and polysaccharides. Consequently, the formulation included not only glycerol as the plasticizer but also water during the formation of doughs, which allows the breakage of hydrogen bonds, increasing processability and allowing thermo-mechanical methods to be employed in the manufacture of the desired bioplastics. Unfortunately, this water content hinders the processing of samples by injection moulding, and thus the dough-like blends were subjected to ageing, obtaining constant and suitable values after 6 days. Once the bioplastics were successfully processed by this technique, the effect of mould temperature was analysed. The obtained results indicate that higher temperatures involved enhanced mechanical properties, which was evidenced by an increase in the elastic moduli (from 3.49×10^7 to 6.01×10^7 Pa for the systems processed at 100 and 150 °C, respectively), as well as an increase in the tensile strength parameters (σ_{max} increased from 0.12 ± 0.04 to 0.26 ± 0.01 MPa for the systems processed at 100 and 150 °C, respectively), which were caused by the development of more protein–protein interactions for higher temperatures. Unfortunately, the increase in the viscoelastic moduli resulted in lower water absorption capacities. In this sense, the greater structuration of the systems coincided with the lower soluble matter loss observed. Lastly, the obtained micrographs confirm the more homogeneous (compact) structure obtained as the temperature increased up to 150 °C, where voids and cracks were absent. This microstructure matches the higher mechanical properties and lower water absorption previously observed, since the continuous structure hindered the water penetration and facilitated the matrix breakdown.

These results confirm the suitability of RB as raw material for the generation of natural bioplastics; however, further research is needed to generate bioplastics for commercial applications.

Supplementary Materials: The following are available online at https://www.mdpi.com/2073-4360/13/3/398/s1, Figure S1: Elastic modulus (E′) during the ageing process of the doughs obtained after mixing (kept in closed containers) versus time.

Author Contributions: Conceptualization, M.A.-G., M.F., A.G. and A.R.; methodology, M.A.-G.; software, M.A.-G.; validation, M.A.-G., M.F. and A.R.; formal analysis, M.A.-G.; investigation,

M.A.-G.; resources, M.A.-G. and A.G.; data curation, M.A.-G.; writing—original draft preparation, M.A.-G.; writing—review and editing, M.A.-G., M.F. and A.R.; visualization, M.A.-G., M.F. and A.R.; supervision, M.F. and A.R.; project administration, M.F. and A.R.; funding acquisition, A.G. All authors have read and agreed to the published version of the manuscript.

Funding: This research was funded by the "Ministerio de Ciencia e Innovación" of the Spanish Government and FEDER (UE), grant number RTI2018-097100-B-C21.

Acknowledgments: The authors acknowledge the University of Seville for the VPPI-US grant (Ref.-II.5) to Manuel Felix and the financial support of the Spanish Government "Ministerio de Ciencia, Innovación y Universidades". The authors also thank CITIUS for granting access to and their assistance with the Microscopy service. Authors also kindly thank Herba Ingredients for providing the raw material used in this study.

Conflicts of Interest: The authors declare no conflict of interest.

References

1. Tábi, T.; Kovács, J.G. Examination of injection moulded thermoplastic maize starch. *Express Polym. Lett.* **2007**, *1*, 804–809. [CrossRef]
2. Urista, C.M.; Fernández, R.Á.; Rodriguez, F.R.; Cuenca, A.A.; Jurado, A.T. Review: Production and functionality of active peptides from milk. *Food Sci. Technol. Int.* **2011**, *17*, 293–317. [CrossRef] [PubMed]
3. Yamada, M.; Morimitsu, S.; Hosono, E.; Yamada, T. Preparation of bioplastic using soy protein. *Int. J. Biol. Macromol.* **2020**, *149*, 1077–1083. [CrossRef] [PubMed]
4. Mostafa, N.A.; Farag, A.A.; Abo-dief, H.M.; Tayeb, A.M. Production of biodegradable plastic from agricultural wastes. *Arab. J. Chem.* **2018**, *11*, 546–553. [CrossRef]
5. Shin, B.Y.; Narayan, R.; Lee, S.I.; Lee, T.J. Morphology and rheological properties of blends of chemically modified thermoplastic starch and polycaprolactone. *Polym. Eng. Sci.* **2008**, *48*, 2126–2133. [CrossRef]
6. Ashter, S.A. *Types of Biodegradable Polymers in Introduction to Bioplastics Engineering*; Elsevier: Amsterdam, The Netherlands, 2016.
7. Felix, M.; Martinez, I.; Romero, A.; Partal, P.; Guerrero, A. Effect of pH and nanoclay content on the morphology and physicochemical properties of soy protein/montmorillonite nanocomposite obtained by extrusion. *Compos. Part. B Eng.* **2018**, *140*, 197–203. [CrossRef]
8. Herald, T.J.; Obuz, E.; Twombly, W.W.; Rausch, K.D. Tensile properties of extruded corn protein low-density polyethylene films. *Cereal Chem.* **2002**, *79*, 261–264. [CrossRef]
9. Chantapet, P.; Kunanopparat, T.; Menut, P.; Siriwattanayotin, S. Extrusion processing of wheat gluten bioplastic: Effect of the addition of kraft lignin. *J. Polym. Environ.* **2013**, *21*, 864–873. [CrossRef]
10. Ramakrishnan, N.; Sharma, S.; Gupta, A.; Alashwal, B.Y. Keratin based bioplastic film from chicken feathers and its characterization. *Int. J. Biol. Macromol.* **2018**, *111*, 352–358. [CrossRef] [PubMed]
11. Verbeek, C.J.R.; Low, A.; Lay, M.C.; Hicks, T.M. Processability and mechanical properties of bioplastics produced from decoloured bloodmeal. *Adv. Polym. Technol.* **2018**, *37*, 2102–2113. [CrossRef]
12. Félix, M.; Martín-Alfonso, J.E.; Romero, A.; Guerrero, A. Development of albumen/soy biobased plastic materials processed by injection molding. *J. Food Eng.* **2014**, *125*, 7–16. [CrossRef]
13. Mekonnen, T.; Mussone, P.; Khalil, H.; Bressler, D. Progress in bio-based plastics and plasticizing modifications. *J. Mater. Chem. A* **2013**, *1*, 13379–13398. [CrossRef]
14. Verbeek, C.J.R.; van den Berg, L.E. Extrusion processing and properties of protein-based thermoplastics. *Macromol. Mater. Eng.* **2010**, *295*, 10–21. [CrossRef]
15. Silviana, S.; Rahayu, P. Central composite design for optimization of starch-based bioplastic with bamboo microfibrillated cellulose as reinforcement assisted by potassium chloride. *J. Phys. Conf. Ser.* **2019**, *1295*, 012073. [CrossRef]
16. Klanwan, Y.; Kunanopparat, T.; Menut, P.; Siriwattanayotin, S. Valorization of industrial by-products through bioplastic production: Defatted rice bran and kraft lignin utilization. *J. Polym. Eng.* **2016**, *36*, 529–536. [CrossRef]
17. Félix, M.; Lucio-Villegas, A.; Romero, A.; Guerrero, A. Development of rice protein bio-based plastic materials processed by injection molding. *Ind. Crops Prod.* **2016**, *79*, 152–159. [CrossRef]
18. AOAC International. *Official Methods of Analysis of AOAC International*; AOAC International: Rockville, MD, USA, 2005.
19. Mariotti, F.; Tomé, D.; Mirand, P.P. Converting nitrogen into protein—Beyond 6.25 and Jones' factors. *Crit. Rev. Food Sci. Nutr.* **2008**, *48*, 177–184. [CrossRef]
20. López-Bascón-Bascon, M.A.; Luque de Castro, M.D. Soxhlet extraction in liquid-phase extraction. In *Liquid-Phase Extraction*; Elsevier: Amsterdam, The Netherlands, 2019; pp. 327–354.
21. Forssell, P.M.; Mikkilä, J.M.; Moates, G.K.; Parker, R. Phase and glass transition behaviour of concentrated barley starch-glycerol-water mixtures, a model for thermoplastic starch. *Carbohydr. Polym.* **1997**, *34*, 275–282. [CrossRef]
22. Rudnik, E. *Compostable Polymer Materials*, 2nd ed.; Elsevier: Amsterdam, The Netherlands, 2019.

23. Felix, M.; Perez-Puyana, V.; Romero, A.; Guerrero, A. Production and characterization of bioplastics obtained by injection moulding of various protein systems. *J. Polym. Environ.* **2017**, *25*, 91–100. [CrossRef]
24. Zárate-Ramírez, L.S.; Romero, A.; Bengoechea, C.; Partal, P.; Guerrero, A. Thermo-mechanical and hydrophilic properties of polysaccharide/gluten-based bioplastics. *Carbohydr. Polym.* **2014**, *112*, 16–23. [CrossRef]
25. International Organization for Standardization. *Plastics—Determination of Tensile Properties—Part 1: General Principles*; ISO 527-1:2012; International Organization for Standardization: Geneva, Switzerland, 2012.
26. ASTM International. *Standard Test. Method for Water Absorption of Plastics*; ASTM D570; ASTM International: West Conshohocken, PA, USA, 2014.
27. Gibbs, J.H.; DiMarzio, E.A. Nature of the glass transition and the glassy state. *J. Chem. Phys.* **1958**, *28*, 373. [CrossRef]
28. Felix, M.; Romero, A.; Cordobes, F.; Guerrero, A. Development of crayfish bio-based plastic materials processed by small-scale injection moulding. *J. Sci. Food Agric.* **2015**, *95*, 679–687. [CrossRef] [PubMed]
29. Fernández-Espada, L.; Bengoechea, C.; Cordobés, F.; Guerrero, A. Protein/glycerol blends and injection-molded bioplastic matrices: Soybean versus egg albumen. *J. Appl. Polym. Sci.* **2016**, *133*, 43524. [CrossRef]
30. Anandito, R.B.K.; Nurhartadi, E.; Iskandar, B.D. Effect of various heat treatment on physical and chemical characteristics of red rice bran (*Oryza nivara* L.) Rojolele. *IOP Conf. Ser. Mater. Sci. Eng.* **2019**, *633*, 012046. [CrossRef]
31. Kunanopparat, T.; Menut, W.; Srichumpoung, W.; Siriwattanayotin, S. Characterization of defatted rice bran properties for biocomposite production. *J. Polym. Environ.* **2014**, *22*, 559–568. [CrossRef]
32. Álvarez-Castillo, E.; Bengoechea, C.; Guerrero, A. Composites from by-products of the food industry for the development of superabsorbent biomaterials. *Food Bioprod. Process.* **2020**, *119*, 296–305. [CrossRef]
33. Bourny, V.; Perez-Puyana, V.; Felix, M.; Romero, A.; Guerrero, A. Evaluation of the injection moulding conditions in soy/nanoclay based composites. *Eur. Polym. J.* **2017**, *95*, 539–546. [CrossRef]
34. Fernández-Espada, L.; Bengoechea, C.; Cordobés, F.; Guerrero, A. Thermomechanical properties and water uptake capacity of soy protein-based bioplastics processed by injection molding. *J. Appl. Polym. Sci.* **2016**, *133*, 43524. [CrossRef]
35. ASTM International. *Standard Test Method for Tensile Properties of Plastics*; ASTM-D638-14; ASTM International: West Conshohocken, PA, USA, 2014.
36. Delgado, M.; Felix, M.; Bengoechea, C. Development of bioplastic materials: From rapeseed oil industry by products to added-value biodegradable biocomposite materials. *Ind. Crops Prod.* **2018**, *125*, 401–407. [CrossRef]
37. Perez, V.; Felix, M.; Romero, A.; Guerrero, A. Characterization of pea protein-based bioplastics processed by injection moulding. *Food Bioprod. Process.* **2016**, *97*, 100–108. [CrossRef]
38. Zárate-Ramírez, L.S.; Martínez, I.; Romero, A.; Partal, P.; Guerrero, A. Wheat gluten-based materials plasticised with glycerol and water by thermoplastic mixing and thermomoulding. *J. Sci. Food Agric.* **2011**, *91*, 625–633. [CrossRef] [PubMed]

Article

Incorporation of ZnO Nanoparticles into Soy Protein-Based Bioplastics to Improve Their Functional Properties

Mercedes Jiménez-Rosado [1,*], Víctor Perez-Puyana [2], Pablo Sánchez-Cid [2], Antonio Guerrero [1] and Alberto Romero [2]

[1] Department of Chemical Engineering, Escuela Politécnica Superior, 41011 Sevilla, Spain; aguerrero@us.es
[2] Department of Chemical Engineering, Facultad de Química, 41012 Sevilla, Spain; vperez11@us.es (V.P.-P.); pabsanbue@alum.us.es (P.S.-C.); alromero@us.es (A.R.)
* Correspondence: mjimenez42@us.es; Tel.: +34-954-557-179

Abstract: The union of nanoscience (nanofertilization) with controlled release bioplastic systems could be a key factor for the improvement of fertilization in horticulture, avoiding excessive contamination and reducing the price of the products found in the current market. In this context, the objective of this work was to incorporate ZnO nanoparticles in soy protein-based bioplastic processed using injection moulding. Thus, the concentration of ZnO nanoparticles (0 wt%, 1.0 wt%, 2.0 wt%, 4.5 wt%) and mould temperature (70 °C, 90 °C and 110 °C) were evaluated through a mechanical (flexural and tensile properties), morphological (microstructure and nanoparticle distribution) and functional (water uptake capacity, micronutrient release and biodegradability) characterization. The results indicate that these parameters play an important role in the final characteristics of the bioplastics, being able to modify them. Ultimately, this study increases the versatility and functionality of the use of bioplastics and nanofertilization in horticulture, helping to prevent the greatest environmental impact caused.

Keywords: bioplastics; nanoparticles; horticulture; soy protein-based

Citation: Jiménez-Rosado, M.; Perez-Puyana, V.; Sánchez-Cid, P.; Guerrero, A.; Romero, A. Incorporation of ZnO Nanoparticles into Soy Protein-Based Bioplastics to Improve Their Functional Properties. *Polymers* **2021**, *13*, 486. https://doi.org/10.3390/polym13040486

Academic Editor: Alina Sionkowska
Received: 14 January 2021
Accepted: 30 January 2021
Published: 4 February 2021

1. Introduction

The use of controlled release devices is increasing in the horticultural sector. In this way, products to release water [1], fertilizers [2] or pesticides [3] can be found in the market, improving their efficiency in crops. Generally, these devices are made using polymeric coatings that contain the substance to be released. This substance is released through the pores of the polymeric matrix due to diffusion between the matrix and the soil, thus reducing the leaching time of the substance and increasing the assimilation efficiency [4,5]. However, these devices have problems derived from the low biodegradability of the plastics used as matrices, which remain in the soil and are difficult to remove. This low biodegradability is due to the non-organic origin of these plastics, not following a natural decomposition process generated by bacteria. For this reason, bioplastic matrices are attracting attention in this application due to their biodegradability, which allows a controlled release of the fertilizer. In addition, this material does not release toxic substances and can be completely degraded without needing to be removed. In this way, these matrices can hold water and encapsulate fertilizers or pesticides. Subsequently, the substance can be controlled released through irrigation water or by biodegradation of the bioplastic matrix generated by different factors (i.e., change in pH or temperature, action of bacteria that degrade the polymeric matrix, irrigation) [6]. Thus, it can improve the assimilation efficiency and solve problems derived from the use of non-biodegradable plastics [7]. To this end, several studies have evaluated the use of bioplastic matrices in horticulture. For example, Olad et al. fabricated starch-based superabsorbent hydrogel from potato peel waste to retain and, subsequently, to control release of water to crops [8]. In this line, Michalik and Wandzinc studied chitosan-based hydrogels [9]. On the other hand, different nutrients have been incorporated into bioplastic matrices, such as sodium and NPK (primary nutrients: nitrogen, phosphorus and potassium) [10–12]. There are also examples of controlled release of pesticides, such as the work of Singh et al. [13], who used imidacloprid to release repellent odours for insects and microorganisms. All these studies use cheap raw materials (i.e., residues and byproducts of agri-food industry) whose costs are in the range 0.03–1.10 €/kg [14], which allows these products to be competitive with conventional plastics (with costs in the range of 0.57–1.59 €/kg) [7]. However, there are still no bioplastic matrices that allow different qualities to be combined, which would enable water, fertilizers and pesticides to be released together. Therefore, bioplastics have opened a new field in the horticultural sector, generating a new reorganization of the sector, increasing the efficiency of crops and reducing the amount of conventional plastics used. However, a lot of research is still needed before they can really compete with conventional plastics and even replace them. For instance, the manufacturing process should be optimized, together with the analysis of possible combinations between different applications in the same device.

In this sense, nanoscience provides a novel possibility to avoid these drawbacks. In this way, it allows the presence of pests and diseases to be detected from the beginning through nanoparticles that can activate a chemical or an electrical signal in the presence of contaminants like bacteria [15]. This signal can be recorded in sensors that allow farmers to act in time and save the crop [16]. In addition, it also allows the correct amount of fertilizers and pesticides that promote productivity to be applied, while ensuring environmental safety (avoiding the excessive use of fertilizers and pesticides), non-contamination of food and greater efficiency in the use of agricultural inputs, which reduces production costs [17]. In this context, new materials based on the use of metallic, polymeric and inorganic nanoparticles have been developed and characterized to increase crop productivity, as well as to improve the uptake and immobilization of nutrients by plants, an area called nanobiofertilization [18,19]. Moreover, these systems minimize the leaching of fertilizers into the subsoil and groundwater, preventing their contamination via nutrient excess. In addition, nanoparticles improve the absorption of nutrients by plants, mitigating eutrophication by reducing the transfer of nitrogen to underground aquifers [20]. Among them, zinc oxide (ZnO) nanoparticles are the most widely used, as they not only allow zinc to be supplied (which is an essential micronutrient for crop growth), but they also have a pesticidal nature, without contaminating the crop, or affecting its quality [21]. In this way, the use of ZnO nanoparticles instead of the salts conventionally used for fertilization generates an improvement in the assimilation efficiency of plants due to their smaller particle size, generating less leaching of nutrients to the subsoil (less contamination). In addition, their greater efficiency means that a lesser amount of product must be incorporated to supply micronutrient deficiencies and it is easier to obtain quality products without nutrient deficit problems. On the other hand, the ZnO nanoparticles generate other benefits such as their antibacterial activity that prevents pesticides from having to be used during cultivation. All this means that, although the commercial price of nanoparticles is higher than that of conventional salts, they are a promising alternative for their replacement [22,23]. These nanoparticles have already been used in other sectors, such as medicine, pharmacology and the food industry due to their antibacterial activity [24,25]. However, farmers are reluctant to use nanoparticles alone, due to their novelty, which does not allow their long-term impact on human health to be understood [26].

Although the incorporation of nanoparticles worsens the mechanical properties of bioplastics without incorporated nanoparticles, their incorporation allows these bioplastics to be used in horticulture for the controlled release of nutrients (in this case, zinc). In addition, it has great advantages over other bioplastics used for this purpose [27,28], to which conventionally used salts are incorporated, making bioplastics have even worse mechanical properties (making them difficult to handle during processing and distribution), further worsening their ability to absorb water and generating a faster release of the micronutrient. In addition, incorporating ZnO nanoparticles generates other additional benefits such as the antimicrobial activity that allows bioplastics can be used as a pesticide. All of this makes these bioplastics a great alternative to currently used devices on the market. For this reason, the objective of this work was to unite both concepts, bioplastics and nanoparticles, to create devices that allow the controlled release of water and fertilizers, while also acting as pesticides. The incorporation of nanoparticles into bioplastic matrices allows the amount of nanoparticles that need to be incorporated into the crop for its growth to be reduced, since they have greater efficiency than conventional fertilizers and would be incorporated in a controlled way in the crop, avoiding their possible leaching. All of this could reduce farmers' reluctance to use them. Furthermore, this could grant bioplastic matrices more than one function, allowing them to compete with plastic devices. In this way, different concentrations of ZnO nanoparticles (0 wt%, 1.0 wt%, 2.0 wt% and 4.5 wt%) were incorporated into soy protein-based bioplastics processed via injection moulding. In addition, different mould temperatures (70 °C, 90 °C and 110 °C) were studied to evaluate the optimal parameters for their processing. To evaluate the different bioplastics, their mechanical, morphological and functional properties were measured.

2. Materials and Methods

2.1. Materials

The bioplastics were manufactured with soy protein isolate (SPI), which is a by-product obtained from soybean oil production (with 91 wt% protein), glycerol (Gly) and zinc oxide nanoparticles (ZnO). SPI and Gly were supplied by Protein Technologies International (Zwaanhofweg, Belgium) and Panreac Química Ltd. (Barcelona, Spain).

ZnO nanoparticles were synthesized from zinc chloride anhydride ($ZnCl_2$) and sodium hydroxide (NaOH), both with 99% purity and provided by Panreac Química Ltd.

2.2. Synthesis of ZnO Nanoparticles

ZnO nanoparticles were synthesized using the colloidal precipitation method [29], using $ZnCl_2$ as precursor and NaOH as reducing agent. For this, 20 mL of both $ZnCl_2$ 0.2 M and NaOH 0.4 M solutions were magnetically mixed at 500 rpm for 2 h at 50 °C to promote reaction 1 (Equation (1)). Then, the zinc hydroxide ($Zn(OH)_2$) was filtered and washed twice with 40 mL of distilled water in order to eliminate residues. Finally, this precipitate was dried in an oven (Memmert B216, Schwabach, Germany) at 100 °C for 8 h and calcined in a muffle furnace (Hobersal HD-230, Barcelona, Spain) at 500 °C for 4 h to promote the synthesis of ZnO nanoparticles according to reaction 2 (Equation (2)). All these conditions were selected to maximize the yield of the synthesis and minimize the particle size (Figure S1), for the nanoparticles to be within the range used in horticulture [22,23]. The assessment was made in a previous study (not yet published) evaluating the two factors that most alter yield and particle size: the concentration of $ZnCl_2$ and $ZnCl_2$:NaOH ratio used. Thus, the lowest concentrations of $ZnCl_2$ are those that obtained the smallest particle sizes, having a higher yield when the $ZnCl_2$:NaOH ratio used was 1:2.

$$ZnCl_{2\,(aq)} + 2\,NaOH_{\,(aq)} \rightarrow Zn(OH)_{2\,(s)} + 2\,NaCl_{\,(aq)} \tag{1}$$

$$Zn(OH)_{2\,(s)} \rightarrow ZnO_{\,(s)} + H_2O_{\,(g)} \tag{2}$$

2.3. Bioplastics Processing Method

Firstly, SPI and Gly (1:1) with different concentrations of ZnO nanoparticles (1.0 wt%, 2.0 wt% and 4.5 wt%) were homogenized in a Polylab QC (Themo Haake, Karlsruhe, Germany) for 30 min at 50 rpm under adiabatic conditions, beginning at 20 °C. During the mixing, the temperature and torque were monitored (Figure S2) to ensure that there was no plasticization in this stage (temperature < 37 °C ($\Delta T < 17\%$) and torque < 10 Nm) and that the blend was homogeneous and correctly mixed (values with a tendency to stabilize).

Later, the dough-like blends were subjected to injection moulding in a MiniJet Piston Moulding System II (Themo Haake, Karlsruhe, Germany). To this end, the blends were introduced in a cylinder at 40 °C, from which they were injected into a mould at a pressure of 600 bar for 20 s. The mould temperature was 70 °C, 90 °C or 110 °C. The bioplastic was kept in the mould for 20 s at 200 bar. The different mould temperatures were selected to evaluate the effect they have in the formation of the bioplastic with nanoparticles included. This temperature induces the formation of a gel-like network among protein chains, mainly through physical interactions and aggregation, which is reinforced by the presence of nanoparticles. Thus, a higher temperature of mould leads to an enhancement of the network.

Finally, bioplastics were subjected to a thermal treatment in a conventional oven at 50 °C for 24 h to strengthen the networks of bioplastics and, thus, maintain their physical integrity. This step has already been implemented in a previous research work for the same purpose [30].

2.4. Characterization of Bioplastics

2.4.1. Mechanical Properties

A minimum mechanical requirement is necessary to guarantee the processability, transport and storage of bioplastics without damaging them. Mechanical properties (both flexural and tensile properties) were measured in rectangular bioplastics ($60 \times 10 \times 1$ mm^3), using a dynamic-mechanical thermal analyser RSA3 (TA Instruments, New Castle, DE, USA) with a dual cantilever or rectangular geometry for flexural and tensile tests respectively.

Flexural Properties

Flexural measurements were performed, following a modification of ISO 178:2019 standard [31], in dynamic mode with a double bending geometry. For this, frequency sweep tests were carried out between 0.02 Hz and 20 Hz at a strain below the critical strain (0.05%) and at room temperature (22 ± 2 °C). Thus, the elastic (E′) and viscous (E″) moduli caused by the application of a small amplitude oscillatory flexural strain were studied as a function of frequency. In addition, the loss tangent (tan δ = E″/E′) was calculated in order to facilitate the evaluation of the behaviour of each bioplastic.

Tensile Properties

Tensile tests were carried out by applying a static axial force at a crosshead speed of 1 mm/min until bioplastic breakage at room temperature. In this way, maximum stress, strain at break and Young's modulus were evaluated as differential parameters of tensile bioplastic behaviour. These tests were performed following a modification of ISO 570-2:1993 standard [32].

2.4.2. Morphological Properties

Fundamentally, the mechanical and functional properties of bioplastics depend on their morphology, with their microstructure and the distribution of their components being of special interest.

Scanning Electron Microscopy (SEM)

Firstly, the bioplastics were subjected to SEM. For this, the samples were previously sputter-coated with palladium-gold and, subsequently, observed using a Zeiss EVO electron microscope (Pleasanton, CA, USA). Two different detectors were used, i.e., secondary electron and scattered electron detectors, to evaluate the microstructure and element distribution, respectively. In both cases, an acceleration voltage of 10 kV was used.

Energy Dispersive X-ray Analysis (EDXA)

In addition to scattered electron, an EDXA complement was coupled to the microscope to analyse the distribution of elemental concentration (Zeiss EVO electron microscope,

Pleasanton, CA, USA). Thus, the different zones obtained in the scattered electron micrographs were evaluated to verify the composition of the bioplastic matrices.

2.4.3. Functional Properties
Water Uptake Capacity (WUC)

One of the purposes of these bioplastics is to absorb and retain water without disintegrating in order to capture rain or irrigation water and supply it to the crop when needed. The ASTM D570-98 standard [33] was used to determine this property. In this way, the bioplastics were submerged in 300 mL of distilled water for 24 h, calculating their water uptake capacity and soluble matter loss as is indicated in previous works [28].

Nanoparticle Release

The study of nanoparticle release is essential for the evaluation of their performance as a controlled release fertilizer and pesticide. Different authors indicate that a quick way to measure this property is by controlling its release in water through conductivity measurements [34]. Thus, an EC-Metro (Crison BASIC 30, Barcelona, Spain) was used to measure the conductivity of the immersion water during the test. The maximum release time was obtained when the conductivity values remained stable for more than 30 min.

Biodegradability

The added value of these bioplastics is their degradation after their use without releasing toxic substances for crops, which eliminates the need for their removal. This quality was evaluated by burying the bioplastics in farmland and irrigating them daily with 20 mL of water (intensive irrigation simulation of 20 L water/m^2). The samples were unearthed at different times, evaluating their disintegration using direct observation.

The used farmland was a special commercial substrate for orchards and fruit trees (Compo, Barcelona, Spain) which contains the ideal ratio of nutrient to the correct crop growth and does not contain microorganisms or pathogens that can alter the tests.

2.5. Statistical Analysis

All specimens were visually analysed prior to testing. In this way, those that presented defects were discarded from the study. The discarded specimens represented less than 10% of the specimens made.

Each analysis was carried out at least three times for each system and all the data were reported with their standard deviation statistically assessed using analysis of variance and Tukey's post hoc test with 95% confidence level ($p < 0.05$) using the statistical package SPSS 18 (Excel, Microsoft, Redmond, WA, USA). The significant differences have been reported with different letters in the corresponding figures.

3. Results and Discussion
3.1. Mechanical Properties
3.1.1. Flexural Properties

Figure 1 shows the flexural properties of different bioplastics. All the systems have similar profiles in elastic modulus (E') and loss tangent (tan δ), regardless of the micronutrient loading and mould temperature used. In this way, E' increases with frequency at a rate that tends to become constant at high frequency, giving rise to a slope lower than 0.19. This behaviour may be due to an extension of the links that recover instantaneously, not leaving the linear range of deformations in the interval studied. This behaviour is similar to those obtained in other works with similar protein-based bioplastics, looking like a typical response from these materials [35,36].

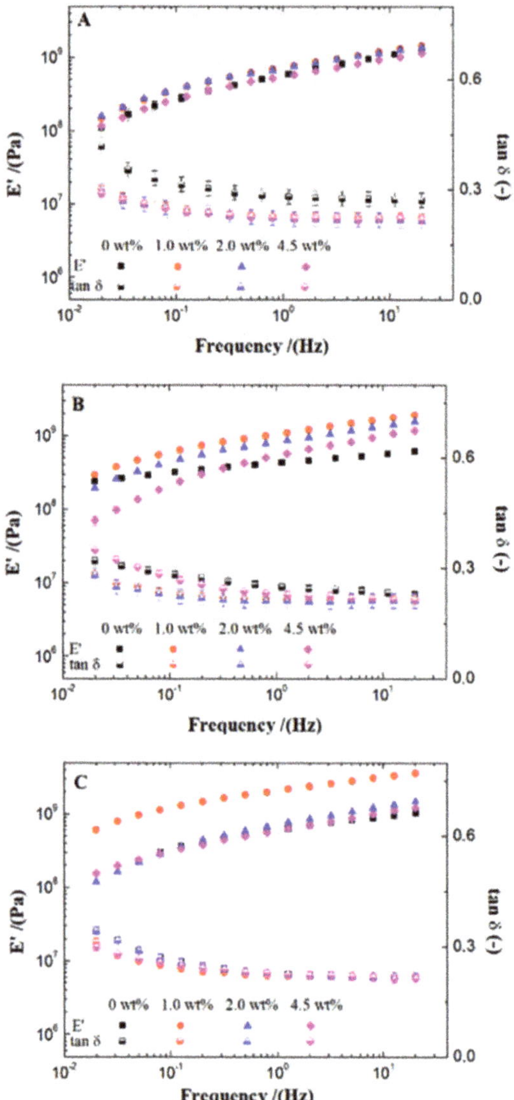

Figure 1. Flexural properties of bioplastics with different ZnO nanoparticle concentrations (0 wt%, 1.0 wt%, 2.0 wt% and 4.5 wt%) processed at different mould temperatures: 70 °C (**A**), 90 °C (**B**) and 110 °C (**C**). Elastic modulus (E') and loss tangent (tan δ) profile in frequency interval.

The effect of the nanoparticle content on the elastic modulus depends on the mould temperature. Among the different mould temperatures used, 70 °C (Figure 1A) showed no significant differences between the different ZnO percentages used. However, this difference was more remarkable when the mould temperature used was 90 °C and 110 °C (Figures 1B and 1C, respectively). In this sense, a low micronutrient load (1.0 wt% and 2.0 wt%) increased the E' values, while higher loads (4.5 wt%) reduced it. In this case, at these temperatures, the incorporation of ZnO nanoparticles always induces an increase in the frequency dependence which is more apparent at 90 °C. The enhancement with micronutrient load, which is modulated by the mould temperature, could be attributed to the interaction between the nanoparticles and the bioplastic, as reported by other authors [37,38]. At low temperatures (70 °C), the nanoparticles, being in small concentrations, did not affect the bioplastics. However, the nanoparticles interacted with each other and with the biopolymeric chains when the temperature increased, improving the mechanical properties of the bioplastics [37]. Nevertheless, this improvement reached a limit, showing no increase in E′ values with 4.5 wt% nanoparticles. It can be kept in mind that a higher content of nanoparticles also involves a reduction in the protein content available that may impair the development of the protein network. Therefore, when the amount of filler material increased, it worsened the crosslinking between the biopolymer chains, thus limiting their mechanical properties. Similar behaviours have already been reported in previous studies, where filler materials improved the mechanical properties up to a certain concentration, reducing them at higher concentrations [39,40]. On the other hand, the bioplastics with 0 wt% and 1.0 wt% nanoparticles presented a slight increase in E′ values when higher mould temperatures are used, which is the common behaviour in these materials [41]. However, this increase is not observed at higher nanoparticle concentrations (2.0 wt% and 4.5 wt% ZnO).

Regarding tan δ, all systems presented similar values, between 0.2 and 0.35. This indicates that all bioplastics had a strong solid character that was enhanced either with the incorporation of nanoparticles or with the increase of temperature. This behaviour is characteristic in protein-based bioplastic, being found in other works. Thus, Yue et al. (2012) also found this behaviour in cottonseed protein [42], and the pea protein-based bioplastics processed by Perez et al. (2016) show this solid character [43]; Gómez-Heincke et al. (2017) obtained similar results with rice, potato and wheat gluten proteins [44].

3.1.2. Tensile Properties

The tensile properties of bioplastics are shown in Figure 2. Firstly, the maximum stress (Figure 2A) increased when higher temperatures were applied, being more notable when the maximum stress started from lower values (0 wt% and 4.5 wt% nanoparticles). Furthermore, 1.0 wt% and 2.0 wt% nanoparticles increased the maximum stress at the same temperature, while 4.5 wt% nanoparticles decreased it. This behaviour is similar to those obtained with flexural properties, although for this parameter it is only significant at the lowest temperature. This evolution reaffirms the hypothesis of some detrimental effect on the development of the protein network caused by an excess of nanoparticles. The strain at break (Figure 2B) shows a similar behaviour as in maximum stress, although in this case, the effect is significant for all temperatures, and 1.0 wt% nanoparticles had significantly the highest values in this case.

On the other hand, Young's modulus presented a different behaviour, which is opposite for the lowest mould temperature. Thus, an increase of mould temperature or content of nanoparticles caused a decrease in Young's modulus, showing no significant differences once the minimum value was reached. This suggests that there is a minimum value of Young's modulus that is not lost regardless of how the bioplastic is processed.

Finally, it is worth mentioning that all bioplastics show strong enough mechanical properties for the suitable transport, storage and distribution of the product, not altering its final functionality and facilitating its production. It is also interesting to point out that these ZnO-containing bioplastics show better mechanical properties than those formulated with zinc sulphate, which is an advantage in this sector [28].

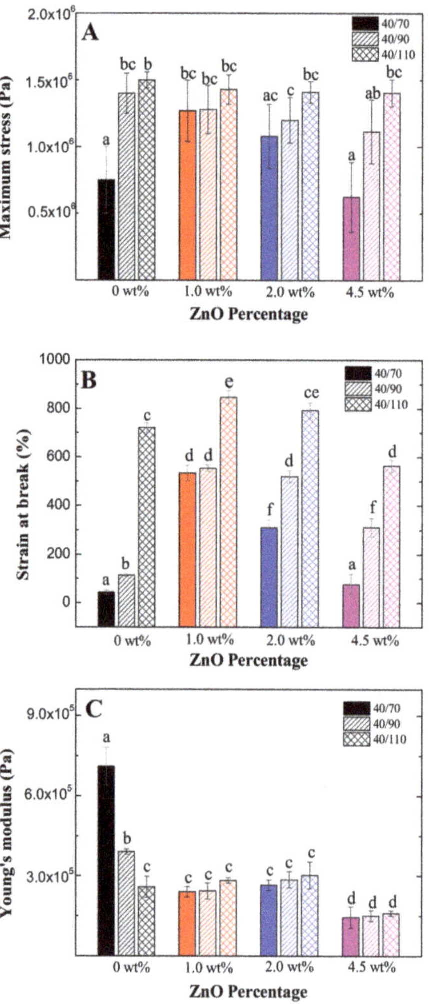

Figure 2. Tensile parameters of bioplastics with different ZnO nanoparticle concentrations (0 wt%, 1.0 wt%, 2.0 wt% and 4.5 wt%) processed at different mould temperatures (70 °C, 90 °C and 110 °C). (**A**): maximum stress. (**B**): strain at break. (**C**). Young's modulus. Different letters in the bars mean that the values are significantly different ($p < 0.05$).

3.2. Morphological Properties

The morphological properties of the bioplastics with 1.0% ZnO nanoparticles are shown in Figure 3 as the representative behaviour of all systems. Nevertheless, Figure S3 shows the morphology of the rest of the systems. Firstly, the macrographic images of the systems (Figure 3A–C) show that all the bioplastics are homogeneous, presenting a colour change with the increase of temperature. This colour change could be attributed to a higher degree of crosslinking generated by Maillard reactions, which colours the systems towards a tanner brown as the temperature increases. This change has already been observed in previous works [45,46]. However, a temperature change not only changes the macrographic appearance of the bioplastic, since differences in the microstructure are also noticed. In this way, structural differences can be seen in micrographic images obtained using a secondary electron detector (Figure 3A′–C′). The bioplastics processed at 70 °C were the only ones that presented porosity in their macrostructure, with cracks appearing in those processed at 90 °C and, above all, at 110 °C. This closure of the microstructure caused by temperature is due to the higher degree of crosslinking generated in these temperatures and has already been reported in previous works [35,41].

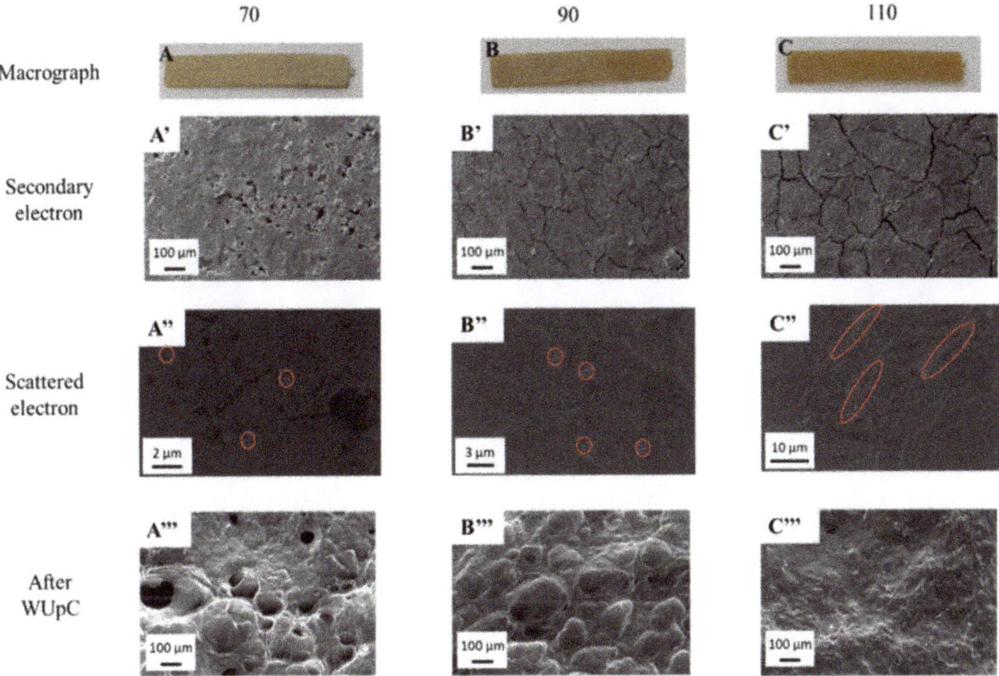

Figure 3. Macrographs (**A–C**) and micrographs of bioplastics with different ZnO nanoparticle concentrations (0 wt%, 1.0 wt%, 2.0 wt% and 4.5 wt%) processed at different mould temperatures (70 °C, 90 °C and 110 °C), using a secondary electron detector and a scattered electron detector before ((**A'–C'**) and (**A"–C"**), respectively) and after water uptake capacity (WUpC) tests (**A'''–C'''**).

Furthermore, the distribution of ZnO nanoparticles in the bioplastics can be seen using a scattered electron detector (Figure 3A"–C"). All the systems present a homogeneous distribution of nanoparticles in the bioplastic, which appears as lighter areas within the dark matrix that makes up the bioplastic. These areas with different tonality were corroborated using EDXA as nanoparticles (white) and protein matrix (black) (Figure S4). As can be observed, the increase of mould temperature generates an effect of nanoparticles sintering, which join together, increasing their size [47], and even forming rings in the direction of injection. This behaviour of the ZnO nanoparticles with temperature could explain the structure observed by the secondary electron detector, since an increase in nanoparticle size makes it difficult to join the biopolymeric chains, causing the observed cracks to appear.

3.3. Functional Properties

3.3.1. Water Uptake Capacity

Figure 4 shows the water uptake capacity (Figure 4A) and soluble matter loss (Figure 4B) of the different bioplastics. As can be seen, the increase in both temperature and ZnO nanoparticles percentage reduced the water uptake capacity of the bioplastic matrices, causing them to lose their superabsorbent quality. This behaviour could be due to the lower free volume and the greater crosslinking of the systems when mould temperature or nanoparticle percentage is increased, reducing the bioplastics' space to interact with water, forming hydrogen bonds that retain it. This is corroborated using the SEM images after the water uptake capacity tests (Figure 3A‴–C‴), which show a decrease in pore size and depth. In addition, other research works have already reported this isolated behaviour when increasing the mould temperature [41] or the amount of additive incorporated [37,48] in similar bioplastics. Moreover, it is also worth mentioning that these bioplastics improve the water uptake capacity generated by systems studied for the same purpose, where microstructure salts were incorporated instead of nanoparticles, improving their functionality [27,28].

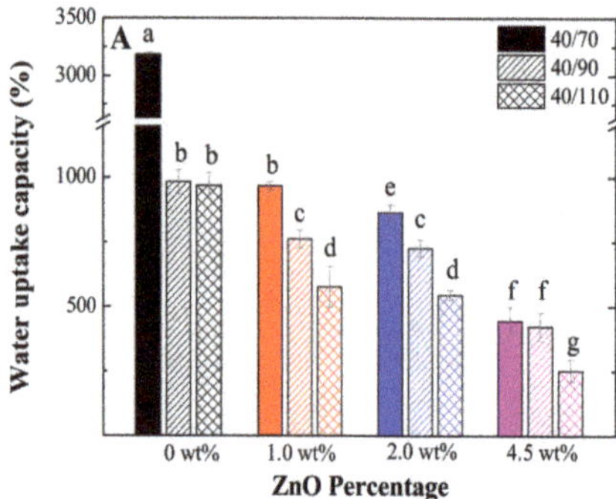
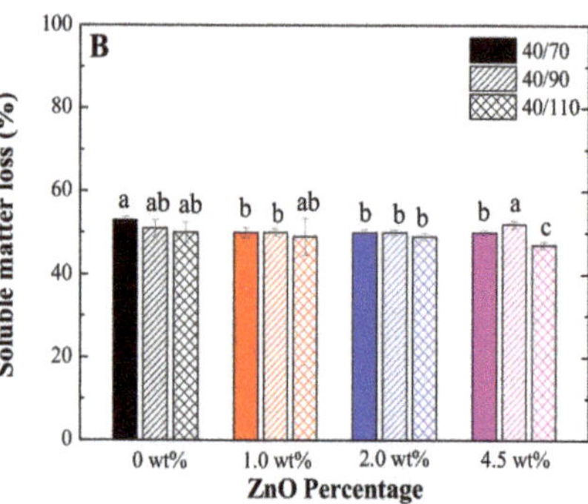

Figure 4. Water uptake capacity (**A**) and soluble matter loss (**B**) of the bioplastics with different ZnO nanoparticle concentrations (0 wt%, 1.0 wt%, 2.0 wt% and 4.5 wt%) processed at different mould temperatures (70 °C, 90 °C and 110 °C). Different letters (a,b, ... , g) in the bars mean that the values are significantly different ($p < 0.05$).

Regarding the soluble matter loss, there were no significant differences between the systems. This indicates that, even if the structure changes, the bioplastics always maintain their integrity by only releasing the incorporated plasticizer (glycerol) and part of the soluble protein.

3.3.2. Nanoparticle Release

The profile of water release of nanoparticles is shown in Figure 5. However, only bioplastics processed with a mould temperature of 90 °C were shown as representative. As can be seen, all the systems present a quick release at short test times, probably due to the greater difference in concentrations between the system and the medium. This release stabilizes over time until reaching the maximum release time. Furthermore, all profiles adapt to the Korsmeyer-Peppas model with an n between 0.1 and 0.3, which indicates that several processes, such as diffusion, disintegration, etc., simultaneously occur during the release, none of them being predominant [49].

Regarding the maximum release time (Table 1), the higher the concentration of nanoparticles, the more prolonged the release over time. This indicates that all the incorporated nanoparticles are released in a controlled way. In addition, this maximum release time is higher than that found when microstructured salts are used instead of nanoparticles [27,50], which indicates that the release is better controlled on this occasion.

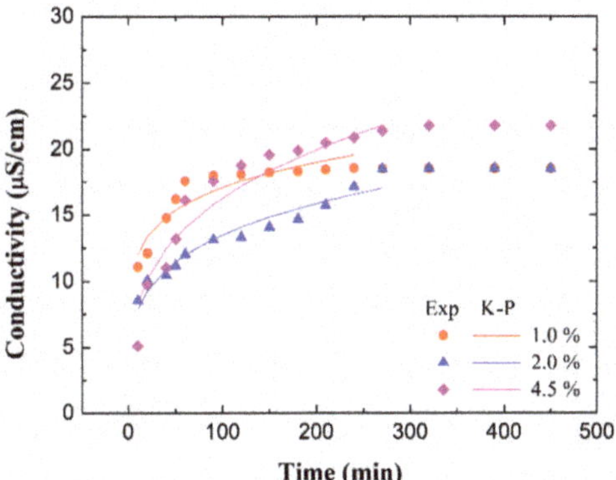

Figure 5. Accumulation of conductivity in the water release tests of bioplastics with different ZnO nanoparticle concentrations (1.0 wt%, 2.0 wt% and 4.5 wt%) processed at a mould temperature of 90 °C.

Table 1. Maximum release time of bioplastics with different ZnO nanoparticle concentrations (0 wt%, 1.0 wt%, 2.0 wt% and 4.5 wt%) processed at a mould temperature of 90 °C and degradation time of bioplastics with different ZnO nanoparticle concentrations (0 wt%, 1.0 wt%, 2.0 wt% and 4.5 wt%) processed at different mould temperatures (70 °C, 90 °C and 110 °C).

	Maximum Release Time (min)	**Degradation Time (Days)**		
		70 °C	**90 °C**	**110 °C**
0 wt%	-	40	60	70
1.0 wt%	240	40	60	70
2.0 wt%	270	20	40	50
4.5 wt%	390	10	20	30

3.3.3. Biodegradability

Finally, the degradation time of each bioplastic matrix is indicated in Table 1. As can be seen, higher mould temperatures lead to more durable bioplastics, probably due to the better mechanical properties observed with increasing temperature, as is reported in previous works [51]. However, the incorporation of nanoparticles in the bioplastics caused this degradation to be faster, except for the systems with 1.0 wt% nanoparticles. This behaviour could be due to the fact that, when the nanoparticles are released, there are more free holes where the bioplastic is more susceptible to degradation, thus accelerating this process. This behaviour has already been reported by Abdullah et al. (2020) [52]. Figure S5 shows the physical appearance of a bioplastic with 1.0 wt% nanoparticles processed at 110 °C. The appearance of bioplastics is similar in all cases, although the lightness of this decomposition changes. It should be noted that, in all cases, the bioplastics decompose into their primary elements (mainly nitrogen), serving as a supplementary fertilizer for the crop and not requiring its removal after use. Furthermore, the degradation time of bioplastics could be modified through the incorporation of nanoparticles and a change in mould temperature, making them very versatile, thus they could be used in all types of horticultural crops.

4. Conclusions

To sum up, soy protein-based bioplastics have shown their great capacity to hold ZnO nanoparticles and release them in a controlled way. In this sense, two novel lines of great interest in horticulture (bioplastics and nanobiofertilization) have been brought together, generating interesting synergies between them, and improving the devices investigated so far. Thus, a controlled release biodegradable device is achieved that presents functionality both to release water and fertilizers, as well as to be used as a long-lasting pesticide, having an enhanced efficiency in plants. In addition, these bioplastics have great versatility to change their characteristics by modifying their composition and processing parameters. In this way, they can be used in different crops, not being necessary to remove them after use. Nevertheless, this is only an initial study, requiring further research to evaluate the functionality of these bioplastics in real and large-scale crops.

Supplementary Materials: The following are available online at https://www.mdpi.com/2073-4360/13/4/486/s1, Figure S1: Yield (K_i, A) and crystal size (B) of nanoparticles processed at different $ZnCl_2$ concentrations (0.2, 0.5 and 1.0 M) and $ZnCl_2$:NaOH ratios (0.5, 1.0 and 2.0); Figure S2: Temperature increment (A) and torque variation (B) of raw materials mixed at different nanoparticle concentrations (0, 1.0, 2.0 and 4.5 wt%); Figure S3: Macro and micrographs of bioplastics processed with 2.0 and 4.5 wt% ZnO nanoparticles at different mould temperatures (70, 90 and 110 °C); Figure S4: EDXA analyses of the different coloured zones in a bioplastic matrix with nanoparticles incorporated; Figure S5: Biodegradability photographs of bioplastics with 1.0 wt% ZnO nanoparticles processed at 110 °C.

Author Contributions: Conceptualization, M.J.-R., V.P.-P. and A.R.; methodology, M.J.-R.; validation, V.P.-P., P.S.-C. and A.R.; formal analysis, M.J.-R.; investigation, M.J.-R.; resources, A.G.; data curation, M.J.-R. and V.P.-P.; writing—original draft preparation, M.J.-R., P.S.-C. and V.P.-P.; writing—review and editing, A.R.; visualization, P.S.-C.; supervision, A.R. and A.G.; project administration, A.R.; funding acquisition, A.G. All authors have read and agreed to the published version of the manuscript.

Funding: This research was funded by the "Ministerio de Ciencia e Innovación" of the Spanish Government and FEDER (UE), grant number RTI2018-097100-B-C21.

Institutional Review Board Statement: Not applicable.

Informed Consent Statement: Not applicable.

Data Availability Statement: The data presented in this study are available on request from the corresponding author.

Acknowledgments: The authors acknowledge the "Ministerio de Educación y Formación Profesional" for the PhD grant of M. Jiménez-Rosado (FPU2017/01718). The authors also thank CITIUS for granting access to and their assistance with the microscopy service.

Conflicts of Interest: The authors declare no conflict of interest.

References

1. Easter, K.W. *Irrigation Innvestment, Technology and Management Strategies for Development*; Routledge: New York, NY, USA, 2019.
2. Ganetri, I.; Essamlali, Y.; Amadine, O.; Danoun, K.; Aboulhrouz, S.; Zahouily, M. Controlling factors of slow or controlled-release fertilizers. In *Controlled Release Fertilizers for Sustainable Agriculture*; Elsevier: Amsterdam, The Netherlands, 2021; pp. 111–129. [CrossRef]
3. Bahadir, M.; Pfister, G. Controlled release formulations of pesticides. In *Chemistry of Plant Protection*; Springer: New York, NY, USA, 1990; pp. 1–64. [CrossRef]
4. Mann, M.; Kruger, J.E.; Andari, F.; McErlean, J.; Gascooke, J.R.; Smith, J.A.; Worthington, M.J.H.; McKinley, C.C.C.; Campbell, J.A.; Lewis, D.A.; et al. Sulfur polymer composites as controlled-release fertilisers. *Org. Biomol. Chem.* **2019**, *17*, 1929–1936. [CrossRef]
5. Beig, B.; Niazi, M.B.K.; Jahan, Z.; Kakar, S.J.; Shah, G.A.; Shahid, M.; Zia, M.; Haq, M.U.; Rashid, M.I. Biodegradable Polymer Coated Granular Urea Slows Down N Release Kinetics and Improves Spinach Productivity. *Polymers* **2020**, *12*, 2623. [CrossRef] [PubMed]
6. Mortain, L.; Dez, I.; Madec, P.-J. Development of new composites materials, carriers of active agents, from biodegradable polymers and wood. *C. R. Chim.* **2004**, *7*, 635–640. [CrossRef]
7. Karan, H.; Funk, C.; Grabert, M.; Oey, M.; Hankamer, B. Green Bioplastics as Part of a Circular Bioeconomy. *Trends Plant Sci.* **2019**, *24*, 237–249. [CrossRef] [PubMed]
8. Olad, A.; Doustdar, F.; Gharekhani, H. Fabrication and characterization of a starch-based superabsorbent hydrogel composite reinforced with cellulose nanocrystals from potato peel waste. *Colloids Surf. Physicochem. Eng. Asp.* **2020**, *601*, 124962. [CrossRef]
9. Michalik, R.; Wandzik, I. A Mini-Review on Chitosan-Based Hydrogels with Potential for Sustainable Agricultural Applications. *Polymers* **2020**, *12*, 2425. [CrossRef]
10. Kartini, I.; Lumbantobing, E.T.; Suyanta, S.; Sutarno, S.; Adnan, R. Bioplastic Composite of Carboxymethyl Cellulose/N-P-K Fertilizer. *Key Eng. Mater.* **2020**, *840*, 156–161. [CrossRef]

11. Bi, S.; Barinelli, V.; Sobkowicz, M.J. Degradable Controlled Release Fertilizer Composite Prepared via Extrusion: Fabrication, Characterization, and Release Mechanisms. *Polymers* **2020**, *12*, 301. [CrossRef]

12. Merino, D.; Gutiérrez, T.J.; Alvarez, V.A. Potential Agricultural Mulch Films Based on Native and Phosphorylated Corn Starch With and Without Surface Functionalization with Chitosan. *J. Polym. Environ.* **2019**, *27*, 97–105. [CrossRef]

13. Singh, A.; Dhiman, N.; Kar, A.K.; Singh, D.; Purohit, M.P.; Ghosh, D.; Patnaik, S. Advances in controlled release pesticide formulations: Prospects to safer integrated pest management and sustainable agriculture. *J. Hazard. Mater.* **2020**, *385*, 121525. [CrossRef]

14. Girotto, F.; Alibardi, L.; Cossu, R. Food waste generation and industrial uses: A review. *Waste Manag.* **2015**, *45*, 32–41. [CrossRef] [PubMed]

15. Ramezani, M.; Ramezani, F.; Gerami, M. Nanoparticles in pest incidences and plant disease control. In *Nanotechnology for Agriculture: Crop Production & Protection*; Springer: Singapore, 2019; pp. 233–272. [CrossRef]

16. Dissanayake, N.M.; Arachchilage, J.S.; Samuels, T.A.; Obare, S.O. Highly sensitive plasmonic metal nanoparticle-based sensors for the detection of organophosphorus pesticides. *Talanta* **2019**, *200*, 218–227. [CrossRef]

17. Nuruzzaman, M.; Rahman, M.M.; Liu, Y.; Naidu, R. Nanoencapsulation, Nano-guard for Pesticides: A New Window for Safe Application. *J. Agric. Food Chem.* **2016**, *64*, 1447–1483. [CrossRef] [PubMed]

18. Zulfiqar, F.; Navarro, M.; Ashraf, M.; Akram, N.A.; Munné-Bosch, S. Nanofertilizer use for sustainable agriculture: Advantages and limitations. *Plant Sci.* **2019**, *289*, 110270. [CrossRef] [PubMed]

19. Rios, J.J.; Yepes-Molina, L.; Martinez-Alonso, A.; Carvajal, M. Nanobiofertilization as a novel technology for highly efficient foliar application of Fe and B in almond trees. *R. Soc. Open Sci.* **2020**, *7*, 200905. [CrossRef]

20. Liu, R.; Lal, R. Potentials of engineered nanoparticles as fertilizers for increasing agronomic productions. *Sci. Total. Environ.* **2015**, *514*, 131–139. [CrossRef] [PubMed]

21. Naderi, M.H.; Danesh-Shahraki, A. Nanofertilizers and Their Roles in Sustainable Agriculture. *Environ. Sci.* **2013**, *5*, 2229–2232.

22. Eichert, T.; Kurtz, A.; Steiner, U.; Goldbach, H.E. Size exclusion limits and lateral heterogeneity of the stomatal foliar uptake pathway for aqueous solutes and water-suspended nanoparticles. *Physiol. Plant.* **2008**, *134*, 151–160. [CrossRef]

23. Martínez-Fernández, D.; Barroso, D.; Komárek, M. Root water transport of *Helianthus annuus* L. under iron oxide nanoparticle exposure. *Environ. Sci. Pollut. Res.* **2016**, *23*, 1732–1741. [CrossRef]

24. Petkova, P.S.P.; Francesko, A.; Perelshtein, I.; Gedanken, A.; Tzanov, T. Simultaneous sonochemical-enzymatic coating of medical textiles with antibacterial ZnO nanoparticles. *Ultrason. Sonochem.* **2016**, *29*, 244–250. [CrossRef]

25. Shi, L.-E.; Li, Z.-H.; Zheng, W.; Zhao, Y.-F.; Jin, Y.-F.; Tang, Z.-X. Synthesis, antibacterial activity, antibacterial mechanism and food applications of ZnO nanoparticles: A review. *Food Addit. Contam. Part A* **2014**, *31*, 173–186. [CrossRef] [PubMed]

26. Strand, R.; Kjolberg, K.L. Regulating Nanoparticles: The Problem of Uncertainty. *Eur. J. Law Technol.* **2011**, *2*, 1–10.

27. Jiménez-Rosado, M.; Perez-Puyana, V.; Rubio-Valle, J.F.; Guerrero, A.; Romero, A. Processing of biodegradable and multifunctional protein-based polymer materials for the potential controlled release of zinc and water in horticulture. *J. Appl. Polym. Sci.* **2020**, *137*, 49419. [CrossRef]

28. Jiménez-Rosado, M.; Pérez-Puyana, V.; Cordobés, F.; Romero, A.; Guerrero, A. Development of soy protein-based matrices containing zinc as micronutrient for horticulture. *Ind. Crop. Prod.* **2018**, *121*, 345–351. [CrossRef]

29. Chen, C.; Liu, P.; Lu, C. Synthesis and characterization of nano-sized ZnO powders by direct precipitation method. *Chem. Eng. J.* **2008**, *144*, 509–513. [CrossRef]

30. Álvarez-Castillo, E.; Del Toro, A.; Aguilar, J.M.; Guerrero, A.; Bengoechea, C. Optimization of a thermal process for the production of superabsorbent materials based on a soy protein isolate. *Ind. Crop. Prod.* **2018**, *125*, 573–581. [CrossRef]

31. ISO 178:2019, Plastics. *Determination of Flexural Properties*; ISO/TC 46; ISO: Geneva, Switzerland, 2019.

32. ISO 570-2:1993, Plastics. *Determination of Tensile Properties: Part 2: Test Conditions for Moulding and Extrusion Plastics*; ISO/TC 46; ISO: Paris, France, 1993.

33. ASTM Interntional. *ASTM D570-98: Standard Test Method for Water Absorption of Plastics*; ASTM Interntional: West Conshohocken, PA, USA, 2005.

34. Essawy, H.A.; Ghazy, M.B.; El-Hai, F.A.; Mohamed, M.F. Superabsorbent hydrogels via graft polymerization of acrylic acid from chitosan-cellulose hybrid and their potential in controlled release of soil nutrients. *Int. J. Biol. Macromol.* **2016**, *89*, 144–151. [CrossRef]

35. Jiménez-Rosado, M.; Bouroudian, E.; Perez-Puyana, V.; Guerrero, A.; Romero, A. Evaluation of different strengthening methods in the mechanical and functional properties of soy protein-based bioplastics. *J. Clean. Prod.* **2020**, *262*, 121517. [CrossRef]

36. Jerez, A.; Partal, P.; Martínez, I.; Gallegos, C.; Guerrero, A. Protein-based bioplastics: Effect of thermo-mechanical processing. *Rheol. Acta* **2007**, *46*, 711–720. [CrossRef]

37. Amjadi, S.; Emaminia, S.; Davudian, S.H.; Pourmohammad, S.; Hamishehkar, H.; Roufegarinejad, L. Preparation and characterization of gelatin-based nanocomposite containing chitosan nanofiber and ZnO nanoparticles. *Carbohydr. Polym.* **2019**, *216*, 376–384. [CrossRef]

38. Ramesan, M.T.; Siji, C.; Kalaprasad, G.; Bahuleyan, B.K.; Al-Maghrabi, M.A. Effect of Silver Doped Zinc Oxide as Nanofiller for the Development of Biopolymer Nanocomposites from Chitin and Cashew Gum. *J. Polym. Environ.* **2018**, *26*, 2983–2991. [CrossRef]

39. Saenghirunwattana, P.; Noomhorm, A.; Rungsardthong, V. Mechanical properties of soy protein based "green" composites reinforced with surface modified cornhusk fiber. *Ind. Crop. Prod.* **2014**, *60*, 144–150. [CrossRef]
40. Fu, H.; Li, X.; Gong, W.; Tian, H.; Zhou, H. Enhanced electrical and dielectric properties of plasticized soy protein bioplastics through incorporation of nanosized carbon black. *Polym. Compos.* **2020**, 25790. [CrossRef]
41. Fernández-Espada, L.; Bengoechea, C.; Cordobés, F.; Guerrero, A. Thermomechanical properties and water uptake capacity of soy protein-based bioplastics processed by injection molding. *J. Appl. Polym. Sci.* **2016**, *133*, 43524. [CrossRef]
42. Yue, H.-B.; Cui, Y.-D.; Shuttleworth, P.S.; Clark, J.H. Preparation and characterisation of bioplastics made from cottonseed protein. *Green Chem.* **2012**, *14*, 2009–2016. [CrossRef]
43. Perez, V.; Felix, M.; Romero, A.; Guerrero, A.; Information, P.E.K.F.C. Characterization of pea protein-based bioplastics processed by injection moulding. *Food Bioprod. Process.* **2016**, *97*, 100–108. [CrossRef]
44. Gómez-Heincke, D.; Martínez, I.; Stading, M.; Gallegos, C.; Partal, P. Improvement of mechanical and water absorption properties of plant protein based bioplastics. *Food Hydrocoll.* **2017**, *73*, 21–29. [CrossRef]
45. Klüver, E.; Meyer, M. Thermoplastic processing, rheology, and extrudate properties of wheat, soy, and pea proteins. *Polym. Eng. Sci.* **2015**, *55*, 1912–1919. [CrossRef]
46. Su, J.-F.; Huang, Z.; Yuan, X.-Y.; Wang, X.-Y.; Li, M. Structure and properties of carboxymethyl cellulose/soy protein isolate blend edible films crosslinked by Maillard reactions. *Carbohydr. Polym.* **2010**, *79*, 145–153. [CrossRef]
47. Pourrahimi, A.M.; Liu, D.; Ström, V.; Hedenqvist, M.S.; Olsson, R.T.; Gedde, U. Heat treatment of ZnO nanoparticles: New methods to achieve high-purity nanoparticles for high-voltage applications. *J. Mater. Chem. A* **2015**, *3*, 17190–17200. [CrossRef]
48. Gamero, S.; Jiménez-Rosado, M.; Romero, A.; Bengoechea, C.; Guerrero, A. Reinforcement of Soy Protein-Based Bioplastics Through Addition of Lignocellulose and Injection Molding Processing Conditions. *J. Polym. Environ.* **2019**, *27*, 1285–1293. [CrossRef]
49. Korsmeyer, R.; Peppas, N. Swelling-controlled delivery systems for pharmaceutical applications: Macromolecular and modelling considerations. In *Controlled Release Delivery Systems*; Marcel Dekker: New York, NY, USA, 1983; pp. 77–90.
50. Perez-Puyana, V.; Felix, M.; Romero, A.; Guerrero, A. Development of eco-friendly biodegradable superabsorbent materials obtained by injection moulding. *J. Clean. Prod.* **2018**, *198*, 312–319. [CrossRef]
51. Jiménez-Rosado, M.; Rubio-Valle, J.F.; Perez-Puyana, V.; Guerrero, A.; Romero, A. Comparison between pea and soy protein-based bioplastics obtained by injection molding. *J. Appl. Polym. Sci.* **2020**, 50412. [CrossRef]
52. Abdullah, A.H.D.; Putri, O.D.; Fikriyyah, A.K.; Nissa, R.C.; Hidayat, S.; Septiyanto, R.F.; Karina, M.; Satoto, R. Harnessing the Excellent Mechanical, Barrier and Antimicrobial Properties of Zinc Oxide (ZnO) to Improve the Performance of Starch-based Bioplastic. *Polym. Technol. Mater.* **2020**, *59*, 1259–1267. [CrossRef]

Article

Strengthening of Porcine Plasma Protein Superabsorbent Materials through a Solubilization-Freeze-Drying Process

Estefanía Álvarez-Castillo *, Carlos Bengoechea and Antonio Guerrero

Escuela Politécnica Superior, Chemical Engineering Department, University of Seville, Calle Virgen de África, 7, 41011 Sevilla, Spain; cbengoechea@us.es (C.B.); aguerrero@us.es (A.G.)
* Correspondence: malvarez43@us.es

Abstract: The replacement of common acrylic derivatives by biodegradable materials in the formulation of superabsorbent materials would lessen the associated environmental impact. Moreover, the use of by-products or biowastes from the food industry that are usually discarded would promote a desired circular economy. The present study deals with the development of superabsorbent materials based on a by-product from the meat industry, namely plasma protein, focusing on the effects of a freeze-drying stage before blending with glycerol and eventual injection molding. More specifically, this freeze-drying stage is carried out either directly on the protein flour or after its solubilization in deionized water (10% w/w). Superabsorbent materials obtained after this solubilization-freeze-drying process display higher Young's modulus and tensile strength values, without affecting their water uptake capacity. As greater water uptake is commonly related to poorer mechanical properties, the proposed solubilization-freeze-drying process is a useful strategy for producing strengthened hydrophilic materials.

Keywords: plasma protein; superabsorbent; protein-based material; freeze-drying; injection molding

Citation: Álvarez-Castillo, E.; Bengoechea, C.; Guerrero, A. Strengthening of Porcine Plasma Protein Superabsorbent Materials through a Solubilization-Freeze-Drying Process. *Polymers* **2021**, *13*, 772. https://doi.org/10.3390/polym13050772

Academic Editors: Andrea Zille and Alexander Böker

Received: 14 January 2021
Accepted: 26 February 2021
Published: 3 March 2021

Publisher's Note: MDPI stays neutral with regard to jurisdictional claims in published maps and institutional affiliations.

1. Introduction

Superabsorbent materials are capable of absorbing and retaining water in quantities higher than ten times their own dry weight [1,2]. Traditionally, superabsorbent materials are based on acrylic derivatives [3,4], which are expensive, toxic, and highly pollutant due to their fairly low biodegradability. These materials are extensively used in the personal care industry, so their use time is relatively short as they are easily disposable, which contributes to the environmental issues caused by poorly biodegradable synthetic plastics. In contrast, some studies have pointed out the feasibility of obtaining superabsorbent materials from natural sources such as porcine plasma protein [5–7], soy protein [8–11], or gluten [12].

The meat industry produces a huge amount of blood, which is rich in proteins [13] and should not be directly disposed of in landfills or effluents due to its high organic charge, which can produce high pollution levels due to the high biochemical oxygen demand (BOD) [14–16]. Therefore, extensive use and revalorization of this by-product would be of great interest to increase the competitiveness and business growth of the meat industry and to promote the integration of circular economy principles. Plasma is the blood fraction that remains after the separation of the red cells and platelets [17] through centrifugation, which can be dried to obtain a porcine plasma protein (PPP) powder. This by-product is already used in the food industry either as an emulsifier, as a water-holding agent [18] in frankfurters [19,20] and sausages [13], or as an alternative to certain other protein ingredients, such as egg [21,22]. Furthermore, the excellent film-forming potential displayed by porcine plasma protein has proven to be useful in the development of films for food packaging, replacing synthetic plastics [23–25]. As mentioned earlier, some studies have also pointed out the potential of PPP for superabsorbent applications, showing water

227

uptake capacity values as high as 3600% [6], which can be attributed to its considerable contents of polar amino acids, such as glutamic and aspartic acid [26].

The superabsorbent capacity of protein-based materials is strongly dependent on the processing conditions, being reduced as temperature increases due to the promotion of crosslinking [5,27]. Furthermore, longer residence times, either in the mold or during storage at relatively high temperatures, lead to an increase in the physical crosslinking within the structure, also hindering the swellability of the samples [5,10,28]. Likewise, the variation of the pH in the raw protein material affect the amount of water that samples can hold [6]. Temperature, time, and pH are crucial factors, as they promote changes in the sample structure, and consequently in the existing interactions between the protein chains [29]. As mechanical properties are mostly inversely related to water absorption, superabsorbent materials commonly possess very poor mechanical properties, sometimes even being solubilized to a certain extent when immersed [5,10]. In an attempt to overcome this drawback, certain strategies have been pursued (e.g., crosslinking agents, acrylic co-polymerization), although at the cost of their ecological character [4,30]. The solubilization and freeze-drying of proteins might impact their conformational structure [31], as has been highlighted before when this procedure was used to modify the pH of PPP [6]. The alteration of the molecular structure through freeze-drying could eventually influence the properties of the material that would be obtained from that protein source after thermal processing (i.e., injection molding) [32–34], even though the economic and environmental impacts should not be neglected if applied industrially.

The present manuscript aims to study the effects of freeze-drying on the properties of porcine-plasma-based superabsorbent materials. For this purpose, the protein source samples are either directly freeze-dried or first solubilized in water and subsequently freeze-dried. To evaluate the differences, rheological measurements, mechanical and water uptake tests, and scanning electron microscopy are carried out.

2. Materials and Methods

2.1. Materials and Sample Preparation

In the present study, porcine plasma protein (PPP) was used as the raw material. The protein flour was kindly supplied by Proanda S.A (AproPork, Essentia Protein, Ankeny, IA, USA), asp: 6.97 g/100 g of protein; Glu: 10.04 g/100 g of protein. The PPP protein content (74% w/w) was determined by multiplying the percentage of nitrogen by 6.25 (the Kejdhal factor for this kind of material). The nitrogen content was estimated using a LECO CHNS-932 nitrogen analyzer (Leco Corporation, St. Joseph, MO, USA). The moisture content was estimated to be around 6% and the ash content was around 17%. Pharma-grade glycerol (Gly), delivered by Panreac Química S.A (Barcelona, Spain), was employed as a plasticizer for all systems.

Untreated PPP flour (UF) in as-received condition (6% humidity) was employed as the reference. Then, the effects of freeze-drying were studied through two different procedures: an initial procedure where the flour was conveniently frozen at −40 °C, then freeze-dried at −80 °C (FD) in a LyoQest freeze-dryer (Telstar Technologies, Barcelona, Spain); and a second procedure where 10 g of PPP was solubilized in 100 mL of deionized water, after which the PPP solution was frozen at −40 °C and subsequently freeze-dried at −80 °C (SFD).

Superabsorbent materials based on PPP have previously been obtained by performing a mild injection molding procedure [5–7], which was also followed in the present study. This procedure started with a blending stage in a Haake Polylab QC two-blade counter-rotating mixer (ThermoHaake, Karlsruhe, Germany), whereby 65 g of PPP and glycerol were intimately mixed at a 50:50 ratio. This stage took place at room temperature for 5 min and at 50 rpm, while the mixing rheometer recorded the torque and temperature in the mixing cavity. Subsequently, 1.5 g of the obtained homogeneous blend was injection molded into a rectangular mold ($1 \times 10 \times 60$ mm^3) using a Minijet Piston Injection Molding System (ThermoHaake, Karlsruhe, Germany). The temperature of the feed cylinder was always 40 °C, while the mold temperature was 60 °C and the pressure employed during the injection and holding stages, which lasted 150 s, was 500 bar.

2.2. Methods

2.2.1. Linear Viscoelastic Properties

Viscoelastic properties were estimated using dynamic mechanical temperature analysis (DMTA) within the linear viscoelastic range (LVR) by carrying out compressional and torsional measurements for the blends and protein-based materials, respectively. A RSA3 rheometer (TA Instruments, New Castle, DE, USA) was used to perform the compression mode tests on blends using a cylindrical geometry measuring 8 mm in. diameter. On the other hand, protein-based materials were tested in a DHR-3 rheometer (TA Instruments, New Castle, DE, USA) in torsion mode. In every case, strain sweep tests (0.001–10%) were initially carried out at 1 Hz to identify the strain amplitudes that defined the LVR. Afterwards, temperature ramp tests were performed by employing a heating rate of 5 °C/min from 25 °C to 140 °C for blends or from −30 °C to 140 °C for the protein-based materials, at a constant frequency (1 Hz) and strain (within the LVR).

2.2.2. Tensile Properties

In order to estimate the mechanical properties of the plastic samples, uniaxial tensile tests were performed until breaking point using a Dynamic Mechanical Analyser RSA3 (TA Instruments, New Castle, DE, USA), with a rectangular tensile geometry (tension mode) at a constant strain rate of 1 mm·min^{-1} at room temperature (≃25 °C). Typical mechanical stress-strain curves were obtained, from which mechanical properties were determined, such as the Young's modulus (E), maximum or ultimate stress (σ_{max}), and strain at break (ε_{max}).

2.2.3. Differential Scanning Calorimetry (DSC)

DSC tests were performed in an 822 calorimeter (Mettler Toledo, Worthington, OH, USA) using Mettler Toledo Star System software. For this purpose, 12–14 mg of biomass were located in hermetically sealed aluminum pans and tests were run at a rate of 10 °C/min from −25 to 300 °C using an empty pan as a reference.

2.2.4. Water Uptake

Water uptake capacity (WUC) values for the obtained samples were determined using a protocol described in previous studies [1,5]. First, the protein-based materials were placed in an oven at 50 °C until constant weight (w_1). Then, they were immersed in deionized water for 24 h and then weighed (w_2). Finally, samples that had been dried for 24 h were weighed again (w_3). The WUC and soluble matter loss (SML) can be calculated using the following equations [5]:

$$\text{WUC (\%)} = 100 \cdot \frac{(w_2 - w_3)}{w_3} \tag{1}$$

$$\text{SML (\%)} = 100 \cdot \frac{(w_1 - w_3)}{w_1} \tag{2}$$

2.2.5. Scanning Electron Microscopy

Following immersion, swollen PPP-based samples were then freeze-dried (−80 °C, 0.01 mbar) and cut into small pieces (2–3 mm). Afterwards, they were gold-coated and observed using scanning electron microscopy (SEM). A ZEISS EVO (Carl Zeiss Microscopy, White Plains, NY, USA) microscope was used to evaluate the microstructure of the selected swollen PPP-based materials. Micrographs were acquired using a beam current of 11–12 pA at a working distance of 6 mm and with an acceleration voltage of 10 kV. Analyses were carried at 60x magnification. In addition, the pore size was studied using a digital processing software (ImageJ, Bethesda, MD, USA). The mean diameter was obtained by measuring several pores in the images obtained for each system.

2.3. Statistical Analysis

In the current study, all measurements were performed in triplicate. Statistical studies were performed using ANOVA comparisons in the Statgraphics software (The Plains, VA, USA). Uncertainty was expressed as mean values ± standard deviations, which were plotted for all calculated parameters.

3. Results and Discussion

3.1. Mixing Stage

In this section, the torque and temperature evolution inside the mixer as the PPP and glycerol were blended are presented in Figure 1. The evolution of both parameters was similar to that observed in previous studies for analogous systems [5,7]. The torque profile initially displayed a sudden increase due to the instantaneous compaction of the raw materials when pressed down by the plunger. The torque value dropped down to 2 N·m immediately after, remaining steady at this value during the whole mixing stage. Regarding the temperature profile (ΔTemperature), explained as the difference between the instantaneous temperature inside the mixer and the initial temperature of the blend, no noticeable changes were observed, as the largest increase of temperature recorded within the cavity of the mixer was about 2.5 °C for the SFD sample, remaining at 1.2 ± 0.3 °C at the end of the mixing stage in all cases. All samples showed the same tendency, which led to the conclusion that no significant mechanical energy dissipation took place inside the mixer during the mixing stage. This means that no important protein reticulation or crosslinking may be expected, as these interactions, which commonly occur when the processing conditions are extreme, typically involve an apparent increase in temperature [35].

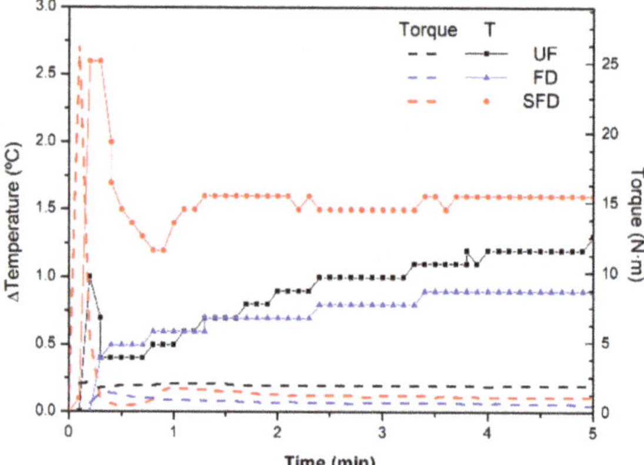

Figure 1. Torque and temperature profile developed within the mixer cavity during the mixing stage for porcine plasma protein-glycerol blends using UF, FD, and SFD protein systems.

3.2. Thermal Characterization of the Systems

3.2.1. Evolution of the Rheological Properties of the Blends with Temperature

The homogeneous blends obtained just after the mixing stage were rheologically characterized (Figure 2) in order to observe the thermal transitions of the PPP systems, which provided valuable information for the subsequent injection procedure [7]. All studied samples showed similar behavior, presenting the same qualitative events: (I) at relatively low temperatures, steady storage moduli (E′) values were observed until a certain temperature (between 45 and 60 °C) was reached; then (II), a temperature-induced drop took place until a minimum was achieved, reaching a decrease of two orders of magnitude in the E′ values; from the temperature where the minimum was located, (III) an increase in E′ took place due to protein aggregation and gelation processes.

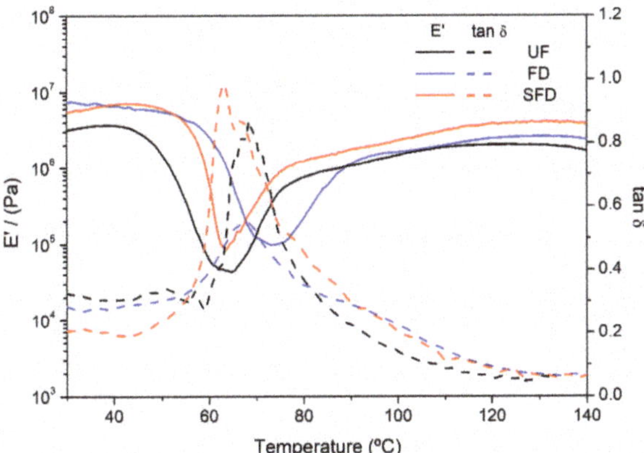

Figure 2. Evolution of the compressional storage modulus (E′) and loss tangent (tan δ) of blends from porcine plasma protein (PPP) and glycerol materials using UF, FD, and SFD protein systems, obtained through temperature sweep tests ranging from 30 to 140 °C at 1 Hz within the lineal viscoelastic range.

The glassy plateau observed in the first stage (I) was kept until a certain temperature, which seemed to be displaced onto higher temperatures when PPP samples were submitted to a freeze-drying step. Thus, the storage moduli of the reference sample (no freeze-drying) started to decrease around 45 °C, while both freeze-dried samples started to decrease at approximately 57 °C, independently of the procedure followed. Likewise, the freeze-drying of the protein flour also seems to quantitatively influence the viscoelastic moduli, showing lower values in the reference (UF) sample than for samples that were freeze-dried, regardless of having been previously solubilized in deionized water (SFD) or not (FD). The observed differences may be associated with the fact that the freeze-dried samples did not contain any moisture, unlike the reference, which contained 6% water, which could play a plasticizer role. Therefore, in spite of all samples possessing the same glycerol content, FD and SFD samples contained lower overall quantities of plasticizer, resulting in reduced mobility between chains, and eventually promoting greater viscoelastic moduli [36]. Moreover, the SFD PPP system could also have been affected by the difference in the ice nucleation history, which may have promoted a difference in the stresses exerted on the protein as the water removal gradually increased the protein concentration in the aqueous solution. As it was freeze-dried, a solution with increasing viscosity was formed, which could increase intermolecular reaction rates, resulting in an alteration in the protein conformation [37].

The decrease observed in the second stage (II) of the thermal treatment of the blends was reported to be associated with the promotion of the mobility of polymeric chains at higher temperatures [6]. At the end of this event, an apparent minimum was observed at a temperature that depended on the procedure followed, as previously observed for the glassy plateau. The reference sample displayed lower viscoelastic moduli values at the minimum point than the freeze-dried samples, which may be connected with the higher amount of plasticizer, as previously mentioned. These minimum values were located at 62.5 °C for the UF and SFD samples and at approximately 73 °C for the FD sample.

The increase in the storage modulus after the minimum (III) occurred as proteins such as albumin [38,39] aggregated. The two different slopes shown during the strengthening may correspond to the different protein fractions present in the PPP [5]. Additionally, similar behavior was presented in previous studies for PPP-Gly blends [5,6]. However, the effects of freeze-drying observed on the rheological properties of the protein source were quite apparent and innovative.

Additionally, as observed in Figure 2, the loss tangent (tan δ = E″/E′) generally showed values below the unity level across the whole temperature range studied, indicating a solid-like behavior for all samples, as the storage moduli (E′) were always higher than the viscous moduli (E″) [7,40]. A remarkable peak in tan δ could always be distinguished, which is typically referred to as the glass transition of the system [41]. Thus, the temperatures at which this peak occurred for the studied samples were between 60 and 65 °C, matching values previously reported for similar samples [6].

3.2.2. Differential Scanning Calorimetry (DSC)

It should be highlighted that the thermal transitions detected in the SFD sample happened at higher temperatures but within a smaller temperature range than the rest. To confirm this, DSC was performed on both the UF reference and the SFD sample (Figure 3). DSC tests confirmed this, which could be explained by certain molecular rearrangements that might have taken place when dispersing the sample in deionized water.

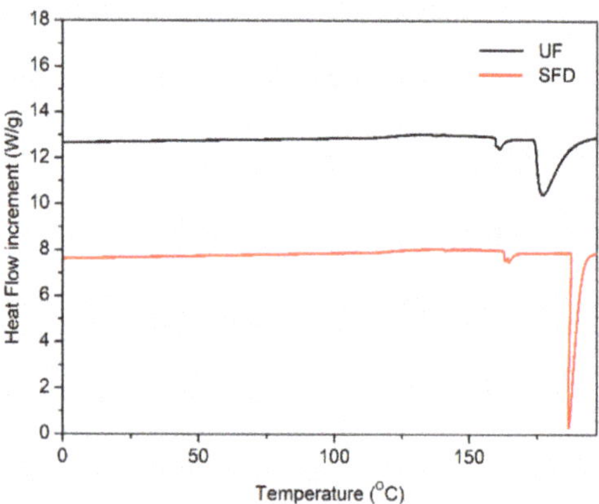

Figure 3. DSC thermograms for the UF and SFD porcine plasma protein systems run at a heating rate of 10 °C/min.

Calorimetric techniques are widely used to identify the thermal transitions in proteins and other biomacromolecules [42,43]. Regarding these results, only two well-differentiated peaks could be distinguished in the thermograms—the first at around 162 and 167 °C and the second located at 177 and 187 °C for the UF and the SFD samples, respectively. The thermal energy values related to the first peak were around 2.4 (W/g)·°C and were quite similar for both samples. On the other hand, in the case of the second endothermal peak, it was slightly higher in the case of the solubilized and freeze-dried sample (27.5 (W/g)·°C) when compared to the reference (23.7 (W/g)·°C). These peaks might correspond to a denaturation point that would favor a greater flowability [44] that would take place in a broader temperature range for the reference (20 °C) than for the SFD sample (10 °C), with a thinner peak. Therefore, these results would confirm the fact that the SFD-containing blend showed thermal transitions across a smaller temperature range, just as observed in the DMTA tests (Figure 2). The differences among samples may be caused by the conformational changes suffered by the proteins after being solubilized in water, which could eventually produce alterations in the protein functionality and stability [45,46].

3.2.3. Evolution of the Rheological Properties of the PPP-Based Materials with Temperature

Injection-molded plastic materials obtained from UF, FD, or SFD systems were submitted to temperature sweep tests in order to identify the influence of temperature on their viscoelastic properties (Figure 4).

As expected, in all cases, the samples showed G′ values higher than those for G″, which resulted in tan δ values below the unity level across the whole temperature range. At lower temperatures, the sample subjected to the SFD process showed higher values for G′ than the rest of the samples, implying the formation of a more strengthened structure, which could be associated with the rearrangements that might have occurred during the solubilization process. As the temperature increased, softening was detected for all samples through a decrease in G′ until a minimum was found, from which G′ significantly increased. This increase in the viscoelastic properties took place at temperatures higher than 60 °C (the molding temperature), as previously reported for similar materials based on PPP and glycerol [5–7,47]. The increase in the viscoelastic properties as the temperature gets higher is typical of the thermosetting potential shown by plastic materials molded in relatively mild conditions, under which plasticized polymers still display thermoplastic behavior. A previous study highlighted the importance of mild processing conditions in producing superabsorbent materials from PPP, as thermal crosslinking hinders water uptake [5]. However, the promotion of the hydrophilic character of the materials is achieved at the expense of a poor strengthening of the structure, which sometimes leads to undesired disintegration of the superabsorbent material when immersed. Moreover, the UF sample displayed the lowest G′ values at the minimum point, which may be explained by the greatest plasticization degree being achieved due to the higher moisture content of the flour. However, samples obtained from freeze-dried flours displayed similar values for G′ at the minimum point, as water was removed from both of them. As the temperature further increased, a tendency toward plateau values was observed for all samples. Nevertheless, the FD sample finally displayed higher values of G′ than the SFD sample, indicating a greater thermosetting potential for this sample.

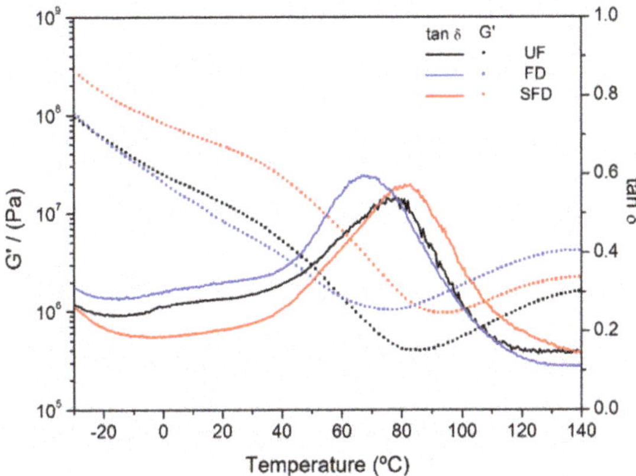

Figure 4. Evolution of the storage (G′) and viscous (G″) moduli in torsion mode for porcine plasma protein-glycerol materials using UF, FD, and SFD protein systems, obtained through temperature sweep tests from 30 to 140 °C at 1 Hz within the lineal viscoelastic range.

3.3. Mechanical Characterization of the PPP-Based Materials

The evolution of the main mechanical properties (E, σ_{max}, ε_{max}) of the protein-based materials obtained from PPP submitted to different procedures is shown in Figure 5. Typical stress-strain curves of uniaxial stress until breaking point were obtained for all samples. At the beginning of the curves, a linear slope characteristic of an initial Hookenian behavior could be distinguished, from which Young's modulus (E) values could be determined. After the yield stress was surpassed, plastic deformation was observed, whereby small stresses resulted in important deformations. The tests ended when materials reached the ultimate stress point (σ_{max}) and underwent rupture at the maximum strain point (ε_{max}).

When the mechanical parameters of the materials obtained from the UF sample were compared to those of the materials from the FD source, no differences were perceived in terms of Young's modulus values. However, a significant although slight decrease could be perceived in σ_{max} values, being more noticeable than ε_{max} values. The lower deformability shown by the FD sample might be related to the reduction in the amount of plasticizer (glycerol + moisture) [48]. On the other hand, more remarkable differences were determined for materials processed from the SFD flour. The SFD stage led to a greater strengthening of the injection-molded sample, as E increased from 0.1×10^5 to 7.8×10^5 Pa. This strengthening was also denoted by a remarkable increase in the tensile strength, as σ_{max} increased from 0.5×10^6 to 1.8×10^6 Pa. As the lower plasticizer content did not have a strong influence on E or σ_{max}, the solubilization in deionized water should be the main reason for this reinforcement. Moreover, ε_{max} dropped from around 135% to 13% when PPP was solubilized, as the reinforcement made the samples more fragile. Some authors have reported that ice formation during freezing may promote protein denaturation through ice-protein interactions, altering the conformational structure of the protein [49]. Zhan et al. showed that unfolding takes place in proteins when they are freeze-dried, which may produce higher amount of reactive sites, promoting the bonding between chains (hydrophobic interactions and hydrogen bonding) (Figure 6) [50]. The results obtained with SFD samples seem to support this hypothesis, whereby a greater exposition of reactive groups along the polymeric chain to ice may lead to a greater reinforcement during the material processing.

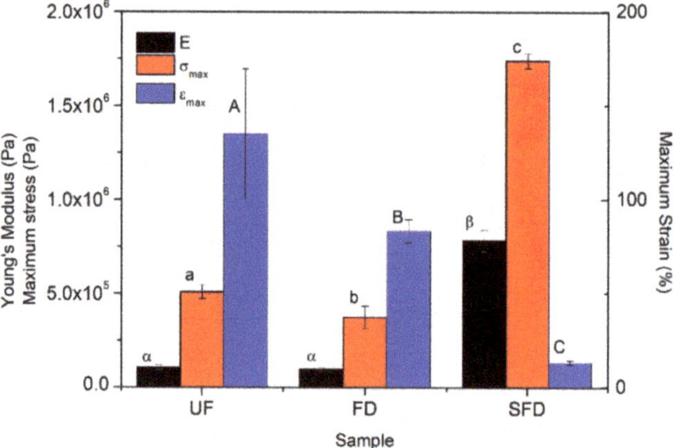

Figure 5. Mechanical parameters of porcine plasma protein-glycerol materials using UF, FD, and SFD protein systems, obtained through uniaxial tensile tests at a deformation rate of 1 mm/s. Average values marked with different lower-case or upper-case Greek letters are statistically different ($p < 0.05$).

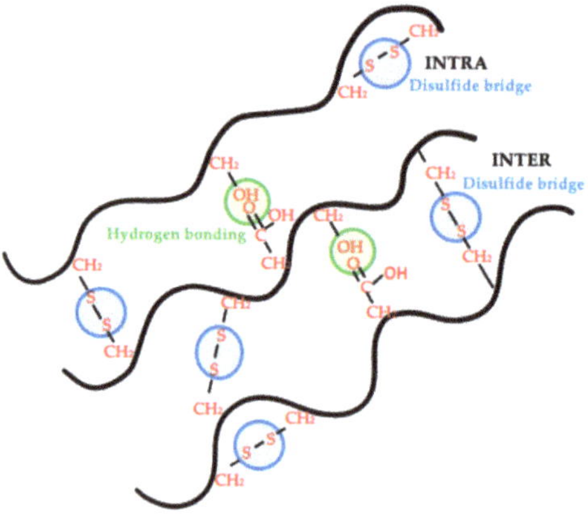

Figure 6. Proposed scheme for the main interactions promoted when protein unfolding takes place.

3.4. Water Uptake Capacity of PPP-Based Materials

The water uptake capacity (WUC) values obtained for the different samples can be observed in Figure 7. Regarding the WUC values of the different samples obtained from different procedures (UF, FD, SFD), it is remarkable that all of them can be considered superabsorbent materials, as their WUC values surpassed the lowest threshold required (1000%) [2]. Superabsorbent materials were previously obtained from porcine plasma protein, as reported in some studies [5–7]. Furthermore, although the SFD sample displayed a higher WUC value than FD or UF samples, no significant differences were found. Thus, the reinforcement in the material achieved by the aqueous solubilization of PPP, as shown by the remarkable increases in the mechanical properties of the samples, did not seem to have any negative consequence in terms of the water absorption capacity.

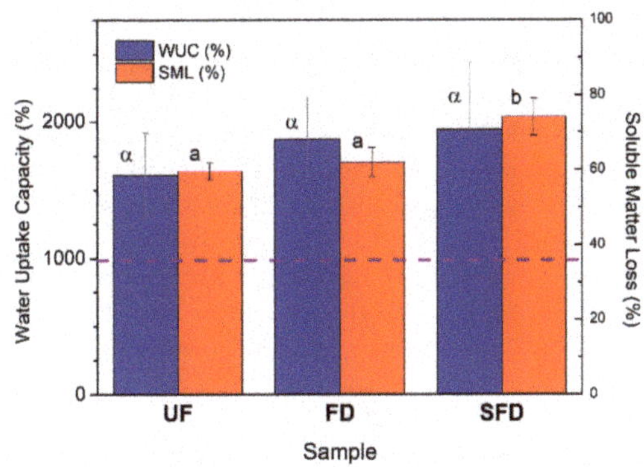

Figure 7. Water uptake capacities of porcine plasma protein-glycerol materials using UF, FD, or SFD protein systems, obtained through deionized water immersion over 24 h. The dashed line indicates the superabsorbent threshold. Average values marked with different lower-case Greek letters are statistically different ($p < 0.05$).

Otherwise, the SFD treatment of the PPP prior to its blending and subsequent injection molding seemed to have a great impact on the soluble matter loss.

3.5. Scanning Electron Microscopy (SEM)

Figure 8 shows the micrographs obtained through SEM of the swollen and freeze-dried matrices of the reference sample (Figure 8A) and the samples submitted to the FD (Figure 8B) and SFD procedures (Figure 8C). The porous structure observed for all samples was caused by the inclusion of water into the polymeric structure during the immersion stage, as glycerol was lost into the immersion media during the 24 h immersion process. This entrapped water was later removed in the freeze-drying stage that took place after swelling, leading to the formation of pores throughout the structure. As can be seen, the UF sample had much larger pores (182 ± 53 μm) than the SFD sample (74 ± 15 μm), which displayed a larger number of smaller pores, while the FD sample contained intermediate pores (110 ± 30 μm). Thus, smaller pore sizes may be the result of the mentioned reinforcement in the structure, as shown by the increases in E and σ_{max} values [7,10] in Figure 5. As mentioned before, a conformational change should take place solubilized and freeze-dried proteins [31], which would influence the protein-protein interactions [42], and consequently the overall structure of the materials. In previous studies, reductions in pore size were achieved through an increase of the mold temperature or through an excessively long molding stage, causing thermal crosslinking, and consequently lower WUC values [5,10]. In the present manuscript, the obvious reduction the pore size was caused by the initial treatment of the raw material, which did not hinder the WUC but did improve the mechanical properties of the final material.

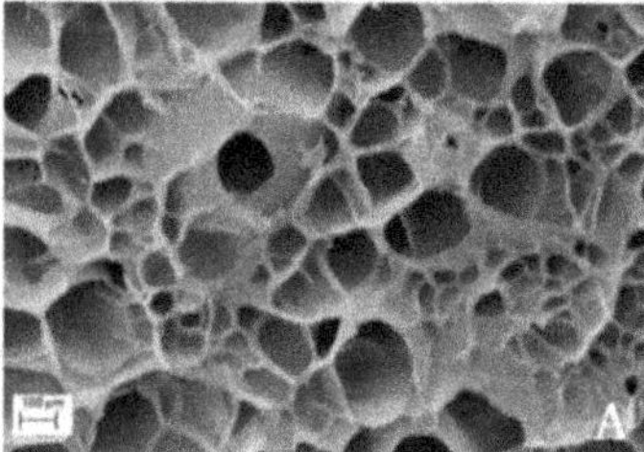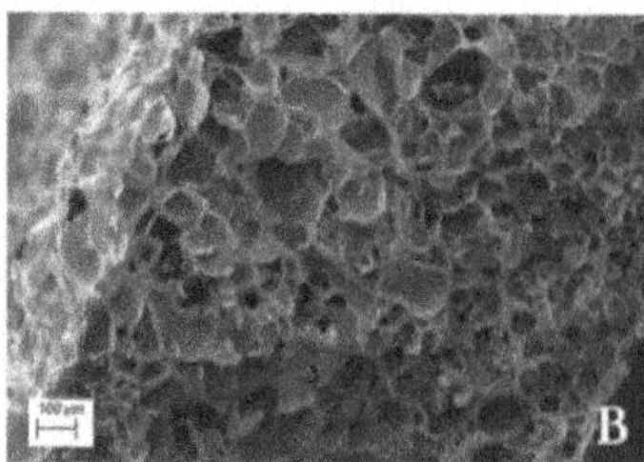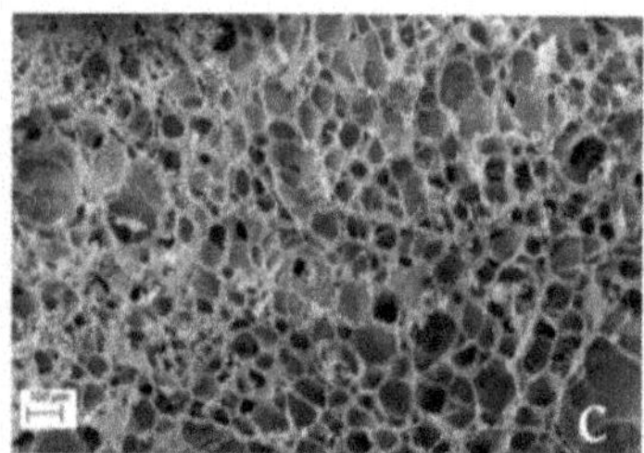

Figure 8. SEM micrographs of matrices obtained after swelling and freeze-drying of injection-molded reference (**A**), FD (**B**), and SFD (**C**) porcine plasma protein samples.

4. Conclusions

The solubilization and freeze-drying processes used in the development of green superabsorbent materials based on porcine plasma protein and glycerol seems to exert a significant influence on the final physicochemical properties.

When the protein source was only freeze-dried prior to blending with the plasticizer, slight changes in the rheological and mechanical properties could be detected, being mainly attributed to its lower plasticizer content due to moisture removal. Otherwise, the addition of a solubilization stage of the porcine plasma protein prior to freeze-drying resulted in greater differences. In this case, samples showed greater viscoelastic moduli across the whole temperature range, either for blends or PPP-based materials. Furthermore, the Young's modulus and maximum stress values of the solubilized-freeze-dried samples were greater, being around 7.5 and 3.5 times higher, respectively. On the other hand, the maximum strain values reduced more than 10-fold when compared to the rest of the samples, making them considerably more fragile. The observed change in the mechanical properties could be supported by the noteworthy decrease in the pore size of the solubilized-freeze-dried samples.

One of the most remarkable facts of the present study is that neither of the two treatments carried out (FD or SFD) led to any significant modification of the water uptake capacity of the UF-containing matrix, with samples surpassing in every case the lowest threshold required to be consider as superabsorbent materials. Thus, improvements in the mechanical properties of superabsorbent materials developed from porcine plasma protein and glycerol could be achieved without needing to use a stronger thermal treatment, submitting the protein source only to freeze-drying (especially if a previous solubilization stage had been previously conducted) before mixing and injection.

Author Contributions: Conceptualization, C.B. and A.G.; methodology, E.Á.-C.; software, E.Á.-C.; validation, E.Á.-C. and C.B.; formal analysis, E.Á.-C.; investigation, E.Á.-C., C.B. and A.G.; resources, C.B. and A.G.; data curation, E.Á.-C.; writing-original draft preparation, E.Á.-C. and A.G.; writing-review and editing, E.Á.-C. and C.B.; supervision, C.B.; project administration, C.B. and A.G.; funding acquisition, C.B. and A.G.; All authors have read and agreed to the published version of the manuscript.

Funding: This research was funded by Spanish Ministerio de Ciencia e Innovación-Agencia Estatal de Investigación (MICINN) and Fondo Europeo de Desarrollo Regional (FEDER) under the project reference RTI2018-097100-B-C21 and by the Spanish Ministerio de Ciencia e Innovación-Ministerio de Universidades through the PhD grant PRE2019-089815.

Institutional Review Board Statement: Not applicable.

Informed Consent Statement: Not applicable.

Data Availability Statement: All the results shown in the manuscript can be provided if requested from the corresponding author.

Acknowledgments: The authors would like to thank Spanish Ministerio de Ciencia e Innovación-Agencia Estatal de Investigación (MICINN), FEDER programs and Spanish Ministerio de Universi-

Polymers **2021**, *13*, 772

dades for the funding suppled. The authors also thank the Microanalysis and Microscopy Service (CITIUS-Universidad de Sevilla) for the full assistance provided in using the Zeiss Evo. The authors are also grateful to the Universidad de Sevilla and the Escuela Politécnica Superior of the Universidad de Sevilla for fully funding the article processing charges.

Conflicts of Interest: All the authors confirm that the manuscript is an original work and that it has not been previously published. The authors also declare that they have no known competing financial interests or personal relationships that could have appeared to influence the work reported in this paper.

References

1. Cuadri, A.A.A.; Romero, A.; Bengoechea, C.; Guerrero, A. The Effect of Carboxyl Group Content on Water Uptake Capacity and Tensile Properties of Functionalized Soy Protein-Based Superabsorbent Plastics. *J. Polym. Environ.* **2018**, *26*, 2934–2944. [CrossRef]
2. Cuadri, A.A.; Romero, A.; Bengoechea, C.; Guerrero, A. Natural superabsorbent plastic materials based on a functionalized soy protein. *Polym. Test.* **2017**, *58*, 126–134. [CrossRef]
3. Ruiz-Hitzky, E.; Darder, M.; Fernandes, F.M.; Wicklein, B.; Alcântara, A.C.S.; Aranda, P. Fibrous clays based bionanocomposites. *Prog. Polym. Sci.* **2013**, *38*, 1392–1414. [CrossRef]
4. Song, W.; Xin, J.; Zhang, J. One-pot synthesis of soy protein (SP)-poly(acrylic acid) (PAA) superabsorbent hydrogels via facile preparation of SP macromonomer. *Ind. Crops Prod.* **2017**, *100*, 117–125. [CrossRef]
5. Álvarez-Castillo, E.; Bengoechea, C.; Rodríguez, N.; Guerrero, A. Development of green superabsorbent materials from a by-product of the meat industry. *J. Clean. Prod.* **2019**, *223*, 651–661. [CrossRef]
6. Álvarez-Castillo, E.; Bengoechea, C.; Guerrero, A. Effect of pH on the properties of porcine plasma-based superabsorbent materials. *Polym. Test.* **2020**, *85*, 106453. [CrossRef]
7. Álvarez-Castillo, E.; Bengoechea, C.; Guerrero, A. Composites from by-products of the food industry for the development of superabsorbent biomaterials. *Food Bioprod. Process.* **2020**, *119*, 296–305. [CrossRef]
8. Zohuriaan-Mehr, M.J.; Pourjavadi, A.; Salimi, H.; Kurdtabar, M. Protein- and homo poly(amino acid)-based hydrogels with super-swelling properties. *Polym. Adv. Technol.* **2009**, *20*, 655–671. [CrossRef]
9. Fernández-Espada, L.; Bengoechea, C.; Cordobés, F.; Guerrero, A. Protein/glycerol blends and injection-molded bioplastic matrices: Soybean versus egg albumen. *J. Appl. Polym. Sci.* **2016**, *133*. [CrossRef]
10. Álvarez-Castillo, E.; del Toro, A.J.; Aguilar, J.M.; Bengoechea, C.; Guerrero, A.; Bengoechea, C.; Del Toro, A.; Aguilar, J.M.; Guerrero, A.; Bengoechea, C. Optimization of a thermal process for the production of superabsorbent materials based on a soy protein isolate. *Ind. Crops Prod.* **2018**, *125*, 573–581. [CrossRef]
11. Álvarez-Castillo, E.; Del Toro, A.J.; Aguilar, J.M.; Guerrero, A.; Bengoechea, C. Formation of soy protein-based superabsorbent materials through optimization o f a thermal processing. *Afinidad* **2019**, *76*, 23–29.
12. Capezza Villa, A.J. *Novel Superabsorbent Materials Obtained from Plant Proteins*; Deparment of Plant Breeding, Swedish University of Agricultural Sciences: Alnarp, Sweden, 2017.
13. Jin, S.-K.; Choi, J.-S.; Kim, G.-D. Effect of porcine plasma hydrolysate on physicochemical, antioxidant, and antimicrobial properties of emulsion-type pork sausage during cold storage. *Meat Sci.* **2021**, *171*, 108293. [CrossRef] [PubMed]
14. Álvarez, C.; Rendueles, M.; Díaz, M. Production of porcine hemoglobin peptides at moderate temperature and medium pressure under a nitrogen stream. Functional and antioxidant properties. *J. Agric. Food Chem.* **2012**, *60*, 5636–5643. [CrossRef]
15. Del Hoyo, P.; Rendueles, M.; Díaz, M. Effect of processing on functional properties of animal blood plasma. *Meat Sci.* **2008**, *78*, 522–528. [CrossRef] [PubMed]
16. Verheijen, L.A.; Wiersema, D.; Hulshoff, P.; de Wit, J. *Management of Waste from Animal Product Processing*; FAO Corporate Document Repository: Wageningen, The Netherlands, 1996.
17. Benitez, B.; Barboza, Y.; Bracho, M.; Izquierdo, P.; Archile, A.; Rangel, L.; Marquez, E. Efecto del pH y concentración de las proteínas sobre la propiedad de gelación de la sangre animal. *Rev. Cient. Fac. Cienc. Vet.* **1999**, *9*, 190–196.
18. Parés, D.; Toldrà, M.; Saguer, E.; Carretero, C. Scale-up of the process to obtain functional ingredients based in plasma protein concentrates from porcine blood. *Meat Sci.* **2014**, *96*, 304–310. [CrossRef] [PubMed]
19. Hurtado, S.; Saguer, E.; Toldrà, M.; Parés, D.; Carretero, C. Porcine plasma as polyphosphate and caseinate replacer in frankfurters. *Meat Sci.* **2012**, *90*, 624–628. [CrossRef]
20. Hurtado, S.; Dagà, I.; Espiqulé, E.; Parés, D.; Saguer, E.; Toldrà, M.; Carretero, C. Use of porcine blood plasma in "phosphate-free frankfurters". *Procedia Food Sci.* **2011**, *1*, 477–482. [CrossRef]
21. Ramos-Clamont, G.; Fernández-Michel, S.; Carrillo-Vargas, L.; Martinez-Calderón, E.; Vázquez-Moreno, L. Functional properties of protein fractions isolated from porcine blood. *J. Food Sci.* **2003**, *68*, 1196–1200. [CrossRef]
22. Nuthong, P.; Benjakul, S.; Prodpran, T. Effect of some factors and pretreatment on the properties of porcine plasma protein-based films. *LWT—Food Sci. Technol.* **2009**, *42*, 1545–1552. [CrossRef]
23. Samsalee, N.; Sothornvit, R. Development and characterization of porcine plasma protein-chitosan blended films. *Food Packag. Shelf Life* **2019**, *22*, 100406. [CrossRef]

24. Sothornvit, R.; Krochta, J.M. Plasticizers in edible films and coatings. In *Innovations in Food Packaging*; Elsevier: Amsterdam, The Netherlands, 2005; pp. 403–433. ISBN 9780123116321.

25. Nuthong, P.; Benjakul, S.; Prodpran, T. Characterization of porcine plasma protein-based films as affected by pretreatment and cross-linking agents. *Int. J. Biol. Macromol.* **2009**, *44*, 143–148. [CrossRef]

26. García, M.C.; Torre, M.; Marina, M.L.L.; Laborda, F.; Rodriguez, A.R.; Garcia, M.C.; Torre, M.; Marina, M.L.L.; Laborda, F.; Rodriquez, A.R.; et al. Composition and characterization of soyabean and related products. *Crit. Rev. Food Sci. Nutr.* **1997**, *37*, 361–391. [CrossRef] [PubMed]

27. Bourny, V.; Perez-Puyana, V.; Felix, M.; Romero, A.; Guerrero, A. Evaluation of the injection moulding conditions in soy/nanoclay based composites. *Eur. Polym. J.* **2017**, *95*, 539–546. [CrossRef]

28. Jiménez-Rosado, M.; Bouroudian, E.; Perez-Puyana, V.; Guerrero, A.; Romero, A. Evaluation of different strengthening methods in the mechanical and functional properties of soy protein-based bioplastics. *J. Clean. Prod.* **2020**, *262*, 121517. [CrossRef]

29. Gómez-Heincke, D.; Martínez, I.; Stading, M.; Gallegos, C.; Partal, P. Improvement of mechanical and water absorption properties of plant protein based bioplastics. *Food Hydrocoll.* **2017**, *73*, 21–29. [CrossRef]

30. Rathna, G.V.N.; Damodaran, S. Swelling behavior of protein-based superabsorbent hydrogels treated with ethanol. *J. Appl. Polym. Sci.* **2001**, *81*, 2190–2196. [CrossRef]

31. Gong, K.-J.; Shi, A.-M.; Liu, H.-Z.; Liu, L.; Hu, H.; Adhikari, B.; Wang, Q. Emulsifying properties and structure changes of spray and freeze-dried peanut protein isolate. *J. Food Eng.* **2016**, *170*, 33–40. [CrossRef]

32. Stärtzel, P.; Gieseler, H.; Gieseler, M.; Abdul-Fattah, A.M.; Adler, M.; Mahler, H.-C.; Goldbach, P. Mannitol/l-Arginine-Based Formulation Systems for Freeze Drying of Protein Pharmaceuticals: Effect of the l-Arginine Counter Ion and Formulation Composition on the Formulation Properties and the Physical State of Mannitol. *J. Pharm. Sci.* **2016**, *105*, 3123–3135. [CrossRef] [PubMed]

33. Costantino, H.R.; Firouzabadian, L.; Wu, C.; Carrasquillo, K.G.; Griebenow, K.; Zale, S.E.; Tracy, M.A. Protein spray freeze drying. 2. Effect of formulation variables on particle size and stability. *J. Pharm. Sci.* **2002**, *91*, 388–395. [CrossRef]

34. Schersch, K.; Betz, O.; Garidel, P.; Muehlau, S.; Bassarab, S.; Winter, G. Systematic investigation of the effect of lyophilizate collapse on pharmaceutically relevant proteins I: Stability after freeze-drying. *J. Pharm. Sci.* **2010**, *99*, 2256–2278. [CrossRef] [PubMed]

35. Apichartsrangkoon, A.; Ledward, D. Dynamic viscoelastic behaviour of high pressure treated gluten–soy mixtures. *Food Chem.* **2002**, *77*, 317–323. [CrossRef]

36. Perez-Puyana, V.; Felix, M.; Romero, A.; Guerrero, A. Characterization of pea protein-based bioplastics processed by injection moulding. *Food Bioprod. Process.* **2016**, *97*, 100–108. [CrossRef]

37. Fang, R.; Bogner, R.H.; Nail, S.L.; Pikal, M.J. Stability of Freeze-Dried Protein Formulations: Contributions of Ice Nucleation Temperature and Residence Time in the Freeze-Concentrate. *J. Pharm. Sci.* **2020**, *109*, 1896–1904. [CrossRef]

38. Aguilar, J.M.; Jaramillo, A.; Cordobés, F.; Guerrerro, A. Influencia del procesado térmico sobre la reología de geles de albumen de huevo. *Afinidad* **2010**, *545*, 28–32.

39. Aguilar, J.M.; Cordobes, F.; Jerez, A.; Guerrero, A. Influence of high pressure processing on the linear viscoelastic properties of egg yolk dispersions. *Rheol. Acta* **2007**, *46*, 731–740. [CrossRef]

40. Zárate-Ramírez, L.S.; Romero, A.; Martínez, I.; Bengoechea, C.; Partal, P.; Guerrero, A. Effect of aldehydes on thermomechanical properties of gluten-based bioplastics. *Food Bioprod. Process.* **2014**, *92*, 20–29. [CrossRef]

41. Bengoechea, C.; Arrachid, A.; Guerrero, A.; Hill, S.E.; Mitchell, J.R. Relationship between the glass transition temperature and the melt flow behavior for gluten, casein and soya. *J. Cereal Sci.* **2007**, *45*, 275–284. [CrossRef]

42. Sedov, I.; Nikiforova, A.; Khaibrakhmanova, D. Evaluation of the binding properties of drugs to albumin from DSC thermograms. *Int. J. Pharm.* **2020**, *583*, 119362. [CrossRef]

43. Johnson, C.M. Differential scanning calorimetry as a tool for protein folding and stability. *Arch. Biochem. Biophys.* **2013**, *531*, 100–109. [CrossRef] [PubMed]

44. Adebisi, A.O.; Kaialy, W.; Hussain, T.; Al-Hamidi, H.; Nokhodchi, A.; Conway, B.R.; Asare-Addo, K. Freeze-dried crystalline dispersions: Solid-state, triboelectrification and simultaneous dissolution improvements. *J. Drug Deliv. Sci. Technol.* **2020**, *61*, 102173. [CrossRef]

45. Pierson, N.A.; Makarov, A.A.; Strulson, C.A.; Mao, Y.; Mao, B. Semi-automated screen for global protein conformational changes in solution by ion mobility spectrometry-massspectrometry combined with size-exclusion chromatography and differential hydrogen-deuterium exchange. *J. Chromatogr. A* **2017**, *1496*, 51–57. [CrossRef] [PubMed]

46. Silva, J.L.; Foguel, D.; Da Poian, A.T.; Prevelige, P.E. The use of hydrostatic pressure as a tool to study viruses and other macromolecular assemblages. *Curr. Opin. Struct. Biol.* **1996**, *6*, 166–175. [CrossRef]

47. Álvarez-Castillo, E.; Oliveira, S.; Bengoechea, C.; Sousa, I.; Raymundo, A.; Guerrero, A. A rheological approach to 3D printing of plasma protein based doughs. *J. Food Eng.* **2020**, *288*, 110255. [CrossRef]

48. Felix, M.; Romero, A.; Cordobes, F.; Guerrero, A. Development of crayfish bio-based plastic materials processed by small-scale injection moulding. *J. Sci. Food Agric.* **2015**, *95*, 679–687. [CrossRef]

49. Arsiccio, A.; Giorsello, P.; Marenco, L.; Pisano, R. Considerations on Protein Stability during Freezing and Its Impact on the Freeze-Drying Cycle: A Design Space Approach. *J. Pharm. Sci.* **2020**, *109*, 464–475. [CrossRef]

50. Zhan, F.; Shi, M.; Wang, Y.; Li, B.; Chen, Y. Effect of freeze-drying on interaction and functional properties of pea protein isolate/soy soluble polysaccharides complexes. *J. Mol. Liq.* **2019**, *285*, 658–667. [CrossRef]

polymers

Article

Analysis of Styrene-Butadiene Based Thermoplastic Magnetorheological Elastomers with Surface-Treated Iron Particles

Arturo Tagliabue, Fernando Eblagon and Frank Clemens *

Laboratory for High Performance Ceramics, Empa, Swiss Federal Laboratories for Materials Science and Technology, Überlandstrasse 129, 8600 Dübendorf, Switzerland; arturo.tagliabue@empa.ch (A.T.); fernandoeblagon@lankhorsteuronete.com (F.E.)
* Correspondence: frank.clemens@empa.ch; Tel.: +41-58-765-4821

Abstract: Magnetorheological elastomers (MRE) are increasing in popularity in many applications because of their ability to change stiffness by applying a magnetic field. Instead of liquid-based 1 K and 2 K silicone, thermoplastic elastomers (TPE), based on styrene-butadiene-styrene block copolymers, have been investigated as matrix material. Three different carbonyl iron particles (CIPs) with different surface treatments were used as magneto active filler material. For the sample fabrication, the thermoplastic pressing method was used, and the MR effect under static and dynamic load was investigated. We show that for filler contents above 40 vol.-%, the linear relationship between powder content and the magnetorheological effect is no longer valid. We showed how the SiO_2 and phosphate coating of the CIPs affects the saturation magnetization and the shear modulus of MRE composites. A combined silica phosphate coating resulted in a higher shear modulus, and therefore, the MR effect decreased, while coating with SiO_2 only improved the MR effect. The highest performance was achieved at low deformations; a static MR effect of 73% and a dynamic MR effect of 126% were recorded. It was also shown that a lower melting viscosity of the TPE matrix helps to increase the static MR effect of anisotropic MREs, while low shear modulus is crucial for achieving high dynamic MR. The knowledge from TPE-based magnetic composites will open up new opportunities for processing such as injection molding, extrusion, and fused deposition modeling (FDM).

Keywords: magnetorheological elastomer; thermoplastic elastomer; magnetorheological effect; static and dynamic mechanical analysis

check for
updates

Citation: Tagliabue, A.; Eblagon, F.; Clemens, F. Analysis of Styrene-Butadiene Based Thermoplastic Magnetorheological Elastomers with Surface-Treated Iron Particles. *Polymers* **2021**, *13*, 1597. https://doi.org/10.3390/polym13101597

Academic Editor: Patrick Ilg

Received: 23 March 2021
Accepted: 13 May 2021
Published: 15 May 2021

Publisher's Note: MDPI stays neutral with regard to jurisdictional claims in published maps and institutional affiliations.

1. Introduction

Magnetorheological elastomers (MRE) are viscoelastic smart composites that show variable stiffness upon application of an external magnetic field. These compounds find applications in dampers for vibration absorption [1,2], in robotics, electronics [3,4], and force/acceleration sensors [5–7]. Depending on how magnetoactive particles are distributed in the matrix, two types of MRE are distinguished in literature: (1) isotropic MREs have homogeneously distributed particles within the matrix. (2) For anisotropic MREs, a magnetic field is applied while the matrix is still liquid; this causes an alignment of magnetoactive particles along the magnetic field. This alignment of the magnetoactive particles often referred to as pre-structuring, which results in a higher magnetorheological effect (MR effect).

Typically, elastomers such as silicon rubber [8] or natural rubber [9,10] are used as matrix material because of their low stiffness and hardness. To a lesser extent, thermoplastic elastomers (TPE) such as polyurethane and styrene block copolymers were investigated [11,12]. Soft magnetic carbonyl iron particles (CIP) are most often used because of the low remanent and high saturation magnetization, as well as the high permeability [13]. Burgaz et al. [14]

addressed the importance of carbonyl coating by comparing bare iron particles (BIP) to CIP and reported higher agglomeration, lower matrix-filler affinity, and increased Payne effect for the former.

The particle size and shape of the ferromagnetic particles affect the MR behavior [15]. The smaller the particle, the greater the surface area, therefore lowering the amount of free polymer, which will result in lower flexibility, expressed in high initial shear moduli, thus lowering the MR effect. The larger the used magnetic particles; however, the greater the interparticle distance that results in a reduced dipole interaction between particles, which finally results in a lower MR effect. The oxygen released from the CIP surface into the TPE can result in a decrease in elastic properties [10]. Therefore, researchers have started to coat the carbonyl iron particles with an oxide layer.

To increase the affinity between matrix and embedded particles, several surface modifications of CIP have been reported [12,16,17]. The potential benefits of a modified CIP surface are the reduced agglomeration and higher matrix-filler affinity. The smaller interfacial thickness between matrix and filler increases flexibility while simultaneously lowering the initial shear modulus affecting the MR effect. However, P. Małecki et al. [17] reported a lower MR effect in SiO_2-coated CIP particles that were uniformly distributed in a styrene-ethylene-butadiene-styrene (SEBS) matrix.

In practice, magnetorheological dampers have to carry a certain weight, which will result in a certain static pre-strain of the MR elastomer. Therefore, in this study, we investigate the effect of static and dynamic strain on the magnetorheological effect of anisotropic MREs based on SEBS TPE and CIP particles. Two thermoplastic elastomers with different shore hardness and carbonyl iron particles with and without surface treatment were used to investigate the MR effect under static strain conditions as well as the dynamic behavior. To avoid heating the samples during the characterization, a dynamic analyzer equipped with permanent magnets was used. Additionally, the magnetization behavior of MR soft composites with various filler content, up to 60 vol.-%, was investigated using a vibrating sample magnetometer (VSM).

2. Materials and Methods

Two SEBS-based materials (KRAIBURG TPE GmbH & Co. KG, Waldkraiburg, Germany) with different shore hardness were used in this study. The relevant properties are shown in Table 1.

Table 1. TPE matrix materials.

	TF1 STL	TF1 STT
Density (g/cm^3)	0.87	0.89
Hardness (Shore A)	7	15

The SEBS is a thermoplastic elastomer (TPE) that contains thermoplastic and elastomeric properties at the same time. TPEs can be divided into six subgroups (ISO 18064), and one of them is styrenic block copolymers, so-called TPS, to which SEBS (styrene-ethylene-butylene-styrene co-block polymer) belongs. SEBS is a styrenic triblock copolymer and consists of soft elastomers and hard thermoplastic blocks. The SEBS of Kraiburg is a composite based on SEBS, PP (polypropylene), fillers, and stabilizers.

Three commercial carbonyl iron particles (BASF, Ludwigshafen, Germany) were used as the soft magnetic fillers for the TPE-based MR elastomers (Table 2). (1) The HS type is a CIP without surface treatment and the mean particle size (d_{50}) of 1.9 μm; (2) the CC type, includes a SiO_2 coating on the particles (d_{50} of 4.7 μm); and (3) the EW-I type, where the carbonyl iron particles are coated with a SiO_2 and phosphate-layer (d_{50} of 3.4 μm).

Table 2. CIP materials with and without surface treatment.

Material Property	HS	CC	EW-I
Density (g/cm^3)	7.73	7.89	7.58
d$_{50}$ (µm)	1.9	4.7	3.4
Surface treatment	No	SiO$_2$	SiO$_2$ + phosphate

The magnetorheological thermoplastic elastomers were mixed in a torque rheometer (Rheomix 600, Thermofisher, Karlsruhe, Germany). For high shear mixing, the TPE material was heated to 170 °C, and the CIP filler was slowly added. Finally, the feedstocks were compounded for 30 min at 30 RPM. The naming convention for the samples consisted of the last three letters of the polymer grade (either STT or STL), followed by the volume percent of the filler and the type of used CIP (either CC, HS, or EW-I). Therefore, a sample made with the TF1 STL copolymer, loaded with 50 vol.-% of HS CIP, will be named STL50HS. After mixing, the density of the feedstock was analyzed by He-pycnometer (AccuPyc 1340, Micrometrics, Norcross, GA, USA).

Cylindrical samples with a thickness of 5 mm were warm-pressed above the melting temperature of the used TPE materials at 220 °C. A 5 kN load was applied for 8 min using a cylindrical die with a diameter of 20 mm. To introduce anisotropic orientation of the CIP particles, warm-pressed samples were placed in a magnetic field of 0.8 T generated by two permanent magnets (Q-51-51-25-N, Webcraft GmbH, Uster, Switzerland) before cooling from 220 °C to room temperature actively under flowing water. Figure 1 shows the sample processing schematically.

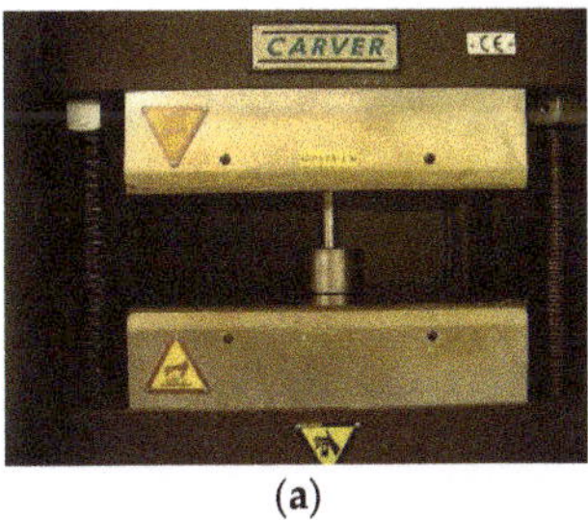
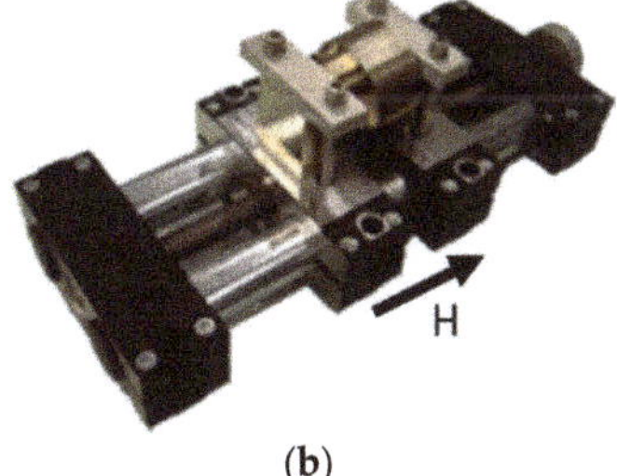

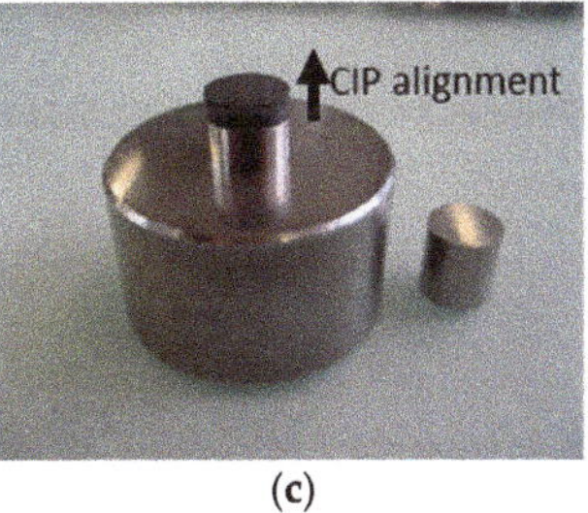

(a)   (b)   (c)

Figure 1. (**a**) Warm-pressing of the MRE samples, (**b**) pre-structuring the MRE sample under magnetic field H, and (**c**) CPI alignment in the TPE-based MR elastomer.

The magnetic properties of the MRE with various filler content and different surface treatments were measured using a vibrating sample magnetometer (VSM) from Quantum Design, USA. A 10^{-2} g sample was used for the VSM measurements. A magnetic field from 0 T to 2.2 T at ~0.1 T/min was applied. Then the magnetic field was ramped down to −2.2 T at the same rate and finally back up to 2.2 T.

An R 2000 rheometer (TA Instruments, New Castle, UK) was used to measure the complex viscosity above the melting temperature (160 °C) with a constant oscillating frequency of 10 Hz and 1% strain amplitude. MR elastomers with a 2 mm thickness were fixed between two parallel plates of 25 mm in diameter. For comparison, the complex viscosity at 170 °C will be reported in the result part for the two different SEBS elastomers filled with 30 vol.-% of carbonyl iron particles, grade HS.

For static and dynamic testing, MRE samples were assembled into a double shear testing structure made of V155 grade steel for the dynamic testing. To achieve good adhesion, the surface of the V155 grade steel was sanded before applying an adhesive paste (Araldite 2011 two-component epoxy).

An Eplexor 500 machine (NETZSCH-Gerätebau GmbH, Selb, Germany) was used for the DMA testing in shear mode. The prepared sample was fixed in the DMA as shown in Figure 2a,b, respectively.

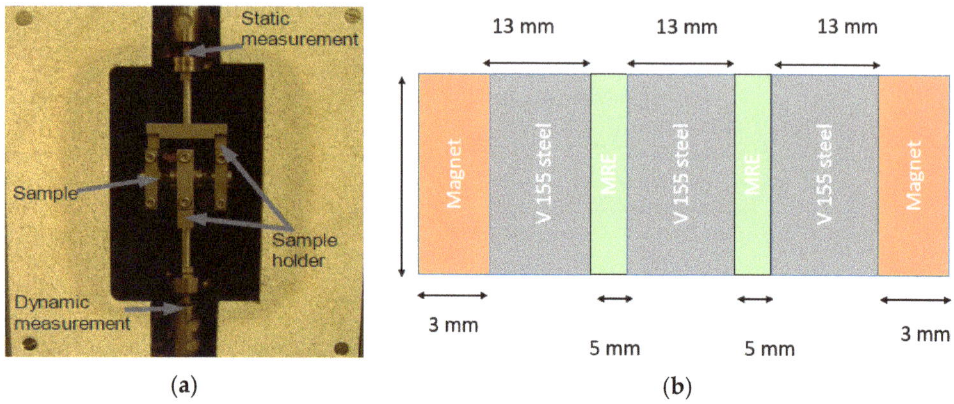

Figure 2. (**a**) Sample preparation in the dynamic mechanical analyzer. (**b**) A sketch of the sample with applied magnets at the edges of the sample.

To induce a magnetic field, disc-shaped permanent magnets with a diameter of 20 mm and thickness of 2 mm (N45, Webcraft GmbH, Uster, Switzerland) were fixed at the faces of the assembled double shear structure. Pairs of 0, 2, 4, and 6 magnets in total were used to induce a magnetic field inside the magnetorheological elastomer composite samples. The intensity and homogeneity of the induced magnetic fields were calculated using finite element analysis (FEMM). In Figure 3, the calculation for one pair of magnets and the results for a higher number of paired permanent magnets is shown. The magnetic field strength calculated for 2 and 3 pairs of magnets is shown in Appendix A.

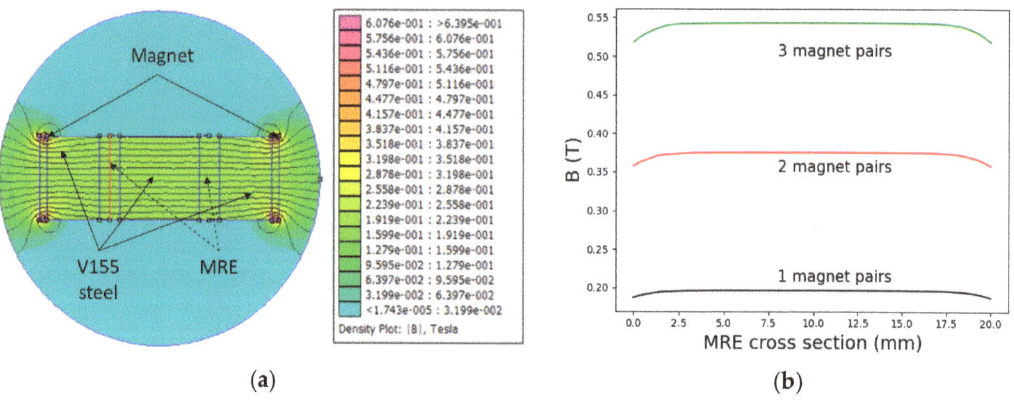

Figure 3. (**a**) Calculated magnetic field intensity using 1 pair of permanent magnets and (**b**) the variability of the field across the MRE.

As already mentioned, in this study, we investigated the static and dynamic magnetorheological behavior of the SEBS-based composites. The composites were cycled with a constant strain amplitude of 1% at different pre-strained levels (referred to as static tests in this report). Additionally, so-called dynamic tests were performed without any pre-strain conditions. Both tests were made with a constant frequency of 10 Hz. The MRE was calculated using:

$$MR_{effect} = \frac{G_s - G_0}{G_0} \; x \; 100\% \tag{1}$$

where G_0 represents the initial shear modulus and G_s the shear modulus upon application of an external magnetic field.

Static strains were set at 0.00%, 0.20%, 0.67%, 1.09%, 1.50%, 1.93%, 2.35%, and 2.75%, whereas the dynamic strains were set at 0.04%, 0.25%, 0.45%, 0.65%, and 0.85%. The test conditions for both static and dynamic experiments are shown in Figure 4.

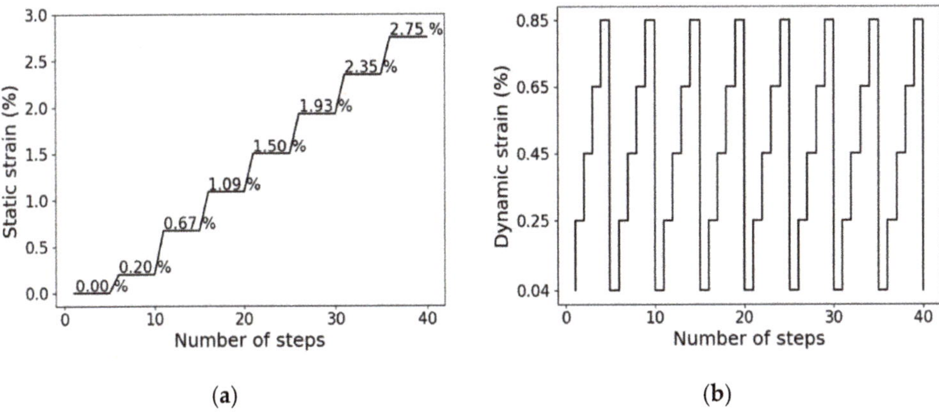

(a) (b)

Figure 4. (**a**) Static and (**b**) dynamic tests procedure to investigate the MR effect of SEBS-based composites.

3. Results

3.1. Effect of CIP Concentration

To investigate the influence of CIP concentration on the MRE STL-based composites, we investigated volume fractions between 10 and 60% of HS. For the static tests, a strain amplitude of 1% and a pre-strain of 1.58% were selected. For the dynamic measurements, a strain amplitude of 0.66% was chosen. All tests were performed with an applied magnetic field of 0.54 T and a constant frequency of 10 Hz. The results of the static and dynamic tests are illustrated in Figure 5. For the static and dynamic measurements, a linear relationship between the MR effect and the CIP content was observed below 50 vol.-%.

As is illustrated in Figure 5, above 50 vol.-%, the magnetorheological effect decreases significantly. This behavior can be explained by the critical particle volume concentration (CPVC), above which linearity approximation, as described by Jolly et al. [18] and Ginder [19], is no longer valid. The point at which there is not enough matrix present to fill all the space between the particles is known as the CPVC. To verify the CPVC point, density measurements can be used (Figure 6). Based on the mixing rule, the theoretical calculated density of a composite linearly increases when filler content increases. At the point where the theoretical and measured density shows a discrepancy, the CPVC point can be defined because air voids in the composite will lower the measured density significantly.

Density measurements of samples with a concentration above 50 vol.-% confirmed this prediction. The theoretical density of the composite with 60 vol.-% CPI was higher than the actual, measured values. Based on the discrepancy, it can be assumed that for 60 vol.-% CIP, air voids between the CIP particles are present (Figure 6). This behavior is in good agreement with the theory of critical particle volume concentration (CPVC). Filling a polymeric material above the CPVC results in a porous structure which causes a significant change in composite properties. Static and dynamic MR effects show similar results [20,21]. In Figure 6, it can be seen that the torque for the 60 vol.-% CIP is almost three times higher in comparison to 50 vol.-%. This implies that the processing behavior (e.g., viscosity) will significantly increase, and significantly higher machine power during thermoplastic shaping is required.

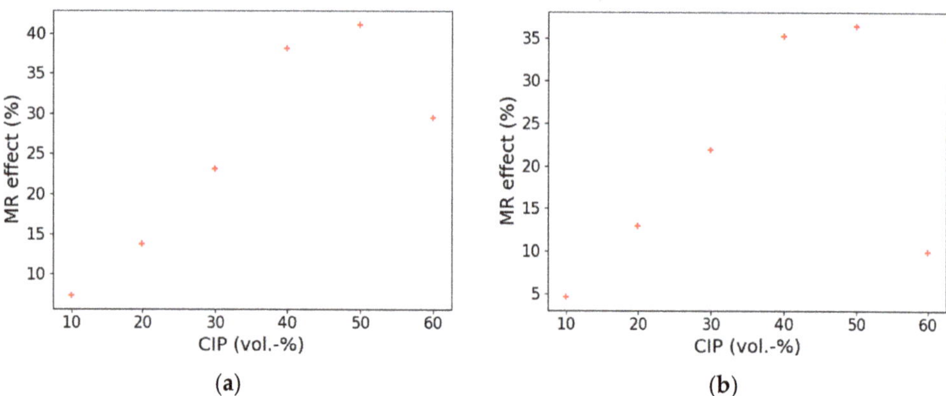

(a)   (b)

Figure 5. The influence of the CIP content on the MR effect (**a**) for the static test, a strain amplitude of 1% and pre-strain of 1.53% was used; (**b**) for the dynamic test, a strain amplitude of 0.66% was investigated. STL matrix with different concentrations of HS-type CIP; a constant frequency of 10 Hz and field strength of 0.54 T was selected for all tests.

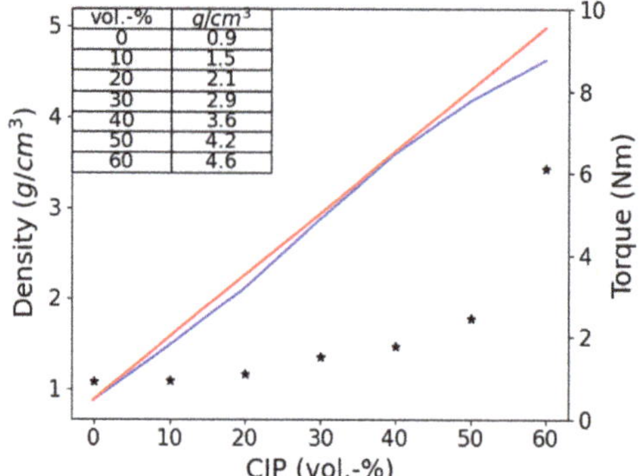

Figure 6. The torque at the end of the mixing process and density of the SEBS-based composite in relation to the volume concentration of CIP. The red line shows the calculated density based on the mixing rule. The blue line presents the measured density, and the stars are torque measured in Nm.

3.2. Flow Behavior at Meting Temperature of TPE Composites

The equation of motion of spherical CIP in a liquid is inversely proportional to the viscosity of the liquid [22]. For comparison reasons, complex viscosity on two different SEBS composites with 30 vol.-% CIP was investigated with a constant oscillating frequency of 10 Hz and 1% strain amplitude. Table 3 shows the complex viscosity values above the melting point of 160 °C.

Table 3. A comparison of the complex viscosity and torque at the end of the mixing process for two different SEBS composites with 30 vol.-% CIP.

	Complex Viscosity at 170 °C (Pas)	Torque at the End of Mixing 170° (Nm)
STT30HS	184	0.1 ± 0.1
STL30HS	30,000	1.6 ± 0.2

The MRE based on STT has a significantly lower complex viscosity in comparison to the one based on STL. The torque after the mixing for both matrix materials confirmed these results. Therefore, it can be expected that the orientation of the carbonyl iron particles in a magnetic field will be less restrained.

3.3. Magnetization Saturation of the CIPs

High saturation magnetization is a crucial factor in maximizing the MR effect, as discussed by Jolly et al. [18]. In Figure 7a, the saturation magnetization in relation to filler content is shown. The results for STL type SEBS composites with three different types of CIP filler (30 vol.-% filler content) are shown in Figure 7b.

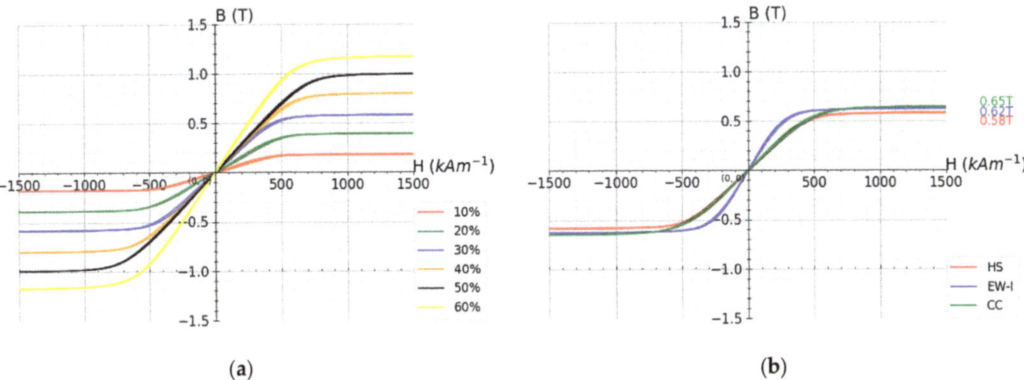

Figure 7. The magnetization loop for (**a**) STL type styrene-ethylene-butadiene-styrene thermoplastic elastomers with HS filler content between 10 and 60 vol.-%, (**b**) 30 vol.-% filler content of three different CIP grades in STL type styrene-ethylene-butadiene-styrene thermoplastic elastomers.

The saturation magnetization for the composites with different filler content is shown in Figure 8. It can be observed that by extrapolation of the composite values, 2 T of pure carbonyl iron can be achieved. This value is very close to the theoretical value of iron (2.2 T).

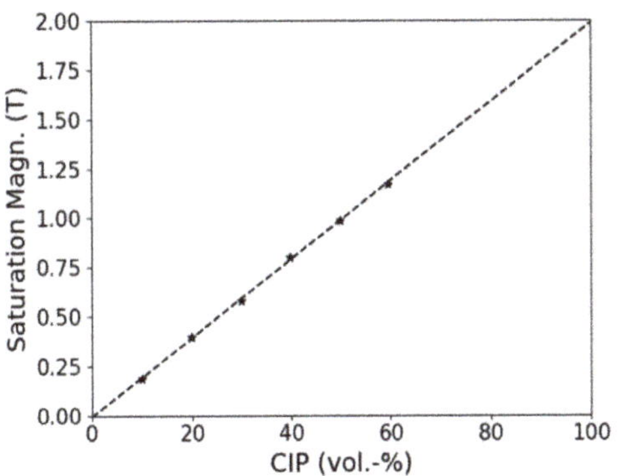

Figure 8. The saturation magnetization in relation to the CIP content and extrapolation of the results to 100% carbonyl iron.

In Figure 7b, a saturation magnetization of 0.58 T, 0.65 T, and 0.62 T for STL30HS, STL30CC, and STL30EW-I were observed, respectively. Therefore, the highest saturation magnetization was obtained for the composites based on the carbonyl iron particles with a SiO_2 coating. This is the CC grade type of CIP used in this study.

3.4. Effect of the Matrix Material

To evaluate the influence of the Shore hardness of styrene-ethylene-butadiene-styrene (SEBS) elastomers on the magnetorheological effect, composites with a constant filler content (30 vol.-%) carbonyl iron particles were selected. Based on the results of the magnetic saturation, the carbonyl iron particles with surface-treatment of SiO_2 (CC grade) were used for this investigation.

According to G. Bossis et al. [22] and Hass [23], the time required to induce alignment of particles along an external field is proportional to the viscosity of the melted matrix. Moreover, alignment of CIP in the MRE has been proven to positively influence the MR effect [22,23]. As shown in Table 3, the viscosity of 30 vol.-% STL is much higher than 30 vol.-% STT, meaning that full alignment of particles is faster in the former composite. This is expressed in the MR effect: MRE sample based on STT showed a larger static MR effect than the one obtained from STL type, as shown in Figure 9a. However, for dynamic strain, the influence of shore hardness of the thermoplastic elastomer matrix becomes more important. Therefore, the softer STL-based composite shows a higher MR effect, as illustrated in Figure 9b.

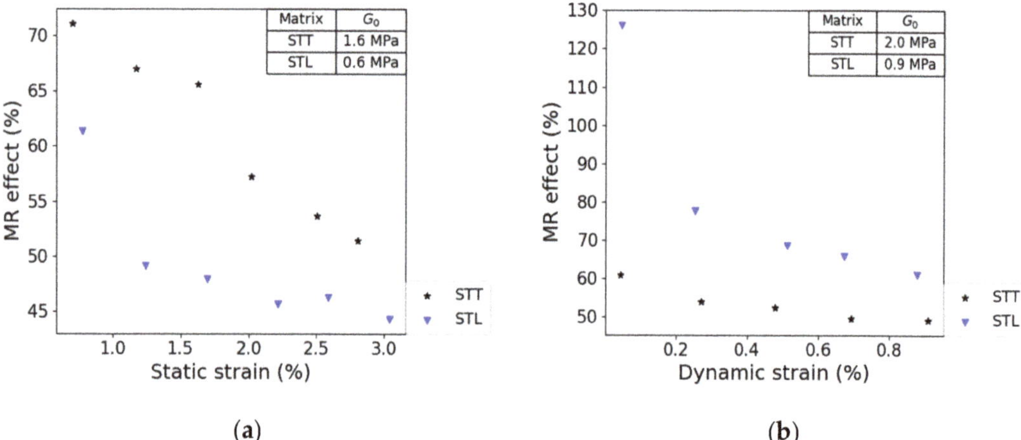

Figure 9. The influence of matrix material during static (**a**) and dynamic (**b**) strain on the MR effect. Composites STL30CC and STT30CC were used for this analysis. A constant strain amplitude of 1%was used for the static tests. A constant frequency of 10 Hz and field strength of 0.54 T were used for both test conditions.

These results indicate that for thermoplastic elastomers, a low melt flow behavior increases the static MR effect thanks to the improved orientation of CIP within the matrix. On the other hand, low matrix Shore hardness resulted in a high MR dynamic effect due to the low initial stiffness of the composite. Based on these results, further experiments were continued with softer STL type styrene-butadiene thermoplastic elastomer.

3.5. Effect of the CIP Surface Treatment

Małecki et al. [17] studied the magnetorheological effect of isotropic MRE based on an SEBS matrix. Coating the CIP with SiO_2 proved to negatively impact the MR effect due to the increased affinity between particles and matrix. This reduced the mobility of CIP within the composite, which in turn hindered the buildup of CIP chains in a magnetic field which results in a lower stiffening effect of the composite.

In our experiments on anisotropic SEBS-based MRE, we observed that a SiO_2-coating on CIP particles increased the MR effect significantly under static and dynamic strain. It can be assumed that for anisotropic MRE, mobility is less important because CIP particles are already aligned in the matrix. Figure 10a,b illustrate the effect of the surface treatment on the MR effect.

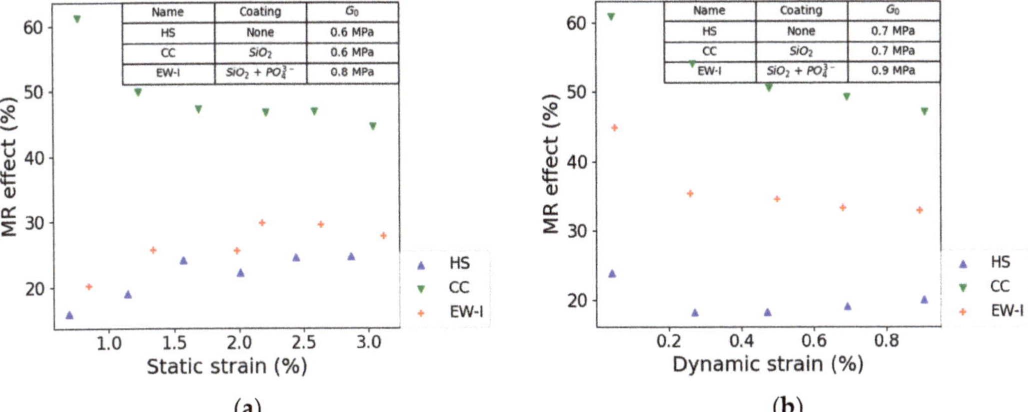

(a) **(b)**

Figure 10. The effect of static (**a**) and dynamic (**b**) strain on MR effect. For this investigation, STL30HS, STL30CC, and STL30EW-I composites were used. All samples contain 30 vol.-% CIP in an STL matrix. A constant strain amplitude of 1%was used for the static tests. A constant frequency of 10 Hz and field strength of 0.54 T were used for both test conditions.

The higher static MR effect could be obtained for the STL30CC (SiO_2-surface) MRE composite. This can be explained by the higher saturation magnetization of these particles (Figure 4).

As already mentioned, the effect of the filler-matrix affinity on the MR effect is important and has been already reported in literature [17]. Both composites STL30HS and STL30CC show low G_0 values of 0.6 MPa, while G_0 of STL30EW-I composite increased to 0.8 MPa. The higher shear modulus can be explained by the higher filler-matrix interfacial adhesion. These results are in agreement with the studies of Wang et al. [24]. Young's modulus of composites is affected by the polymer-filler interfacial adhesion. When the polymer–filler interfacial adhesion is weak, the composites exhibit lower modulus. Therefore, it can be assumed that the phosphate coating on top of the silica layer results in a higher interfacial adhesion. This results in a higher G_0. Based on Figure 10, we can conclude surface treatment of CIP of static and dynamic strain result in slightly different results, especially for the combined SiO_2 and phosphate coating.

4. Conclusions

Styrene-ethylene-butylene-styrene, also known as SEBS, is an important thermoplastic soft elastomer (TPE) that behaves like rubber without undergoing vulcanization. In this study, we reported the magnetorheological effect (MR effect) of two different SEBS-based composites filled with different kinds of carbonyl iron fillers. Carbonyl iron powder without surface treatment and with SiO_2 and SiO_2-phosphate surface treatment were used as magneto active fillers. The main application of MR elastomers is found in the damping structures. The structure will generate shear stress on the MR elastomers, which results in a static shear strain. Based on this background, we investigated the MR effect under static and dynamic strain. The source of SEBS and carbonyl iron filler affected the static and dynamic magnetorheological performance of the magnetorheological elastomer composite (MREC) significantly.

Increasing the carbonyl iron content from 10 to 60 vol.-%, we could observe a significant drop in the MR effect above 50 vol.-% of filler. This behavior is in good agreement with the theory of critical particle volume concentration (CPVC). Filling a polymeric material above the CPVC results in a porous structure which causes a significant change in composite properties. Static and dynamic MR effects show similar results.

For anisotropic MR elastomers, changing the SEBS matrix affected the static MR effect differently. Higher static MR properties can be achieved with low complex viscosity above the melting point, and the shore hardness of the SEBS plays a minor role. With a lower complex viscosity, the resistance to generate alignment of the carbonyl iron particles is lower; thus, the chain formation under a magnetic field is faster. For the dynamic MR effect, the shore hardness of the SEBS is an important parameter, as expected.

The surface treatment of the CIPs resulted in a higher magnetization saturation of MRECs with a filler content of 30 vol.-%. The highest MR effect could be obtained with a surface treatment of SiO_2 coating. We assume that higher static and dynamic MR effect with SiO_2 coating is caused by two phenomena: (1) SiO_2 coated CPI show higher saturation magnetization, which will result in higher MR-effect. (2) Higher matrix-filler affinity reduced the initial shear modulus, positively affecting the MR-effect.

Author Contributions: The idea for this work came from F.C.; The work mixing and characterization were done by F.E.; the manuscript was written by A.T. and F.C.; presentation of the results and the draft structure were discussed by all authors. All authors have read and agreed to the published version of the manuscript.

Funding: This project has received funding from the European Union's Horizon 2020 research and innovation programme under the Marie Sklodowska-Curie grant agreement No 860108 (SMART PROJECT).

Institutional Review Board Statement: The study was conducted according to the guidelines of the Declaration of Helsinki and comply with the research integrity guidelines of EMPA (https://www.empa.ch/web/empa/integrity, accessed on 10 May 2021).

Data Availability Statement: The data presented in this study are available on request from the corresponding author.

Conflicts of Interest: The authors declare no conflict of interest.

Appendix A

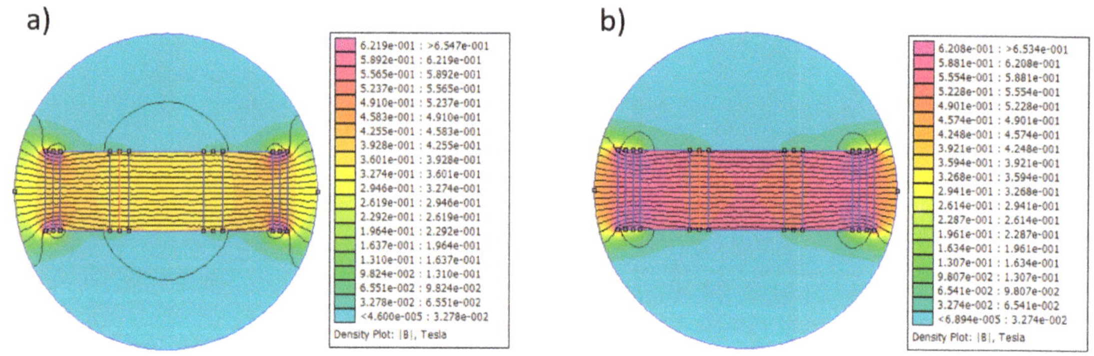

Figure A1. Calculated magnetic field intensity using (**a**) 2 pairs of permanent magnets and (**b**) 3 pairs of permanent magnets.

References

1. Molchanov, V.S.; Stepanov, G.V.; Vasiliev, V.G.; Kramarenko, E.Y.; Khohlov, A.R.; Xu, Z.D.; Guo, Y.Q. Viscoelastic properties of magnetorheological elastomers for damping applications. *Macromol. Mater. Eng.* **2014**, *299*, 1116–1125. [CrossRef]
2. Lee, C.W.; Kim, I.H.; Jung, H.J. Fabrication and Characterization of Natural Rubber-Based Magnetorheological Elastomers at Large Strain for Base Isolators. *Shock Vib.* **2018**, *2018*. [CrossRef]
3. Yuan, L.; Sun, S.; Pan, Z.; Ding, D.; Gienke, O.; Li, W. Mode coupling chatter suppression for robotic machining using semi-active magnetorheological elastomers absorber. *Mech. Syst. Signal Process.* **2019**, *117*, 221–237. [CrossRef]
4. Christie, M.D.; Sun, S.S.; Ning, D.H.; Du, H.; Zhang, S.W.; Li, W.H. A torsional MRE joint for a C-shaped robotic leg. *Smart Mater. Struct.* **2017**, *26*. [CrossRef]

5.   Tian, T.F.; Li, W.H.; Alici, G. Study of magnetorheology and sensing capabilities of MR elastomers. *J. Phys. Conf. Ser.* **2013**, *412*, 012037. [CrossRef]

6.   Li, W.; Kostidis, K.; Zhang, X.; Zhou, Y. Development of a Force Sensor Working with MR Elastomers. *IEEE/ASME Int. Conf. Adv. Intell. Mechatron* **2009**, 233–238. [CrossRef]

7.   Phan, K.L.; Mauritz, A.; Homburgb, F.G.A. A novel elastomer-based magnetoresistive accelerometer. *Sens. Actuators A Phys.* **2008**, *145*. [CrossRef]

8.   Lokander, M. Performance of Magnetorheological Rubber Materials. Ph.D. Thesis, KTH, Fiber and Polymer Technology, Stockholm, Sweden, 2004; p. 54.

9.   Davis, L.C. Model of magnetorheological elastomers. *J. Appl. Phys.* **1999**, *85*, 3348–3351. [CrossRef]

10.   Mattias Lokander, B.S.; Reitbergerb, T. Oxidation of natural rubber-based magnetorheological elastomers. *Polym. Degrad. Stab.* **2004**, *86*, 467–471. [CrossRef]

11.   Lu, X.; Qiao, X.; Watanabe, H.; Gong, X.; Yang, T.; Li, W.; Sun, K.; Li, M.; Yang, K.; Xie, H.; et al. Mechanical and structural investigation of isotropic and anisotropic thermoplastic magnetorheological elastomer composites based on poly(styrene-b-ethylene-co-butylene-b-styrene) (SEBS). *Rheol. Acta* **2012**, *51*, 37–50. [CrossRef]

12.   Yu, M.; Qi, S.; Fu, J.; Zhu, M.; Chen, D. Understanding the reinforcing behaviors of polyaniline-modified carbonyl iron particles in magnetorheological elastomer based on polyurethane/epoxy resin IPNs matrix. *Compos. Sci. Technol.* **2017**, *139*, 36–46. [CrossRef]

13.   Zajac, P.; Kaleta, J.; Lewandowski, D.; Gasperowicz, A. Isotropic magnetorheological elastomers with thermoplastic matrices: Structure, damping properties and testing. *Smart Mater. Struct.* **2010**. [CrossRef]

14.   Burgaz, E.; Goksuzoglu, M. Effects of magnetic particles and carbon black on structure and properties of magnetorheological elastomers. *Polym. Test.* **2020**, *81*, 106233. [CrossRef]

15.   Jin, Q.; Xu, Y.G.; Di, Y.; Fan, H. Influence of the particle size on the rheology of magnetorheological elastomer. *Mater. Sci. Forum* **2015**, *809*, 757–763. [CrossRef]

16.   Li, W.; Chen, J.; Gong, X.; Yang, T.; Sun, K.; Chen, X.; Li, W. Microstructure and magnetorheological properties of the thermoplastic magnetorheological elastomer composites containing modified carbonyl iron particles and poly(styrene-b-ethylene-ethylenepropylene-b-styrene) matrix. *Smart Mater. Struct.* **2012**. [CrossRef]

17.   Małecki, P.; Królewicz, M.; Krzak, J.; Kaleta, J.; Pigłowski, J. Dynamic mechanical analysis of magnetorheological composites containing silica-coated carbonyl iron powder. *J. Intell. Mater. Syst. Stuct.* **2015**. [CrossRef]

18.   Jolly, M.R.; Carlson, J.D.; Muñoz, B.C. A model of the behaviour of magnetorheological materials. *Smart Mater. Struct.* **1996**, *5*, 607–614. [CrossRef]

19.   Ginder, J.M. Rheology Controlled by Magnetic Fields. *Digit. Encycl. Appl. Phys.* **2003**. [CrossRef]

20.   Howard, K.W.; Hodgson, K.T. Influence of pigment packing behavior on the adhesive requirements of aqueous paper coatings. *J. Coat. Technol. Res.* **2015**, *12*, 237–245. [CrossRef]

21.   Khorassani, M.; Pourmahdian, S.; Afshar-Taromi, F.; Nourhani, A. Estimation of critical pigment volume concentration in latex paint systems using gas permeation. *Iran. Polym. J.* **2005**, *14*, 1000–1007.

22.   Bossis, G.; Lançon, P.; Meunier, A.; Iskakova, L.; Kostenko, V.; Zubarev, A. Kinetics of internal structures growth in magnetic suspensions. *Phys. A.* **2013**, *392*. [CrossRef]

23.   Hass, K.C. Computer simulations of nonequilibrium structure formation in electrorheological fluids. *Phys. Rev. E* **1993**, *47*, 3362–3373. [CrossRef] [PubMed]

24.   Wang, K.; Wu, J.; Ye, L.; Zeng, H. Mechanical properties and toughening mechanisms of polypropylene/barium sulfate composites. *Compos. Part A Appl. Sci. Manuf.* **2003**, *34*, 1199–1205. [CrossRef]

sustainability

MDPI

Article

KEYme: Multifunctional Smart Toy for Children with Autism Spectrum Disorder

Raquel Cañete, Sonia López and M. Estela Peralta *

Department of Design Engineering, Higher Polytechnic School, University of Seville, 41011 Seville, Spain; raqcanyaq@alum.us.es (R.C.); sonlopgar@alum.us.es (S.L.)
* Correspondence: mperalta1@us.es; Tel.: +34-954552827

Abstract: The role that design engineering plays in the quality of life and well-being of people with autism spectrum disorder around the world is extremely relevant; products are highly helpful when used as "intermediaries" in social interactions, as well as in the reinforcement of cognitive, motor and sensory skills. One of the most significant challenges engineers have to face lies in the complexity of defining those functional requirements of objects that will efficiently satisfy the specific needs of children with autism within a single product. Furthermore, despite the growing trends that point toward the integration of new technologies in the creation of toys for typically developing children, the variety of specialized smart products aimed at children with autism spectrum disorder is very limited. Based on this evidence the KEYme project was created, where a multifunctional smart toy is developed as a reinforcement system for multiple needs which is adaptable to different kinds of autism for therapies, educational centers or family environments. This approach involves the knowledge transfer from the latest neuroscience, medicine and psychology contributions to the engineering and industrial design field.

Keywords: social sustainability; smart product; autism spectrum disorder; inclusive design; therapy toys; design for ASD

Citation: Cañete, R.; López, S.; Peralta, M.E. KEYme: Multifunctional Smart Toy for Children with Autism Spectrum Disorder. *Sustainability* **2021**, *13*, 4010. https://doi.org/ 10.3390/su13074010

Academic Editors: Yadir Torres Hernández, Manuel Félix Ángel, Ana María Beltrán Custodio and Francisco Garcia Moreno

Received: 13 January 2021
Accepted: 30 March 2021
Published: 3 April 2021

Publisher's Note: MDPI stays neutral with regard to jurisdictional claims in published maps and institutional affiliations.

1. Introduction

Autism spectrum disorder, henceforth ASD [1], is a complex neurological disorder; it involves different types of needs, which makes it difficult to create product solutions that are adaptable to all of them. ASD presents a set of social restrictions associated with the alteration of social interaction (problems in verbal and non-verbal communication, in social and emotional reciprocity), sensory sensitivity, mental dysfunctions and various behaviors (such as aggressiveness, self-harm, restricted interests, hyperactivity or passivity). These symptoms of autism appear at an early age and remain until adulthood, thus making it difficult for such behaviors to disappear [2,3]. Furthermore, the disorders mentioned above lead to other indirect impacts such as harassment, social exclusion and school dropout.

It is important to highlight that society actively coexists with ASD: one out of every 100 births is diagnosed [4]. However, the availability of products that act as facilitators for the development of the needs of children with ASD is limited; parents, guardians and other professionals involved in therapy and education are forced to use not adapted products. After a market and patent research carried out on more than 1000 products [5], it was concluded that few large companies have adapted products for children with disabilities in their portfolios. Furthermore, research, development and innovation in this type of products do not move forward at the same rate as they do in toys for typically developing children. For this, the current trend is the introduction of new technologies to integrate the property of "smart" in the products.

On the other hand, advances in therapies for children with ASD reflect the importance of having facilitators (many of them technological) that allow enhancing the child's interaction with their outside world; technological products are a means of learning for physical, psychological and social development, and they function as tools that stimulate attention, following orders and socialization [6]. Different studies have demonstrated the role that technological products currently play in therapies; they are generally classified as (1) visual, tactile and auditory stimulation devices, (2) video-based instruction and feedback, (3) computer and ICT-assisted actions and instructions, (4) virtual reality and (5) robotics [7–12]. However, they are in the early research and development phase, meaning that the demand for specialized assistive technology in ASD is not covered by the current supply available in the market.

Therefore, this research introduces the KEYme project, an assistive technology design for ASD. The project uses a specific design framework for adaptive assistive technology that makes it possible to cover the set of needs included in the ASD classification [1]. Using this methodology, a multifunctional toy with interactive and smart properties is developed. The combination of multiple requirements allows the creation of game modes adaptable to different autism pictures and contexts of use (therapy, family, didactic). KEYme can help the child to develop those skills in which he has difficulties (like social communication and interaction challenges, repetitive and restrictive behavior, over or under sensitivity to perceptual factors). In addition, it includes those changes, trends and new concepts that are currently being applied in the market within products for typically developing children.

Lastly, the KEYme project contributes to the improvement of products and their accessibility in the field of human development and social sustainability (equity). It works within the scope of childhood and disability, specifically, the wide range of signs and symptoms of assistive technology ASD. The development of quality assistive technology, its universal design and its accessibility (supply and price) is one of the main research challenges identified by the World Health Organization [13].

To do this, the paper is structured as follows. Section 2 contains the background and the market study that analyzes the current situation of the toy sector dedicated to typically developing children and children with special needs, including assistive technology for ASD, product categories and relevant research. Section 3 exposes the methods used for the development of the KEYme project, which are later applied in Section 4, where the design and development of the proposal are explained. Finally, Section 5 summarizes the results and discussions of this research.

2. Background: Interactive Smart Toys Dedicated to Children with ASD

To date, the known causes, as well as the developmental and behavioral disorders associated with ASD, have not been demonstrated categorically. Scientific evidence establishes different demonstrable hypotheses and in many cases can jointly influence from neurobiological bases (genetic influence), obstetric complications, cerebral and cognitive structural alterations (cerebellum, hippocampus, mirror neurons or mamillary bodies) to the connection to other factors such as education, context and environment [14]. These reasons determine the importance acquired by the integrating initiatives of people with ASD in society.

Currently, there are some solutions that allow those people with ASD to minimize problems and satisfy certain needs that improve their quality of life and independence: didactic, educational, behavioral and occupational therapies. They are usually carried out from the detection of the disorder; the early interventions in children are the ones that most efficiently improve the social, emotional, communication and motor skills. In such interventions, assistive technology (AT) is useful [13]. Its correct configuration provides the scientific and technical basis for the development of smart products specialized in ASD. The AT group includes any device, software or equipment that helps overcome certain challenges, specifically for people who require assistance to carry out activities in their daily lives independently.

In this field, the quality of AT and its universal design are some of the main challenges of the World Health Organization [13]. To design a product as AT, emerging technologies are required (such as ICT, sensors, connectivity or the Internet of Things, among others) to create "smart properties" and incorporate specific functions that improve user connection and interaction, allowing imitative activities, stimulation, recognition and understanding of the environment to be faster and more intuitive.

When AT is intended for children, it usually makes use of game strategies to enhance the interaction between child and playmate. In this way, *AT configured as a toy* makes it possible to improve, develop and work on certain skills of the child while being fun, safe, age-appropriate and attractive. In addition to this, through play time, growth is enhanced from the anthropological, psychological and social points of view.

However, all of the above becomes more complicated when AT must be adapted to children with ASD. The heterogeneous group of signs and symptoms of autism, as well as the wide range of needs grouped in (i) communication and social interaction and (ii) flexibility of thought and behavior, makes it difficult to define and select the functional requirements and design variables suitable for product design. This circumstance makes it necessary to either integrate multifunctionality in products or to make AT adaptive, that is, to adapt flexibly to the specific needs or to the resolution of specific problems of each type of disorder and child. This complexity has had repercussions in recent years resulting in a low supply and availability of AT products for children with ASD in the market.

At the European level, Switzerland and Spain are the first two powers in R&D within the toy sector [15]. On the one hand, the Swedish market has a long tradition and recognition record of its products. Despite the great difficulties caused by the health crisis, sales continue to increase due to the introduction of key trends in their toys [16]. On the other hand, the Spanish market is characterized by a high number of toy companies (most of them micro or small companies) generating an annual turnover of around 1600 million euros [17]. Spanish toys are exported to more than 100 countries (the most important being Germany, France, United Kingdom, Italy and Portugal) although representing only 2% of the international market share [15], led by China in mass production and in exports and followed by the United States [18].

Having analyzed the sales of 2019, we concluded that (i) dolls, soft toys (interactive toys that simulate human interaction and affectivity), (ii) technical toys, educational toys, action toys, (iii) electronic toys and (iv) games, books, learning and experimenting were the best-selling toy categories [19]. On the other hand, the trends of the toy sector for the coming years are highly influenced by the introduction of smart features with seven key development categories: (i) movie industry, advertising and digitization (toys based on television series and movies), (ii) STEAM toys (focused on science, technology, engineering, arts and mathematics), (iii) unboxing (the game begins with packaging), (iv) RC toys (remote-controlled), (v) virtual reality (for a full immersion experience), (vi) toys and unisex gender roles and (vii) intended for kid-adults (new versions of old toys for nostalgic adults).

However, most of the toys on the market are targeted toward typically developing children. Those specialized in ASD imply less variety in shapes, interfaces and features; in addition, their availability is much more limited. On top of that, they do not follow the same trends. In general, they are marketed according to the needs which they are aimed for, thus classified into (1) cause-effect toys, (2) toys that improve gross motor skills, (3) toys that improve fine motor skills, (4) to facilitate language and communication skills and (5) favor symbolic and functional play. Most of the toys available within these five categories do not use technology and cannot be considered high-level AT, despite the fact that their benefits as a means of learning for physical, psychological and social development have been evaluated and demonstrated in different studies [6–12]. In particular, the use of SARs (socially assistive robotics) as collaborators has beneficial effects on the development of social skills in children with ASD, especially in areas where they present the most difficulties [20], such as attention, verbal communication or **imitation**, as well as in reducing stereotypic behaviors, where they present an immediate positive effect. Some of the most famous robots on the market are Kaspar Robot [21], Robota [22] or Sprite [23]; among the newly developed ones, we can mention Keepon [24], Nao [25] and Leka [26].

Most of these high-level AT products are off the market; many of them are just projects from research centers and the academic field. On the contrary, the ones that are being commercialized are neither affordable nor universally available because of their high price [27]. Finally, analyzing the design variables used in these products, there is a wide margin for improvement compared to existing toys for typically developing children. In addition to this, each available toy is adapted to a very specific need, so that the same child with ASD requires a large variety of products to meet his/her set of special needs.

As a result of this research, the KEYme project has the main goal of solving these problems through the design of assistive technology based on game strategies. Moreover, KEYme (1) follows current market trends in the toy sector (with smart properties), (2) can be adapted to different special needs and, finally, (3) makes use of affordable and open-source resources to promote universal design.

3. Methods

The design and development of adaptive and specialized assistive technology in special populations is a challenge. The design requirements must comply with a universal and adaptative design as a principle, facilitating **physical accessibility** (adapted mobility characteristics), **cognitive accessibility** (adapted processing, understanding, learning and decision-making characteristics in the task) and **sensory accessibility** (characteristics that assure the correct perception of environmental factors and information about the environment).

If we analyze the frameworks, methodologies and tools that are currently available and specify design principles and fundamentals, it is possible to conclude that the bibliography is very limited. There are general frameworks for applying universal design [28] or design processes for people with special needs [29]; there is also a set of tools focused on participatory and collaborative design to involve children with ASD and their families in the design process [30–34]. However, there is no exclusive agreed methodology for designing assistive technology specialized in ASD, and not all of the available ones can be applied to intelligent and multifunctional products. For this reason, in this research, the USERfit tool within user-centered design [35] was used in the first place. It is oriented to the system and promotes interactive design. Its objective is to facilitate the definition of the best product development framework by integrating various techniques adapted to the problem. This methodology is appropriate and has been applied in successful assistive technology projects, such as [36–38].

Specifically, USERfit is a user-centered design methodology applied to assistive technology. It is based on a design philosophy that can be described as user-centered, system-oriented and promoting interactive design. This approach to design implies a greater commitment to user requirements. The objective is to learn how to use certain design guidelines in order to determine whether or not they are suitable for our product dedicated to a user with specific needs. To get there, this methodology provides a framework divided into four phases: (1) check if the guidelines are relevant and appropriate for our product, (2) choose those guidelines that best suit our situation, (3) check if the guidelines are consistent with each other and (4) be able to manage a high number of guidelines [35]. These steps allow one to define which principles and design guides are relevant for our target user, in this case, the child with ASD.

Its application in the KEYme project starts with the identification of special needs of the user; through an iterative process (Figure 1), different methods were analyzed, selected and integrated: conventional design (like the Functional Analysis Screening Tool (FAST), Quality Function Deployment (QFD), Analytic Hierarchy Process (AHP) and Hierarchical Task Analysis (HTA), among others), specialized guidelines in universal design [28] and inclusive design [39], participatory design [40], collaborative design [41] or design methodologies for people with special needs [29] as well as smart toy design guidelines [42], concluding in the definition of the design methodology for the KEYme project.

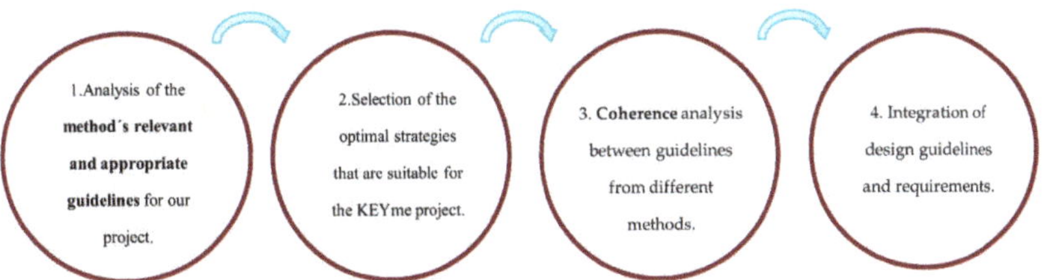

Figure 1. Application of USERfit in design project planning.

In this way, a specific design framework is defined for adaptive AT. It would make it possible to cover the set of disorders included in the ASD classification [1] and would help in making decisions about functional requirements (FRs) and their adequate translation to design variables (DVs) to be included in an interactive and intelligent product. Figure 2 shows the methodology proposed in the KEYme project for product design, which will be applied later in Section 4.

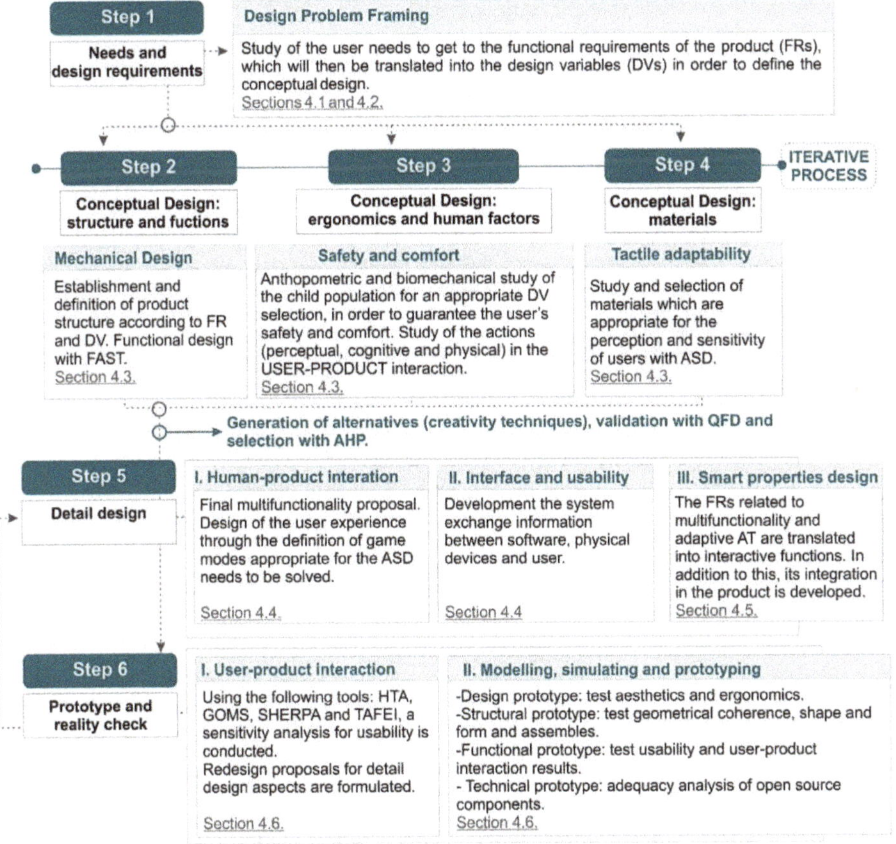

Figure 2. Product design methodology.

4. KEYme: Design and Development

This section covers the development of adaptive assistive technology for ASD. Following the design model of the applied methodology (Figure 2), this section analyzes the design problem, describes the development of the concept and synthesizes the detailed design of the proposal.

4.1. Design Problem Analysis

The design problem, which is our starting point of the KEYme project, raises the need to develop AT configured as a toy adaptable to any user and context of use (therapy, family, didactic environment). This device helps the child with ASD to develop those skills in which he/she has more difficulties; on top of this, it includes the changes, trends and new concepts that we are seeing in toys targeted at typically developing children. To make AT adaptive, first, it must be multifunctional, interactive, and smart; this allows for different game modalities to be included in the same object that would be able to cover a wide range of ASD needs grouped in (i) communication and social interaction and (ii) flexibility of thought and behavior. Moreover, and in order to make this AT universal and accessible to everyone, it must make use of affordable, open-source resources, as well as simple development and manufacturing processes.

Therefore, the project focused on the following design objectives:

1. Design a product as game-based assistive technology. It should enhance and work on different skills of children with autism while being fun, safe, age-appropriate and attractive. Target market and user group: children aged 3–12 years old with ASD and potential playmates (parents, guardians, therapists, teachers or typically developing children).
2. Develop a multifunctional toy, including in the same product the satisfaction of various needs of ASD grouped in (i) communication and social interaction and (ii) flexibility of thought and behavior. This will allow the product to be flexibly adapted to solve specific problems of each type of disorder and child, or what is the same, it can be used by users of different ages and degrees of autism [1].
3. Develop an interactive and smart product that allows for mutual interaction between object and users (children with ASD and potential playmates—parents, guardians, therapists, teachers or typically developing children) and encourages purposeful tasks. The product must be self-contained, as it embeds the interactions within the physical structure, creating an interactive environment of its own.
4. Develop game sequences that integrate sensory, motor (sensory-motor stage) and cognitive activities (preoperational and concrete operation stages) in the same device. The objective is to work on different skills, specifically, to improve non-verbal (sensory-motor stage) and verbal language through social and symbolic imitative play (preoperational stage), improve social interaction (preoperational and concrete operations stages), to work on emotional reciprocity and flexibility in routines (concrete operations stage) and to reduce frustration due to failure during the game.
5. Prioritize the use of imitation in play sequences to improve the interaction between the child with ASD and the playmate.
6. Develop a collaborative product to be used as a facilitator of social interaction and to improve the cognitive, motor and sensory skills of children with special needs.
7. Develop an adaptable product to different contexts of use: indoor (game rooms, home or school) and outdoor environments.

4.2. Needs and Functions Definition

As a result from the analysis of the ASD disorder according to its description in the DSM-V by the American Psychiatric Association [1] and within the heterogeneous group of signs and symptoms, a broad-spectrum set of primary needs was selected:

1. Improve playmate-child with ASD relationships (N1).
2. Sensory stimulation (N2).

3. Improve gross and fine motor skills (N3).
4. Promote shared actions linked to feelings of enjoyment, interest and common goals (N4).
5. Reduce levels of frustration in the game (N5).
6. Increase emotional and social reciprocity (N6).
7. Facilitate learning about changes in game turns (N7).

Subsequently, these needs, as well as the way of working with them, were evaluated in detail; they were subdivided and defined by analyzing early intervention therapies, specifically, applied behavior analysis [43], Denver [44], pivotal response treatment [45], sensory integration [46], physical therapy [47], auditory integration [48], music therapy [49] and use of high assistive technology [50]; other techniques to improve attention with imitation tasks and environmental resources were also analyzed and considered; their efficiency is contrasted with the neural activation of the subject [51–55]. Due to the heterogeneous group of signs and symptoms linked to sub-needs, it was necessary to direct the design of assistive technology toward multifunctionality; this implies including in the same product the satisfaction of various needs, making it possible for it to adapt to the disorder in a flexible way and therefore being adequate for a greater number of users (different ages and degrees of autism [1]). This exhaustive research study on psychological, physiological and neuronal factors resulted in conclusions on the most suitable design variables (motor, visual, auditory, tactile, olfactory, imitation, interaction, context, cognitive and temporal) contrasted by their relationship with (1) neuronal activity, (2) the subject's adequate and comfortable reaction to environmental stimuli and (3) the use of imitation tasks to improve the child's attention. In addition, the most appropriate functions and design variables were defined (see synthesis in Table 1), since the relationships between the need and the design variables were evaluated as "excellent, medium, low or null", and only those relationships rated as "excellent" were taken into account in the conceptual design.

It must be taken into account that, unlike typically developing children, a user with ASD may present deficiencies in social interaction, communication, behavior or executive function. This implies that the execution of tasks while using the product implies an added difficulty when having to pay attention to the object, to other people or to the actions that they carry out. To improve usability, comfort, reduce frustration and minimize errors, the toy should have motivational and behavioral elements or facilitators as secondary functional requirements. They will make the product adapt to the user in three dimensions:

(1) Physical accessibility: it includes all anthropometric adaptations (static or dimensional and dynamic or mobility).
(2) Cognitive accessibility: functions and elements of the product are adapted to the processing, understanding, semiotics, learning and decision-making of a user with ASD.
(3) Sensory accessibility: it assures that the perception of the product is comfortable, intuitive and adapted to the perceptual peculiarities of a user with ASD.

Table 2 details and classifies the set of functional requirements necessary to improve the accessibility of the product; they are derived from the general requirements defined in Table 1 (numbered from 1 to 10). These allowed to carry out the conceptual design in the project were defined.

Table 1. Analysis of needs and functional requirements.

Functional Requirements		Num	N1	N2	N3	N4	N5	N6	N7	Smart Product
Major	Secondary									
STRUCTURE should allow	Your own mobility and displacement.	①			X					The child must be able to change the environment/modality based on his/her wishes.
	User movements on it (swing, tunnel, climbing) with the aim of improving gross motor skills.	②			X					Feedback from the toy should be clear, explicit and easily understood by the child.
	Included components should allow for multiple movements to improve fine motor skills.	③			X					
MATERIALS	Stimulate the perceptual system with different textures adapted to hyper and hypo sensitivity.	④		X	X		X			Visual design according to the content, objectives and goals.
GAME MODES should include	Collaborative, aimlessness, not-sexist and not-violent.	⑤	X	X	X	X	X	X	X	Optimal number of game modes: too many variations affect attention and concentration. Toys should allow children to play collaboratively.
	Cooperative sequences for problem resolution.	⑥	X	X	X		X	X	X	The child must be aware of the results obtained from the toy. Avoid dead time; reduces concentration.
User-product INTERACTION INTERFACE Information should be reinforced simultaneously by visual and auditory stimuli	Safe and comfortable stimuli adapted to the child's perceptual characteristics.	⑦					X			Feedback given by the toy should be clear, explicit and easily understood by children.
	Musical sounds and lights combined with body movements and force generation (key pressure) to reinforce imitation tasks.	⑧	X	X	X	X	X	X	X	Background music can optionally be used to attract attention. Show dynamic content. "Help" screen before starting to play. The virtual environment must be interactive, and children must be able to change the environment based on their wishes. It must be easily controlled. There must be immediate feedback from the toy.
	Informative and intuitive interface.	⑨		X				X	X	
	Reinforcement stimuli for the performance of tasks.	⑩				X	X	X		

Table 2. Detail of functional and technical requirements for adaptability.

General Requirements	Specific Requirements
Physical accessibility (adapted mobility characteristics) ① ② ③ ⑩	Comfort and ergonomic adaptability for an age range 3–12 years old through a variety of body positions and ranges in the use of the product (crawling, sitting, standing, sliding or kneeling).
	Simple instructions on information exchange.
	The interface should include control elements grouped according to function, easily distinguishable by shape, color, size and movement (press, turn or slide).
	Safety: the dimensions and shapes of the product allow balancing actions while preventing tipping. The structure of the product ensures a non-slip grip to the ground and compressive strength. The surface of the product does not pose a risk of shocks, cuts or other damages.
Sensory accessibility (characteristics that assure the correct perception of environmental factors and information about the environment) ④ ⑦ ⑨ ⑩	Tactile, visual, auditory and proprioceptive stimulation through a variety of elements of different sizes, shapes and materials.
	Comfortable, safe stimulation adapted to hyper-hypo sensitivity.
	Improve attention processes through keyboard-guided imitation tasks (with mirrored keys).
	Reinforcement of the action sequence on the keys through pressure, auditory and visual stimuli.
Cognitive accessibility (adapted processing, understanding, learning and decision-making characteristics in the task) ⑤ ⑥ ⑦ ⑧ ⑨ ⑩	Adaptation to different ages according to learning level and degrees of autism with game modes I to IV.
	Custom game modes designed for ASD (I to IV).
	Reinforcement of information processing (success in tasks) with an automatic response from the product to user failures.
	Facilitate the game sequence with an automatic response from the product for each change in game turns.
	Facilitate the interpretation of the game through multisensoriality.
	Automatic collection of relevant data from the game experience.
Technical and economic feasibility	Affordable product.
	Low cost of electronic platform.
	Free-use electronic platform.
	Reduced manufacturing costs and times.
	Affordable product.

4.3. Conceptual and Detail Design

In design engineering, the basic stages to achieve an optimal result are [56] the generation of candidate design alternatives and selection of the most optimal one (through a prior quantitative evaluation). The generation of alternatives is carried out with the conceptualization of the system, that is, recognizing viable solutions for the target user. In this project, three different alternatives destined to satisfy the needs of Table 1 (and requirements of Table 2) were created (Figure 3); they were developed using creative methods (brainstorming and mind map) and the quality of the solutions was validated with QFD [57]. Specifically, QFD allowed us to transform qualitative user demands into quantitative parameters, that is, translate the needs and demands of the users (child with ASD, typically developing children, adult playmates such as parents, guardians or therapists) to the product specifications (definition of functions and selection of design parameters).

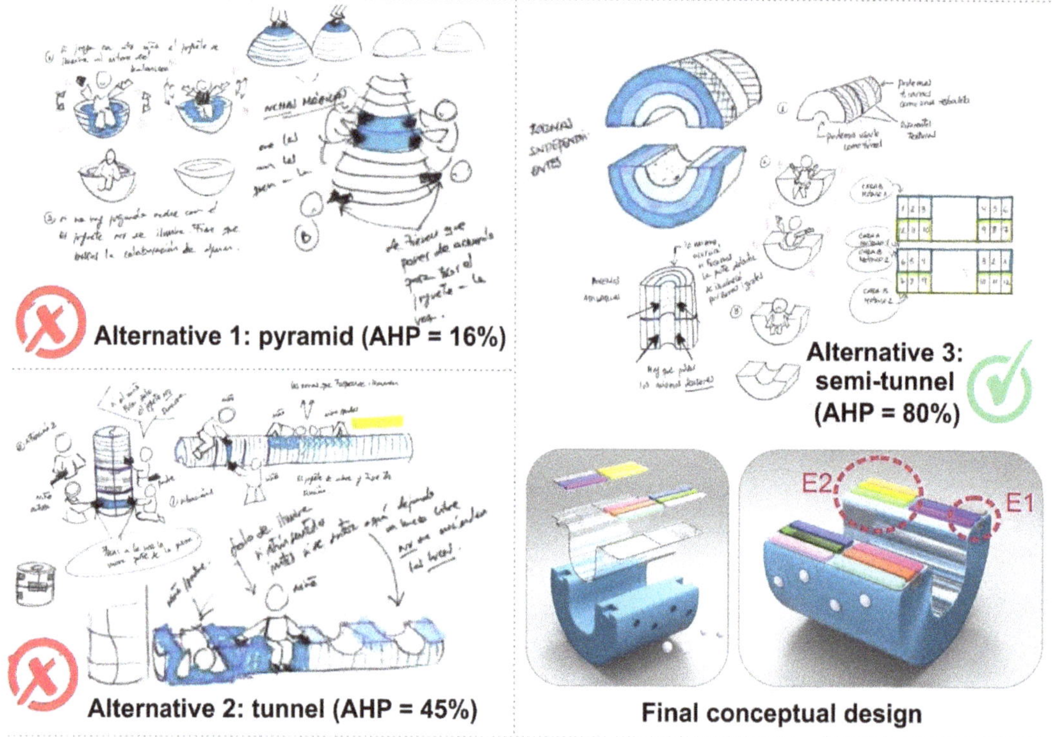

Figure 3. Generation and selection of design alternatives.

As shown in Figure 3, three alternatives were developed that combine *gross motor skills, fine motor skills and sensory stimulation*. Alternative 1 is a pyramidal structure, made up of stackable components; alternative 2 is a modular tunnel that can be used horizontally and vertically; and alternative 3 is a semi-tunnel formed by a single module. All the alternatives have an interactive interface that combines sensory stimuli (motor, tactile, visual and auditory).

Subsequently, the most suitable one was selected with the AHP method (analytic hierarchy process) [58], which assesses each alternative according to the fulfillment of different criteria and sub-criteria. In this case, the AHP hierarchy was structured modeling the evaluation with (1) the design goals 1 to 7, (2) alternatives 1, 2 and 3 for reaching the goals and (3) a group of criteria and sub-criteria that relate the alternatives and goals: (i) suitability for sensory stimulation (somatic, vestibular, visual, auditory, tactile, cognitive, motor and language); (ii) formal design (structure, materials, innovation, environmental impact and integration in space); (iii) safety and comfort (stability, toxicity, anthropometry and biomechanics) and (iv) usability (affordance, visibility, mapping and feedback). After carrying out the pair-wise ranking, calculating the *numerical dominance matrix, obtaining priority concepts and checking the consistency of the priority concepts*, the scope of the design objectives for each alternative was quantified: 16%, 45% and 80% for alternatives 1, 2 and 3, respectively.

Consequently, alternative 3 of the semi-tunnel obtained better results in all the analysis criteria. Specifically, it presents better sensory stimulation by focusing the user's attention on the objective of the game and making the sequence of tasks more intuitive; this is because alternative 3 combines different stimuli (motor, tactile, visual and auditory) in the interface feedback, compared to alternative 1 (visual, motor and auditory) or alternative 2 (visual and tactile). In addition, due to the design of the structure, alternative 3 combines different components with concave and convex shapes, the possibility of turning, interchangeable components and keyboard, allowing to exercise gross and fine motor skills in a more complete way: crawl, climb, swing, squeezing, pressing, pushing, grasping and pinching, compared to alternative 1 (climb, squeezing, pressing and pushing) and alternative 2 (crawl, pressing, pushing and grasping). On the other hand, the distribution of the elements and controllers (interface) of alternative 3 makes the players face the front, a more ideal position for observing and repeating their partner's actions in real time (compared to alternative 1, in a circle, or alternative 2, side to side). In terms of safety and comfort in the use of the product, alternative 3 presented better results. The manipulation of alternatives 1 or 2 is more complicated since to activate the different game modes it is necessary to stack several modules and build pyramidal (alternative 1) or tubular (alternative 2) structures. In contrast, alternative 3 has a single module that integrates all the game modes in the same structure. Although alternatives 1 and 2 are suitable for outdoor environments, the size of alternative 3 is the most suitable for indoor environments.

Despite the positive results compared to alternatives 1 and 2, alternative 3 showed signs of deficiencies in (ii) formal design, specifically, the structure of the basic module was not adequate (see Figure 3, E1) since the keys came into contact with the ground when the product was turned over, which could cause premature deterioration of the keyboard and damage to the circuit if the user left the toy on; and (iv) usability, specifically, in the chromatic selection of the keyboard (Figure 3, see E2) that presented similar colors in keys with close locations, which could generate confusion, errors in the game and therefore increase the frustration of the child [59].

After the final adjustments, the detailed design of alternative 3 was subsequently carried out. To meet the requirements defined in Table 2, this alternative includes four game modes that, as will be explained later in Section 4.3, convert the KEYme into adaptive AT:

- Mode I: Seesaw.
- Mode II: Multikey. Keyboard games with visual and auditory reinforcement, making use of colored lights and musical sounds:
    - Submodality I: sequence and simultaneous pressing game, where one player imitates another by pressing keys of the same color.
    - Submodality II: turn-based sequence and memory game.
    - Submodality III: musical game, each key produces a sound.
- Mode III: Tunnel with visual reinforcement.
- Mode IV: mobility enhanced with stimulation using touch spheres of different textures.

The basic principles of toy design were applied to design game-based assistive technology. These principles guide the design of play sequences that (1) allow the acquisition or exercise of certain skills of the child and (2) are fun, safe, age-appropriate and attractive. For this, all KEYme game sequences are collaborative "aimlessness", between equals, non-sexist and non-violent. In particular, the different primary modes (from I to IV) were configured so that the user experience was linked to (1) promoting cooperation, that is, promoting from an early age the enjoyment of sharing interests in free and equal spaces; (2) replace competitive interaction strategies—rivalry—with cooperation between equals to achieve a common goal (that is, "aimlessness" where the focus is on the process and on the interest in the user's activity and not on a final result of the winner); and (3) develop feelings and affections with the playmate, work on emotions and tackle challenges in a constructive and creative way. Usability and user experience are described in detail in Section 4.3.

To integrate these game modes, the structure of the product is made up of a single module with the components shown in Table 3; these components were designed according to the functional requirements established in the design problem (Tables 1 and 2). Figure 4 shows a simplified exploded view.

Table 3. Product structure.

Component and Number (See Figure 4)	Units	Material	Function
Base module (1)	1	HDPE. PVC-coated foam	Main structure.
Support module (2)	1	HDPE	Covers tunnel wiring; keyboard structural element and trim.
Keys: structure (3) and cover (4)	18 18	PET PET	Mirrored according to color. Structural element for the led strip and push buttons.
Safety cover for battery (5)	1	HDPE	Allows access to battery.
Spheres (6) interchangeable sets according to type of sensitivity	8	Nylon, cotton velvet, beechwood, cork, artificial grass, silicone, leather, felt, polyurethane foam	Tactile stimulation; support elements for climbing and sliding.
Countersunk screw M6, 12	58	Steel	Joining elements.
LED strips: 25 cm (7)	20	-	Light actuators activated with buttons (9); light up: keyboard in mode 2 and sides in sub-game 2.2.
LED strips: 1 m (8)	2	-	Light actuators: they are activated when the product is turned over and when the photoresistor (7) registers a light intensity lower than the set value.
Buttons on keys (9)	18	-	Touch sensors located under the keys (4); activate the LEDs (7) and passive buzzer (12).
On/off button (10)	1	-	Turns the product on and off. Input touch sensor. The ON position activates the rest of the sensors (9, 11, 13, 14).
Game selection switch (11)	1	-	Game selection in mode 2.
Passive buzzer (12)	1	-	Sound actuator. Loudspeaker that plays the tunes on, off and melodies from game mode 2.
Light sensor: fotorresistor (13)	1	-	Analog input light sensor; detects incident light intensity on the surface of the legs; when the product is turned over, it sends a signal to activate the tunnel LEDs (8) in mode 3.
ESP32 board (14)	1	-	Board for programming the toy. Requires programming software.
Stabilizing feet (15)	4	HDPE	Anti-tip and anti-slip stabilizing feet.

Table 3 lists the materials that were selected during the components development according to their suitability to the design of toys [60]. The following technical conditions were taken into account: suitability to create hollow structures, impact resistance, suitability for additive manufacturing, resistance to environmental conditions in outdoor contexts, low toxicity and compatibility with the sensitivity of children with ASD. Additive manufacturing allows fast and economical reproduction of any geometry, custom parts, easy maintenance, short series manufacturing and more environmentally sustainable results (reducing raw material by up to 40% compared to conventional processes). Using 3D printing and open-source electronic platforms allows AT to be affordable and replicable in an agile way; this is a great advantage for users (associations, training centers, therapy centers, etc.), as they have completely personalized and adapted products at their fingertips.

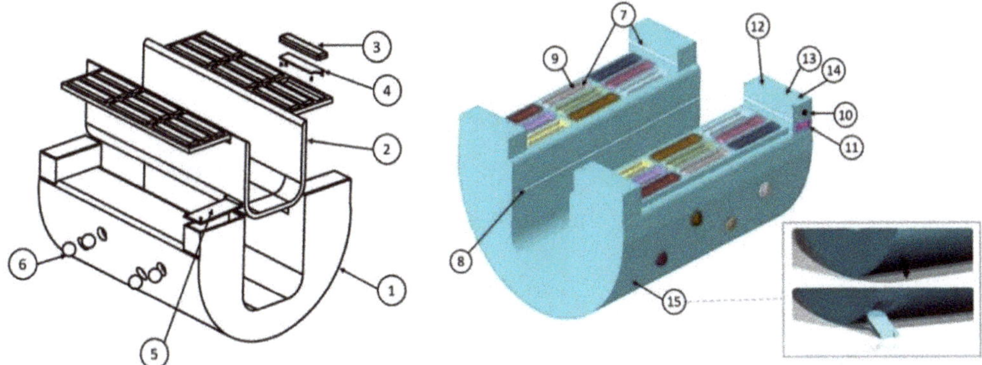

Figure 4. Product Structure.

Finally, KEYme is an interactive smart toy that allows for an autonomous interaction between the user and the product. For the implementation of the functions related to the smart object, the product's structure integrates (i) a set of sensors that capture the child's actions in the game and the information from the environment (Table 3: 9, 10, 11, 13 elements); and (ii) a set of actuators that provide the autonomous response from the product (Table 3: 7, 8, 12 items). The joint action of these elements closes the information loop in user-product interaction.

The dimensioning of the toy was established according to the defined user target: child population aged between 3 and 12 years. The selection of this wide age range is due to the fact that the goal of the project was to create adaptive AT for different users, and therefore one of the strategies used was to organize the game modes by age: (i) modes I, III and IV suitable for younger children (3–7 years), and (ii) mode II adapted to older children (7–12 years). In addition to this, the product as a whole must allow the user to stand and sit while playing, perform movements on it (slide, climb, crawl) and allow for smaller comfortable movements (grasp, press, pinch, squeeze, etc.) with adequate access and visibility of the interface.

With these requirements, an anthropometric study of the child population was carried out, and the product was sized according to each anthropometric variable (Figure 5):

1. Height: the keyboard games are played standing or sitting; the product must allow players to reach the keyboard. The 5th percentile of girls was selected.
2. Hip width, buttock-popliteal length and shoulder width: since the seesaw and tunnel are intended for younger children, maximum measurements were selected for 7 years, in this case, the 95th percentile of girls.
3. Weight: older children must be able to sit or climb on the toy; moreover, the product must also support older playmates. The 50th percentile of weight was selected for 12-year-old children.

4.4. Interface Design, Human-Product Interaction and Usability

The interface design is the main priority to produce high-quality AT. With the functional requirements described in Table 1, the KEYme interface consists of a keyboard made up of two mirror groups of 9 colored keys arranged, an ON and OFF button, a game selection switch and different elements that provide information about the implementation of the game (visual, auditive, tactile and proprioceptive stimuli). These elements improve the user's attention and focus and adapt the environment to the special needs of the child with ASD. Figure 6, as well as Tables 4 and 5, describes in detail the elements of the interface and analyzes the usability of the game modes.

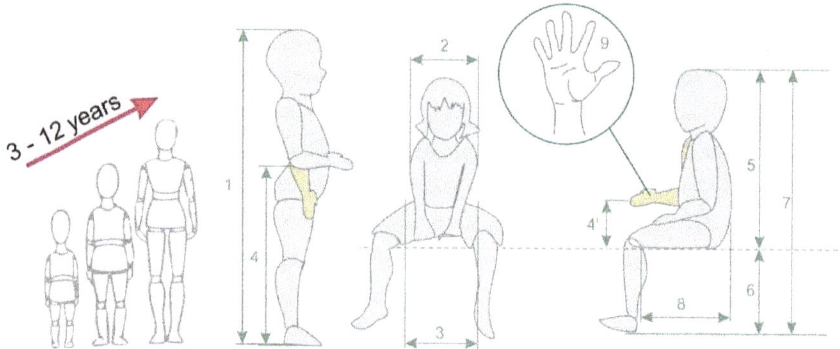

Figure 5. Anthropometric requirements: (**1**) stature (body height); (**2**) shoulder (bideltoid) breadth; (**3**) hip breadth, sitting; (**4**) elbow height, erect and sitting; (**5**) sitting height (erect); (**6**) lower leg length (popliteal height); (**7**) body height, sitting; (**8**) buttock-popliteal length; (**9**) hand measurement (circumferences, breadths, palm length, middle finger length, proximal phalanx length of middle finger, finger grip and hand grasp); and body mass (weight).

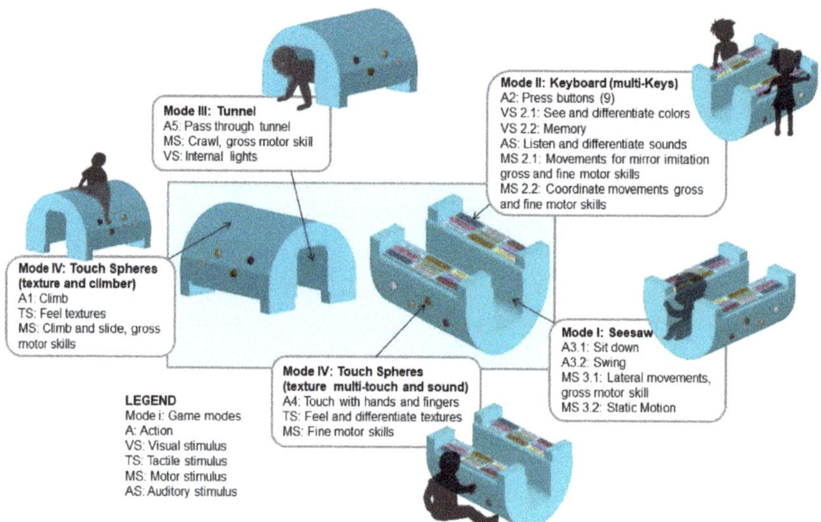

Figure 6. Interface components and usability analysis of the different game modes.

Table 4. Game mode description (part 1).

Mode	Objective
Mode I. Seesaw	The child sits on the module with the keys facing up and swings side to side. This mode will favor gross motor skills.
Mode III. Tunnel	Turning the toy over and pressing the on/off button will turn the inside lights on. As long as the module is in this position, the rest of the modes are automatically off.
Mode IV. Touch Spheres	Two interchangeable sets of spheres according to the type of sensitivity (hypo or hyper) for sensory stimulation and development of fine motor skills. In addition to this, the spheres allow easy climbing and sliding on the base module structure.

Table 5. Game mode description (part 2).

Sub-Game Modes	Interaction with Smart Product
Simultaneous sequence (position 1 on the interface). A child plays the leading role, and the partner must imitate him/her. It enhances the understanding of role changes in the game, among others.	The user selects option 1 on the interface (a). One of the players presses a key (b), while the partner imitates those moves. In case of a hit (both players press the same color and position), both keys will light up and a melody will be played (c); in case of failure, the melody changes, and there will be no visual reinforcement.
Sequence and memory (position 2 on the interface). The two playmates will sequentially press different keys that they must remember in their next turns, creating a sequence of keystrokes that will become more complex as the game moves forward. It helps to identify shared feelings of enjoyment, interest and goals.	The user selects option 2 on the interface (a). The turns are identified by tuning on lights on the side of each player's interface (b). In turns, each playmate presses a key (which activates lighting and a musical sound). In the following turns, each player must remember the sequence and add a new key (c). If the sequence is incorrect, a melody will be played, and the player will have a number of chances to get it right. The game ends with the restart or change of mode.
Musical (position 3 on the interface). Turns the toy into a musical and light keyboard. It favors the ability to improvise, among others.	The user selects option 3 on the interface. The keyboard produces musical sounds and with each keystroke, the key lights up. This mode can be individual or collaborative; there are no turns, and players can use the keyboards simultaneously if they wish.
4 non-slip and anti-tip feet can be set up by releasing the snap-fit. These elements ensure that the structure does not move during mode II.	

Modes I, III and IV work on motor skills with different game sequences (Table 4). Each of them has elements that allow the user to perform various motor tasks: crawl, climb, swing, squeezing, pressing, pushing, grasping, and pinching. The combination of movements with sensory stimulation (visual, auditory and tactile) improves coordination, attention to the task, mood, relaxation and social interaction. The modes are adapted to the needs of sensory seeking and both hyper-sensitivities (over-responsiveness) and hypo-sensitivities (under-responsiveness). Mode I is meant for the user to use the toy as a seesaw, allowing the child to swing to the sides. The players must place the module with their concave part facing upward, like an armchair, to sit down and begin with the lateral movements. Mode III converts the toy into a tunnel. The child can turn the toy upside down so that the lights inside the tunnel light on. In this way, the child can use the toy as a tunnel and crawl inside it. On the other hand, mode III allows you to create a safe space to be alone. Mode IV is meant for the child to interact with the sets of touch spheres. All the spheres are made with different textures, including hard and soft material; the spheres have the function of facilitating the child's climb to the highest part of the module and being able to slide back down.

Mode II is based on imitation and sensory stimulation tasks. It includes 3 submodalities (sequence, memory and musical), which collaboratively and individually work on all the needs described in the analysis of the design problem. Due to the complexity of the game's implementation, it integrates most of KEYme's smart properties. In addition, it is designed to collect interesting data from the gaming experience, such as the number of hits/misses, playing time or sustaining attention.

As stated in Tables 3 and 4, all the modalities make use of visual, proprioceptive and auditory stimuli to improve user-product interaction. Regarding visual stimuli, the following specific considerations related to ASD needs were taken into account: (1) use geometric shapes, (2) use neutral or pastel colors, (3) use striking colors, but not excessively bright (optimal luminosity) and (4) easily distinguishable colors. Figure 7 shows the color selection for the product, as well as the final visual result.

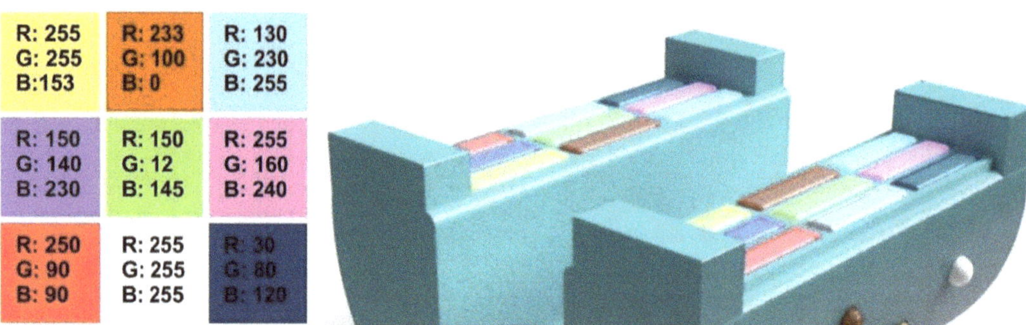

Figure 7. Color Selection.

On the other hand, regarding auditory stimuli, it must be taken into account that the majority of children with ASD present auditory hypersensitivity [61]. This means that, whereas for a typically developing peer the comfort limit is 70–80 dB, for a child with autism it will be 40 dB [62]. Moreover, strident sounds are especially annoying, the most appropriate being the use of musical instruments that emit natural sounds (wood, string or wind) such as piano, violin, flute, guitar and bongos. Lastly, children with ASD have difficulty hearing frequencies of 1–2 kHz; therefore, the notes and frequencies selected ranged from Do_3/C_3 (261.626 Hz) to Do_4/C_4 (523.251 Hz).

Finally, Table 6 summarizes the user-product interaction. The primary and secondary functions (subfunctions) derived from the requirements defined in the analysis of the design problem are listed (Table 1). The functions and the states of the product will be activated or modified with different actions of the user (child with ASD or playmate). The type of user activity will depend on the stage in the game sequence and the objective to be achieved, defining (1) motor actions to change the status of the product or function; (2) retrieval, that is, reading and interpretation of information of the user on the function and status of the product; (3) communication of information to the product by the user for the change of a function or status of the device; (4) checking or verification of a task or a status of the product; and (5) selection of options and alternatives between various functions or product states. All these user activities involve cognitive processing and are omitted for brevity (see legend Table 6).

4.5. Smart Product

According to its characteristics and functionalities, KEYme can be considered (1) an interactive and (2) a smart toy. First, as stated by Moggridge [63], interaction design is concerned with subjective and qualitative values and "would start from the needs and desires of the people who use a product or service, and strive to create designs that would give aesthetic pleasure as well as lasting satisfaction and enjoyment". By taking a closer look at the roles of interactivity in products stated by Clint Heyer, an interactive product focuses on controlling or accessing something unavailable, providing sensuous experience and pleasure and allowing for expression and communication. KEYme project emphasizes this last two roles of interactivity, which are being addressed following dialogue: (1) it works in a conversation-like way as the user and product wait for the next inputs; and (2) implicit styles of interaction: the product reacts to human and other phenomena, for example, the light that the photoresistor detects [64]. Second, as stated by Kara et al. [42], the concept of a "smart toy refers to a technologically developed physical toy constructed with a meaningful purpose". This concept is further explained by Cagiltay et al. in *Smart Toy Based Learning*, in which it is argued that "smart toys are new forms of toys that incorporate tangible objects and electronic chips to provide two-way interactions leading to purposeful tasks with behavioral or cognitive merit" [65] and they allow for an interactive environment to develop cognitive, social and behavioral abilities by means of the toys' dynamic structure. Following this definition, KEYme's different game modalities allow for a very dynamic play that would be mostly focused on improving the child's cognitive, social and behavioral skills. Every game mode is focused on specific needs and is destined for a specific set of skills. Game mode and its consequent actions have a very clear purpose (as explained in Section 4.4). On top of this, these smart toys can be categorized based on the kinds of tasks initiated (behavioral or cognitive tasks), which in the case of KEYme's game modes would be divided into (1) behavioral tasks (Mode I: Seesaw, Mode III: Tunnel and Mode IV: Touch spheres) and (2) cognitive tasks (Mode II: Keyboard games).

Another essential characteristic of smart toys according to Cagiltay is that smart toys should raise the awareness of children about their activities. Building on this statement, KEYme requires the child's full attention to play the different sub-games of the game Mode II. The lights and sounds given by the object help the child understand their performance in the game.

- Submodality I: sequence and simultaneous pressing game, where one player imitates another by pressing keys of the same color: being a game based on empathy, the child must be fully aware of his actions as well as his playmate's to be able to press the same color at the same time.
- Submodality II: turn-based sequence and memory game: the child must again not only be fully aware of his own moves but also his playmate's. The awareness is achieved through memory.
- Submodality III: musical game, each key produces a sound: different sounds are used together with lights responding to keystrokes. This helps the child become aware of their actions.

Table 6. Product functions and user activities relationships.

Product Functions		User Activities
Functions	Subfunctions/Functional Requirement (Table 1)	
Turn toy ON	Toy ON/ ③⑨	Press ON/OFF button
	"Hello" melody sounds ⑦ ⑨	Check if toy is ON
Game selection	Game mode selection	Choose game mode
	Sub-game selection ③⑨	Choose sub-game using the switch
Game Mode I: Seesaw	Allows for sitting ①⑨	Sit down
		Check posture
	Allows for swing ① ② ⑨	Swing
Sub-Game 1: Simultaneous sequence.	Allows for keystrokes ① ③ ④ ⑨	Stand up
		Verify posture
		Decide on roles
		Press key
	Recognizes if players press at the same time ⑤⑥	Imitate playmates' moves
		Check if we are pressing the correct key
	Two keys of the same color light up ⑦⑧⑨⑩	If both players press at the same time
	"Correct" melody sounds ⑦⑧⑨⑩	
	"Miss" melody sounds ⑦⑧⑨⑩	I players don't press at the same time
	Turn LEDs light up ⑤⑥⑦⑨	Check whose turn it is
		Wait for playmate to press a key
Sub-Game 2: Memory game	Allows for Keystrokes ① ③ ④ ⑨	Stand up
		Verify posture
		Press key
	Memorizes the sequence ⑤⑥	Remember sequence
	Key lights up ⑦⑧⑨⑩	If sequence was correctly entered
	"Correct" melody sounds ⑦⑧⑨⑩	
	"Miss" melody sounds ⑦⑧⑨⑩	If sequence was wrongly entered

Table 6. *Cont.*

Product Functions		User Activities
Functions	Subfunctions/Functional Requirement (Table 1)	
Sub-Game 3: Musical	Allows for Keystrokes ① ③ ④ ⑨	Stand up Verify posture Decide which key to press Press key
	Key lights up ⑦ ⑨	
	A music note sounds ⑦ ⑨	
Game Mode III: Tunnel	Allows for turning toy upside down ① ⑨	Turn toy upside down
	Recognizes when the toy is upside down through incident light	Check position
	Tunnel lights up ⑦ ⑨	Check lights
	Allows for crawling inside ① ② ⑨	Verify posture Crawl
	Recognizes when the toy is facing up through incident light	Check position
	Tunnel lights off ⑨	Check lights
Game Mode IV: Touch Spheres	Allows for the child to feel different textures ① ③ ④ ⑦ ⑨	Touch spheres with hands and fingers (squeezing, pressing, pushing and grasping)
	Allows for climbing ① ② ⑨	Climb
	Allows for sliding down ① ② ⑨	Slide down
Turn toy OFF	Turn OFF manually ③	Push ON/OFF button
	Turns off automatically after 15 min	Wait for toy to turn off
	"Goodbye" melody sounds ⑦	Check toy is OFF
LEGEND		
Categorization of User Activities		
Action: include attention, perception, decision making and action		
Checking: include attention, perception and evaluation of own task and product state		
Retrieval information of product: include attention, perception, decode information and interpretation		
Communication (with product): include attention, perception, decision making and data communication		
Selection: include attention, perception and decision making between two or more options		

In smart toys, child–toy interaction is not only important to raise awareness of children about their actions but also in terms of technological components and instructional activities. In this way, KEYme has different modalities of interaction depending on the game mode: (1) *Child-Child interaction*: Mode II—Submodality I and II; (2) *Child-toy interaction*: Mode I; Mode II—Submodality III; Mode III; and Mode IV. It also counts with different types of feedback (light, movement and sound) to guide the child along the games.

According to Cagiltay [65], "the relationship between the characteristics of smart toys and the developmental stages of children needs to be analyzed to develop effective smart toy learning environments". In this way, it is important that smart toys are created according to the different developmental stages in children:

- Sensory-motor stage: In this first stage, simple reflexes, actions and movements are the main activities of children. Language is yet not present, and object permanence is not developed. KEYme's game modes that belong to this stage are Mode I and Mode IV.
- Preoperational stage: Symbolic functioning and language acquisition are the main characteristics. Therefore, cognitive processes are mostly emphasized, though behavioral processes can also play an important role at first. In this stage, creativity and imagination should also be emphasized for children. Using smart toys for collaborative purposes should be implemented from this stage forward. KEYme's game modes that belong to this stage are Mode II—Submodality III and Mode III.
- Concrete operations stage: The concrete operations developmental stage involves children's ability to engage in calculations, rational relations and numerical activities. This is also the stage at which children become capable of classifying objects according to similarities and differences and serializing according to size and weight. KEYme's game modes that belong to this stage are Mode II—Submodality I and II.

Smart toys must therefore also have specific functions making the toy easy to use for play. In this case, KEYme gives feedback for the child to understand how the toy works and how he/she is doing in the game: (1) sounds to express changes of state in the product; (2) lights and sound to express hits and misses; (3) lights to help the child understand the turns; and (4) tunnel lights to reinforce how the toy must be used.

Finally, as stated by Malone and Lepper [66], learning experiences should be intrinsically motivated and, therefore, define toys as objects guided by the internal goal. In their article, they state which are the main dimensions for intrinsic motivation. In this way, Cagiltay [65] goes further, saying that smart toys must explore the different intrinsic motivation components concerning learning experiences:

- Challenge: Activities should challenge learners in order to motivate them intrinsically [66]. Smart toys that provoke behavioral or cognitive tasks provide possibilities for challenging and motivating children. This dimension is mostly explored in the concrete operations stage, more specifically in Mode II—Submodality I, as it requires higher concentration and coordination from the child, and Submodality II, as it is the one that requires more memory.
- Curiosity: It is the most effective component in motivating learners intrinsically. Several smart toys provide open-ended features that allow children to explore new facets of play and increase curiosity. This dimension is explored in all the different developmental stages through the different activities, thus implemented in every game, as all of these modalities allow for a different result every time they are played, depending on how the game develops.
- Control: Activities should give a powerful sense of control to learners to provide a successful learning experience. Some smart toys allow the child to take control of the toy itself to conduct purposeful tasks [66]. This dimension is mostly explored in the preoperational and concrete operations stages more specifically addressed in Mode II—Submodality I and II, in which the games require the user to make decisions and take the leading role.
- Fantasy: According to Cassell and Ryokai, "Fantasy play allows children to explore different possibilities in their life without the risk of failure and frustration from

unexpected events" [67]. This dimension is mostly explored in the sensory-motor and preoperational stages. In this case, it is explored through the multifunctionality of the product and the multiple positions in which it can be placed, as well as through Mode II—Submodality III, in which the child can improvise.

We could therefore argue that KEYme meets all the above objectives, being a self-contained smart toy, as it embeds the interactions within the physical structure, creating an interactive environment of its own. Table 7 summarizes the different characteristics that have been mentioned and how they are implemented in KEYme.

Table 7. Smart functions and properties of KEYme.

Characteristics of Smart Toys		Functions and Characteristics Embedded in KEYme
Purposeful tasks as the main function.		Multifunctional toy. Focused on cognitive, social and behavioral skills.
Tasks can be categorized	Behavioral	Modes I, III, IV.
	Cognitive	Mode II.
Raise awareness about actions.		Submodality I: empathy and imitation as basis of the game; lights and sounds. Submodality II: memory; lights and sounds. Submodality III: lights and sounds.
Interaction modalities	Child-child interaction	Mode II—Submodality I and II.
	Child-toy interaction	Mode I, III, IV. Mode II—Submodality III.
Reinforce instructions with interactivity.		Feedback: lights, movement and sound to guide the child.
Focus on different developmental stages	Sensory-motor stage	Modes I, IV.
	Preoperational stage	Mode II—Submodality III. Mode III.
	Concrete operations stage	Mode II—Submodalities I, II.
Specific functions: easy to use for play.		(1) Sounds to express changes of state in the product. (2) Lights and sound to express hits and misses. (3) Lights to help the child understand the turns. (4) Tunnel lights to reinforce how the toy must be used.
Intrinsic motivation components	Challenge	Mode II—Submodality I and II: concentration and memory.
	Curiosity	All different modalities: different results every time.
	Control	Mode II—Submodality I, II: make decisions, leading roles.
	Fantasy	Multifunctionality: multiple positions.Mode II—Submodality III: improvisation.

Finally, in smart toy play, learning through interaction can be defined as learning several concepts or skills combined with purposeful tasks that are accomplished by interacting with fun technological and instructional components. This statement clearly summarizes the purpose of the KEYme project, which was created with the goal of supporting children with ASD with specific needs in a fun and enjoyable manner using interactive interfaces. Smart toys should be focused on children's learning and cognition, and in this case, the KEYme project is strictly aimed toward the learning process of the child with ASD. The aim is to help children with those skills that can become more difficult. It is worth noting that smart toys provide children with richer experiences by combining physical and digital elements. KEYme is an attractive physical product enhanced with electronic components to achieve interactivity through a reality environment. The different game modes encourage the child to pay attention to both, the physical elements (object, playmate) and the digital elements. In this case, technology is used to reinforce and guide the physical actions of the child.

4.6. Design Verification

The aim of the last phase of the KEYme project development process was to verify the design concept. First, a prototype was manufactured at 1:6 scale that integrated the following scopes: (1) design to validate aesthetic and ergonomic aspects, (2) structural to verify geometric consistency, shape and assemblies between components, (3) functional to verify the detail design and the results of use and interaction and (4) technical to verify the suitability of the smart features of the product and its adequacy to the game modes and interface. The prototype was manufactured using 3D printing, PLA plastic and the Ultimaker Cura 4.5 software, with which adjustments were made to the final design in order to optimize material, time and manufacturing costs. Figure 8 includes some images of the prototype that show details such as the integration of the electronic circuit, the touch spheres or the keys. The next stage, included in the future lines of work, will be focused on the validation of the design. A working full-scale prototype will be tested on a group of target users.

Figure 8. KEYme validation prototype.

As an example, and to illustrate the implementation of the smart properties in the prototype, Figure 9 shows the sub-game 2.1. with the schematics and wiring diagrams for all components. The Arduino IDE software was used to make the programming code for the prototype. Sub-game 2.1. was selected to verify the detailed design and the results of use and interaction in the functional prototype. The developed code was entered on the selected board (ESP32). Since the loudspeaker that was implemented had difficulty reproducing the frequencies of the real design (too low), in the functional prototype all the notes were raised one octave, the following being the frequencies implemented: C [C: 523,251; Mi: 659,255; Sol: 783,991; Do: 1046.50]; C # [C #: 554,365; Fa: 698,456; Sol #: 830,609; C #: 1046.50]; and Re [Re: 587,330; Fa #: 739,989; A: 880,000: Re: 1174.66], only valid for the prototype because, as stated in Section 4.2., musical sounds should not exceed a frequency of 1 KH for users with ASD. On the other hand, the rest of the game modes that include smart properties in user-product interaction were validated in the technical prototype using Autodesk's Tinkercad software.

Finally, the usability of the product was analyzed in order to validate and optimize the user-centered design. After modeling and simulating human interaction with the tools HTA [68], GOMS [69], Sherpa [70], TAFEI [71] and user experience analysis [11], a set of results was obtained that allowed us to implement different improvements in the final design. Table 8 summarizes the obtained results. The usability requirements of the table (column 1, "basis") correspond to the basic analysis criteria taken into account in the simulation and evaluation of user-product interaction:

- Visibility: the interface elements are visible, recognizable and accessible.
- Feedback: real-time synchronization between user and product; adequate information on the current and future sequences of actions.
- Affordance: perceived and actual properties of an object that give clues to its operation.
- Mapping: clear, obvious and intuitive relationships between controllers and their effect on product performance.
- Constraints: the controllers and interface design reduce error by adequately preventing some movements.

- Consistency: interface guides the learning process with similar operation and use of elements for achieving similar tasks.
- Reinforcements: adequacy of interface feedback stimuli.

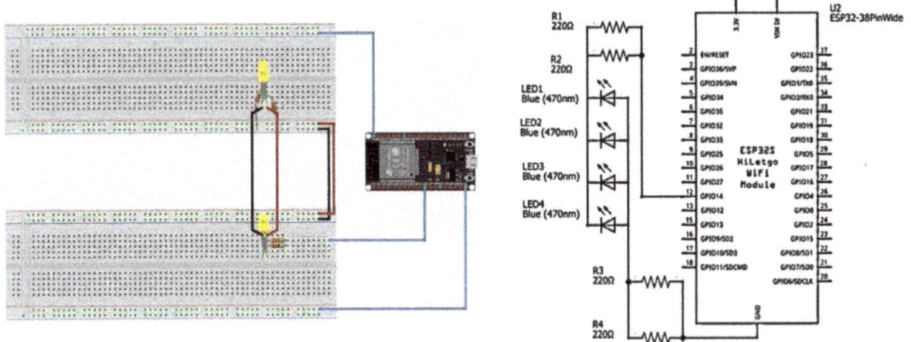

Figure 9. Wiring diagrams for sub-game 2.1.

Table 8. Usability sensitivity analysis and improvements carried out.

Basis	Usability Analysis	Improvements on Final Design
Visibility	All interface elements are visible while the toy is facing up. When turned over, the keys are hidden to prevent them from being pressed in tunnel mode. The design is safe.	
Feedback	It is done with visual, perceptual and proprioceptive stimuli; feedback is correct except for (1) on/off, (2) game turns and (3) game misses.	Add sound and light for power on. LED strip is added as a "shift in turns" indicator. Melodies are added for error identification.
Affordance	All the affordance used has an intuitive meaning that is easy for the user to understand.	
Mapping	(1) There may be errors during turn on and selection of game mode. (2) The action-reaction consistency of musical sounds is not adequate.	(1) A 3-position touch switch is added for game mode II; this is independent of the on/off switch. (2) Pitch rise (musical sound) is set from left to right and bottom to top on the keyboard.
Constraints	Correct. They reduce the probability of human error. The interface helps to focus attention on the tasks and the product with different simultaneous stimuli; it avoids unwanted actions thanks to the design of buttons with mechanical restriction and helps to reduce errors by restricting the type of interactions that the user can carry out.	
Consistency	Aesthetic, functional, internal and external consistency is correct. The grip, movement and stereotype patterns are correct.	
Reinforcements	Correct. Through visual (light and color), auditory (musical sounds) and proprioceptive (textures, movements, grip, pulsation, pressure and force generation) stimulation.	
Working memory use	Game mode II is the most complex and requires user training (sequence, memory and music). As its objective is to improve the capacity of sequential memory (remembering, while imitating and respecting the given order), in this mode short-term memory makes more effort (it stores more information as the game turns progress). For this reason, user-product interaction was reinforced with visual, auditory and proprioceptive stimuli.	
Automatic shutdown	To avoid pressing keys by mistake in mode I, in addition to manual shutdown, the product automatically shuts down as long as no activity is detected.	

5. Discussion

The results of the KEYme are consistent with the development of *a game-based AT* and adaptable to any user and context of use (therapy, family, didactic) that helps children with ASD to develop those skills in which they have difficulties; and, in addition to this, it includes changes, trends and new concepts of toys intended for typically developing children. The properties of multifunctionality, smart features and interactivity allow for different game modes to be included in the same object to cover a wide range of ASD needs grouped in (i) communication and social interaction and (ii) flexibility of thought and behavior.

KEYme is not a toy for individual use; it is a collaborative product to be used as a facilitator of social interaction and to improve cognitive, motor and sensory skills of children with special needs; in particular, seven broad-spectrum needs: (i) playmate-child with ASD relationships, (ii) sensory stimulation, (iii) motor skills, (iv) shared actions related to feelings of enjoyment, interest and common goals, (v) levels of frustration in the game, (vi) emotional and social reciprocity and (vii) learning about changes in game turns.

The balanced implementation of motor, visual, sound and tactile stimuli results in an increase in neuronal activity, which in turn facilitates the achievement of objectives within the aforementioned skills. More specifically, the improvement of psychomotor skills is achieved (referring to the actions and motor skills that involve the movement of the muscles in the body). First, with the development of gross motor skills orienting and guiding the necessary movements (such as crawling, climbing, swaying or maintaining balance) to finish each game sequence; and second, with the development of fine motor skills, thanks to the performance of movements such as *squeezing, pressing, pushing, grasping or pinching (touch spheres)*, all accompanied by visual and auditory stimuli used as reinforcement or positive feedback, which improve hand-eye coordination. It should be noted that the spatial distribution of the interactive functions restricts movements to mirror imitation, which increases neuronal activity four times more compared to anatomical imitation. This type of imitation helps to understand the consequences of each partner's actions, thanks to the experience acquired and the codification of the playmate's intentions.

The improvement of **<u>social behavior</u>** is done through physical contact "child> product> playmate". The motor stimulus has a strong emotional component; it helps to create a bond between playmates increasing the enjoyment of sharing interests. Moreover, during the interaction, the child observes the playmate performing actions in real time; this favors neuronal activity, achieving better results compared to viewing the same actions in videos or images. On the other hand, the frustration of failure during the game is worked through the coordination and synchronization between actions, as well as with the dynamics of the change of roles in mode II, where the two playmates assume an active or passive role in performing imitative tasks (depending on the game turns). In addition to this, sequential memory work contributes to the learning of everyday tasks and independence in daily life. Recalling information while following an order of tasks allows the child to develop skills in everyday tasks (school, didactic and domestic tasks).

Finally, KEYme allows for the interaction between several pairs of players, creating spaces for collaborative play. The structural design of the product makes the result adaptable to different contexts of use. Figure 10 shows the integration of KEYme in indoor (game rooms, home or school) and outdoor environments.

KEYme proposes **inclusive and accessible AT for ASD**. It is an attractive physical product enhanced with electronic components to achieve interactivity through a reality environment. The different game modes encourage the child to pay attention to both the physical elements (object, playmate) and the digital elements. In this case, technology is used to reinforce and guide the physical actions of the child. It is developed with an open-source electronic platform, with versatile and easy-to-use software and hardware. The use of additive manufacturing [72] allows the reproduction of any geometry, the manufacture of completely customized parts in a short period of time and at a low cost, easy product maintenance, as well as more environmentally sustainable results. The integration of additive manufacturing and open-source electronic platforms makes assistive technology affordable, replicable in an agile way and with reduced costs. This is a great advantage for users (associations, families, school environments) since they have at their disposal the possibility of manufacturing fully customized products adapted to different needs.

Figure 10. Integration of KEYme in different contexts of use.

Future developments in this project will focus on transforming the *smart toy* into a *smart toy product-service system* [73]. This involves the KEYme project moving toward interactions with external computers, in order to provide reports of results that would allow for medical monitoring and personalization. The technology used in KEYme makes it possible to configure it for the collection and storage of data, being the most interesting *the number of hits and misses in the game, the reaction time of the child when interacting with his partner or the game time dedicated to each game mode*. The data, collected for further processing, would be sent to therapists and parents for monitoring and observing the child's evolution. On top of this, and to build on the smart characteristics of the design, this information collected and memorized by KEYme could be used for the personalization of the different sub-games, automatically adapting the game's difficulty to the performance of the child. That is, the different game modes would have different levels of play classified by difficulty, using additional feedback (lights, sounds, labels) to help the child in the first stages and removing these reinforcement stimuli as he or she improves those skills. Some of the personalization elements that will be considered in future work are defined in Table 9. Furthermore, the data collected could be used to analyze the user experience and draw conclusions about the child's motivation, attention and interest in the toy, relevant information for future redesigns, integration of new game modes and improvements in the product.

The second major line of future work focuses on the development of a framework for the interactive and intelligent design of assistive technology for children with ASD. This framework will allow for the design of products adapted to the specific needs and problems of each disorder (according to the DSM-V classification [1]). It will take into account the level of neuronal activity (mild, moderate or severe) that different stimuli and imitation actions provoke, which will be directly linked to the design variables to be used in the development of the concept. Directly relating the needs of the child and the design variables (related to sensory activation stimuli) helps in the interpretation of the needs of each disorder and their translation into functional requirements and smart properties of assistive technology.

Table 9. Examples of custom elements.

Game Mode	First Stages	Later Stages
General Mode II	Colors are labeled.	Labels are removed.
Mode II, submodality 1: Simultaneous sequence	KEYme chooses a color by lighting up a key that both players must press at the same time.	Players decide the roles they want to take and the color that is going to be pressed at the same time.
Mode II, submodality 2: Sequence and memory	KEYme helps the players to memorize the sequence using light cues.	Players must memorize the sequence without any cues.
Mode III: Luminous tunnel with music touch wall	KEYme detects if the user is moving or still.	If the user stops, KEYme activates calming sounds and multi-colored LED lights when he/she presses the walls. Mode III becomes a luminous tunnel with a musical touch wall.

Finally, the KEYme project contributes to the improvement of products and their accessibility in the field of human development and social sustainability (equity). It works within the scope of childhood and disability, specifically, the wide range of signs and symptoms of assistive technology ASD. The development of quality assistive technology, its universal design and its accessibility (sufficient supply and price) is one of the main research challenges identified by the World Health Organization [25]. This project involves the knowledge transfer from the latest neuroscience, medicine and psychology contributions to the engineering and industrial design field. In addition to this, it takes advantage of new trends in accessible toys and resources (such as additive manufacturing and open-source electronic platforms) to create products available to everyone.

6. Conclusions

This research is an open proposal for the design of products developed as game-based assistive technology. This article presents the design of KEYme, a single and multifunctional module, although the project, the proposed methodology and the design strategy are open to the development of new products, with the aim of contributing to the challenge and need to transform more inclusive, innovative and innovative societies. reflective. Introducing new emerging technologies (related to open source electronic prototypes to create interactive objects, rapid prototyping techniques and 3D printing) into everyday products is of great interest. These solutions can improve the health and social well-being of currently underprivileged groups, helping engineering advance the development of science with and for society. KEYme also aims to drive a shift in the toy industry toward a more inclusive approach; it is important to expand its platforms and portfolios with adapted products that follow the same trends demanded by typically developing children, but integrating the resolution and satisfaction of the needs of special populations.

Author Contributions: Conceptualization, R.C., S.L. and M.E.P.; formal analysis, R.C.; investigation, R.C. and M.E.P.; methodology, R.C., S.L. and M.E.P.; software, R.C.; supervision, M.E.P.; validation, M.E.P.; writing—original draft, R.C. and M.E.P.; writing—review and editing, M.E.P. All authors have read and agreed to the published version of the manuscript.

Funding: This research received no external funding.

Institutional Review Board Statement: Not applicable.

Informed Consent Statement: Not applicable.

Data Availability Statement: Data sharing not applicable.

Conflicts of Interest: The authors declare no conflict of interest.

References

1. American Psychiatric Association. *Diagnostic and Statistical Manual of Mental Disorders: DSM-5*; American Psychiatric Association: Washington, DC, USA, 2013.
2. Berkell Zager, D.; Cihak, D.F.; Stone-MacDonald, A. *Autism Spectrum Disorders: Identification, Education, and Treatment*; Routledge: London, UK, 2004.
3. Matson, J.L. *Applied Behavior Analysis for Children with Autism Spectrum Disorders*; Springer: Berlin/Heidelberg, Germany, 2009.
4. World Health Organization. Autism Spectrum Disorders. Available online: https://www.who.int/news-room/fact-sheets/detail/autism-spectrum-disorders (accessed on 10 December 2020).
5. Hernández, A.; Peralta, M.E. *Market. Research for Special Needs Products*; University of Seville: Seville, Spain, 2020.
6. Ricks, D.J.; Colton, M.B. Trends and considerations in robot-assisted autism therapy. In Proceedings of the 2010 IEEE International Conference on Robotics and Automation, Anchorage, AL, USA, 3–7 May 2010; pp. 4354–4359. [CrossRef]
7. Bradley, R.; Newbutt, N.; Bradley, R.; Newbutt, N. Autism and virtual reality head-mounted displays: A state of the art systematic review. *J. Enabling Technol.* **2018**, *12*, 101–113. Available online: http://eprints.uwe.ac.uk/35706/ (accessed on 16 May 2018). [CrossRef]
8. Brown Lofland, K. *The Use of Technology in Treatment of Autism Spectrum Disorders*; Indiana Resource Center for Autism: Bloomington, IN, USA, 2019.
9. Chia, G.L.C.; Anderson, A.; McLean, L.A. Use of Technology to Support Self-Management in Individuals with Autism: Systematic Review. *Rev. J. Autism Dev. Disord.* **2018**, *5*, 142–155. [CrossRef]
10. Goldsmith, T.R.; LeBlanc, L.A. Use of technology in interventions for children with autism. *J. Early Intensive Behav. Interv.* **2004**, *1*, 166–178. [CrossRef]
11. Moore, D.; Yufang Cheng, Y.; McGrath, P.; Powell, N.J. Collaborative Virtual Environment Technology for People With Autism. *Focus Autism Other Dev. Disabl.* **2005**, *20*, 231–243. [CrossRef]
12. Sherer, M.; Pierce, K.L.; Paredes, S.; Kisacky, K.L.; Ingersoll, B.; Schreibman, L. Enhancing Conversation Skills in Children with Autism Via Video Technology. *Behav. Modif.* **2001**, *25*, 140–158. [CrossRef]
13. World Health Organization. Global Priority Research Agenda for Improving Access to High-Quality Affordable Assistive Technology. Available online: https://apps.who.int/iris/handle/10665/254660 (accessed on 30 September 2020).
14. Currenti, S.A. Understanding and Determining the Etiology of Autism. *Cell Mol. Neurobiol.* **2010**, *30*, 161–171. [CrossRef]
15. Solunion. Radiografía y Tendencias de la Industria Juguetera Para 2019. Solunion. Available online: https://www.solunion.es/blog/radiografia-tendencias-industria-juguetera-para-2019/ (accessed on 4 January 2021).
16. Euromonitor International. *Toys and Games in Switzerland*; Euromonitor International: London, UK, 2020.
17. Calleja, P. Los Fabricantes de Juguetes se Frotan las Manos. EL PAÍS. 2019. Available online: https://elpais.com/economia/2019/12/18/actualidad/1576669686_782345.html (accessed on 15 September 2020).
18. Bernal, I. Las Exportaciones Españolas de Juguetes Crecen un 6.8% en un Mercado Dominado por Lego. El Correo. 2019. Available online: https://www.elcorreo.com/economia/tu-economia/exportaciones-espanolas-juguetes-20191227181415-nt.html (accessed on 15 September 2020).
19. Spielwarenmesse: Exhibition Center, Halls & Products. 2020. Available online: https://www.spielwarenmesse.de/en/fair/products-halls (accessed on 16 September 2020).
20. Syriopoulou-Delli, C.K.; Gkiolnta, E. Review of assistive technology in the training of children with autism spectrum disorders. *Int. J. Dev. Disabil.* **2020**. [CrossRef]
21. Wainer, J.; Dautenhahn, K.; Robins, B.; Amirabdollahian, F. A Pilot Study with a Novel Setup for Collaborative Play of the Humanoid Robot KASPAR with Children with Autism. *Int. J. Soc. Robot.* **2014**, *6*, 45–65. [CrossRef]
22. Autonomous Systems Lab and Ecole Polytechnique Fédérale de Lausanne. *Robot. Gossip: Robots to Reach Autistic Kids*; Robot Gossip: New York, NY, USA, 2006.
23. Clabaugh, C.; Mahajan, K.; Jain, S.; Pakkar, R.; Becerra, D.; Shi, Z.; Deng, E.; Lee, R.; Ragusa, G.; Matarić, M. Long-Term Personalization of an In-Home Socially Assistive Robot for Children With Autism Spectrum Disorders. *Front. Robot. AI* **2019**, *6*, 110. [CrossRef] [PubMed]
24. Kozima, H.; Marek, P.; Nakagawa, C.; Michalowski, M.P.; Keepon, A. Playful Robot for Research, Therapy, and Entertainment. *Int. J. Soc. Robot.* **2009**, *1*, 3–18. [CrossRef]
25. NAO Autism Pack. 2020. Available online: https://www.robotlab.com/store/robotlab-nao-autism-pack (accessed on 15 September 2020).
26. De Toldi, L. Three years on: An Update from Leka, Robot Launch Winner. *Robohub*, 15 March 2017.
27. Waltz, E. Therapy Robot Teaches Social Skills to Children With Autism. IEEE Spectrum 2018. Available online: https://spectrum.ieee.org/the-human-os/biomedical/devices/robot-therapy-for-autism (accessed on 23 February 2021).
28. Iwarsson, S.; Stahl, A. Accessibility, usability and universal design—Positioning and definition of concepts describing person-environment relationships. *Disabil. Rehabil.* **2003**, *25*, 57–66. [CrossRef]
29. Chew, S. *An Approach to Design with People Who Have Special Needs*; Springer: Berlin, Heidelberg, Germany, 2013; pp. 221–225.
30. Benton, L.; Johnson, H.; Ashwin, E.; Brosnan, M.; Grawemeyer, B. Developing IDEAS: Supporting children with autism within a participatory design team. In Proceedings of the 2012 ACM Annual Conference on Human Factors in Computing Systems CHI'12, Austin, TX, USA, 5–10 May 2012; p. 2599. [CrossRef]

31. Fails, J.A.; Guha, M.L.; Druin, A. Methods and Techniques for Involving Children in the Design of New Technology for Children. *Found. Trends Human Computer Interact.* **2012**, *6*, 85–166. [CrossRef]

32. Khare, R.; Mullick, A. Educational spaces for children with autism; design development process. In Proceedings of the Building Comfortable and Liveable Environments for All, International Meeting 'Education And Training', Atlanta, GA, USA, 15–16 May 2008; pp. 66–75.

33. Millen, L.; Cobb, S.; Patel, H. A method for involving children with autism in design. In Proceedings of the 10th International Conference on Interaction Design and Children—IDC'11, Ann Arbor, MI, USA, 19–23 June 2011; pp. 185–188. [CrossRef]

34. Williams, J.H.G. Self–other relations in social development and autism: Multiple roles for mirror neurons and other brain bases. *Autism Res.* **2008**, *1*, 73–90. [CrossRef] [PubMed]

35. Abascal, J.; Nicolle, C. The Application of USERfit Methodology to Teach Usability Guidelines. In *Tools for Working with Guidelines*; Springer: London, UK, 2001; pp. 209–216.

36. Chambers, M.; Connor, S.L. User-friendly technology to help family carers cope. *J. Adv. Nurs.* **2002**, *40*, 568–577. [CrossRef] [PubMed]

37. Lannen, T.; Brown, D.J.; Standen, P.J. Design of virtual environment input devices for people with moderate to severe learning difficulties-a user-centred approach. In Proceedings of the International Conference on Disability, Virtual Reality and Associated Technologies, Veszprém, Hungary, 18–20 September 2002.

38. Porrero, P.; de la Bellacasa, P. *Assistive Technology: From Virtuality to Reality*; IOS Press: Amsterdam, The Netherlands, 1995.

39. Clarkson, J. *Inclusive Design: Design for the Whole Population*; Springer: Berlin/Heidelberg, Germany, 2003.

40. Common Ground Publishing. *Design Principles and Practices: An International Journal*; Common Ground Research Networks: Champaign, IL, USA, 2007.

41. Mccrickard, D.S.; Abel, T.D.; Scarpa, A.; Wang, Y.; Niu, S. Collaborative Design for Young Children with Autism: Design Tools and a User Study. In Proceedings of the 2015 International Conference on Collaboration Technologies and Systems (CTS), Atlanta, GA, USA, 1–5 June 2015; pp. 175–182. [CrossRef]

42. Kara, N.; Cagiltay, K. Smart toys for preschool children: A design and development research. *Electron. Commer. Res. Appl.* **2020**, *39*, 100909. [CrossRef]

43. Alves, F.J.; De Carvalho, E.A.; Aguilar, J.; De Brito, L.L.; Bastos, G.S. Applied Behavior Analysis for the Treatment of Autism: A Systematic Review of Assistive Technologies. *IEEE Access* **2020**, *8*, 118664–118672. [CrossRef]

44. Rogers, S.J.; Hayden, D.; Hepburn, S.; Charlifue-Smith, R.; Hall, T.; Hayes, A. Teaching Young Nonverbal Children with Autism Useful Speech: A Pilot Study of the Denver Model and PROMPT Interventions. *J. Autism Dev. Disord.* **2006**, *36*, 1007–1024. [CrossRef] [PubMed]

45. Smith, I.M.; Flanagan, H.E.; Garon, N.; Bryson, S.E. Effectiveness of Community-Based Early Intervention Based on Pivotal Response Treatment. *J. Autism Dev. Disord.* **2015**, *45*, 1858–1872. [CrossRef]

46. Case-Smith, J.; Weaver, L.L.; Fristad, M.A. A systematic review of sensory processing interventions for children with autism spectrum disorders. *Autism* **2015**, *19*, 133–148. [CrossRef] [PubMed]

47. Downey, R.; Rapport, M.J.K. Motor Activity in Children With Autism. *Pediatr. Phys. Ther.* **2012**, *24*, 2–20. [CrossRef]

48. Dawson, G.; Watling, R. Interventions to Facilitate Auditory, Visual, and Motor Integration in Autism: A Review of the Evidence. *J. Autism Dev. Disord.* **2000**, *30*, 415–421. [CrossRef]

49. Bharathi, G.; Venugopal, A.; Vellingiri, B. Music therapy as a therapeutic tool in improving the social skills of autistic children. *Egypt. J. Neurol. Psychiatry Neurosurg.* **2019**, *55*, 1–6. [CrossRef]

50. Dautenhahn, K.; Werry, I. Towards interactive robots in autism therapy: Background, motivation and challenges. *Pragmat. Cogn.* **2004**, *12*, 1–35. [CrossRef]

51. Rizzolatti, G.; Fabbri-Destro, M. Mirror neurons: From discovery to autism. *Exp. Brain Res.* **2010**, *200*, 223–237. [CrossRef]

52. Vivanti, G.; Rogers, S.J. Autism and the mirror neuron system: Insights from learning and teaching. *Philos. Trans. R. Soc. B Biol. Sci.* **2014**, *369*, 20130184. [CrossRef]

53. Schulte-Rüther, M.; Markowitsch, H.J.; Fink, G.R.; Piefke, M. Mirror Neuron and Theory of Mind Mechanisms Involved in Face-to-Face Interactions: A Functional Magnetic Resonance Imaging Approach to Empathy. *J. Cogn. Neurosci.* **2007**, *19*, 1354–1372. [CrossRef] [PubMed]

54. Hamilton, A.F.D.C.; Brindley, R.M.; Frith, U. Imitation and action understanding in autistic spectrum disorders: How valid is the hypothesis of a deficit in the mirror neuron system? *Neuropsychologia* **2007**, *45*, 1859–1868. [CrossRef] [PubMed]

55. Iacoboni, M.; Dapretto, M. The mirror neuron system and the consequences of its dysfunction. *Nat. Rev. Neurosci.* **2006**, *7*, 942–951. [CrossRef]

56. Hazelrigg, G.A. Validation of engineering design alternative selection methods. *Eng. Optim.* **2003**, *35*, 103–120. [CrossRef]

57. Bouchereau, V.; Rowlands, H. Methods and techniques to help quality function deployment (QFD). *Benchmarking Int. J.* **2000**, *7*, 8–20. [CrossRef]

58. Ishizaka, A.; Nemery, P. Analytic hierarchy process. In *Multi-Criteria Decision Analysis*; John Wiley & Sons Ltd.: Chichester, UK, 2013; pp. 11–58.

59. Peralta, M.E.; Lopez, S.; Aguayo, F.; Lama, J. *Neuro-Juguete Orientado a la Satisfacción de las Necesidades de Usuarios con Discapacidad*; University of Seville: Seville, Spain, 2017.

60. Grynkiewicz-Bylina, B. *Designing, Prototyping and Manufacture of Safe. Toys Made of Plastics*; Oficyna Wydawnicza Polskiego Towarzystwa Zarządzania Produkcją: Opole, Poland, 2012.
61. Nieves Bouza, A.; Carreiro Prieto, P.; Arceo Tourís, M. Comprendo mi Entorno. 2020. Available online: http://www.autismo.org.es/sites/default/files/comprendo_mi_entorno._manual_de_accesibilidad_cognitiva_para_personas_con_tea.pdf (accessed on 15 October 2020).
62. Khalfa, S.; Bruneau, N.; Rogé, B.; Georgieff, N.; Veuillet, E.; Adrien, J.-L.; Barthélémy, C.; Collet, L. Increased perception of loudness in autism. *Hear. Res.* **2004**, *198*, 87–92. [CrossRef]
63. Moggridge, B. *Designing Interactions*; Footprint Books: Warriewood, Australia, 2007.
64. Heyer, C. Lecture: What Is Interaction? 2020. Available online: http://clintio.us/papers/ (accessed on 15 November 2020).
65. Cagiltay, K.; Kara, N.; Cigdem, C. *Smart Toy Based Learning: Handbook of Research on Educational Communications and Technology*, 4th ed.; Springer: Berlin/Heidelberg, Germany, 2014; pp. 1–1005. [CrossRef]
66. Malone, T.W.; Lepper, M.R. Making Learning Fun: A Taxonomy of Intrinsic Motivations for Learning. In *Aptitude, Learning, and Instruction. Volume 3: Conative and Affective Process Analyses*; Lawrence Erlbaum Associates: London, UK; Hillsdale, NJ, USA, 1987; pp. 223–253.
67. Cassell, J.; Ryokai, K. Making space for voice: Technologies to support children's fantasy and storytelling. *Pers. Ubiquitous Comput.* **2001**, *5*, 169–190. [CrossRef]
68. Stanton, N.A. Hierarchical task analysis: Developments, applications, and extensions. *Appl. Ergon.* **2006**, *37*, 55–79. [CrossRef]
69. Kieras, D. A Guide to GOMS Model Usability Evaluation using NGOMSL. In *Handbook of Human-Computer Interaction*; Elsevier: Amsterdam, The Netherlands, 1997; pp. 733–766.
70. Harris, D.; Stanton, N.A.; Marshall, A.; Young, M.S.; Demagalski, J.; Salmon, P. Using SHERPA to Predict Design-Induced Error on the Flight Deck. *Aerosp. Sci. Technol.* **2005**, *9*, 525–532. [CrossRef]
71. Barber, C.; Stanton, N.A. Task analysis for error identification: A methodology for designing error-tolerant consumer products. *Ergonomics* **1994**, *37*, 1923–1941. [CrossRef]
72. Carfagni, M.; Fiorineschi, L.; Furferi, R.; Governi, L.; Rotini, F. The role of additive technologies in the prototyping issues of design. *Rapid Prototyp. J.* **2018**, *24*, 1101–1116. [CrossRef]
73. Valencia, A.; Mugge, R.; Schoormans, J.; Schifferstein, H. The Design of Smart Product-Service Systems (PSSs): An Exploration of Design Characteristics. *Int. J. Des.* **2015**, *1*, 13–28.

Article

Experimental and Numerical Study on the Flexural Performance of Assembled Steel-Wood Composite Slab

Guodong Li, Zhibin Liu, Wenjia Tang, Dongpo He and Wei Shan *

School of Civil Engineering, Northeast Forestry University, Harbin 150040, China; ldlgd@163.com (G.L.); liu_zhibin2020@163.com (Z.L.); tang_wenjia@163.com (W.T.); dongpo_nefu@163.com (D.H.)
* Correspondence: shanwei1456@163.com; Tel.: +86-155-6182-1696

Abstract: This paper presents research on a new type of fabricated steel–wood composite floor material in the style of a slab-embedded beam flange, using test methods and finite element numerical analysis to study the flexural load-bearing performance of the composite slabs. Through experimental phenomena, the failure process and mechanism of the composite floor are analyzed, and the deformation performance and ultimate bearing capacity of the composite floor material are assessed. Through numerical analysis of the finite element model, the influence of the connection mode of the floor and the composite beam, the type and number of connectors, and the width of the flange of the composite beam on the bending performance of the composite beam–slab system is studied. The research results show that the fabricated steel–wood composite floor slab has good load-bearing and deformation performance. The self-tapping screw connection of the floor slab is better than the ordinary steel nail connection, and the reasonable screw spacing is 100–150 mm. Increasing the flange width of the composite beam can significantly improve the load-bearing capacity of the steel–wood composite floor component.

Keywords: composite structure; steel-wood combination; experimental research; finite element analysis

Citation: Li, G.; Liu, Z.; Tang, W.; He, D.; Shan, W. Experimental and Numerical Study on the Flexural Performance of Assembled Steel-Wood Composite Slab. *Sustainability* **2021**, *13*, 3814. https://doi.org/10.3390/su13073814

Academic Editor: Yadir Torres Hernández

Received: 1 February 2021
Accepted: 25 March 2021
Published: 30 March 2021

Publisher's Note: MDPI stays neutral with regard to jurisdictional claims in published maps and institutional affiliations.

1. Introduction

The production process of traditional building materials consumes a lot of energy and causes serious pollution problems. The development and promotion of environmentally friendly and energy-saving building materials are of great significance to the sustainable development of human society. Compared to other materials, wood has the advantages of lightweight, high strength, easy processing, low carbon, environmental protection, etc. It is a typical green building material and is favorably used in residential buildings.

Assembly timber structure refers to a structure constructed by factory-prefabricated wooden structure components and parts and using on-site assembly as the main method [1,2]. Cold-formed thin-walled steel is lightweight steel that is bent in a cold state to obtain a variety of economical cross-sections. Combining wood and cold-formed thin-walled steel in a certain way can form steel-wood composite components with various cross-sectional forms, which can not only obtain a larger section radius of gyration and moment of inertia and improve the rigidity and load-bearing capacity, but also, with the help of wood's restraint on thin-walled steel, the problem of premature buckling of thin-walled steel can be effectively solved, and the good effect of saving steel and reducing cost can be realized, which reflects the superiority in a new high-performance combination structure [3].

Several studies have recently conducted research on steel–wood composite structures. Loss and Buick Davidson [4] conducted an experimental study on the failure modes of composite slabs under horizontal loads. The results show that the main deformation of the slab is concentrated at the beam–beam joints and the steel–wood composite material has good elastic properties without fracture. Xia Yonghui [5] studied the influence of steel plate thickness and other factors on the natural vibration and bending characteristics of composite slabs. The results show that increasing the thickness of steel and plywood can significantly improve the load-bearing capacity and vibration serviceability of composite panels. The test results of Li Yushun [6] show that the overall working performance of the profiled steel plate–bamboo plywood composite slab is excellent; the bamboo plywood board and the steel plate have a good combined effect and have higher load-bearing capacity and rigidity. Yang Yue [7] studied the flexural performance and failure mode of a steel–concrete composite slab and gave the calculation formula for the flexural bearing capacity. L. Xu and F.M. Tangorra [8] carried out a recent study on the vibration characteristics of cold-formed steel-supported lightweight residential floor systems. On-site tests were also conducted to evaluate the actual vibration performance of the cold-formed steel-supported lightweight residential floor systems. Zhou Xuhong [9] conducted a monotonic static load test on the bearing capacity of a thin-walled steel beam-oriented strand board (OSB) composite slab. The results show that the cold-formed steel joists-OSB composite floor has higher bearing capacity and lower deformation; reducing the screw spacing can improve the bearing capacity of the composite slab. The abovementioned scientific research and [10–13] have detailed beneficial explorations and obtained valuable conclusions on the combination methods, materials and bending performance of steel–wood composite structures.

The assembly and construction process of multi-story and high-rise wooden structure houses is column (wall) → beam → slab. The size of the floor lap and installation interface provided by ordinary rectangular cross-section wooden beams is small, and the assembly quality of the slab and beam is not easy to control. Under the action of large horizontal load, the shear force of the connecting parts (screws, bolts, etc.) between the plate and the beam is relatively large, which is likely to cause the failure of the force transmission of the beam and plate. This paper proposes a beam–slab system in the form of slab-embedded beams. The floor slab can be directly embedded into the board grid formed by prefabricated beams during assembly. The accuracy of installation is high, and the limiting effect of the beam on the floor can reduce the stress level of the assembly parts and provide the reliability of the beam–slab system. In the system, the prefabricated beam is a thin-walled steel–glulam composite beam, which can greatly reduce the weight of the beam. The lower part of the beam is a double "C"-shaped cold-formed thin-walled steel composite section, and the upper part adopts a special shaped cross-section with a middle protrusion (the height of the protrusion is the same as the thickness of the floor), which can form a groove between the beams during the installation of the floor. The wide and thick flanges of the composite beams provide sufficient space for the assembly and connection of the floor slabs and ensure good assembly performance of the floor slab and the composite beams. This form of beam can be adapted to medium- and small-span OSB floor slabs and large-span cross-laminated timber plate.

At present, there is no research on steel-wood composite beam–slab structure systems in the literature, and the performance of beam–slab collaboration in fabricated beam–slab systems is not clear. This paper uses the methods of experiments and finite element numerical analysis to study the assembly performance of composite beams and slabs; the flexural load-bearing capacity and the failure mode of the special shaped cross-section steel-wood composite beam and steel–wood composite slabs composed of composite beams and floor slabs. Moreover, this paper uses the finite element analysis method to explore the key factors affecting the bearing capacity of the steel-wood composite floor.

2. Test Overview

2.1. Specimen Design

The multi-span continuous slab–beam system in the project is shown in Figure 1a. Due to the limitation of the test loading equipment, the test piece shown in Figure 1a cannot be loaded in a laboratory. In the experimental research, the test model is correspondingly simplified. The simplified model shown in Figure 1b can characterize the stress state of the middle section of the continuous floor; therefore, the design and processing of the test model have been carried out.

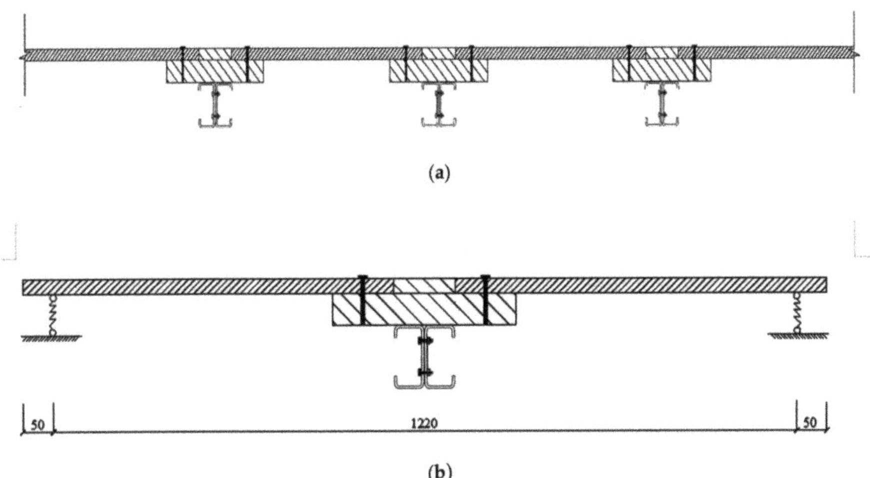

Figure 1. Schematic diagram of multi-span continuous plate–beam system and simplified model. (**a**) 1-1 Multi-span continuous beam-and-slab system; (**b**) simplified model (mm).

The cross-section design of the steel–wood composite beam used in the test and analysis is shown in Figure 2. The outer profile size of the lower double "C"-shaped cold-formed thin-walled steel composite cross-section was 100 × 100 mm, and the upper laminated material was 100 mm plywood (single layer thickness is 25 mm; 3 layers in total); the size of the partially protruding glulam (glued laminated timber) was 100 mm × 25 mm and the size of the extended flanges on both sides was 100 mm × 50 mm. The wooden floor choice was a 25-mm-thick OSB.

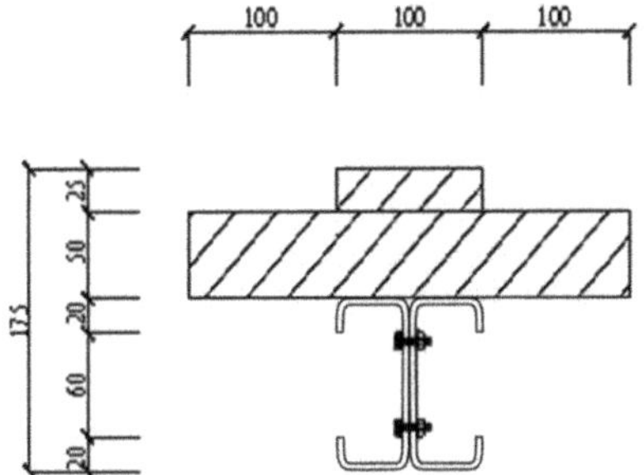

Figure 2. Schematic diagram of steel–wood composite beam section (mm).

The two composed sets of experimental samples were processed and made by professional factories; the dimensions of the parts are shown in Table 1. The two groups of specimens were named ZHLB-1 and ZHLB-2. Between them, ZHLB-1 utilized iron nails with a length of 80mm as the fixing parts of the slab on the flange of the composite beam; ZHLB-2 used 80mm self-tapping screws produced by Shanghai Meigu Chengfan Fasteners Co., Ltd. as the fixing parts of the slab on the flange of the laminated beam. The grouping of test pieces is shown in Table 1.

Table 1. Grouping table of composite floor test pieces. OSB—oriented strand board.

Specimen	Glulam Section Size Length × Width × Height (mm)	Steel Beam Size (mm)	OSB Size (mm)	Connection between Plate and Beam	Superimposed Beam Cross-Section Diagram
ZHLB-1	2500 × 300 × 75	2500 mm long C100 × 50 × 20 × 2.4	2440 × 610 × 25	80 mm nails 80 mm self-tapping screws	Picture 3
ZHLB-2					

To implement the test and meet the boundary conditions of vertical free deformation of the unconnected section of the floor slab in Figure 3b, an elasticity support composed of a rectangular extruded foam board and glulam with a cross-sectional size of 100 mm × 50 mm was set in the test. During the test, the elastic support did not produce an excessive reaction force to the floor slab that hinders the loading of the floor slab. The test pieces of the composite floor are shown in Figure 3.

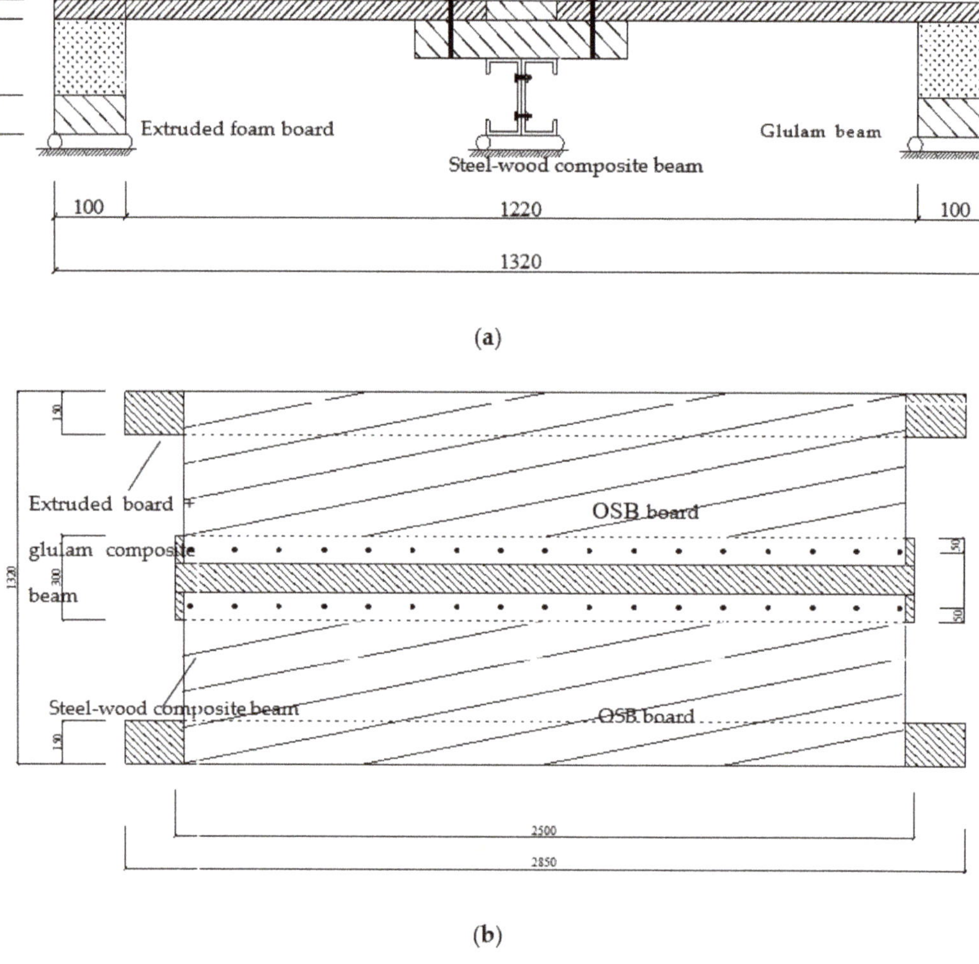

Figure 3. (**a**) 1-1 section picture of the prefabricated steel–wood composite floor test piece; (**b**) structure drawing of the composite floor test piece (mm).

According to the calculation of "Wood Structure Design Standard" (GB/T50005-2017) [14], the distance between the connecting pieces was 150mm; the size and structure of the test pieces are shown in Figure 3b.

2.2. Material Properties Text

In accordance with the "Test Method for Tensile Strength of Wood Along the Grain" (GB/T 1938-2009) and the "Metallic Material Tensile Test-Part 1 Room Temperature Test Method" (GB/T 228.1-2010), the wood and steel used to make the test pieces were tested. According to the along-grain compression test of 15 prismatic glulam specimens (Figure 4a), the along-grain compressive strength of the glulam was 50.10 MPa, and the compressive elastic modulus of the glulam was 10,722.13 MPa. According to the along-grain tensile test of the glulam specimens (Figure 4b), the tensile strength along the grain was 74.61 MPa, and the tensile elastic modulus was 9912.00 MPa. According to the material property test of the steel (Figure 4c), the elastic modulus was taken as 195,687.65 MPa; the Poisson's ratio was taken as 0.3; the yield stress was taken as 225.41 MPa and the tensile strength was 332.72 MPa. The material parameters of OSB [15], extruded board [16], steel glue [17] and other components are shown in Table 2.

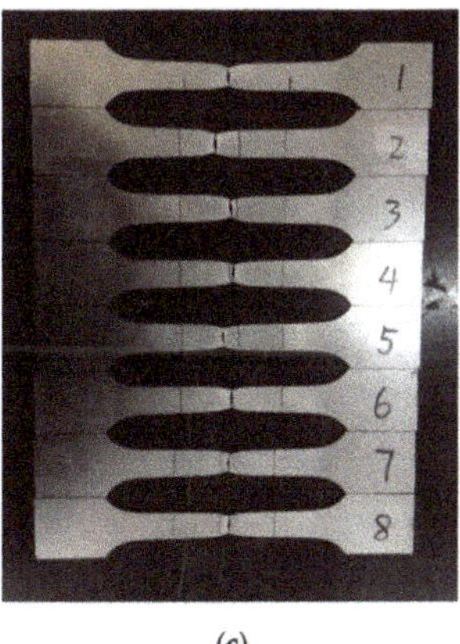

(a)   (b)   (c)

Figure 4. (**a**) Diagram of loading device for compression test of glulam prism. (**b**) Tensile test of glulam specimen. (**c**) Destruction phenomenon of steel tensile test.

Table 2. Material parameters of OSB, extruded board and steel glue.

Material	Elastic Modulus (N/mm^2)	Yield Strength (N/mm^2)	Poisson's Ratio
OSB	4760	9.83	0.4
Extruded board	6.6	0.15	0.27
Steel sticking glue	4000	55	0.15

The test adopted the symmetrical four-point loading method. The loading equipment required was a 32t jack and a distribution beam, 1000 mm long. I-beams were arranged at the three points of the specimen for uniform loading. This test was a simple support test piece. For failure loading, loading started from F0 and increased step by step with 10% of the estimated ultimate load. After loading to 50% of the estimated ultimate load, the load of each level increased by 5% of the estimated ultimate load until the specimen failed. After stopping the loading, the damage pattern was observed and recorded.

Two strain sensors with a size of 100 mm × 3 mm were set at the top of the OSB plate in the middle of the span and the middle of the height along the section in the middle of the span. In order to measure the strain change of the mid-span section of the thin-walled steel–glulam composite beam, 1 strain sensor was set on the top of the mid-span beam and 3 strain sensors with a specification of 100 mm × 3 mm were set on the three-layer glued laminate. Three strain sensors were set along the height of the steel beam section on the I-beam on the same side, sized 2 mm × 3 mm, to measure the strain of the mid-span section of the composite beam. These strain sensors were in the middle, upper edge and lower edge of the web. At the same time, the corresponding temperature compensation sheet was set for OSB, glulam and thin-walled steel. Displacement sensors were set at the mid-span of the laminated beam with a displacement sensor with a range of 200 mm, two displacement sensors with a range of 150 mm at the third point and two displacement sensors with a range of 100 mm at the support. The specific location and loading device are shown in Figure 5. The measuring instrument used a 50t pressure sensor, a 60-channel JM3813 static strain test system and a 10-channel DH3818 static strain tester.

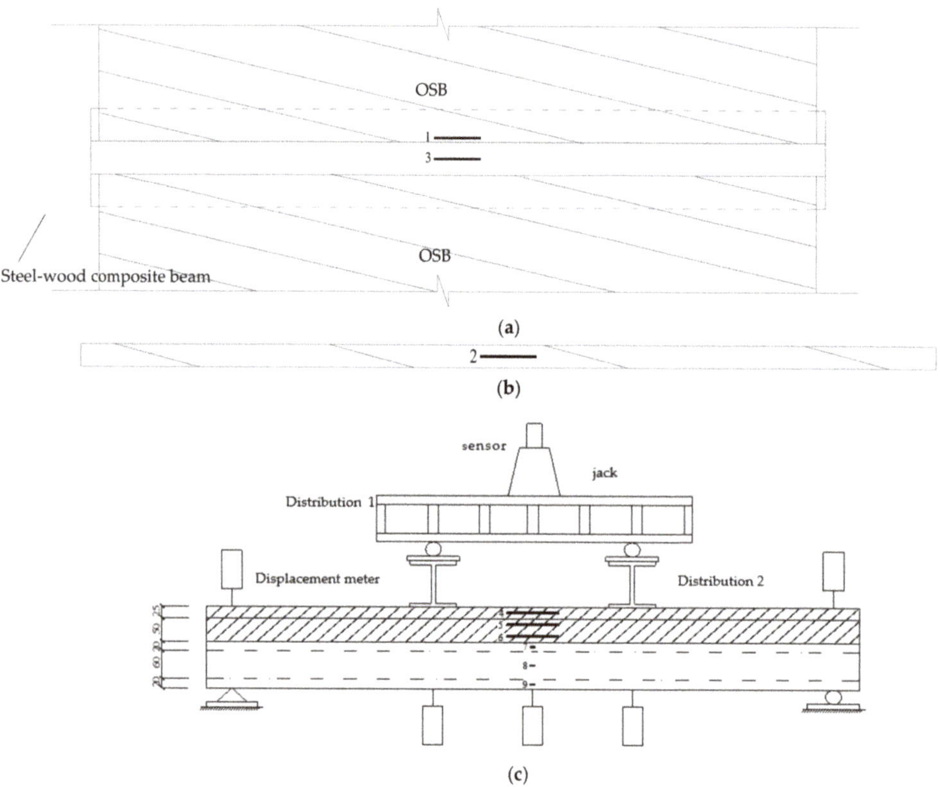

Figure 5. Layout of measuring points for bending test of steel–wood composite floor. The strain sensors were attached to the measuring point tightly and the displacement sensors were fixed at the measuring point. (**a**) Plan view of combined floor measuring points; (**b**) elevation view of OSB measuring points. (**c**) Elevation view of composite beam measuring point (mm).

3. Test Results and Analysis

3.1. Test Phenomenon

The failure mode of the ZHLB-1 group of specimens can be roughly divided into the cracking of the glue layer of the glulam laminate and the fracture of the upper flange of the steel–wood beam, as shown in Figure 6. In the failure loading stage, as the load increases, the bending deformation of the composite beam gradually increases (as shown in Figure 6a) and vertical cracks appear on one flange of the transverse tension glulam (as shown in Figure 6b). When the load reaches about 80% of the maximum bearing capacity, the vertical crack width here gradually expands, and the glued wood laminate is opened at the mid-span part and extends to the support part (as shown in Figure 6c). When the load reaches the ultimate bearing capacity, the glulam laminate is completely degummed along the length of the beam and the iron nails are pulled out (as shown in Figure 6d). Finally, the entire glulam laminate is completely degummed and destroyed along the length of the beam. At this time, the pressure sensor indicator drops below 50% of the maximum bearing capacity, and the bearing capacity of the composite floor is reduced and the load cannot be continued. At this time, the composite floor is regarded as broken.

(a)

(b)

(c)

(d)

Figure 6. Failure phenomenon of ZHLB-1 test. (**a**) Bending deformation of steel–wood composite floor; (**b**) vertical cracks appear on the left flange of glulam; (**c**) glued wood laminates are broken; (**d**) the iron nails connected to the glulam laminate are pulled out.

The failure mode of the ZHLB-2 specimen is roughly that the glued surface of the thin-walled steel and glulam is broken and the flange on one side of the glulam is broken, as shown in Figure 7. In the failure loading stage, as the load increases, the bending deformation of the composite floor gradually increases. The cemented surface of the steel–wood beam mid-span is broken, and the flange on the side of the glulam has vertical cracks (as shown in Figure 7a). The OSB above the foam beam of the plastic board is sinking and deformed downward. When the load reaches about 60 kN, the cemented surface of the mid-span part of the laminated beam is degummed, and the bottom of the glulam beam changes from a compressed state to a tensioned state, causing the bottom of the glulam beam to be broken (as shown in Figure 7b). When the load reaches about 80% of the maximum bearing capacity, an oblique crack appears at the glulam bearing position; the load continues to increase, and the oblique crack gradually expands laterally (as shown in Figure 7c). This is because when the specimen is approaching the ultimate load, the surface load on the OSB plate is transformed into a larger linear load and added to the flange of the composite beam; the shear stress in the flange of the laminated beam gradually increases until the main tensile stress in the beam exceeds the tensile strength of the wood's transverse grain, forming a beam crack of about 45°. At this time, the glued layer of glulam and the glued surface of thin-walled steel also experienced degumming failure, which accelerated the lateral expansion of diagonal cracks. When an oblique crack appeared on the flange of the composite beam, damage occurred on the composite beam from the middle of the span to the bonding surface between the glulam and the thin-walled steel of the support, and finally, the thin-walled steel and the glulam were completely peeled off. At this time, the pressure sensor reading suddenly dropped to about 50% of the maximum bearing capacity, the composite floor could not continue to be loaded and the steel beam underwent very large plastic bending deformation after unloading (as shown in Figure 7d).

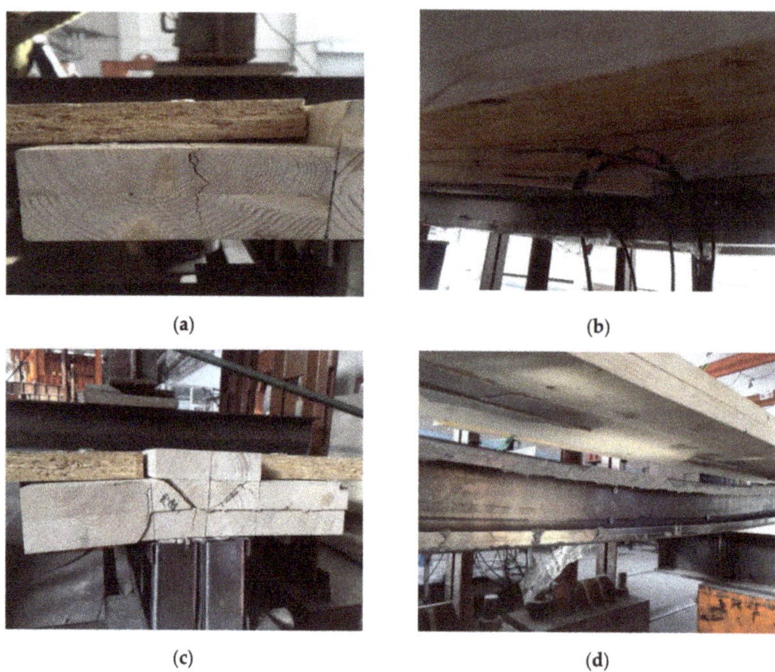

(a)

(b)

(c)

(d)

Figure 7. Failure phenomenon of ZHLB-2 test. (**a**) Vertical cracks appear in the flange of glued wood; (**b**) the flange on one side of the glulam span is pulled off along the plywood; (**c**) cracks and deformation of glulam bearings; (**d**) gluing failure of steel-wood cementing surface as a whole.

3.2. Mid-Span Load–Deflection Curve

By analyzing the experimental data, it can be concluded that the ultimate bearing capacity of the specimen ZHLB-2 is 18.31% higher than that of the specimen ZHLB-1; the corresponding mid-span deflection deformation is 42.41% higher and the ultimate bending moment is 21.43% higher. After the two groups of tests were completed, the mid-span deflection limit, ultimate load and bending moment of each group of specimens were summarized and are presented in Table 3.

Table 3. Test results of composite floor specimens.

Specimen Number	Mid-Span Deflection Limit(mm)	Maximum Deflection/Calculation Span	Ultimate Load (kN)	Bending Moment (kN·m)
ZHLB-1	40.46	1/59	71	28
ZHLB-2	57.62	1/42	84	34

It can be seen from Figure 8 that the deformation of the test beam with the increase in the load manifests in two stages: (1) The elastic stage, where the load value in this stage reaches 2/3~1/2 of the ultimate bearing capacity and its mid-span deflection value increases linearly with the increase in the load; (2) The elastic–plastic stage—With the increase in the load value, the slope of the mid-span load–strain curve changes nonlinearly, and the stiffness of the member gradually decreases until the member fails. When the specimen finally fails, the composite floor shows good ductility. When the ultimate load is reached, the load drops to less than 50% of the maximum bearing capacity. At the initial stage of loading, the composite floor works well, the load–deflection curve maintains a linear relationship growth and the composite floor is in an elastic working state; when the load of the ZHLB-1 specimen reaches about 65% of the ultimate load, the ZHLB-2 specimen reaches about 60% of the ultimate load, the lower flange of the steel beam yields, the load–deflection curve grows nonlinearly and the composite floor is in the elasto-plastic stage; as the load continues to increase, the deflection growth rate of the composite floor becomes larger and the slope of the mid-span load–deflection curve decreases continuously, the stiffness of the specimen gradually decreases. When the load reached the maximum value, a full-length crack along the length of the beam was observed on the top of the glulam beam, indicating that as the deflection of the top of the glulam beam increases, the beam eventually reaches the ultimate compressive strain and brittle failure occurs. The cemented surface of the steel–wood beam also undergoes brittle failure due to the accumulation of deformation; the thin-walled steel undergoes plastic deformation. At this time, the bearing capacity drops below 50% of the maximum, and the destruction shows obvious brittleness.

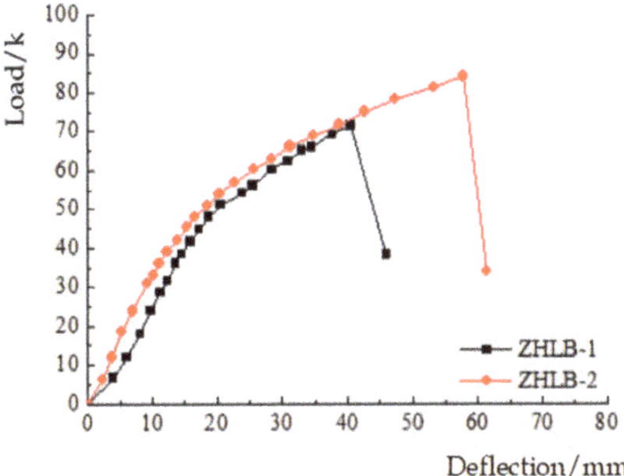

Figure 8. Mid-span load–deflection curve of composite floor specimen.

3.3. Section Mid-Span Height–Strain Curve

The changes in the strain distribution of the mid-span section of the two specimens ZHLB-1 and ZHLB-2 along the section height are shown in Figure 9. Analysis of the data in the figure shows that during the test, the OSB and glulam are mainly compressed, and the thin-walled steel is mainly tensioned. There is only one neutral axis in the cross-section of the composite beam of ZHLB-1 and ZHLB-2 before failure. The strain value has a linear relationship with the section height. At the beginning of loading, the neutral axis of ZHLB-1 and ZHLB-2 is located in the section of thin-walled steel. As the load increases, the central axis slowly moves up, and at the end of the loading stage, the neutral axis moves up to the glulam section. The height of the mid-span section of the composite slab is linearly distributed, and the OSB slab and the thin-walled steel–glulam composite beam also have a gradient change. A few curves have abrupt changes, which are caused by the warping of the strain gauges during the test.

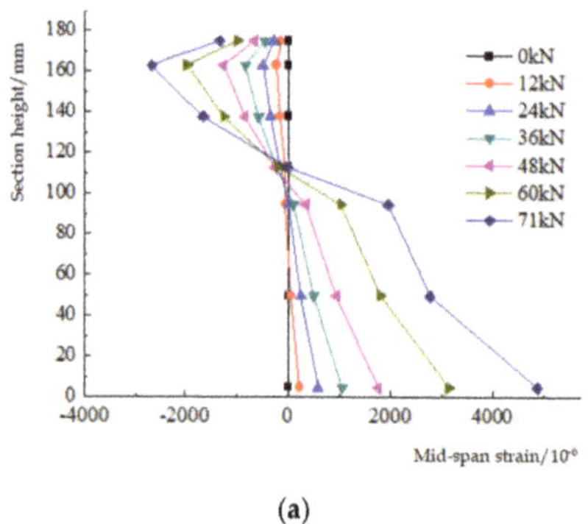
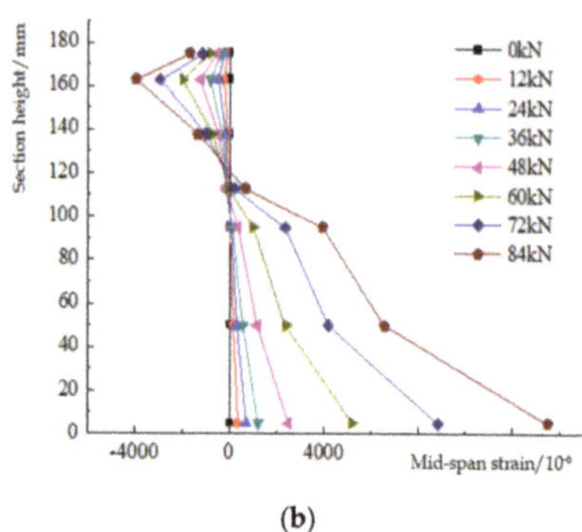

Figure 9. Steel–wood composite floor section mid–span height–strain curve. (**a**) ZHLB–1 section mid–span height–strain curve; (**b**) ZHLB–2 section mid–span height–strain curve.

3.4. Load–Strain Curve

According to the two sets of test data, the load–strain curve of the composite floor was drawn as shown in Figures 10 and 11. The analysis of the data shows that the strain value of each measuring point at the initial stage of loading increases linearly with the increase in the load. As the load increases, there is a tendency to deviate outwards. This is due to the elastoplastic deformation of OSB panels, glulam and thin-walled steel, which presents nonlinear changes. When the two sets of steel–wood composite slabs fail, the compressive strains on top of the OSB slab are 1332 and 1702 $\mu\varepsilon$, respectively; the compressive strains on the top of the glulam beam are 3275 and 4498 $\mu\varepsilon$; the tensile strains of the thin-walled steel are 4884 and 13,503 $\mu\varepsilon$, respectively. When the ZHLB-1 and ZHLB-2 specimens are damaged, the full cross-section of the OSB is compressed, the amount of plywood in the compression zone reaches two layers and the full cross-section of the thin-walled steel is tensioned. Measuring point OSB2 and measuring point M4 are the OSB plate strain change measuring point and the glulam strain change measuring point, respectively, located at the same section height. The deformation of the two measuring points is similar at the beginning of loading, but as the load increases, because the OSB plate has vertical deformation, the deformation of the two members is inconsistent, and the strain of the two members with the same section height is quite different.

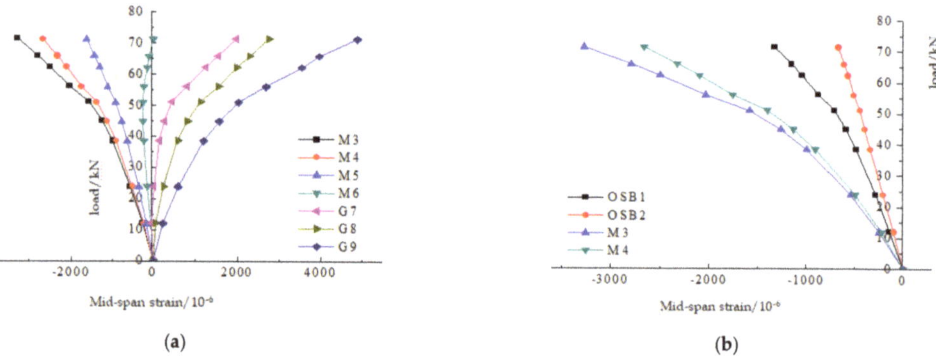

Figure 10. ZHLB–1 mid–span load–strain curve. (**a**) Load–strain curve of steel–wood composite beam; (**b**) load–strain curve of OSB and glulam beam. G7, G8, G9: Corresponding curves of strain measuring points 7, 8 and 9 for thin–walled steel; M3, M4, M5, M6: Corresponding curve of glulam strain measuring points 3, 4, 5 and 6; OSB1, OSB2: Corresponding curve of OSB plate strain measuring points 1 and 2.

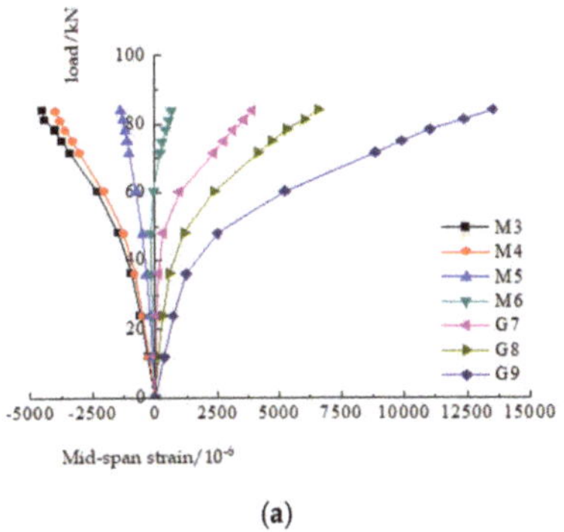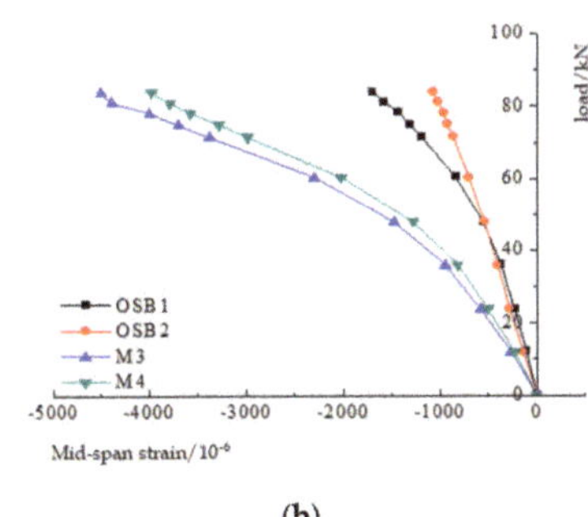

(a)   **(b)**

Figure 11. ZHLB–2 mid–span load–strain curve. (**a**) Mid–span load–strain curve of steel and wood composite beams; (**b**) mid–span load–strain curve of OSB slab and glued timber beam. G7, G8, G9: Corresponding curves of strain measuring points 7, 8 and 9 for thin–walled steel; M3, M4, M5, M6: Corresponding curve of glulam strain measuring points 3, 4, 5 and 6; OSB1, OSB2: Corresponding curve of OSB plate strain measuring points 1 and 2.

4. Finite Element Model Analysis

4.1. Establishment of Finite Element Model

On the basis of the material property test and the test of the composite floor test piece, a finite element model of the test piece was established using ABAQUS software, and in the modeling of the component, the finite element numerical model was processed as follows: (1) The glulam beam is glued by SPF (Spruce-Pine-Fir) boards, so first, build a model of 25-mm glulam laminates for each layer, and then, assemble to obtain it; (2) Assume that the glue layer thickness in the glue layer model of the steel sticking glue is 2mm; (3) According to the literature [15], the joints of the OSB and the glulam flange are treated by coupling degrees of freedom at the joint.

Wood is an anisotropic material. In the finite element analysis, wood was defined as an orthotropic material. According to the compression test results of prismatic glulam, nine engineering constants in the orthotropic stiffness matrix of the glulam material were calculated. For details, see Table 4. The material properties of other parts were obtained based on experimental data. The elastic modulus of steel was taken as 195,687.65 MPa; Poisson's ratio was taken as 0.3; the yield stress was taken as 225.41 MPa and the tensile strength was taken as 332.72 MPa. For rigid parts such as self-tapping screws, iron nails and cushion blocks, since the elastic deformation during the finite element simulation process is not considered, the elastic modulus was taken as 2.1×1011 N/m^2, and Poisson's ratio was 0.3. These data generally conform to the ideal elastoplastic constitutive law that we set in the finite element.

Table 4. Nine engineering constants in the elastic stage of glulam.

D_{1111} (N/mm^2)	$D_{2222} =$ D_{3333}(N/mm^2)	$D_{1122} = D_{1133}$ (N/mm^2)	D_{2233} (N/mm^2)	$D_{1212} =$ D_{1313}(N/mm^2)	D_{2323} (N/mm^2)
$11{,}532.52 \times 10^6$	938.98×10^6	620.28×10^6	611.72×10^6	723.74×10^6	193.00×10^6

According to the experimental phenomenon and the actual force characteristics of the steel-wood composite beam, the interaction of the finite element components is reasonably defined: (1) Assuming that the glulam laminate does not undergo relative slippage, the interaction between the laminates is set as a "binding" constraint; (2) The interaction of thin-walled steel and steel glue layer, glulam and steel glue layer is set as a "binding" constraint; (3) A C-shaped steel web splits the area of the same size as the bolt, and the bolt position is defined by the "binding" constraint relationship when modeling; (4) For the self-tapping screw connecting the OSB and the glulam flange, the connection is handled by the method of joint coupling degree of freedom; (5) Set the lower surface of OSB to "contact" with the upper surface of the extruded board; the contact surface of the elastic bearing extruded plate and glulam, I-beam and the contact surface of the cushion block is set as the "binding" constraint.

The element type of this model adopts the reduced integral of the C3D20R 20-node quadratic hexahedron element. The grid seed density of glulam, OSB and extruded board is 0.02, the grid seed density of thin-walled steel is 0.01 and the grid seed density of I-beam and steel pad is 0.05. The grid division of the composite slab is shown in Figure 12.

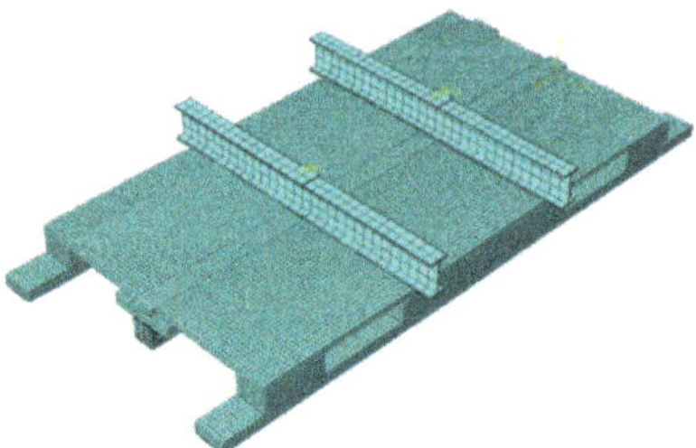

Figure 12. Grid division of composite floor.

4.2. Verification of Finite Element Model of ZHLB-2 Specimen

4.2.1. Comparison of Test Phenomena of ZHLB-2 Specimen

The along-grain stress cloud diagram of "S33" wood is shown in Figure 13. It can be seen from Figure 13 that when the ultimate load is reached, most of the glulam at the pure bending section enters plasticity, and the maximum compressive stress at the top of the glulam beam in the middle span is 37.84 MPa, while the maximum tensile stress at the bottom reaches 14 MPa. Most thin-walled sections located in the pure bending section enter the plastic stage, and the bottom of the mid-span beam reached the ultimate tensile stress. The OSB plate also enters the yield state, and the maximum compressive stress located at the mid-span part of the connection with the glued wood flange is 11.45 mpa. Through calculation of the experimental data for ZHLB-2 test pieces, the two OSB slabs bear 8% of the bearing capacity of the composite floor. The adhesive layer is about to reach the plastic state, which corresponds to the opening of the bonding surface of the steel and wood of the ZHLB-2 specimen. The extruded board + glulam composite beam does not yield, corresponding to the ZHLB-2 specimen; the elastic support is not damaged and is still in the elastic stage. From the stress cloud diagram of the finite element simulation analysis, the composite slab of the simulation analysis is basically consistent with the experimental phenomenon, which verifies the feasibility of the model.

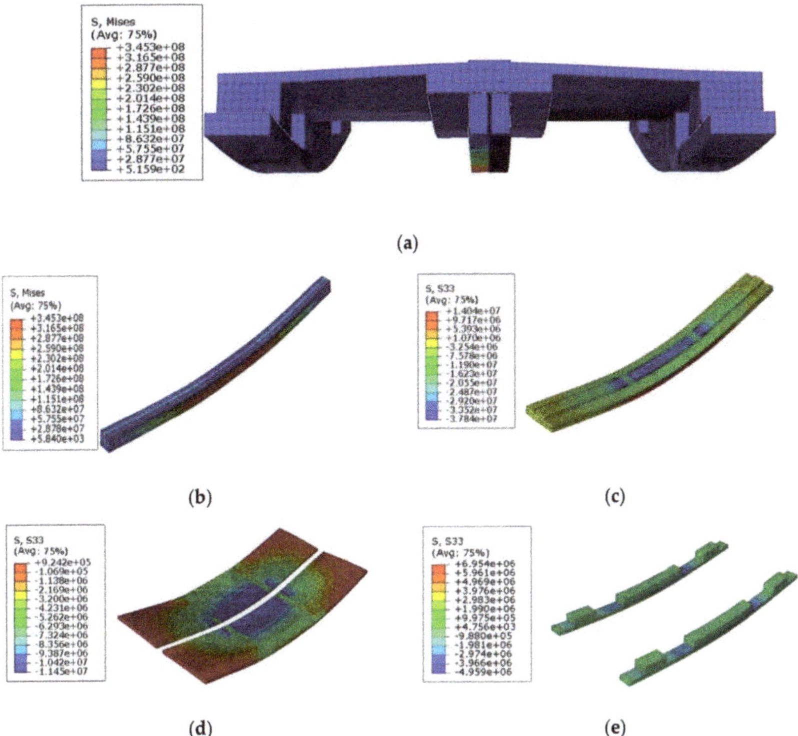

Figure 13. Stress–field nephogram of ZHLB–2 specimen. (**a**) Combined floor deformation stress–field nephogram. (**b**) Stress–field nephogram of thin–walled steel beam; (**c**) stress–field nephogram of glulam beams; (**d**) OSB plate stress–field nephogram; (**e**) extruded board + glued wood stress–field nephogram.

4.2.2. Comparison of Test Results of ZHLB–2 Specimen

The comparison between the finite element analysis results and the test results is shown in Table 5. The data were extracted from the model and a load–deflection curve was drawn for comparison with the test results, as shown in Figure 14.

Table 5. ZHLB-2 specimen test value and finite element comparison.

Sub Option	Ultimate Load/kN	Maximum Deflection/mm	Bending Moment/kN·m
The test results	84.03	57.62	33.61
Finite element value	93.67	63.16	37.47
Error	11.47%	9.6%	11.48%

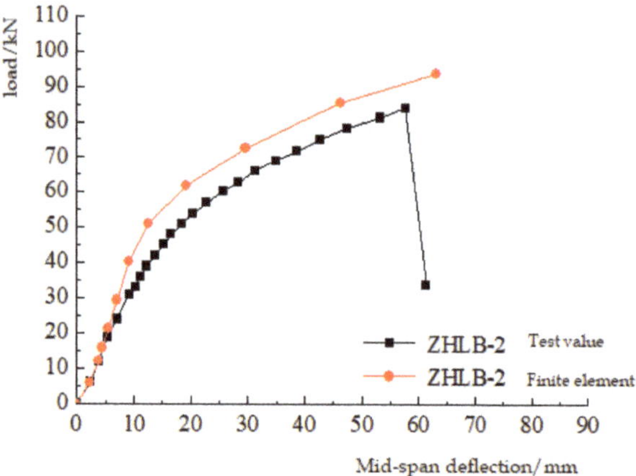

Figure 14. Comparison of ZHLB-2 specimen and finite element test results.

Analyzing Figure 14, it can be seen that the finite element simulation results are basically close to the initial slope and curve trend of the test results, indicating that the initial stiffness of the finite element and the force and deformation of the elastoplastic stage are comparable to the test. The bearing capacity error is small and the curves are in good agreement, which proves that the ZHLB-2 specimen model established by the finite element method is reasonable and feasible. Among them, the difference in ultimate load is 11.47%, and the difference in deflection and deformation at mid-span is 9.6%; the overall stiffness of the finite element model is greater than that of the ZHLB-2 specimen, and the bearing capacity and deformation capacity are greater than the test values. The main reasons for the error in the analysis and comparison test results are as follows: (1) The quality of the thin-walled steel–glulam composite beam during the test was affected by knots, cracks and the strength of the laminate glue layer; (2) the adhesive layer was subject to uncontrollable human factors such as the surrounding environment, the bonding strength and the time that the test plan is shelved after the production is completed. The above two factors will affect the bearing capacity and deformation of composite floor specimens, but they do not exist in the finite element analysis.

4.3. Factors Affecting the Bearing Capacity of Steel-Wood Composite Slabs

4.3.1. The Influence of Self-Tapping Screw Spacing

Under the condition that the steel-wood composite beams and OSB slabs remain unchanged, the spacing of the self-tapping screws connecting the composite slabs and the OSB slab was set to 100, 150 and 200 mm, and mid-span load–deflection curves were drawn and a comparative analysis performed for the composite floors with different screw spacings, as shown in Figure 15.

The ultimate load simulation values of different self-tapping screw spacings are summarized in Table 6. From the table, it can be seen that when the self-tapping screw spacing is reduced from 200 to 150 mm, the bearing capacity of the composite floor is increased by 7.14%; when the tapping screw spacing is reduced from 100 to 50 mm, the bearing capacity of the composite floor is only increased by 2.5%. It can be seen that as the spacing of self-tapping screws continues to decrease, the increase in the bearing capacity of the steel–wood composite floor gradually decreases.

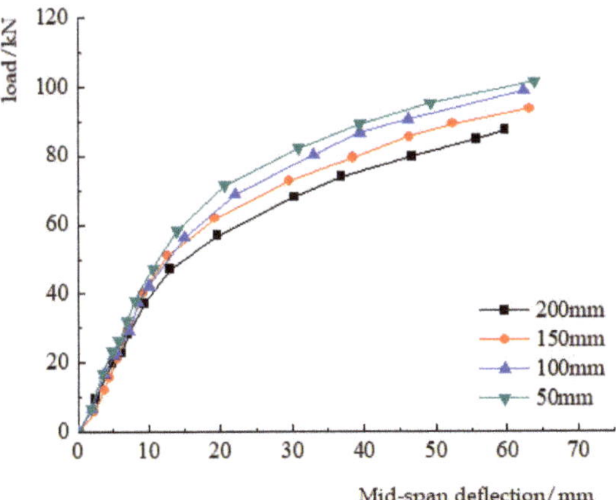

Figure 15. Comparison of finite element test results of different self-tapping screw spacings.

Table 6. Finite element analysis results of composite floor with different self-tapping screw spacings.

Self-Tapping Screw Spacing (mm)	Ultimate Load Value (kN)	Compared with the Benchmark	Increase with Reduction
200	87.43	−6.67%	—
150	93.67	—	7.14%
100	98.92	5.60%	5.60%
50	101.39	8.24%	2.5%

4.3.2. The Influence of the Width of Laminated Beam Glulam

Under the condition that the screw spacing remains unchanged, the flange width of the laminated beam was set to 60 and 80 mm and mid-span load–deflection curves of different glulam flange widths were drawn, compared and analyzed, as shown in Figure 16.

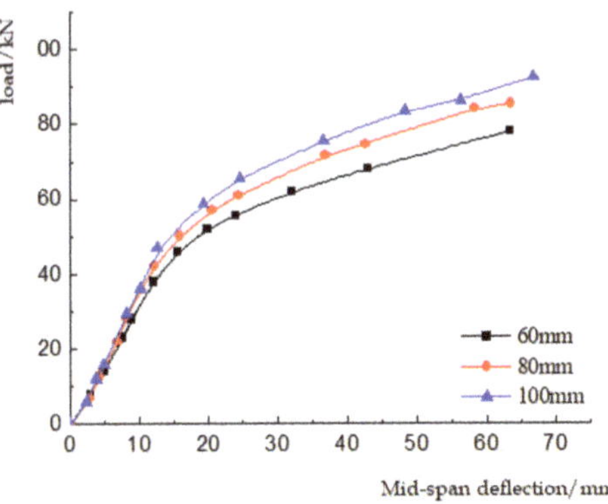

Figure 16. Comparison of finite element test results of different glulam flange widths.

The ultimate load simulation values of different glulam flange widths are summarized in Table 7. The analysis data in the table show that when the glulam flange width is increased from 60 to 80 mm, the bearing capacity of the composite floor increases by 9.21%; when the glulam flange width is increased from 80 to 100 mm, the bearing capacity of the composite floor is increased by 9.90%. The analysis data in the table show that with every increase of 20 mm in the width of the glulam flange, the bending bearing capacity of the composite floor can be increased by about 10%, which verifies that the flange width of the glulam beam has a significant influence on the bearing capacity of the composite floor.

Table 7. Finite element analysis results of composite slabs with different glulam flange widths.

Glulam Flange Width (mm)	Ultimate Load Value (kN)	Increase Compared to the Benchmark	Increase by One Step
60	78.04	−16.69%	—
80	85.23	−9.01%	9.21%
100	93.67	—	9.90%

5. Conclusions

In this paper, flexural performance tests of two groups of fabricated steel–wood composite slabs were carried out, and the failure modes, bearing capacity and stiffness of each specimen were compared and analyzed. The conclusions are as follows:

(1) For the thin-walled steel–glulam composite beams connected by glue, the glulam was mainly compressed and the thin-walled steel was mainly tensioned during the test, and the two materials entered the plastic stage before the specimen was damaged, indicating that the steel and wood can work together, thereby increasing the material efficiency of the composite beam.

(2) The ultimate bearing capacity of the composite slab assembled with self-tapping screws was 18.31% higher than that of the composite slab assembled with iron nails; the corresponding mid-span deflection deformation increased by 42.41% and the ultimate bending moment also increased by 21.43%. The steel–wood composite slab assembled using self-tapping screws can superimpose the OSB and the thin-walled steel–glulam so that the material strength can be fully utilized, thereby improving the bend-bearing capacity and overall rigidity of the composite floor.

(3) Based on the finite element analysis, a comparative analysis of the self-tapping screw spacing and the flange width of the laminated beam glulam which affect the bearing capacity of the steel–wood composite floor was carried out. The analysis results show that reducing the spacing of self-tapping screws can improve the bearing capacity of the composite floor, and the reasonable screw spacing is between 100 and 150mm. Increasing the flange width of the steel–wood composite beam can significantly improve the bend-bearing capacity of the composite floor.

Author Contributions: Conceptualization, G.L. and W.T.; methodology, G.L. and W.T.; software, G.L. and W.T.; validation, G.L., W.T. and D.H.; formal analysis, G.L. and W.T.; investigation, G.L. and W.T.; resources, G.L., W.S.; data curation, G.L., W.T., Z.L. and Z.L.; writing—original draft preparation, G.L., W.T. and Z.L.; writing—review and editing, G.L.; visualization, G.L., W.T., Z.L. and D.H.; supervision, G.L.; project administration, G.L.; funding acquisition, G.L. All authors have read and agreed to the published version of the manuscript.

Funding: This research was funded by the Fundamental Research Funds for the Central Universities, grant number 2572019BJ03, the Harbin science and technology innovation talent fund project, grant number 2017RAQXJ086 and the Natural Science Foundation of Heilongjiang Province, grant numbers LH2019E005 and LH2020E009.

Institutional Review Board Statement: Not applicable.

Informed Consent Statement: Not applicable.

Data Availability Statement: Data available in a publicly accessible repository.

Conflicts of Interest: The authors declare no conflict of interest.

References

1. Yang, X. Development and Application of Prefabricated Wood Structure Building System. *Constr. Sci. Technol.* **2017**, *19*, 57–62.
2. GB/T 51233-2016. *Technical Standards for Prefabricated Wood Structure Buildings*; China Building Industry Press: Beijing, China, 2016.
3. Wang, X.; Chen, Z.; Bai, J.; Li, S.; An, Q. Research status and development prospects of steel-wood composite structures. In Proceedings of the 2011 National Steel Structure Academic Year Conference, Yinchuan, China, 17 October 2011; pp. 654–657.
4. Cristiano, L.; Buick, D. Innovative composite steel-timber floors with prefabricated modular components. *Eng. Struct.* **2017**, *132*, 695–713.
5. Xia, Y.; Zheng, X.; Wei, J. Study on vibration and bending performance of plywood-thin-walled steel composite floor. *Wood Ind.* **2020**, *34*, 18–22.
6. Li, Y.; Shan, W.; Huang, Z.; Ge, B.; Wu, Y. Experimental study on mechanical properties of profiled steel sheet-bamboo sheet composite slab. *J. Build. Struct.* **2008**, *96*, 102–111.
7. Yang, Y.; Liu, J.; Fan, J.; Nie, X. Experimental study on flexural performance of steel plate-concrete composite plate. *J. Archit. Struct. Sci.* **2013**, *34*, 24–31.
8. Xu, L.; Tangorra, F.M. Experimental investigation of lightweight residential floors supported by cold-formed steel C-shape joists. *J. Constr. Steel Res.* **2006**, *63*, 422–435. [CrossRef]
9. Zhou, X.; Li, Z.; Wang, R.; Shi, Y. Research on the Flexural Bearing Capacity of Cold-formed Thin-walled Steel Beam-OSB Slate Composite Floor. *China Civ. Eng. J.* **2013**, *46*, 1–11.
10. Li, G.; Yang, B.; Shan, W.; Guo, N. Experimental research on the cold-formed thin-walled steel-laminated timber composite beam. *Prog. Eng. Sci.* **2017**, *8*, 177–181.
11. Guo, N.; Zhang, P.; Zuo, Y.; Zuo, H. Bending performance of glue-lumber beam reinforced by bamboo plyboard. *J. Jilin Univ. (Eng. Technol. Ed.)* **2017**, *47*, 778–788.
12. Zuo, H.; Liu, H.; Lu, J. Effect of New Self-tapping Screw Reinforcement Measures on Bending Performance of Glulam Beams. *J. Northeast. Univ.* **2020**, *48*, 112–116, 121.
13. Guo, N.; Chen, H.; Zhang, P.; Zuo, H. The research of parallel to the grain compression performance test of laminated glued bamboo-wood composites. *Tech. Gaz.* **2016**, *23*, 129–135.
14. GB/T 50005-2017. *Wood Structure Design Standard*; China Building Industry Press: Beijing, China, 2017.
15. Jiang, H. Research on the Control Effect of Prestressed Glulam String Beams. Master's Thesis, Northeast Forestry University, Harbin, China, 2018.
16. Yang, X. Research on Bending Performance and Design Method of Steel Plate Reinforced Glulam Beam. Ph.D. Thesis, Northeast Forestry University, Harbin, China, 2016.
17. GB 50009-2012. *Building Structure Load Code*; China Building Industry Press: Beijing, China, 2012.

sustainability

Article

Instrumentalization of a Model for the Evaluation of the Level of Satisfaction of Graduates under an *E-Learning* Methodology: A Case Analysis Oriented to Postgraduate Studies in the Environmental Field

Eduardo García Villena [1,2,*], Silvia Pueyo-Villa [3,4], Irene Delgado Noya [1,5,*], Kilian Tutusaus Pifarré [1,2,*], Roberto Ruíz Salces [1] and Alina Pascual Barrera [5,6]

Citation: García Villena, E.; Pueyo-Villa, S.; Delgado Noya, I.; Tutusaus Pifarré, K.; Ruíz Salces, R.; Pascual Barrera, A. Instrumentalization of a Model for the Evaluation of the Level of Satisfaction of Graduates under an *E-Learning* Methodology: A Case Analysis Oriented to Postgraduate Studies in the Environmental Field. *Sustainability* **2021**, *13*, 5112. https://doi.org/10.3390/su13095112

Academic Editor: Yadir Torres Hernández

Received: 31 March 2021
Accepted: 28 April 2021
Published: 3 May 2021

Publisher's Note: MDPI stays neutral with regard to jurisdictional claims in published maps and institutional affiliations.

1. Higher Polytechnic School, Universidad Europea del Atlántico (UNEATLANTICO), Isabel Torres 21, 39011 Santander, Spain; roberto.ruiz@uneatlantico.es
2. Department of Environment and Sustainability, Universidad Internacional Iberoamericana (UNINI-PR), Arecibo 00613, Puerto Rico
3. Department of Languages and Education, Universidad Europea del Atlántico (UNEATLANTICO), Isabel Torres 21, 39011 Santander, Spain; silvia.pueyo@uneatlantico.es
4. Department of Language, Education and Communications Sciences, Universidad Internacional Iberoamericana (UNINI-PR), Arecibo 00613, Puerto Rico
5. Department of Project Management, Universidad Internacional Iberoamericana (UNINI-MX), Campeche 24560, Mexico; alina.pascual@unini.edu.mx
6. Area of Environmental Engineering, Universidade Internacional do Cuanza, Rua Padre Fidalgo s/n, Municipio do Cuito, Bié, Bairro Sede (6335,09 km), Angola
* Correspondence: eduardo.garcia@uneatlantico.es (E.G.V.); irene.delgado@uneatlantico.es (I.D.N.); kilian.tutusaus@uneatlantico.es (K.T.P.)

Abstract: The purpose of this article was to evaluate the level of satisfaction of a sample of graduates in relation to different online postgraduate programs in the environmental area, as part of the process of continuous improvement in which the educational institution was immersed for the renewal of its accreditation before the corresponding official bodies. Based on the bibliographic review of a series of models and tools, a Likert scale measurement instrument was developed. This instrument, once applied and validated, showed a good level of reliability, with more than three quarters of the participants having a positive evaluation of satisfaction. Likewise, to facilitate the relational study, and after confirming the suitability of performing a factor analysis, four variable grouping factors were determined, which explained a good part of the variability of the instrument's items. As a result of the analysis, it was found that there were significant values of low satisfaction in graduates from the Eurasian area, mainly in terms of organizational issues and academic expectations. On the other hand, it was observed that the methodological aspects of the "Auditing" and "Biodiversity" programs showed higher levels of dissatisfaction than the rest, with no statistically significant relationships between gender, entry profile or age groups. The methodology followed and the rigor in determining the validity and reliability of the instrument, as well as the subsequent analysis of the results, endorsed by the review of the documented information, suggest that the instrument can be applied to other multidisciplinary programs for decision making with guarantees in the educational field.

Keywords: Sustainable Development Goals; environment; student satisfaction survey; measuring instrument; factor analysis; postgraduate degrees

1. Introduction

Mejías and Martínez [1] refer to satisfaction as the level of students' state of mind towards their institution, and which results from comparing the perception they have regarding the fulfillment of their needs, expectations and requirements.

According to González, Tinoco and Torres [2], the abundant current literature on the relationship between the satisfaction of graduates and the degree of educational quality in obtaining different academic degrees, as well as its quantification, demonstrates the importance of this topic for universities interested in attracting new generations of students to the different modalities of higher education.

Pichardo and García [3] consider there to be an increase in the interest in knowing the needs those university students have about the contexts to improve their higher education process.

In parallel, studies on student satisfaction are required by national and international university evaluation bodies [4] and are indicative of areas of improvement for better positioning in academic performance among higher education institutions [5].

However, educational quality is often confused with the quality of the service offered. Indeed, according to Mejías and Martínez [1], the objective is to find out what needs the student has and not so much how efficient the service provided is, despite the obvious positive correlation between student satisfaction and the quality of the service provided [4–6].

Most of the models found in the bibliography are directed to face-to-face teaching and, to a lesser extent, to the distance modality, being a common practice in the former to attribute little relevance to new technologies, which, in case they appear, usually does so as one more element of the teaching itself, as it occurs in the proposal of Mejías and Martínez [1], instead of granting them the separate protagonism they deserve.

Virtual education plays a preponderant role in satisfying the right (and the need) that everyone has to access education [7]. In this sense, already at the end of the 20th century, and in the first decade of the 21st century, there were two approaches to virtual education: a partial one, where the training activity, training materials, platforms and the cost/benefit ratio were evaluated [8–10], and a global one, referring to evaluation systems centered on models and/or standards of total quality and on the practice of benchmarking. In this context, according to Pereira and Gelvez [11], as in face-to-face education programs, in the systemic models of quality evaluation in virtual education, there is no clear concept of what quality is and that determines the criteria to be assessed, so it is important to establish standards for the quality of *e-learning*.

One of these standards is "Quality Management. Quality of Virtual Training" [12], which determines the three factors involved in meeting the needs and expectations of students: recognition of training for employability, learning methodology and accessibility.

In this context, the lack of knowledge of the methodology involved in *e-learning* and the digital divide represent two major challenges, since not all students feel comfortable with the procedures or have equal opportunities in accessing technologies, respectively [13].

Precisely, the pandemic caused by the appearance of the SARS COV-2 coronavirus at the end of 2019 and the first quarter of 2020 has only deepened the problem of inequalities, since, on the one hand, the lack of access to education with the help of ICTs by some families became evident, either due to a lack of resources or a lack of technical support [14]; on the other hand, confinement and teleworking led to a significant increase in the number of accesses to the programs offered by educational institutions (Figure 1).

Other standards are the EFQM model and the ISO 9000 family of standards, which indicate, as a requirement for a quality management system, the need to establish a process for measuring customer or user satisfaction [4].

The reason for measuring student satisfaction lies in the fact that students are the main factor and guarantee of the existence and maintenance of educational organizations. Students are the recipients of education; they are the ones who can best value it and, although they have a partial vision, their opinion provides a reference that should be taken into account [4,15].

According to Mejías and Martínez [1], "measuring student satisfaction in a consistent, permanent and adequate manner would guide the right decisions to increase their strengths and remedy their weaknesses."

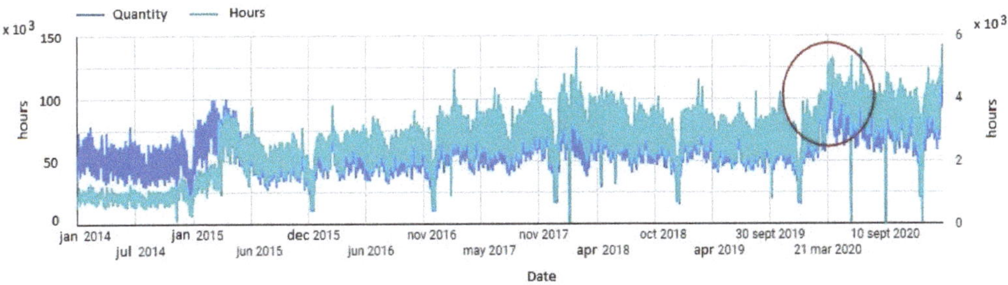

Figure 1. Number of accesses and connection hours of different programs of an educational institution. Note. The overprinted circle illustrates the increase in accesses and connection hours to an educational institution's online programs, starting in the first quarter of 2020.

Romo, Mendoza and Flores [16] emphasize the importance of measuring student satisfaction through tests of contrasted validity and reliability, in order to establish plans for improving educational quality.

The validity of the measurement instrument can be content, criterion and construct validity. In general, it is customary in models to carry out content validity by means of a literature review and consultation with panels of experts [1].

Reliability is given, among others, by the statistical parameter "Cronbach's alpha," widely used in the university environment [17], as a simple way of measuring how reliable the measuring instrument is on a scale where the approximation to unity is the most desirable.

The importance of having valid and reliable instruments (from a statistical point of view) is therefore evident, given the implications that may arise from their use [18].

Surdez et al. [5] used as an evaluative instrument a questionnaire to estimate university students' satisfaction with education (SEUE), proposed by [15], and adapted by the authors themselves (Table 1).

Table 1. Variables and associated dimensions used as an evaluative instrument.

Variables	Dimensions
Section I: demographic variables	School situation
	Individual profile
Section II: student satisfaction	Teaching–learning profile
	Respectful treatment of the people with whom he/she must interact to achieve his/her academic goals
	Teaching–learning space infrastructure
	Self-realization

Note. Own elaboration based on Surdez et al. [5].

In this context, Fainholc [19] suggests that the traditional application of business quality models or others based on management rather than on the teaching–learning process to measure user satisfaction in distance education based on the completion of an opinion questionnaire can lead to misunderstandings if the institution and its technological-educational proposal are not known in depth.

Such a proposal should be based on indicators related, among others, to the fulfillment of the Sustainable Development Goals (SDGs) and the needs and expectations of stakeholders in the university context [20].

In the scope of this research, the delivery of environmental programs is fundamental to develop competencies in graduates that contribute to sustainability within the framework of the SDGs. In this sense, the sustainability model should be one of the guiding principles to be included transversally in the different curricula of an institution [21].

Based on the background described above, the purpose of this research was to develop a model for an educational institution to evaluate the degree of satisfaction of the graduates of several postgraduate programs in the field of the environment, through a valid and reliable instrument, applied to a test directed to a sample of 150 participants. It is hoped that the results of this report will contribute to the continuous improvement of the quality of training, to the promotion of new university programs and to an increase in the competency profile of the participants, which will undoubtedly result in better performance at the professional level to develop sustainable policies in the future.

2. Materials and Methods

The methodology followed in this work is based on a cross-section survey suited to the purposes of a descriptive and relational type of research, with a quantitative, non-experimental, transect approach, due to the fact that no hypotheses were posed and no variables were manipulated, but indeed "data were measured, evaluated or collected on various aspects, dimensions or components of the phenomenon to be investigated [in its natural working environment and in a single time]" [4,22].

As it can be seen from the purpose that guided this research, the methodological scope comprised, broadly speaking, four distinct stages: firstly, based on the literature review, a Likert scale instrument was developed. Secondly, the validity and reliability of this instrument was determined thanks to the data collected from a quantitative approach [22] on the satisfaction of a sample of 150 graduates of various online postgraduate programs in the environmental field and in relation to different aspects of training. Thirdly, a factor analysis was carried out with the aim of grouping variables or items of similar characteristics into a reduced number of factors and, in this way, obtaining a simplified model. Finally, a study was carried out to determine possible relationships based on the results obtained.

2.1. Population and Sample

The target population consisted of a total of 215 graduates from different online graduate programs in the environmental area. To determine the necessary sample size, and given that the intention was to estimate distributions or percentages when working with qualitative variables, the following sample size calculation formula for a finite population was used [23]:

$$n \geq \frac{N * Z^2_{1-\frac{\alpha}{2}} * (p * q)}{(N-1) * \varepsilon^2 + Z^2_{1-\frac{\alpha}{2}} * (p * q)}$$

where:

- n = required sample size;
- N = population size;
- $Z_{1-\alpha/2}$ = 1.96 (Z-statistic calculated at a 95% confidence level);
- $p = q = 0.5$ (typical values for worst-case conditions);
- Error (epsilon) = 0.05.

Substituting the values in the formula resulted in the sample size required for the study to be at least $n \geq 138$.

The sampling procedure was non-probability convenience sampling.

2.2. Data Collection

At the end of the corresponding postgraduate program, the final project director sent the final grade along with an email and an access link, inviting the 215 recent graduates of the postgraduate programs between the dates of 01 January 2016 and 31 December 2018 to voluntarily participate in the survey.

The data were collected through the "Google Forms" platform, which allows administering questionnaires over the Internet.

On the other hand, participants were assured that their contribution to the research was confidential.

During the analyzed period, survey data were received from a total of 168 students (78%). Of these, 18 were discarded when deficiencies were detected in the filling out of the form. SPSS version 26 statistical software was used for data analysis.

2.3. Methodology

In particular, the following activities were carried out:

(1) Selection of theories, models and tools for training evaluation, through bibliographic review techniques and analysis of the educational institution's processes.
(2) Finding indicators based on the models, literature review and authors' experience.
(3) Selection of nominal qualitative variables (gender, origin and program) and ordinal variables (age group, entry profile and graduate satisfaction).
(4) Development of a measurement instrument or questionnaire for the variable "graduate satisfaction" on a Likert scale ("1. Strongly disagree" to "4. Strongly agree"), using Microsoft Word text editing software.
(5) Determination of the validity of the instrument by requesting a panel of experts.
(6) Application of the instrument to a final sample of 150 graduates belonging to different *online* environmental postgraduate programs.
(7) Findings of the reliability of the proposed measuring instrument using Cronbach's alpha statistic in SPSS version 26 statistical software.
(8) Testing the adequacy of a factor analysis by finding the Kaiser–Meyer–Olkin (KMO) sampling adequacy statistic and Bartlett's test of sphericity.
(9) Execution of factor analysis and determination of factors or homogeneous groups of variables.
(10) Finding of possible significant relationships between variables by means of the chi-square statistical parameter in the SPSS version 26 statistical software, Infostat 2020 and Excel spreadsheet.
(11) Interpretation of results for decision making.

3. Results

3.1. Selection of Nominal and Ordinal Qualitative Variables

The nominal qualitative variables considered during the research were "gender," "origin" and "program" studied by the graduate. Each of them was assigned a categorical value to differentiate their categories. The assignment of these characteristics was exhaustive and mutually exclusive, i.e., it did not have any idea of quantity, order or hierarchy associated with them [18].

On the other hand, the chosen ordinal qualitative variables did have an implicit ordering of the attribute. In this case, the "age group," "entry profile" and "graduate satisfaction" variables were considered.

3.2. General Characteristics of the Graduates

As shown as Table 2, it was observed that 70.7% were male and the remaining 29.3% were female; 37.3% came from South America, 32% from North America, 24% from Central America and 6.7% from Eurasia. In relation to age, the 30–39 years age group accounted for 40.7%, the 20–29 years age group for 22%, the 40–49 years age group for 20%, the 50–59 years age group for 13.3% and the 60–69 years age group for 4%. In reference to previous studies, 66% completed a degree/diploma/bachelor, 15.3% a master's degree, 14.7% a postgraduate degree and 4% a doctorate. Finally, 44.7% studied the Climate Change program, 19.3% the Marine Sciences program, another 16% the Water program and, finally, 4% the Biodiversity program.

Table 2. Educational and demographic characteristics of the graduates.

	Attribute	Category	*n*	%
Gender	1	Male	106	70.7
	2	Female	44	29.3
Origin	1	North America	48	32
	2	Central America	36	24
	3	South America	56	37.3
	4	Eurasia	10	6.7
Program	1	Audits	15	10
	2	ISO 14001	9	6
	3	Biodiversity	6	4
	4	Waters	24	16
	5	Marine Sciences	29	19.3
	6	Climate Change	67	44.7
Age group (years)	1	20–29	33	22
	2	30–39	61	40.7
	3	40–49	30	20.0
	4	50–59	20	13.3
	5	60–69	6	4
Entry profile	1	PhD	6	4
	2	Master's degree	23	15.3
	3	Postgraduate	22	14.7
	4	Degree/Dip./Bachelor	99	66

3.3. Determination of the Indicators of the "Graduate Satisfaction" Variable

Based on the literature review, a total of ten dimensions with their corresponding indicators were identified (Table 3).

Table 3. Dimensions and indicators of the "graduate satisfaction" variable.

Dimensions	Indicators
Self-realization	Level of achieved academic satisfaction
Relevance for training	Degree of learning achieved
Academic program requirements	Time pressures, attention effort, complexity of the tasks …
Economic and social context of the participant	Number of scholarships granted, payment facilities …
Interaction between participants, tutors and other interested parties	Degree of satisfaction with the channels established for external communication
Academic program	Degree of satisfaction with the curriculum design
Continuous evaluation	Degree of satisfaction with continuous evaluation activities
Teacher evaluation	Degree of satisfaction with academic tutors
	Degree of satisfaction with the tutor of the final project
Accessibility to the product and other services offered by the institution	Number of shipments of teaching material, reception times …
Virtual campus	Degree of ease of use of the virtual campus
	Technical support and number of incidents

3.4. Instrument Design Based on the Measurement Criteria

Once the list of indicators was available, the corresponding measurement criteria for the diagnosis were proposed. To this end, a bibliographic search and the contribution of a panel of experts were used.

In this way, thirteen items or measurement criteria were obtained, which provided the basis for a Likert scale questionnaire with scale categories that gradually ranged from "1. Strongly disagree" to "4. Strongly agree" to measure the variable of graduate satisfaction in relation to the reference postgraduate programs (Table 4).

Table 4. Likert scale questionnaire for measuring graduate satisfaction with postgraduate programs.

Item No.	Measurement Criteria
1	Delivery of didactic materials has been punctual and on time.
2	My tutor has gone out of his/her way to help me
3	I am satisfied with the attention I received prior to enrollment
4	The handling of the virtual campus has been very user friendly
5	Overall, I am satisfied with the program
6	My assessment of the organization of the program is very satisfactory
7	The Institution has provided me with facilities to carry out the study
8	My Final project manager has been accessible
9	This program will be relevant to my professional training and performance
10	I found the information provided during the program to be sufficient
11	The academic program has met my initial expectations
12	I found the contents of the program interesting
13	My assessment of the continuous evaluation is very satisfactory

3.5. Validity and Reliability of the Measurement Instrument

The validity of the measurement instrument was determined based on the relevance, pertinence and clarity of each of the items by a panel of experts [4].

Ultimately, the aim was to see what proportion of the measurement criteria was accepted by each of the experts. To do this, each expert filled in a table in which he or she considered whether that question was of significant interest (relevant), whether it was suitable (feasible, novel, ethical and interesting) and, finally, whether it was unambiguously written (clear).

For example, issues were considered ambiguous, such as "The tutor has adequately complied with the teaching plan and the qualification of exercises", if they included two activities for only one answer, or if they made reference to assumptions, for example, "I consider that my classmates adequately value the activities of the tutor".

Once the tables were completed, the frequencies were determined and weighted with a test proportion of 85%, using the binomial in the SPSS program. In this way, significance levels were determined that allowed us to accept in all cases the null hypothesis that the accepted criteria ratio in the instrument was 85%.

The reliability of the instrument was based on the determination of Cronbach's alpha and included all the ordinal qualitative variables or items associated with the "graduate satisfaction" variable.

Cronbach's alpha provided a result of 0.834, considered a good value for internal consistency [18]. Consequently, we proceeded to the study of the information collected.

Table 5 shows the statistics associated with Cronbach's alpha. In the last column, it can be seen that all are greater than 0.808 and that the elimination of item 3 could improve the value obtained. However, it was considered that the small improvement in the statistic (from 0.834 to 0.836) was not significant enough to compensate for the loss of information that the exclusion of the item from the analysis would entail, so it was decided to leave it.

3.6. Measures of Central Tendency and Dispersion for Ordinal Qualitative Variables

Table 6 shows the measures of central tendency (mean) and dispersion (standard deviation) for the ordinal qualitative variables.

It could be observed that the mean of age group was in the range of 30–39 years old and that the mean of entry profile was located between undergraduate and postgraduate.

Table 5. Statistics associated with Cronbach's alpha.

Item	Scaling Average If the Element Has Been Suppressed	Scale Variance If the Element Has Been Suppressed	Total Correlation of Corrected Elements	Cronbach's Alpha If the Item Has Been Deleted
1	44.36	14.433	0.674	0.810
2	44.35	15.478	0.371	0.829
3	44.58	15.024	0.331	0.836
4	44.47	14.264	0.568	0.815
...	...	...	...	...
13	44.28	15.156	0.414	0.827

Table 6. Central tendency and dispersion statistics for ordinal qualitative variables.

Variable		M	SD
Graduate Satisfaction Item			
	1	3.72	0.493
	2	3.73	0.504
	3	3.50	0.673
	4	3.61	0.601
	5	3.71	0.651
	6	3.63	0.550
	7	3.75	0.480
	8	3.73	0.473
	9	3.85	0.408
	10	3.65	0.567
	11	3.70	0.576
	12	3.71	0.595
	13	3.80	0.543
Age Group			
		2.37	1.089
Entry Profile			
		3.43	0.893

Note. M and SD represent mean and standard deviation, respectively.

Therefore, the mean and the standard error of the mean were as follows (Table 7):

Table 7. Mean and standard error of the mean of graduate satisfaction variable.

Variable	M	SE
Graduate Satisfaction	3.72	0.02

Note. M and SE represent mean and standard error of the mean, respectively.

3.7. Exploratory Factor Analysis

The proposal is to group the items into a reduced number of factors, capable of explaining the greatest proportion of the total variability contained in them. To this end, the variables or items must meet the assumptions of normality, homoscedasticity (equality of variances) and multicollinearity (correlation between variables).

From the statistical analysis, it was observed that the data did not follow a normal distribution. However, according to Hair et al. [24], these three assumptions can be ignored, as factor analysis is more conceptual than statistical; in fact, these checks are rarely carried out.

Aiquipa [25] suggests the need to use the Kaiser–Meyer–Olkin (KMO) sampling adequacy coefficient, complemented by Bartlett's test of sphericity, to check whether factor analysis is applicable.

As shown in Table 8, being 0.793 ≥ 0.5, it was observed that there was a strong correlation between the items; therefore, factor analysis was applicable [26].

Table 8. KMO and Bartlett's test.

Kaiser–Meyer–Olkin Measure of Sampling Adequacy		0.793
Bartlett's test for sphericity	Approx. Chi-Square	601.695
	df	78
	Sig.	0.001

Likewise, the *p*-value of Bartlett's test of sphericity, which is very sensitive to the sample size, turned out to be lower than the significance level in the social sciences (0.05), thus confirming, with more consistency, the suitability of performing factor analysis.

It is also convenient to review the data of the diagonal of the anti-image matrix to see to what extent they are close to unity or, if necessary, if it is necessary to eliminate any variable or item.

In this sense, Hair et al. [24] recommend eliminating, by default, the variable with the lowest value, which, as shown in Table 9, corresponded to item No. 3: "I am satisfied with the attention received prior to enrollment".

Table 9. Anti-image correlation matrix.

	Item 1	Item 2	Item 3	Item 4	Item 5	Item 6	Item 7	Item 8	Item 9	Item 10	Item 11	Item 12	Item 13
Item 1	0.813 [a]	−0.290	−0.111	0.110	−0.154	0.150	−0.232	−0.149	−0.200	−0.072	−0.029	−0.412	0.015
Item 2	−0.290	0.766 [a]	0.070	−0.189	0.050	0.007	0.017	−0.223	0.002	0.068	−0.064	0.169	−0.071
Item 3	−0.111	0.070	0.655 [a]	−0.509	0.019	−0.012	−0.143	−0.018	0.044	−0.089	0.053	0.103	0.155
Item 4	0.110	−0.189	−0.509	0.742 [a]	0.014	−0.083	0.011	0.016	−0.089	0.035	−0.003	−0.394	−0.096
Item 5	−0.154	0.050	0.019	0.014	0.830 [a]	−0.272	0.111	0.051	−0.082	−0.100	0.020	−0.090	−0.053
Item 6	0.150	0.007	−0.012	−0.083	−0.272	0.706 [a]	−0.215	−0.164	−0.159	0.285	−0.310	−0.035	−0.059
Item 7	−0.232	0.017	−0.143	0.011	0.111	−0.215	0.851 [a]	0.032	0.094	−0.155	−0.087	−0.046	0.019
Item 8	−0.149	−0.223	−0.018	0.016	0.051	−0.164	0.032	0.892 [a]	−0.015	−0.205	−0.115	−0.070	0.010
Item 9	−0.200	0.002	0.044	−0.089	−0.082	−0.159	0.094	−0.015	0.853 [a]	−0.118	−0.109	0.126	−0.089
Item 10	−0.072	0.068	−0.089	0.035	−0.100	0.285	−0.155	−0.205	−0.118	0.764 [a]	−0.392	0.029	−0.113
Item 11	−0.029	−0.064	0.053	−0.003	0.020	−0.310	−0.087	−0.115	−0.109	−0.392	0.830 [a]	−0.128	0.115
Item 12	−0.412	0.169	0.103	−0.394	−0.090	−0.035	−0.046	−0.070	0.126	0.029	−0.128	0.780 [a]	−0.352
Item 13	0.015	−0.071	0.155	−0.096	−0.053	−0.059	0.019	0.010	−0.089	−0.113	0.115	−0.352	0.812 [a]

[a] Measures of sampling adequacy (MSA).

Once this item was eliminated, factor analysis was performed again, showing an improvement in the KMO statistic (Table 10).

Table 10. KMO and Bartlett's test without item 3.

Kaiser–Meyer–Olkin Measure of Sampling Adequacy		0.807
Bartlett's test for sphericity	Approx. Chi-Square	538.981
	df	66
	Sig.	0.001

Table 11 shows how the anti-image matrix also generally yields values closer to unity than in the previous case.

The communalities matrix also improved in this case. Table 12 shows that the model is able to reproduce 65.1% of the variability of the first item, 68.5% of the second item, etc.

On the other hand, despite the fact that the sedimentation graph recommended selecting three factors for the analysis, i.e., those with eigenvalues above unity, it was decided to choose a number of four, based on the literature review, which recommends reaching an approximate value of 60–65% in social sciences for the rotated cumulative explained variance [27].

As shown in Table 13, after the rotation, there was a redistribution of the variability among the factors, with the four components explaining 63.095% of the total variability, which is an appropriate percentage in the context of the research.

Table 11. Anti-image correlation matrix without item No. 3.

	Item 1	Item 2	Item 4	Item 5	Item 6	Item 7	Item 8	Item 9	Item 10	Item 11	Item 12	Item 13
Item 1	0.814 [a]	−0.285	0.062	−0.153	0.150	−0.252	−0.152	−0.196	−0.083	−0.023	−0.405	0.033
Item 2	−0.285	0.771 [a]	−0.178	0.049	0.008	0.028	−0.222	−0.001	0.075	−0.068	0.163	−0.083
Item 4	0.062	−0.178	0.835 [a]	0.028	−0.103	−0.073	0.008	−0.078	−0.012	0.028	−0.399	−0.021
Item 5	−0.153	0.049	0.028	0.827 [a]	−0.272	0.115	0.052	−0.083	−0.099	0.019	−0.093	−0.056
Item 6	0.150	0.008	−0.103	−0.272	0.698 [a]	−0.219	−0.164	−0.159	0.285	−0.31	−0.034	−0.058
Item 7	−0.252	0.028	−0.073	0.115	−0.219	0.837 [a]	0.030	0.101	−0.17	−0.08	−0.032	0.042
Item 8	−0.152	−0.222	0.008	0.052	−0.164	0.03	0.889 [a]	−0.014	−0.207	−0.115	−0.069	0.013
Item 9	−0.196	−0.001	−0.078	−0.083	−0.159	0.101	−0.014	0.855 [a]	−0.114	−0.111	0.122	−0.097
Item 10	−0.083	0.075	−0.012	−0.099	0.285	−0.17	−0.207	−0.114	0.761 [a]	−0.390	0.039	−0.101
Item 11	−0.023	−0.068	0.028	0.019	−0.31	−0.08	−0.115	−0.111	−0.39	0.829 [a]	−0.134	0.108
Item 12	−0.405	0.163	−0.399	−0.093	−0.034	−0.032	−0.069	0.122	0.039	−0.134	0.774 [a]	−0.374
Item 13	0.033	−0.083	−0.021	−0.056	−0.058	0.042	0.013	−0.097	−0.101	0.108	−0.374	0.823 [a]

[a] Measures of sampling adequacy (MSA). Once item 3 has been removed.

Table 12. Communalities matrix.

No. Item	Initial	Extraction	No. Item	Initial	Extraction
1	1.0	0.651	8	1.0	0.574
2	1.0	0.685	9	1.0	0.577
4	1.0	0.572	10	1.0	0.578
5	1.0	0.591	11	1.0	0.611
6	1.0	0.682	12	1.0	0.776
7	1.0	0.700	13	1.0	0.574

Table 13. Total variance explained.

Component	Extraction Sums of Squared Loadings			Rotation Sums of Squared Loadings		
	Total	Variance %	Accumulated %	Total	Variance %	Accumulated %
1	4.427	36.892	36.892	2.171	18.089	18.089
2	1.160	9.664	46.557	2.093	17.440	35.529
3	1.075	8.960	55.517	1.705	14.207	49.736
4	0.909	7.579	63.095	1.603	13.360	63.095

Note. Extraction method: principal component analysis.

The matrix of rotated components provided the items corresponding to each of the factors (Table 14).

Table 14 shows that factor 1 refers to the methodology followed during the program. Factor 2 is related to the organization. Factor 3 refers to the fulfillment of the academic expectations of the graduate. Finally, factor 4 is related to the work of the teaching staff.

Table 15 shows the measurement criteria grouped by factors and other indicators.

3.8. Analysis of Variance (ANOVA)

ANOVA assumes normality and homoscedasticity.

Table 14. Matrix of rotated components.

Component	1	2	3	4
Item 12	0.793	0.308		
Item 4	0.716			
Item 13	0.715			
Item 1	0.483	0.422		0.466
Item 7		0.777		
Item 6		0.675	0.397	
Item 10		0.665		0.343
Item 5			0.724	
Item 11			0.693	
Item 9			0.616	0.431
Item 2				0.792
Item 8		0.440		0.564

Note. Values below 0.3 have been eliminated. Components or Items have been shaded for each factor.

Table 16 shows that residuals were normally distributed (*p*-value > 0.05).

Figure 2 illustrates that the residuals do not have any information, i.e., they do not follow a definite trend, such as a "funnel shape," in which case the hypothesis of equal variances would not be fulfilled. However, this is only a graphical test, which must then be confirmed with Levene's statistic (Table 17).

Table 15. Measurement criteria grouped by factors and measures of central tendency.

Factor	Item No.	Mean Item	Mean Factor	Measurement Criteria
Methodology	Item 12	3.71	3.71	I found the contents of the program interesting
	Item 4	3.61		The handling of the virtual campus has been very user friendly
	Item 13	3.8		My assessment of the continuous evaluation is very satisfactory
	Item 1	3.72		Delivery of didactic materials has been punctual and on time
Organization	Item 7	3.75	3.68	The Institution has provided me with facilities to carry out the study
	Item 6	3.63		My assessment of the organization of the program is very satisfactory
	Item 10	3.65		I found the information provided during the program to be sufficient
Academic expectations	Item 5	3.71	3.75	Overall, I am satisfied with the program
	Item 11	3.7		The academic program has met my initial expectations
	Item 9	3.85		This program will be relevant to my professional training and performance
Teaching work	Item 2	3.73	3.73	My tutor has gone out of his/her way to help me
	Item 8	3.73		My Master's Final project director has been accessible

Table 16. Test of Normality. Shapiro–Wilk Test (modified *).

Variable	n	*M*	*SD*	*W* *	*p*-Value (Unilateral D)
Residual Satisfaction	12	0.00	0.06	0.91	0.3917

Note. (*) indicates that it is a modified version of Shapiro-Wilk Test. M and SD represent mean and standard deviation, respectively.

Table 17. Test of homogeneity of variances.

	Levene Statistic	df_1	df_2	Sig.
Satisfaction	1.303	3	8	0.339

Levene's homogeneity test effectively confirms that analysis of variance can be performed (Significance Level > 0.05).

Analysis of variance confirmed Significance Level > 0.05, which means that differences in the mean factors were attributed to random chance (Table 18).

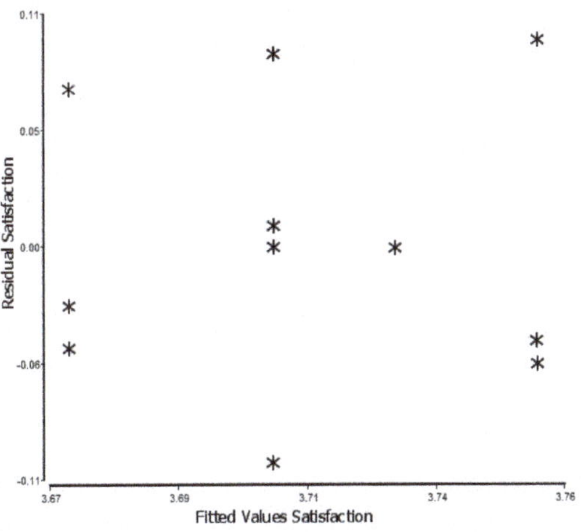

Figure 2. Residuals vs. fitted values.

Table 18. Table of analysis of variance.

F.V.	SS	*df*	MS	F	Sig.
Between	0.009	3	0.003	0.616	0.624
Within	0.041	8	0.005		
Total	0.05	11			

3.9. Categorization of Factors

Based on the four factors previously found, as many new variables were created as a result of the sum of the corresponding items for each of the 150 participants.

These values were then classified into different categories: "very low", "low", "medium" and "high," which are intended to give an idea of the level of satisfaction corresponding to each factor.

3.10. Overall Relationship between Variables

Before studying whether there are significant relationships between the factors and the rest of the variables, it is useful to approach this same objective from a global perspective.

3.10.1. Creation of the "Level of Satisfaction" Variable

To this end, a new global variable called "level of satisfaction" was created from the sum of the items for each of the 150 participants.

The descriptive data of central tendency and dispersion of this new variable are shown in Table 19.

3.10.2. Normality Test of the Data of the Variable "Satisfaction Level"

In order to determine whether the "satisfaction level" variable was distributed according to a normal distribution, and given that the sample was larger than 50 individuals, the Kolmogorov–Smirnov test was carried out in the SPSS software version 26 [28].

Table 19. Basic statistics of the level of satisfaction variable.

	N	Minimum	Maximum	*M*	*SD*
Level of Satisfaction	150	29	48	44.58	3.876

Note. *M* and *SD* represent mean and standard deviation, respectively.

The result is shown in Table 20.

Table 20. Kolmogorov–Smirnov normality test of the Satisfaction Level variable.

	Normality Tests					
Variable	Kolmogorov–Smirnov [a]			Shapiro–Wilk		
	Statistical	*df*	Sig.	Statistical	*df*	Sig.
Satisfaction Level	0.190	150	0.001	0.810	150	0.001

[a] Lilliefors correction.

Since the resulting p-value (0.001) is less than 0.05, the null hypothesis of a normal distribution of the data is rejected, so these values do not follow a normal distribution. This conditioned the notion that the method to be used later to determine a possible significant relationship between different variables would be nonparametric.

3.10.3. Categorization of the Variable "Level of Satisfaction"

Next, the "level of satisfaction" variable was grouped into a range of three categories: low, medium and high. For this purpose, two cut-off points were sought to establish these categories.

The following were considered:

$$44.58 - 0.75 * 3.876 \cong 42$$

$$44.58 + 0.75 * 3.876 \cong 47$$

Figure 3 shows the frequencies of the "graduate satisfaction" variable with the lower and upper cut-off values.

As a result, the ranges shown in Table 21 were obtained, to which a categorical value and level of satisfaction were assigned.

Table 21. Grouping ranges for evaluating graduate satisfaction levels.

Value	Range	Frequency	%	Level of Satisfaction
1	Values $\leq$ 42	34	23	Low
2	Values between 43 and 47	80	53	Medium
3	Values $\geq$ 48	36	24	High

Table 21 shows that 53% of the graduates had a medium level of satisfaction with the institution's online environmental postgraduate programs, 23% had a low level of satisfaction and the remaining 24% were highly satisfied after completing the corresponding postgraduate program.

Since the assumptions of normality of the grouped "level of satisfaction" variable were not met, parametric tests such as Pearson's test could not be applied, so it was necessary to resort to nonparametric, a priori, less robust ones.

Since there are nominal variables, it was not possible to apply Spearman's correlation; therefore, it was justified to resort to the chi-square test.

The results of applying the chi-square test to each of the five variables of the model against the grouped "level of satisfaction" variable gave rise to as many contingency tables, which are summarized in Table 22.

Since, in all cases, the *p*-value was $\geq$0.05, it was concluded to accept the null hypothesis of there being no relationship between variables in all cases.

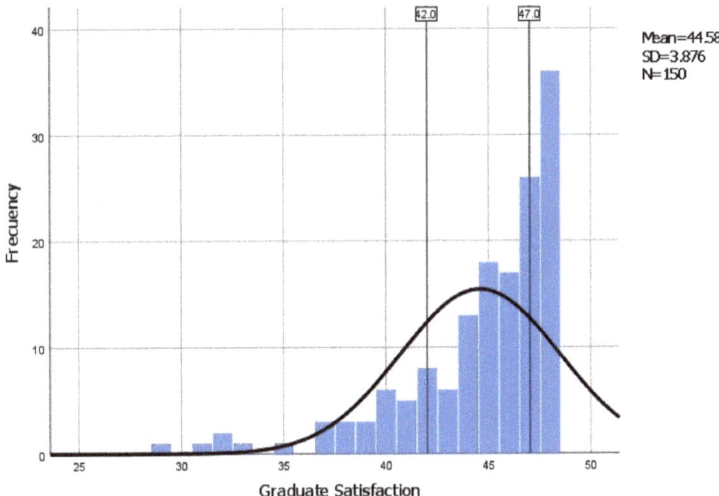

Figure 3. Frequency distribution of the "level of satisfaction" variable and normal distribution curve (mean = 44.58 and standard deviation = 3.876) overprinted. Note. The lower (42) and upper (47) cut-off values for determining the categories are shown.

Table 22. Summary of contingency tables of chi-square test values ("Grouped Level of Satisfaction" vs. "Other Variables").

Variables	Grouped Level of Satisfaction(p-Value) **
Gender	0.858
Age group	0.666
Origin	0.059
Entry profile	0.778
Program	0.327

Note. ** Variable statistically significant (p-value < 0.05).

3.11. Relationship between the Factors and the Rest of the Variables of the Model

As in the previous case, the results of applying the chi-square test to each of the four factors of the model vs. the rest of the variables gave rise to a set of contingency tables summarized in Table 23.

In this case, significant relationships between variables do appear. In scenarios where the p-value is < 0.05, it is concluded to reject the null hypothesis of there being no relationship between variables. A stacked bar chart is then justified in these cases.

This case is proof of how important and useful it is to perform a factor analysis, since by analyzing each factor separately, information unnoticed when approaching the problem from a global perspective comes to light.

Figure 4 shows that, in general, there is a medium to high level of satisfaction with the methodology followed; however, the "Biodiversity" and "Audits" programs show high levels of dissatisfaction with this factor.

Table 23. Summary of contingency tables of chi-square test values ("Level of Satisfaction by Factors" vs. "Other Variables").

Variables	Methodology	Organization	Academic Expectations	Teaching Work
Gender	0.738	0.369	0.072	0.482
Age group	0.927	0.750	0.984	0.462
Origin	0.166	0.003 **	0.045 **	0.187
Entry profile	0.369	0.954	0.831	0.811
Program	0.002 **	0.104	0.073	0.920

Note. ** Variable statistically significant (p-value < 0.05).

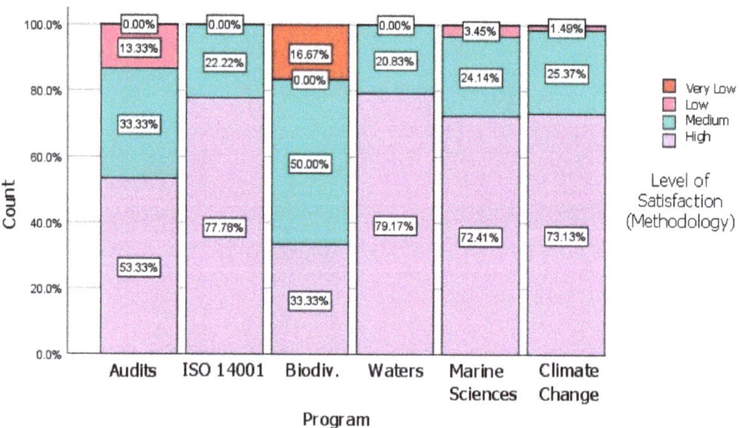

Figure 4. Level of satisfaction referring to methodology by program.

Figure 5 shows a fairly homogeneous level of satisfaction with the organization depending on the country of origin, except in the case of Eurasia, where graduates have very significant values of dissatisfaction, around 40%. There are also significant levels of dissatisfaction among graduates from South and Central America.

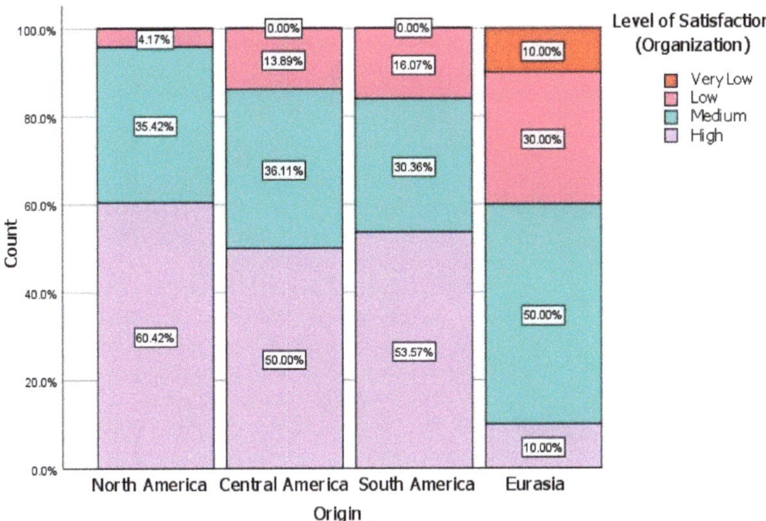

Figure 5. Level of satisfaction referring to the organization by origin.

Figure 6 shows that, with regard to academic expectations, there is a high level of satisfaction by areas of origin, except in the case of graduates from Eurasia, whose values contrast once again with the rest.

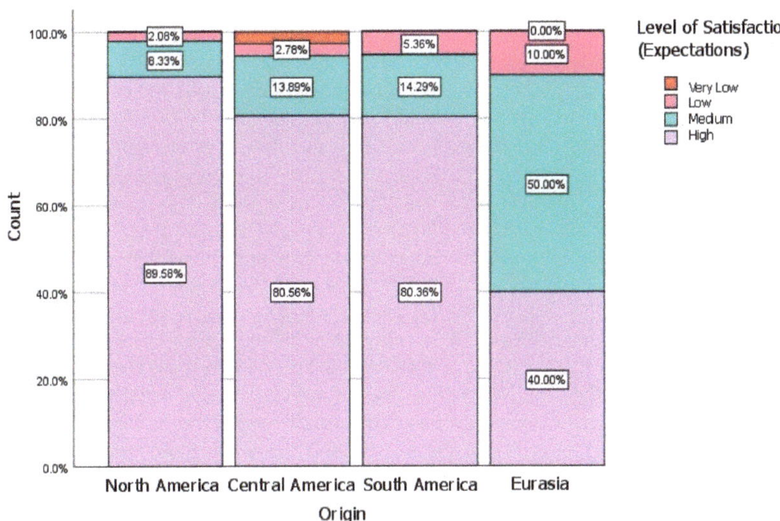

Figure 6. Level of satisfaction referring to academic expectations by origin.

4. Discussion

In this research article, the need to develop a model was proposed based on a series of variables collected in the bibliography to evaluate the satisfaction of 150 graduates of various online postgraduate programs in the environmental field, and to study possible relationships between them.

In this research, the estimated global mean of the variable "graduate satisfaction" was 3.72, very close to the maximum value of the Likert scale, with a standard error of 0.02, which means that the potential error made in the estimation with respect to the true average does not exceed 0.04 (with 95% confidence).

The objective was to implement the model in a Likert scale questionnaire ("1. Strongly disagree" to "4. Strongly agree"), in order to measure the variable of "graduate satisfaction" in relation to the reference postgraduate programs. This result is endorsed in most of the consulted literature references, which obtain the items of the measurement instrument, starting from indicators already contemplated in research by one or more authors, even when they differ in the number of used scales [5,6,16]. For example, Álvarez et al. [6] rely on the studies of [15]. In this sense, it is concluded that it is possible to obtain the items of the instrument from the indicators of other models, regardless of the number of scales contemplated.

With the aim of establishing its validity and reliability, the results yielded a value of Cronbach's alpha statistic of 0.836, which indicates that the measurement instrument is consistent over time and is within the range of values of good reliability. These results are similar to those found in most of the literature, where the values of this coefficient are above 0.80 [5,16], which hints at good to excellent reliability [18]. In most models, the validity of the content of the instrument is verified by means of a literature review and an expert panel consultation. If it is verified to be possible from a statistical point of view, an exploratory factor analysis is carried out, in order to find the factors based on the grouping of items [4,5]. In other cases, a criterion and construct validity is also performed [1].

In order to group the homogeneous variables into factors and simplify the complexity of the instrument, the feasibility of an exploratory factor analysis was investigated, resulting in a sampling adequacy KMO (Kaiser–Meyer–Olkin) of 0.807 and a Bartlett's test of sphericity, whose p-value was 0.000. Thus, the suitability of the application of factor analysis was demonstrated. The exploratory factor analysis yielded a total of four factors: methodology, organization, academic expectations and teaching work, which explained 63% of the total variance. The results are in agreement with some studies, such as Hair et al. [24], who allude to the elimination, by default, of the item with the lowest coefficient of the diagonal of the anti-image matrix, or Pardo and Ruíz [26], who use KMO values higher than 0.5 and a Bartlett's p-value of null sphericity to provide a good fit and thus find a reduced number of homogeneous groups or factors that can be crossed, in turn, with other variables to establish possible relationships between them [29]. As a result, it is concluded that with this data reduction technique, in addition to obtaining an instrument with good internal consistency, the operation with the data matrix is considerably simplified, as we work with only four factors after regrouping.

In this research, when determining the level of satisfaction with environmental postgraduate programs in general, it was found that 77% of the graduates had a medium-high level of satisfaction. In this sense, it was found that there were no statistically significant differences between the averages of the factors of the "graduate satisfaction" variable, so the estimated differences between averages were attributed to random chance.

Regarding the objective of finding some type of relationship between the level of graduate satisfaction and the rest of the variables (gender, entry profile, origin, program and age group), the results of the chi-square test applied globally did not yield significant results. Some research works corroborating these results are those of [4,5,16,30], who also found no significant differences between satisfaction and some dimensions such as gender, average years at university and school cycle. However, Kuo et al. [31], after conducting a preliminary test with a set of 111 students from the United States to measure their satisfaction in an online course, showed that satisfaction was conditioned by the handling of ICTs and that there were differences between gender, academic level (undergraduate and graduate) and time spent. However, the chi-square test applied to each individual factor did show significant relationships in the aspects of methodology, organization and expectations in relation to the programs studied and the origin of the graduate. In this sense, significant levels of dissatisfaction were found in the methodology of the "Auditing" and "Biodiversity" programs and from the point of view of organization and academic expectations in graduates from Central America, South America and the Eurasia zone, respectively. It is worth mentioning that, from an overall point of view, no significant differences were found among the variables, which provides an idea of the importance of performing factor analysis to discover findings that would otherwise go unnoticed. Finally, the fact that no other significant relationships were found between other variables does not mean that the relationship does not exist, but simply that it could not be demonstrated with a significance level of 0.05 and the number of study units with which it worked.

5. Conclusions and Recommendations

Throughout the investigation, the following points were found:

- The importance for educational institutions to know the degree of satisfaction of postgraduate graduates, either to attract new generations of students [2], to respond to third parties [4] or, fundamentally, to find out their needs [1].
- The existence of a great heterogeneity of models for measuring student satisfaction in general and graduate satisfaction in particular, some of them quite complex.
- The existence of partial and global approach models for virtual education, as well as standards for this training modality.
- The possibility of developing a Likert scale instrument to measure the satisfaction of 150 graduates of various online postgraduate programs in the environmental area.
- The validity and reliability of the measurement instrument, thanks to the assessment of a team of experts and the determination of a Cronbach's alpha value of 0.834, respectively. This value, close to unity, guarantees good internal consistency.
- The adequacy of factor analysis by providing KMO sampling adequacy values of 0.793, greater than 0.5, and a p-value close to or equal to zero for Bartlett's test of sphericity.
- An improvement in the KMO coefficient (from 0.793 to 0.807) when removing one of the variables or items from the study, as well as a very insignificant increase in Cronbach's alpha value (from 0.834 to 0.836).
- The existence of four new underlying variables or factors as a result of the factor analysis: methodology, organization, academic expectations and teaching work, which together explain 63% of the total variance.
- A medium-high level of satisfaction in the order of 77% with environmental online postgraduate programs.
- It was found that there were no statistically significant differences between the averages of the dimensions of the "graduate satisfaction" variable, so the estimated differences between averages were attributed to random chance.
- That the results of the chi-square test indicate that there is no significant overall relationship ($\alpha = 0.05$) between the level of satisfaction and the rest of the variables (gender, entry profile, origin, program and age group).
- That, at the level of individual factors, it is possible to establish significant relationships between methodology, organization and academic expectations with the programs and the origin of the graduate.
- That graduates from Central and South America have higher dissatisfaction values than the rest in the organizational sphere.
- That graduates from the Eurasian zone have higher dissatisfaction values than the rest in organizational aspects and academic expectations.
- That the "Audit" and "Biodiversity" programs present higher levels of dissatisfaction than the rest in relation to methodological issues.

To conclude, some recommendations would be the following:

- Find out what causes dissatisfaction rates of graduates from Central and South America, but especially from Eurasia, and afterwards review organizational processes and academic expectations.
- Review the methodological issues of the "Audit" and "Biodiversity" programs.

- Perform a comparison with other online, face-to-face postgraduate or undergraduate degrees [4].
- Complement the results obtained with the opinions of the teaching staff [32].
- Expand the sample with more participants from Europe and Asia.
- Improve the institution's policy to facilitate access to training according to the participants' social and economic context.

Author Contributions: Conceptualization, R.R.S. and A.P.B.; data curation, A.P.B.; formal analysis, R.R.S.; investigation, S.P.-V.; methodology, E.G.V.; project administration, I.D.N.; resources, K.T.P.; software, I.D.N.; supervision, S.P.-V.; validation research and measurement instrument validity, E.G.V., S.P.-V. and I.D.N.; visualization, S.P.-V.; writing—original draft, E.G.V.; writing—review and editing, I.D.N. All authors have read and agreed to the published version of the manuscript.

Funding: This research is part of European Project Erasmus + Lovedistance (Reference: 609949-EPP-1-2019-1-PTEPPKA2-CBHE-JP), funded by SODERCAN (Society for Regional Development of Cantabria), in conjunction with CITICAN (Investigation and Technology Centre of Cantabria).

Institutional Review Board Statement: Not applicable.

Informed Consent Statement: Informed consent was obtained from all subjects involved in the study.

Data Availability Statement: The data presented in this study are available on request from the corresponding author. The data are not publicly available due to the conditions of the project contract with the funder (Society for Regional Development of Cantabria).

Acknowledgments: The authors would like to thank to the Centro de Investigación y Tecnología Industrial de Cantabria (CITICAN) and the Universidad Europea del Atlántico for their valuable collaboration.

Conflicts of Interest: The authors declare no conflict of interest.

References

1. Mejías, A.; Martínez, D. Desarrollo de un instrumento para medir la satisfacción estudiantil en educación superior. *Docencia Univ.* 2009, 10. Available online: http://saber.ucv.ve/ojs/index.php/rev_docu/article/view/3704 (accessed on 15 March 2021).
2. González, R.; Tinoco, M.; Torres, V. Análisis de la satisfacción de la experiencia universitaria de los egresados en 2015 de la Universidad de Colima. *Paradig. Económico* **2017**, *8*, 59–84.
3. Pichardo, M.; García, B.A. El estudio de las expectativas en la universidad: Análisis de trabajos empíricos y futuras líneas de investigación. *REDIE: Rev. Electrónica Investig. Educ.* **2007**, *9*, 1–16.
4. Pérez, F.J.; Martínez, P.; Martínez, M. Satisfacción del estudiante universitario con la tutoría. Diseño y validación de un instrumento de medida. *Estud. Educ.* **2015**, *29*, 81–101.
5. Surdez, E.G.; Sandoval, M.d.C.; Lamoyi, C.L. Satisfacción estudiantil en la valoración de la calidad educativa universitaria. *Educ. Educ.* **2018**, *21*, 9–26. [CrossRef]
6. Álvarez, J.; Chaparro, E.M.; Reyes, D.E. Estudio de la satisfacción de los estudiantes con los servicios educativos brindados por instituciones de educación superior del Valle de Toluca. *REICE. Rev. Iberoam. Calid. Efic. Cambio Educ.* **2015**, *13*. Available online: https://revistas.uam.es/reice/article/view/2788 (accessed on 20 March 2021).
7. García, L. *La Educación a Distancia: De la Teoría a la Práctica*; Ariel: Barcelona, Spain, 2002.
8. Vann Slyke, C.; Kittner, M.; Belanger, F. Identifying Candidates for Distance education: A telecommuting perspective. In Proceedings of the America's Conference on Information System, Baltimore, MD, USA; 1998; pp. 666–668.
9. McArdle, G.E. *Training Design and Delivery*; American Society for Training and Development: Alexandria, VA, USA, 1999.
10. Kirkpatrick, D.L. *Evaluating Training Programs: The Four Levels*; Berret Koehler Publishers: San Francisco, CA, USA, 1994.
11. Pereira, A.; Gelvez, L.N. Propuesta de un modelo latinoamericano para apoyar la gestión de calidad de la educación virtual. Un enfoque dinámico sistémico. Available online: https://reposital.cuaed.unam.mx:8443/xmlui/bitstream/handle/20.500.12579/5314/VEAR18.0426.pdf?sequence=1&isAllowed=y (accessed on 15 March 2021).
12. Asociación Española de Normalización y Certificación, *Gestión de la Calidad. Calidad de la Formación Virtual*, (UNE 66181:2012). 2012.
13. Cabero-Almenara, J.; del-Carmen Llorente-Cejudo, M.; Puentes-Puente, A. Online Students' Satisfaction with Blended Learning. *Comunicar* **2009**, *18*, 149–157. [CrossRef]
14. García, N.; Rivero, M.L.; Ricis, J. Brecha digital en tiempo del COVID-19. *Rev. Educ. HEKADEMOS* **2020**, *28*, 76–85.
15. Gento, S.; Vivas, M. EL SEUE: Un Instrumento para Conocer la Satisfacción de los Estudiantes Universitarios con su Educación. *Acción Pedagógica* **2003**, *12*, 16–27.
16. Romo, J.R.; Mendoza, G.; Flores, G. Relaciones conceptuales entre calidad educativa y satisfacción estudiantil, evaluadas con ecuaciones estructurales: El caso de la facultad de filosofía y letras de la Universidad Autónoma de Chihuahua. 2012. Available online: http://cie.uach.mx/cd/docs/area_04/a4p11.pdf (accessed on 7 March 2021).

17. González, J.; Pazmiño, M. Cálculo e interpretación del Alfa de Cronbach para el caso de validación de la consistencia interna de un cuestionario, con dos posibles escalas tipo Likert. *Rev. Publicando* **2015**, *2*, 62–67.
18. Rodríguez, J.; Reguant, M. Calcular la fiabilidad de un cuestionario o escala mediante el SPSS: El coeficiente alfa de Cronbach. *REIRE Rev. d'Innovació Recer. Educ.* **2020**, *13*, 1–13. [CrossRef]
19. Fainholc, B. La calidad en la educación a distancia continúa siendo un tema muy complejo. *Rev. Educ. Distancia* **2004**. Available online: https://revistas.um.es/red/article/view/25311 (accessed on 11 March 2021).
20. Perero, G.; Isaac, C.L.; Díaz, S.; Ramos, Y. Propuesta de indicadores valorativos de la sostenibilidad de universidades ecuatorianas. *Ing. Ind.* **2020**, *41*, e4125.
21. Piza-Flores, V.; Aparicio, J.L.; Rodríguez, C.; Beltrán, J. Transversalidad del eje "Medio ambiente" en educación superior: Un diagnóstico de la Licenciatura en Contaduría de la UAGro. *RIDE. Rev. Iberoam. Investig. Desarro. Educ.* **2018**, *8*, 598–621. [CrossRef]
22. Hernández, R.; Fernández, C.; Baptista, P. *Metodología de la Investigación*, 3rd ed.; McGraw-Hill: New York, NY, USA, 2003.
23. Torres, M.; Karim, P. Tamaño de una muestra para una investigación de mercado. Facultad de Ingeniería. Universidad Rafael Landívar. *Boletín Electrónico* **2021**, *2*. Available online: https://docplayer.es/424351-Tamano-de-una-muestra-para-una-investigacion-de-mercado.html (accessed on 13 March 2021).
24. Hair, J.; Anderson, R.; Tatham, R.; Black, W. *Análisis Multivariante*, 5th ed.; Prentice Hall: Madrid, Spain, 1999.
25. Aiquipa, J. Diseño y validación del inventario de dependencia emocional. *Rev. Investig. Psicol.* **2012**, *15*, 133–145. [CrossRef]
26. Pardo, A.; Ruiz, M. *SPSS11. Guía para el Análisis de Datos*, 1st ed.; McGraw Hill: Madrid, Spain, 2002.
27. Vailati, P. Alfa de Cronbach y Análisis Factorial en SPSS—Investigación de Mercados II UADE. 2020. Available online: https://www.youtube.com/watch?v=PjZZeajjZYU&t=1603s (accessed on 23 March 2021).
28. De la Garza, J.; Morales, B.N.; González, B.A. *Análisis Estadístico Multivariante*; McGraw Hill: New York, NY, USA, 2013.
29. Johnson, D. *Métodos Multivariados Aplicados al Análisis de Datos*; International Thomson Editores: London, UK.
30. Troyano, Y.; García, A.J. Expectativas del alumnado sobre el profesorado tutor en el contexto del Espacio Europeo de Educación Superior. *Boletín RED-U* **2009**, *7*. [CrossRef]
31. Kuo, Y.C.; Walker, A.E.; Belland, B.R.; Schroder, K.E.E. A predictive study of student satisfaction in online education programs. *Int. Rev. Res. Open Distrib. Learn.* **2013**, *14*, 16–39. [CrossRef]
32. Llorent, V.; Cobano, V. Análisis crítico de las encuestas universitarias de satisfacción docente. *Rev. Educ.* **2019**, *385*, 91–117. [CrossRef]

Article

Total Saponins Isolated from Corni Fructus via Ultrasonic Microwave-Assisted Extraction Attenuate Diabetes in Mice

Shujing An [1], Dou Niu [1], Ting Wang [1], Binkai Han [1], Changfen He [1], Xiaolin Yang [1], Haoqiang Sun [1], Ke Zhao [2], Jiefang Kang [1,*] and Xiaochang Xue [2,*]

[1] National Engineering Laboratory for Resource Development of Endangered Crude Drugs in Northwest China, College of Life Sciences, Shaanxi Normal University, Xi'an 710119, China; Asjcyp@163.com (S.A.); kristin@snnu.edu.cn (D.N.); wangting86@snnu.edu.cn (T.W.); hanbinkaisnnu@163.com (B.H.); hcf14416106@163.com (C.H.); yangxiaolin163@126.com (X.Y.); sunhaoqiang32@163.com (H.S.)

[2] The Key Laboratory of Medicinal Resources and Natural Pharmaceutical Chemistry, The Ministry of Education, College of Life Sciences, Shaanxi Normal University, Xi'an 710119, China; zhaoke_1999@163.com

* Correspondence: kangjiefang@snnu.edu.cn (J.K.); xuexch@snnu.edu.cn (X.X.)

Citation: An, S.; Niu, D.; Wang, T.; Han, B.; He, C.; Yang, X.; Sun, H.; Zhao, K.; Kang, J.; Xue, X. Total Saponins Isolated from Corni Fructus via Ultrasonic Microwave-Assisted Extraction Attenuate Diabetes in Mice. *Foods* **2021**, *10*, 670. https://doi.org/10.3390/foods10030670

Academic Editors: Yadir Torres Hernández, Manuel Félix Ángel and Ana María Beltrán Custodio

Received: 12 February 2021
Accepted: 18 March 2021
Published: 22 March 2021

Publisher's Note: MDPI stays neutral with regard to jurisdictional claims in published maps and institutional affiliations.

Abstract: Saponins have been extensively used in the food and pharmaceutical industries because of their potent bioactive and pharmacological functions including hypolipidemic, anti-inflammatory, expectorant, antiulcer and androgenic properties. A lot of saponins-containing foods are recommended as nutritional supplements for diabetic patients. As a medicine and food homologous material, Corni Fructus (CF) contains various active ingredients and has the effect of treating diabetes. However, whether and how CF saponins attenuate diabetes is still largely unknown. Here, we isolated total saponins from CF (TSCF) using ultrasonic microwave-assisted extraction combined with response surface methodology. The extract was further purified by a nonpolar copolymer styrene type macroporous resin (HPD-300), with the yield of TSCF elevated to 13.96 mg/g compared to 10.87 mg/g obtained via unassisted extraction. When used to treat high-fat diet and streptozotocin-induced diabetic mice, TSCF significantly improved the glucose and lipid metabolisms of T2DM mice. Additionally, TSCF clearly ameliorated inflammation and oxidative stress as well as pancreas and liver damages in the diabetic mice. Mechanistically, TSCF potently regulated insulin receptor (INSR)-, glucose transporter 4 (GLUT4)-, phosphatidylinositol 3-kinase (PI3K)-, and protein kinase B (PKB/AKT)-associated signaling pathways. Thus, our data collectively demonstrated that TSCF could be a promising functional food ingredient for diabetes improvement.

Keywords: *Cornus officinalis*; saponins; ultrasonic extraction; microwave extraction; response surface methodology; type 2 diabetes mellitus

1. Introduction

Corni Fructus (CF), as a traditional Chinese medicine and food homologous plant, possesses a wide spectrum of biological activities including anti-hyperglycemia, antineoplastic, antimicrobial, anti-aging and antioxidant properties, as well as hepatoprotective and immune regulation effects [1–3]. As a result, CF has been extensively used in fruit wine, vinegar, jam, and health drinks to meet the increasing demand for production of functional foods that claim to provide health benefits [4,5]. CF is rich in morroniside, polyphenols, loganin, ursolic acid, iridoid glycosides, gallic acid and vitamin C [6,7]. Accumulating evidence has demonstrated that these bioactive constituents of CF could improve metabolic disorders in diabetes [8,9]. Therefore, CF has beneficial effects on treating and preventing diabetes and its complications.

Saponins, also referred to as triterpene glycosides, are a subclass of terpenoids. The amphipathic nature of saponins supplies them with the property as surfactants that can be used for the development of food, cosmetics and drugs [10]. There is enormous, commercially driven promotion of saponins as dietary supplements and functional foods. For example, the activity of β-galactosidase, one of the most important and classical biotechnological enzymes used in the food industry, can be increased by Quillaja bark saponin [10]. It is well known that saponins exhibit antidiabetic activity. Total saponins isolated from *Aralia taibaiensis* alleviated polydipsia, polyuria, polyphagia and weight loss of diabetic rats in a dose-dependent manner through increasing the levels of serum insulin, superoxide dismutase and reducing glutathione [11]. The hypoglycemic mechanism of total saponins from *Schisandra chinensis (Turcz.) Baill* was at least partially due to the activation of GLUT4, which is regulated by the IRS-1/PI3K/AKT pathway [12]. Astragalus total saponins and curcumin can achieve significant protective effects on diabetic nephropathy rats by improving the glycometabolism, insulin resistance (IR), lipid metabolism, oxidative stress levels, and pathological changes [13]. However, extremely few studies reported the isolation of the total saponins from CF (TSCF), and the role of TSCF on diabetes is almost completely unknown [14,15]. Thus, it is of great significance to study the extraction of TSCF and uncover its role in T2DM.

Unfortunately, the research on CF mainly focuses on the extraction technology of iridoid glycosides, polysaccharides, triterpenoid acids and tannins, and their anti-inflammatory and thirst-quenching properties, while the extraction technology and pharmacological effects of TSCF are rarely reported. Currently, commercially available saponins-containing products are very limited and predominantly derived from *Saponaria officinalis* and *Quillaja saponaria* by using ethanol, macroporous adsorption resin or n-butanol for extraction and polished purification. For example, the *Sanguisorba officinalis* ethanol extract, which mainly contained saponins, can inhibit bacterial toxin production at low concentrations [16]. However, the saponins in CF are more complicated and at least composed of eight glycosides [17]. In addition, the content of TS in CF is lower than that in *Saponaria officinalis* [18,19]. As a result, although various techniques including ethanol extraction, membrane separation [20], ultra-high pressure extraction [21], and chromatography were used to extract and purify TS from CF, extraction of TSCF with high yield is still difficult to achieve.

Compared to conventional extraction methods, which usually mean longer extraction time, higher temperature but with lower extraction efficiency, ultrasonic microwave-assisted extraction (UMAE), a technology that utilizes the advantages of microwave and sonochemistry to optimize the rate and efficiency of extraction, has been extensively used in the preparation of active ingredients of Chinese herbal medicine. It works well under mild conditions and thus, avoids the destruction of the components of interest [22]. As a statistical method to solve multivariable problems with high precision of regression equation and favorable fitting accuracy, response surface methodology (RSM) has been used in food, natural medicine and other fields [23]. These two methods can be combined to optimize the extraction technique with higher yields. We have reported recently that the total triterpenoid acids in CF can be extracted with high yields by UMAE optimized by RSM [24]. We wonder whether UAME can be used to extract TSCF with higher efficiency.

In this study, UMAE optimized by RSM was employed to extract TSCF, which was further purified with HPD-300 macroporous resin. By means of which the yield of TS elevated to 13.96 mg/g compared to 10.87 mg/g with unassisted extraction. In addition, we established T2DM model via feeding mice with high-fat diet (HFD) followed by streptozotocin (STZ) administration and assessed the therapeutic effects of TSCF on glucose metabolism, lipid metabolism, inflammation and oxidative stress in the diabetic mice.

2. Materials and Methods

2.1. Plant Materials

The mature fruit of CF was collected from Foping, Shaanxi Province (longitude 107.41° E to 108.10° E, latitude 33.16° N to 33.45° N, elevation 1120 m) in October 2017 and approved by Yaping Xiao, a plant identification expert from Shaanxi Normal University. CF was treated as previously described [24]. In brief, after removing the inner ripe core, the outer pulp of CF was dried, and the sarcocarp was crushed and passed through a 40-mesh sieve. The powder was stored at −20 °C for later use.

2.2. Chemicals and Reagents

Oleanolic acid standard (Purity $\geq$ 98%) was purchased from Chengdu Manster Biotechnology Co. (Chengdu, China). Streptozotocin (STZ) was from Sigma Chemical Co. (St. Louis, MO, USA). HPD-300 macroporous resin was from Shaanxi Shenlan Technology Co. Ltd. (Hanzhong, China). All the other chemicals used in the present work were at least of analytical grade.

2.3. Standard Curve of TS

As TS is a multi-ingredient mixture, oleanolic acid (OA) is used to draw the standard curve for TS quantification according to the method previously described [24]. Briefly, 0.2 mg/mL OA standard solution was firstly prepared by dissolving 2.0 mg OA standard substance in 10 mL methanol. Then, the solution was divided into a series of different volumes in the test tubes (concentration range: 0.2–1.0 µg/mL). After removing the methanol via rotary evaporation, 0.2 mL vanillin–glacial acetic acid solution (5%) and 0.8 mL perchloric acid solution were sequentially added to OA samples followed by a 15-min water bath at 70 °C and cooled to room temperature in flowing water. Finally, 5 mL glacial acetic acid solution was added and the absorbance value of each sample was detected at 554 nm with an ultra-micro spectrophotometer (Thermo Fisher Scientific, Waltham, MA, USA). The standard curve was drawn with the quantity of OA (Y = abscissa) and absorbance value (X = ordinate). The regression line for OA was Y = 0.0379X − 0.0209 (R^2 = 0.9992).

2.4. Single Factor Experimental Design for Extraction of Total Saponins from CF (TSCF) with Ultrasonic-Microwave Assisted Extraction (UMAE)

The values of each single factor in the TSCF extraction process were screened using an XO-200 UMAE system by changing one single factor and keeping the others unchanged. In brief, one gram of CF powder was sealed with 70% aqueous ethanol with solid/liquid ratios of 1:10, 1:15, 1:20, 1:25, and 1:30 g/mL, respectively. The extraction time was set for 3–15 min. The ultrasonic powers ranged in 240–480 W, while the microwave powers ranged in 200–600 W. After extracting at room temperature for the indicated time, samples were centrifuged and the supernatant was gathered for absorbance value measurement as mentioned in Section 2.3. TSCF content was finally calculated with the equation:

$$\text{content of TSCF (mg/g)} = \frac{CeVe}{Me} \tag{1}$$

in which *Ce* (mg/mL), *Ve* (mL), and *Me* (g) are TSCF concentration in the extract, volume of the extract, and weight of the dried CF powder, respectively.

2.5. Optimization of TSCF Extraction by RSM

To optimize the extraction parameters by RSM, a four-variable (X_1, X_2, X_3 and X_4) three-level (high (+1), intermediate (0), and low (−1), respectively) Box–Behnken design was employed to determine the collective influence of them on the extraction of TSCF. As shown in Table 1, X_1 (solid/liquid ratio), X_2 (microwave power), X_3 (ultrasonic power), and X_4 (extraction time) are the coded variables. Thus, 29 experimental conditions were obtained and each of them was conducted in triplicate. All data were fitted to the following quadratic second degree polynomial model:

$$Y = \beta_0 + \sum_{i=1}^{4} \beta_{ii}X_i^2 + \sum_{i<j=2}^{4} \beta_{ij}X_iX_j \tag{2}$$

where Y is the predicted response; β_0 is the model constant; β_{ii} and β_{ij} are the quadratic and cross-product coefficients; X_i and X_j are independent variables. The experimental design analysis and prediction were conducted by using Design-Expert® software (8.0.5b version).

Table 1. Results of central composite experimentation.

No.	X_1 (g/mL)	X_2 (W)	X_3 (W)	X_4 (min)	Theoretical Yield of TS (mg/g)	Actual Yield of TS (mg/g)
1	0 (1/20)	−1 (300	1 (420)	0 (6)	11.38	11.42
2	0 (1/20)	1 (500)	−1 (300)	0 (6)	11.56	11.48
3	0 (1/20)	1 (500)	0 (360)	−1 (3)	12.66	12.69
4	1 (1/25)	0 (400)	0 (360)	1 (9)	10.64	10.38
5	0 (1/20)	−1 (300)	0 (360)	1 (9)	9.71	9.76
6	−1 (1/15)	−1 (300)	0 (360)	0 (6)	11.65	11.54
7	0 (1/20)	0 (400)	1 (420)	−1 (3)	12.60	12.31
8	0 (1/ 20)	0 (400)	0 (360)	0 (6)	13.76	13.94
9	0 (1/ 20)	0 (400)	0 (360)	0 (6)	13.76	13.75
10	0 (1/ 20)	0 (400)	−1 (300)	1 (9)	9.83	10.08
11	0 (1/ 20)	−1 (300)	0 (360)	−1 (3)	12.29	12.57
12	−1 (1/15)	1 (500)	0 (360)	0 (6)	11.65	12.02
13	−1 (1/15)	0 (400)	1 (420)	0 (6)	11.65	11.52
14	1 (1/ 25)	1 (500)	0 (360)	0 (6)	12.83	12.91
15	0 (1/ 20)	0 (400)	−1 (300)	−1 (3)	12.70	12.42
16	0 (1/ 20)	0 (400)	1 (420)	1 (9)	10.58	10.82
17	0 (1/ 20)	1 (500)	1 (420)	0 (6)	12.71	12.51
18	0 (1/ 20)	0 (400)	0 (360)	0 (6)	13.76	13.57
19	0 (1/ 20)	0 (400)	0 (360)	0 (6)	13.76	13.85
20	0 (1/ 20)	0 (400)	0 (360)	0 (6)	13.76	13.71
21	1 (1/ 25)	0 (400)	1 (420)	0 (6)	13.01	13.34
22	0 (1/ 20)	−1 (300)	−1 (300)	0 (6)	11.89	12.05
23	1 (1/ 25)	0 (400)	−1 (300)	0 (6)	12.00	12.21
24	−1 (1/15)	0 (400)	−1 (300)	0 (6)	12.00	11.75
25	−1 (1/15)	0 (400)	0 (360)	1 (9)	9.98	9.89
26	0 (1/ 20)	1 (500)	0 (360)	1 (9)	10.35	10.15
27	1 (1/ 25)	−1 (300)	0 (360)	0 (6)	11.83	11.43
28	1 (1/ 25)	0 (400)	0 (360)	−1 (3)	13.11	13.15
29	−1 (1/15)	0 (400)	0 (360)	−1 (3)	12.40	12.62

2.6. Purification of Crude TSCF

The crude TSCF extracts were further purified by chromatography. In brief, the sample was concentrated and re-suspended with H_2O to 3.5 mg/mL and loaded onto a column packaged with HPD-300 resin and pre-equilibrated with distilled water at a flow rate of 2 bed volume (BV)/h. After removing the unbound remnants with 3 BV of distilled water, TS were finally eluted with 8.5 BV of 50% ethanol at a flow rate of 1.5 mL/min.

To determine the TS extracted from CF, the extracted products were dissolved in aqueous ethanol to a serial-diluted concentration so that it fell within the linear range of the standard curve. Then, 0.2 mL extract solution was used for TS determination by the method in Section 2.3.

2.7. Establishment of T2DM Model in Mice

Sixty male ICR mice (5 weeks old) weighed 16 ± 2 g were purchased from the Animal Center of College of Medicine, Xi'an Jiaotong University (Xi'an, China). All mice were kept under controlled temperatures (22–23 °C) and relative humidity (40–60%) on a 12 h light/dark cycle and with free access to water and normal diet for 7 days before the experiment. Animal studies were carried out according to the National Institutes of Health's Guide for the Care and Use of Laboratory Animals and were approved by the Animal Care and Use Committee of Shaanxi Normal University.

The T2DM mouse model was established according to the method we described previously [24]. In brief, the mice of the normal control (NC) group were fed with normal chow diet (11% kcal fat), whereas all the others were given a high-fat diet (HFD) (58% kcal fat, Slac Laboratory Animal, Shanghai, China) for 4 weeks. After fasting for 12 h, all HFD-fed mice received intraperitoneally administered STZ at a dosage of 60 mg/kg once every three days, four times, and equal volume of citrate buffer was administered for NC mice. Then, the mice were fasted for another 12 h and blood samples were promptly collected from the tail vein for the determination of fasting blood glucose (FBG) values. Mice with FBG $\geq$ 7.8 mmol/L were confirmed as T2DM for further studies.

2.8. Treat T2DM Mice with TSCF

The confirmed diabetic mice were divided into 5 groups: T2DM model control (DM), positive control (PC) and three TSCF-treated groups. The experimental design flow chart is shown in Figure 1. For TSCF treatment, diabetic mice were intragastrically (i.g.) administered with 50 (TSCF-L), 100 (TSCF-M), and 200 mg/kg (TSCF-H) of TSCF once a day for 4 weeks. Body weight was measured once a week. All mice were fasted for 12 h after the last administration, and blood samples were collected from the retro-orbital sinus for serum preparation and biochemical analysis. Then, the mice were killed and tissues (liver, kidney, and spleen) were immediately removed and weighed. One part of the tissues was stored at −80 °C for quantitative PCR analysis, while the other was fixed in 4% paraformaldehyde solution for histopathological examination. The organ index was calculated as the organ weight divided by the body weight.

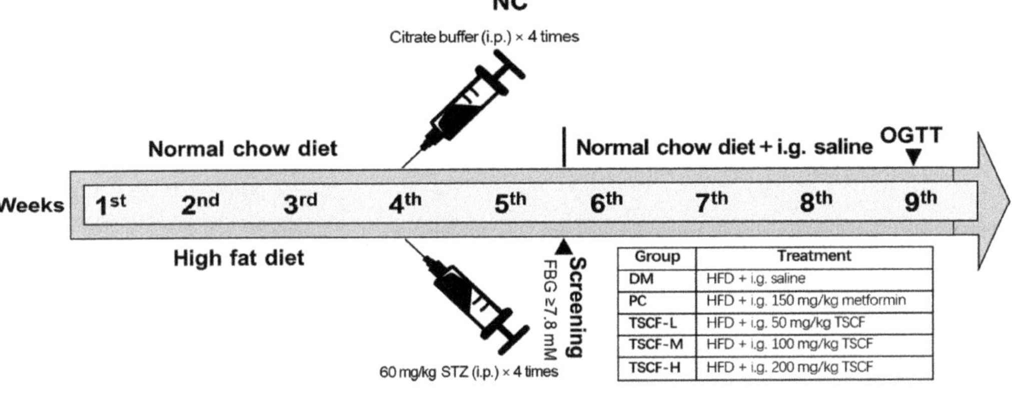

Figure 1. Schematic diagram of the experimental process. i.p., intraperitoneal injection; i.g., intragastric administration; HFD, high-fat diet; OGTT, oral glucose tolerance test; FBG, fasting blood glucose; STZ, streptozotocin; NC, normal control group; DM, diabetes mellitus model group; PC, positive (metformin-treated) control group; TSCF, total saponins from Corni Fructus; TSCF-L, low-dose TSCF group; TSCF-M, middle-dose TSCF group; TSCF-H, high-dose TSCF group.

2.9. Fasting Blood Glucose (FBG) and Oral Glucose Tolerance Test (OGTT)

The blood samples were obtained weekly from the tail vein of the mice after a 12-h fast, and FBG was measured by using a One-Touch glucometer (HMD Biotechnology Co., Ltd., Qingdao, China). OGTT was performed after four weeks of TSCF administration. In brief, the mice were firstly fasted for 12 h and FBG was determined as the 0 min blood glucose level. Then, glucose solution (2 g/kg) was administered to the mice by oral gavage and blood glucose (BG) values were determined at 30, 60, and 120 min, respectively. The OGTT, presented via a complete area under the curve (AUC) from 0 to 120 min, was calculated by the formula: AUC = (FBG + BG$_{30min}$) × 1/4 + (BG$_{30min}$ + BG$_{60min}$) × 1/4 + (BG$_{60min}$ + BG$_{120min}$) × 1/2.

2.10. Biochemical Assay

The plasma total cholesterol (TC), total triglyceride (TG), low-density lipoprotein cholesterol (LDL-C) and high-density lipoprotein cholesterol (HDL-C), and the superoxide dismutase (SOD) activity and malondialdehyde (MDA) content of liver homogenate were measured using marketing diagnostic kits (Nanjing Jiancheng Bioengineering Institute, Nanjing, China). The inflammatory cytokines tumor necrosis factor (TNF)-α, interleukin (IL)-6, and C-reactive protein (CRP) in the liver homogenate, and fasting insulin (FINS) levels in the serum, were tested by ELISA kits (Tianjin Anoric Biotechnology Co., Ltd., Tianjin, China). All measurements were performed according to the manufacturer's protocols. The homeostasis model assessment (HOMA)-insulin sensitivity (HOMA-IS) index and HOMA-insulin resistance (HOMA-IR) index were calculated following the formulas: HOMA-IS = ln [1/(FINS × FBG)], HOMA-IR = (FBG × FINS)/22.5 [25].

2.11. Histopathological Examinations

Histology of the livers and pancreas was studied using hematoxylin and eosin (H&E) staining following the standard method. In brief, fresh isolated liver and pancreas samples were sequentially fixed in 4% paraformaldehyde solution, embedded in paraffin, and cut into 5-μm thick sections. Sections were stained with H&E and were detected by light microscopy. Finally, the images were examined and evaluated with a CX23 biomicroscope (Olympus, Kyoto, Japan).

2.12. RNA Quantitation by Real-Time qPCR

Real-time PCR analysis of AKT, GLUT4 (glucose transport 4), INSR (insulin receptor), and PI3K mRNA expression was performed using an ABI 7500 PCR System (Applied Biosystems, Carlsbad, CA, USA). Primers for AKT were: forward 5′-TT TGGGAA GGTGATTCTGGTG-3′; reverse 5′-CGTAAGGAAGGGATGCCTAGA GTT-3′. Primers for INSR were: forward 5′-CAAGAAATGATTCAGATGACAGCAG-3′; reverse 5′-AGA CTCCATCCTTCAGGGACTCA-3′. Primers for GLUT4 were: forward 5′-CCCCATTCCC TGGTTCATT-3′; reverse 5′-GACCCATAGCATCCGCAAC-3′. Primers for PI3K were: forward 5′-GACCAATACTTGATGTGGCTGACG-3′; reverse 5′-CTCGCAATAGGTTCT CCGCTTT-3′. Primers for the control housekeeping gene β-ACTIN were: forward 5′-GC CTTCCTTCTTGGGTAT-3′; reverse 5′-GGCATAGAGGTCTTTACGG-3′. The fold change of target genes was calculated with the $2^{-\Delta\Delta Ct}$ method.

2.13. Statistical Analysis

All experiments were repeated in triplicate and the data were expressed as means $\pm$ standard deviation. Statistical analysis was performed by one-way analysis of variance (ANOVA) and Duncan's multiple range tests and $p < 0.05$ was considered statistically significant.

3. Results and Discussion

3.1. Optimization of UMAE Conditions by RSM for TSCF Extraction

To determine which factor in UMAE mainly influences TSCF yield and the range for RSM optimization, variable factors such as solid/liquid ratio, etc., were screened by single factor experiments. As shown in Figure 2, although these factors have different effects on TSCF yield, they do have the optimal value: solid/liquid ratio = 1:20 g/mL, ultrasonic power = 360 W, extraction time = 6 min, and microwave power = 400 W. The effect of ethanol concentration on TSCF extraction was also investigated, but no apparent effects were found (data not shown). Therefore, the constant 70% ethanol was used for all experiments. Interestingly, the origin of CF has a much significant effect on TSCF extraction because of various TS contents. We previously screened CF from 15 origins and found that the TS content in Foping CF is about 25.81 $\pm$ 1.33 mg/g, which is an ideal material for TSCF extraction (unpublished Chinese article).

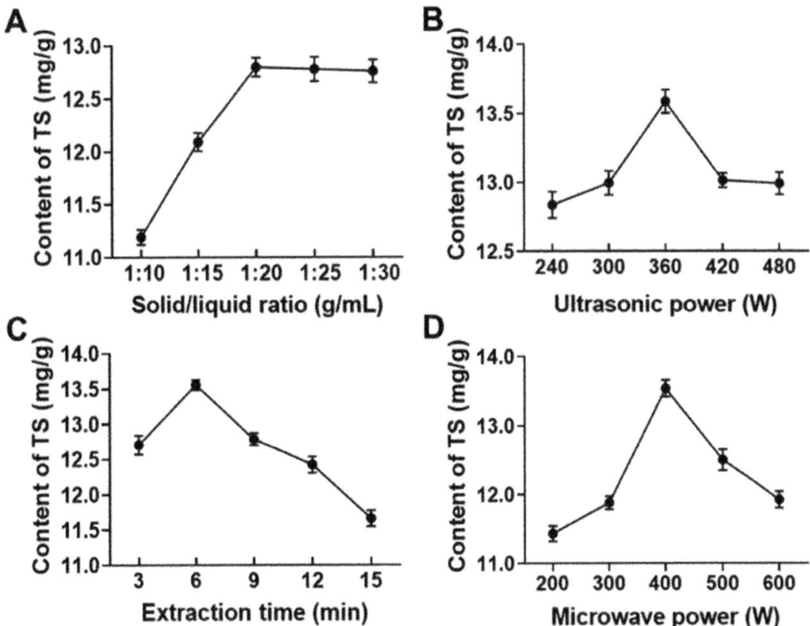

Figure 2. Screening of the variable factors that affect TSCF extraction yield. TSCF was extracted under certain conditions, including solid/liquid ratio (**A**), ultrasonic power (**B**), microwave power (**C**), and extraction time (**D**), and the extraction yield was determined.

Then, we confirmed the most influential factors and their possible interactions via optimizing the UMAE conditions by RSM. In order to avoid losing the optimal value, a wider range including the best condition was selected for further RSM studies in which 29 experiments were designed involving three levels of four solitary factors (Table 1). The maximum yield (13.94 mg/g) of TSCF was found in the condition of X_1 = 1:20 g/mL, X_2 = 400 W, X_3 = 360 W, and X_4 = 6 min. Quadratic regression analysis of the data was carried out with Design-Expert 8.05b software and fitted with the following second-order polynomial equation:

$$Y = 13.76 + 0.34X_1 + 0.25X_2 + 0.16X_3 - 1.22X_4 + 0.25X_1X_2 + 0.34X_1X_3$$
$$-0.01X_1X_4 + 0.42X_2X_3 + 0.068X_2X_4 + 0.21X_3X_4 \qquad (3)$$
$$-0.75X_1{}^2 - 1.02X_2{}^2 - 0.85X_3{}^2 - 1.49X_4{}^2$$

where R1 is the yield of TSCF; X_1, X_2, X_3 and X_4 represent the factors of solid/liquid ratio, microwave power, ultrasonic power, and extraction time, respectively.

The statistical significance of the regression model was measured by F-test and *p*-value, and the analysis of variance (ANOVA) for the response surface quadratic model is shown in Table S1. The high F-value, very low *p*-value, the coefficient of determination, the adjusted coefficient of determination, non-significant lack of fit, and the adequate precision collectively demonstrated that the model was significantly accurate for prediction of response within the range of experimental variables. This was further confirmed by the fact that the predicted yields of TSCF were fitted well to the actual values (Table 1).

To determine the effects of these variables, individually and in combination, on TSCF production, three-dimensional response surface plots were generated. As shown in Figure 3, the response (TSCF production) was plotted on the Z-axis against any two independent variables, while keeping other variables at a fixed optimal level. The data were well fitted into the foregoing equation, and the most favorable level of each variable was determined to be as follows: X_1 = 1:21.47 g/mL, X_2 = 417.37 W, X_3 = 368.78 W, and

X_4 = 4.81 min. For ease of operation, these parameters were adjusted and confirmed by experiments to be 1:21 g/mL, 417 W, 369 W, and 5 min, respectively. Under these conditions, the actual yield of TSCF was 13.96 ± 0.14 mg/g, which was highly consistent with the theoretical value. These data demonstrated that RSM was suitable and sufficient for TSCF extraction optimization. However, as an energy consumption technology, further studies are still needed to confirm whether UMAE may bring obstacles in the future commercial development of TSCF products.

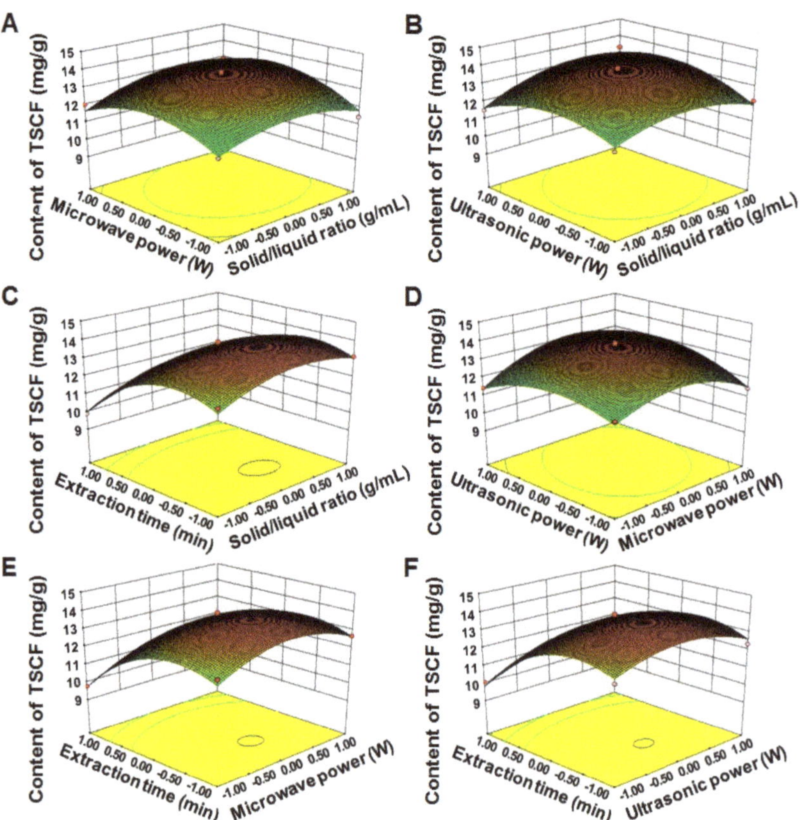

Figure 3. Three-dimensional response surface plot for TSCF production. The contour plots showed the interactive effects of the solid/liquid ratio and microwave power (**A**), solid/liquid ratio and ultrasonic power (**B**), solid/liquid ratio and extraction time (**C**), microwave power and ultrasonic power (**D**), microwave power and extraction time (**E**), and ultrasonic power and extraction time (**F**), respectively.

3.2. Purification of Crude TSCF

After UMAE, the extraction solution was filtrated and the ethanol was removed by rotary evaporation. The insoluble residue was removed by water-saturated n-butanol two-phase extraction until the n-butanol phase was colorless. Then, n-butanol was removed by evaporation under reduced pressure, and the liquid was fixed to 20 mL with 70% ethanol for TSCF quantification. The purity of the crude TSCF is only about 6.23%, which indicated that there are still a lot of impurities in the products. We found that X-5 macroporous resin is suitable for purifying total triterpenoid acids from CF, while Zhao et al. reported that HPD-300 macroporous resin had the best adsorption and desorption properties for extracting TS from CF as compared with other eight different types of macroporous resins. Then, the TSCF crude extract was dissolved in water and loaded onto an HPD-300 macroporous resin column for further polished purification. The purity of TSCF increased from 6.23% to 37.36%, which was six times higher than before. The eluted TSCF was finally lyophilized and stored at $-20\ ^\circ$C for later use.

At present, there is no legal standard substance for TSCF quantification. Considering the single saponin (such as ginsenoside) and the aglycone (oleanolic acid) share similar structure to TSCF, they are usually selected to draw the standard curve, and TSCF can be calculated by colorimetry. The reaction conditions should be strictly controlled because the reaction of TSCF with vanillin glacial acetic acid reagent is sensitive to temperature and time.

3.3. The Influence of TSCF on Body Weight and Organ Index of Diabetic Mice

We wondered whether TSCF has protective effects on diabetic mice; therefore, TSCF was used to treat T2DM mice established by HFD-feeding and STZ injection. There was no significant difference in the initial body weight of the mice in each group before TSCF administration. At the end of the study, the weight of DM mice was significantly lower than that of the NC group ($p < 0.05$). However, after 4 weeks of gavage administration of TSCF, the body weight loss of diabetic mice was significantly ameliorated and showed a relatively stable state, in contrast to the steady loss observed in the DM group ($p < 0.05$), whereas no significant difference was found between TSCF-treated and NC groups (Table 2).

Table 2. Effect of TSCF on the body weight of diabetic mice.

Group	Body Weight (g)				
	0 w	1 w	2 w	3 w	4 w
NC	33.33 ± 2.78	33.68 ± 2.70	34.04 ± 1.80 $^\triangle$	33.66 ± 1.85 $^\triangle$	32.71 ± 1.63 $^\triangle$
DM	33.29 ± 2.69	31.35 ± 2.76	30.89 ± 1.89 *	29.01 ± 2.13 **	28.65 ± 1.48 *
PC	32.81 ± 1.74	32.18 ± 2.35	32.41 ± 2.28	32.73 ± 2.32 $^\triangle$	32.42 ± 1.56 $^\triangle$
TSCF-L	33.23 ± 1.98	32.04 ± 2.28	32.47 ± 0.81	32.47 ± 0.81 $^\triangle$	32.15 ± 1.57 $^\triangle$
TSCF-M	33.65 ± 2.67	33.07 ± 2.65	32.54 ± 1.46	33.26 ± 2.22 $^{\triangle\triangle}$	32.41 ± 1.17 $^\triangle$
TSCF-H	32.48 ± 2.38	32.58 ± 1.66	32.27 ± 1.10	33.33 ± 1.77 $^{\triangle\triangle}$	32.25 ± 1.63 $^\triangle$

Data were expressed as means ± SD (n = 10). * $p < 0.05$, ** $p < 0.01$ vs. NC group; $^\triangle$ $p < 0.05$, $^{\triangle\triangle}$ $p < 0.01$ vs. DM group.

The organ index can be used to determine the health of the internal organs of T2DM mice. As shown in Figure 4, kidney, liver and spleen indexes of mice in the DM group significantly increased when compared with the NC mice ($p < 0.05$ or $p < 0.01$). Actually, the liver and spleen indexes of the DM mice were nearly double those of the NC mice, which indicated that diabetes seriously damaged the liver and kidneys. However, TSCF potently suppressed these trends and all the organ indexes were restored to the normal level. In particular, no obvious differences can be found for liver, kidney and spleen indexes between TSCF-H and NC groups ($p > 0.05$). These data suggested that TSCF could significantly lessen the damage of organs in diabetic mice.

3.4. The Effects of TSCF on Glucose Metabolism of T2DM Mice

Impaired glycemic control is a main feature of diabetes. FBG and OGTT are the most commonly used indexes to detect diabetes mellitus, which reflects the function of islet β cells, and the ability of the body to regulate blood glucose. As shown in Figure 5A, FBG of T2DM mice elevated remarkably as compared with NC mice ($p < 0.01$). Both metformin and TSCF administration potently reduced FBG levels in T2DM mice ($p < 0.01$), although the levels were still higher than those of NC mice. In addition, the effect of high dosage of TSCF on FBG reduction was even better than metformin. As to OGTT, the blood glucose of all mice reached peak level at 30 min following oral glucose challenge. Thereafter, metformin, intermediate and high dosage of TSCF significantly suppressed the increase in blood glucose, and reached a much lower level at 120 min when compared with the DM mice (Figure 5B). Consistently, both TSCF-H and PC mice had the most significantly low levels of AUC as compared with the DM mice (Figure 5C, $p < 0.01$). These data revealed that TSCF is beneficial for blood glucose maintenance.

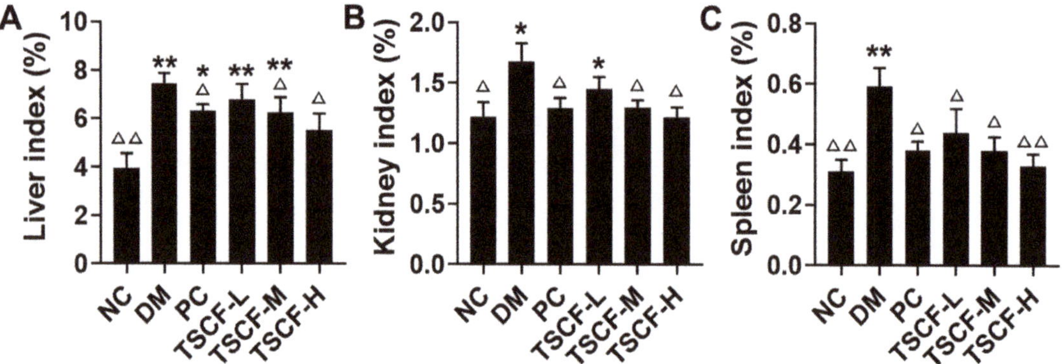

Figure 4. Effects of TSCF on the organ indexes of diabetic mice. (**A**) Liver index; (**B**) kidney index; (**C**) spleen index. NC, normal control group; DM, diabetic mice treated with PBS; PC, diabetic mice treated with 100 mg/kg metformin each day; TSCF-L, TSCF-M, and TSCF-H, diabetic mice treated with 50, 100, and 200 mg/kg TSCF each day. * $p < 0.05$, ** $p < 0.01$ compared with the NC group; $^\Delta$ $p < 0.05$, $^{\Delta\Delta}$ $p < 0.01$, compared with the DM group.

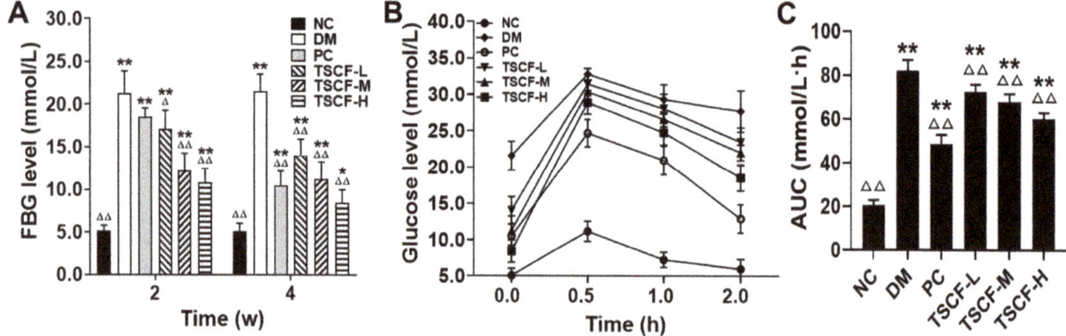

Figure 5. Effects of TSCF on glycemic modulation of T2DM mice. (**A**) FBG; (**B**) OGTT; (**C**) AUC; * $p < 0.05$, ** $p < 0.01$ compared with the NC group; $^\Delta$ $p < 0.05$, $^{\Delta\Delta}$ $p < 0.01$ compared with the DM group.

IR, which is indispensable for the development of T2DM, constitutes a pathophysiological state, where insulin fails to regulate glucose homeostasis in peripheral tissues [26]. In order to quantify the IR and β-cell function in diabetic mice, FINS, HOMA-IS, and HOMA-IR levels were determined. We found that FINS and HOMA-IS were downregulated, whereas HOMA-IR was upregulated in the mice of the DM group as compared with the mice of the NC group ($p < 0.01$) (Table 3), which indicated damaged islet cells and obvious IR. However, this condition was greatly relieved by TSCF treatment and the FINS level was clearly restored to the normal level in the TSCF-M and TSCF-H groups; there were no significant differences between them and the mice of NC group ($p > 0.05$). Similarly, TSCF clearly improved the insulin sensitivity in T2DM mice as indicated by the HOMA-IS and HOMA-IR levels ($p < 0.05$ or $p < 0.01$ vs. DM group), and all of them were restored to the normal baseline under TSCF treatment. Notably, the effect of TSCF-H was comparable to, if not more potent than, that of the positive metformin. Collectively, all these data confirmed that TSCF could increase insulin secretion and sensitivity and relieve IR.

Table 3. Effect of TSCF on FINS, HOMA-IS, and HOMA-IR of diabetic mice.

Group	FINS (mIU/L)	HOMA-IS	HOMA-IR
NC	9.46 ± 1.71 $^{\triangle\triangle}$	-3.84 ± 0.25 $^{\triangle\triangle}$	2.13 ± 0.55 $^{\triangle\triangle}$
DM	5.89 ± 0.73 **	-4.93 ± 0.15 **	6.19 ± 1.02 **
PC	8.40 ± 1.92 $^{\triangle\triangle}$	-4.46 ± 0.22 **,$^{\triangle\triangle}$	3.92 ± 0.90 **,$^{\triangle\triangle}$
TSCF-L	7.03 ± 0.97 *	-4.53 ± 0.19 **,$^{\triangle\triangle}$	4.38 ± 0.61 **,$^{\triangle\triangle}$
TSCF-M	8.26 ± 0.69 $^{\triangle\triangle}$	-4.45 ± 0.15 **,$^{\triangle\triangle}$	3.84 ± 0.57 **,$^{\triangle\triangle}$
TSCF-H	9.43 ± 1.53 $^{\triangle\triangle}$	-4.31 ± 0.08 **,$^{\triangle\triangle}$	3.52 ± 0.81 *,$^{\triangle\triangle}$

Data were expressed as means $\pm$ SD (n = 10). * $p < 0.05$, * $p < 0.01$ vs. NC group; $^{\triangle}$ $p < 0.05$, $^{\triangle\triangle}$ $p < 0.01$ vs. DM group.

3.5. The Effects of TSCF on Lipid Metabolism in T2DM Mice

The metabolic profile of T2DM includes not only impaired glucose metabolism, but also dyslipidemia combined with elevated lipid profile levels [27]. Table 4 shows the effect of TSCF on the TG, TC, LDL-C and HDL-C levels in diabetic mice. Compared with NC mice, DM mice exhibited significantly higher levels of TG, TC and LDL-C; lower levels of HDL-C. TSCF treatment for 4 weeks led to a great decrease in TG and LDL-C ($p < 0.01$) and a significant increase in HDL-C ($p < 0.01$) as compared with the mice of the DM group. A number of recent studies have claimed that diabetes is strongly associated with increased lipid accumulation, and elevated blood glucose was accompanied by increases in serum TC and TG in diabetic rats [28]. The interesting findings of this study are that treatment with TSCF reduced not only FBG but also total cholesterol and TG levels in T2DM mice. At the same time, administration of TSCF increased HDL-C but decreased LDL-C. These results revealed that TSCF could improve lipid metabolism in T2DM mice.

Table 4. Effect of TSCF on the lipid levels in diabetic mice ($\overline{x} \pm s$, mmol/L).

Group	TC	TG	HDL-C	LDL-C
NC	2.49 ± 0.11 $^{\triangle\triangle}$	0.67 ± 0.08 $^{\triangle\triangle}$	1.68 ± 0.13 $^{\triangle\triangle}$	2.98 ± 0.79 $^{\triangle\triangle}$
DM	2.89 ± 0.28 **	0.93 ± 0.11 **	1.01 ± 0.96 **	4.81 ± 0.71 **
PC	2.66 ± 0.13	0.68 ± 0.11 $^{\triangle\triangle}$	1.66 ± 0.11 $^{\triangle\triangle}$	4.48 ± 0.71 **
TSCF-L	2.50 ± 0.23 $^{\triangle\triangle}$	0.73 ± 0.09 $^{\triangle\triangle}$	1.35 ± 0.09 **,$^{\triangle\triangle}$	3.70 ± 0.30 *,$^{\triangle\triangle}$
TSCF-M	2.69 ± 0.16	0.70 ± 0.06 $^{\triangle\triangle}$	1.39 ± 0.12 **,$^{\triangle\triangle}$	3.45 ± 0.25 $^{\triangle\triangle}$
TSCF-H	2.68 ± 0.23	0.67 ± 0.10 $^{\triangle\triangle}$	1.46 ± 0.11 *,$^{\triangle\triangle}$	3.21 ± 0.43 $^{\triangle\triangle}$

Data were expressed as means $\pm$ SD (n = 10). * $p < 0.05$, ** $p < 0.01$ compared with NC group; $^{\triangle}$ $p < 0.05$, $^{\triangle\triangle}$ $p < 0.01$ compared with DM group.

3.6. Effects of TSCF on Inflammation and Oxidative Stress in T2DM Mice

Inflammation is the body's defensive response to stimulation, and inflammatory factors play critical roles in regulating the insulin signaling pathway and the structure and function of islet β cells [29]. IL-6, TNF-α, CRP and other inflammatory factors secreted by various cells can induce a cytokine signal to activate c-Jun amino terminal kinase (JNK), reducing the activation of downstream PI3K, thus inhibiting the expression and synthesis of GLUT4, and leading to β cell dysfunction and IR [30,31]. We found that IL-6, TNF-α, and CRP levels were greatly upregulated in the liver tissues of T2DM mice compared to the NC mice. After treating these T2DM mice with TSCF for 4 weeks, all these cytokines were decreased significantly ($p < 0.01$) in a TSCF dose-dependent manner (Figure 6A–C).

Considering the levels of MDA and SOD in liver tissues reflect the severity of free radical attack and the ability of scavenging free radicals, they can be used to detect the damage of liver function. As shown in Figure 6D,E, MDA clearly increased, whereas SOD activities significantly decreased in the diabetic mice as compared with the normal mice ($p < 0.01$). In contrast, both MDA levels and SOD activities tended to restore to normal levels as compared to DM mice after a 4-week treatment with TSCF. Additionally, TSCF dose-dependently recovered SOD levels of the T2DM mice. It has been reported that hepatic inflammation and oxidative stress are closely connected with diabetes [32]. Thus, all these data suggested that TSCF potently attenuated inflammation and oxidative stress in T2DM mice.

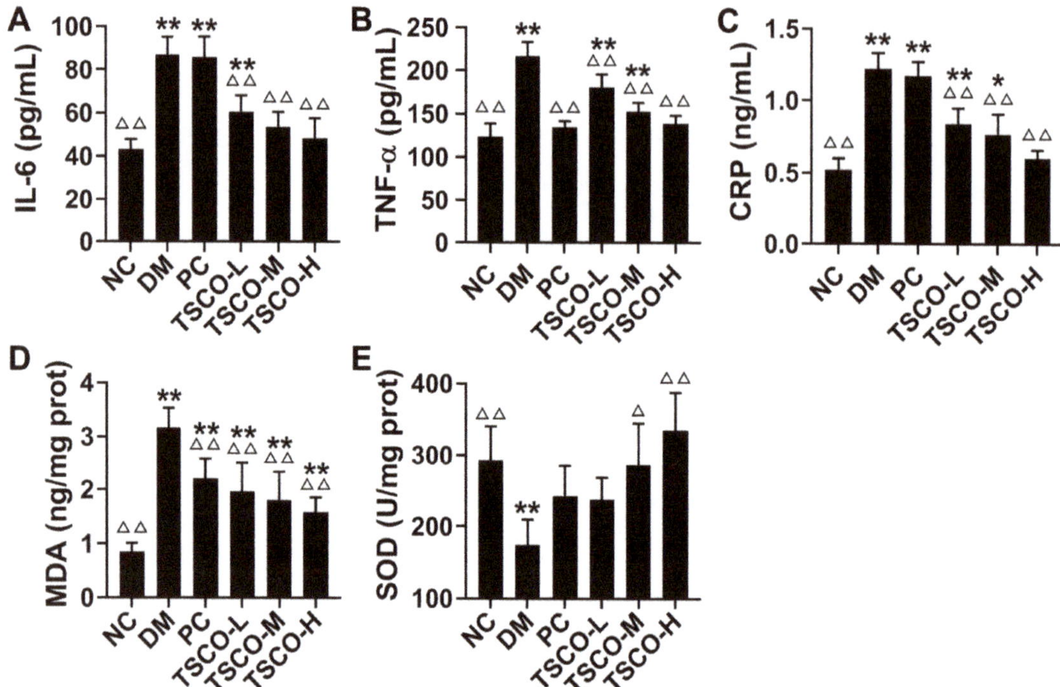

Figure 6. Effects of TSCF on inflammation and oxidative stress in T2DM mice. Sera of diabetic mice were prepared and (**A**) IL-6, (**B**) TNF-α, (**C**) CRP, (**D**) MDA, and (**E**) SOD levels were measured. NC, normal control group; DM, T2DM model group; PC, Metformin group; TSCF-L, TSCF-M, and TSCF-H, diabetes mice treated with 50 mg/kg, 100 mg/kg and 200 mg/kg TSCF per day, respectively. * $p < 0.05$, ** $p < 0.01$ compared with NC group; $^{\triangle}$ $p < 0.05$, $^{\triangle\triangle}$ $p < 0.01$ compared with DM group.

3.7. TSCF Ameliorated Pancreas and Liver Damages in T2DM Mice

The trends of organ index, glucose metabolism, inflammatory cytokines production and oxidative stress suggested that TSCF could improve organ damages in T2DM mice. To confirm this, histopathological assays were performed. As shown in Figure 7, HFD and STZ triggered severe injury to the pancreas and liver of the diabetic mice. Intact pancreatic structure and neatly aligned islet β cells can be seen in NC mice, while the pancreatic cells were severely damaged, with dilated ducts and accumulated fat tissue, represented by big and optically empty cells filled with lipids, in the DM and TSCF-L groups. However, all these features were significantly reduced in TSCF-M, TSCF-H and PC groups (Figure 7A). As to liver tissues, NC mice showed intact normal hepatic histology, but HFD-STZ-induced DM mice exhibited numerous cytotoxic and inflammatory alterations including steatosis with some affected cells, vacuolar and hydropic degenerations, hepatocellular necrosis, and leukocyte infiltrations. In contrast, a significant decline in the severity and frequency of hepatic lesions was observed in the livers of TSCF-treated mice (Figure 7B). Meanwhile, the protective effect of TSCF on liver can also be found by organ index (Figure 4A). All these data revealed that TSCF could ameliorate the hepatic and pancreatic injuries in T2DM mice, and therefore, may be a promising candidate for diabetes treatment.

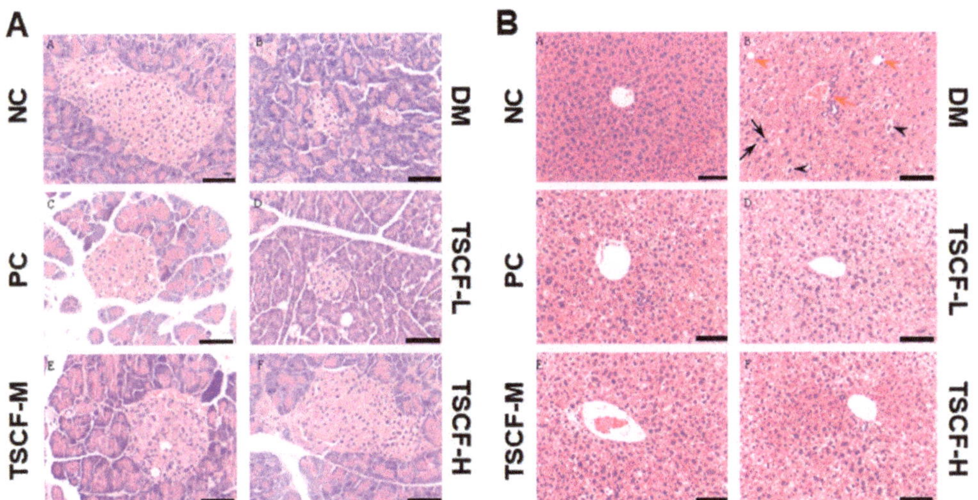

Figure 7. Histopathological changes of diabetic mice treated with TSCF. Pancreas (**A**) and liver (**B**) tissues were isolated from diabetic mice and H&E staining was performed according to standard methods. Apoptotic hepatocytes (black arrow), focal mononuclear cell aggregation (red arrow), diffuse hydropic degeneration (black arrowheads), fatty change (red arrowheads). NC, normal control group; DM, T2DM model group; PC, Metformin group; TSCF-L, TSCF-M, and TSCF-H, diabetes mice treated with 50, 100 and 200 mg/kg TSCF per day, respectively. Bar = 100 μm.

3.8. TSCF Regulates INSR, GLUT4, PI3K and AKT Signaling Pathways in Skeletal Muscle of T2DM Mice

It is well known that skeletal muscle is the principal site for glucose regulation and metabolism, and muscle insulin resistance is pivotal to abnormal glucose metabolism in diabetes. Considering that INSR-, GLUT4-, PI3K- and AKT-associated signaling pathways play critical roles in insulin-glucose regulation and homeostasis in skeletal muscle, we determined the effects of high-dosage TSCF on these signaling pathways to elucidate the possible molecular mechanisms of TSCF on T2DM mice. We found that T2DM induction greatly decreased ($p < 0.01$), whereas high-dose TSCF and metformin administration remarkably restored the level of INSR, GLUT4, PI3K and AKT in skeletal muscle of mice (Figure 8). In addition, the effect of TSCF-H on the expression of GLUT4 in the skeletal muscle of mice was seven times greater than that of the metformin ($p < 0.05$). These data suggested that TSCF suppressed hyperlipidemia and hyperglycemia in diabetic mice by regulating the INSR-, GLUT4-, PI3K- and AKT-associated signaling pathways to maintain the homeostasis of insulin-glucose metabolism.

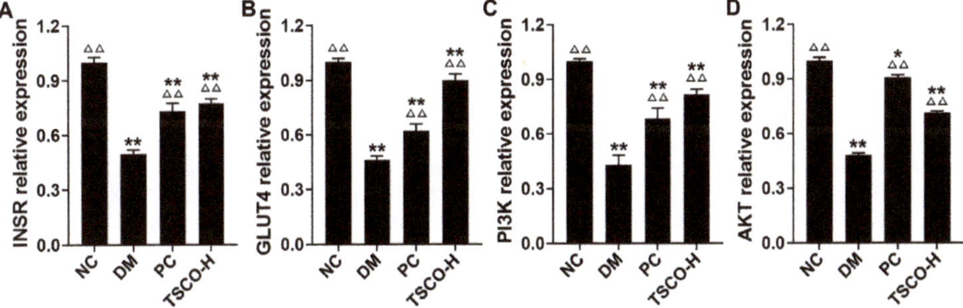

Figure 8. The effects of TSCF on MDA (**A**) INSR, (**B**) GLUT4, (**C**) PI3K, and (**D**) AKT signaling pathways in T2DM mice. * $p < 0.05$, ** $p < 0.01$ compared with NC group; $^{\triangle}$ $p < 0.05$, $^{\triangle\triangle}$ $p < 0.01$ compared with DM group.

4. Conclusions

T2DM is a complex metabolic disorder associated with pancreas dysfunction and varying degrees of insulin resistance, accounting for over 90% of all individuals diagnosed with diabetes. CF has great potential as a drug homologous food for the treatment of diabetes. Optimizing the processing technology is an effective way to improve the yield of active ingredients for the industrial production of medicinal and food homologous plants. In this study, RSM was used to optimize the UMAE of TSCF and the yield of TSCF increased from 10.87 up to 13.96 mg/g. If administered to HFD-TSZ-induced T2DM mice, TSCF significantly ameliorated the syndrome of diabetic mice as indicated by improved glucose and lipid metabolism, reduced inflammation and oxidative stress, and ameliorated pancreas and liver damages in the mice. Mechanistically, TSCF may alleviate hyperlipidemia and hyperglycemia in diabetic mice by regulating the INSR-, GLUT4-, PI3K- and AKT-associated signaling pathways. In summary, TSCF has a therapeutic effect on T2DM and can be used as a potential natural product to treat diabetes.

Supplementary Materials: The following are available online at https://www.mdpi.com/2304-8158/10/3/670/s1, Table S1: Variance analysis for the regression model.

Author Contributions: Conceptualization, J.K.; methodology, J.K.; software, J.K. and X.X.; validation, J.K. and X.X.; formal analysis, S.A., J.K. and X.X.; investigation, S.A., D.N., T.W., B.H., C.H., X.Y., H.S. and K.Z.; resources, J.K.; data curation, J.K. and X.X.; writing—original draft preparation, S.A. and J.K.; writing—review and editing, X.X.; visualization, J.K. and X.X.; supervision, J.K. and X.X.; project administration, J.K. and X.X.; funding acquisition, J.K. All authors have read and agreed to the published version of the manuscript.

Funding: This work was funded by the National Key Technologies Research and Development Program for Modernization of Traditional Chinese Medicine (Grant No. 2017YFC1701302), the Fundamental Research Funds for the Central Universities (Grant No. GK201706012), the major Project of Shaanxi Province, China (Grant No. 2020ZDLSF05-10), and the Central University Innovation Team Project (GK202001006).

Data Availability Statement: Not applicable.

Conflicts of Interest: The authors declare no conflict of interest.

References

1. Park, C.H.; Noh, J.S.; Kim, J.H.; Tanaka, T.; Zhao, Q.; Matsumoto, K.; Shibahara, N.; Yokozawa, T. Evaluation of Morroniside, Iridoid Glycoside from Corni Fructus, on Diabetes-Induced Al-terations such as Oxidative Stress, Inflammation, and Apoptosis in the Liver of Type 2 Diabetic db/db Mice. *Biol. Pharm. Bull.* **2011**, *34*, 1559–1565. [CrossRef] [PubMed]
2. Wang, D.; Li, C.; Fan, W.; Yi, T.; Wei, A.; Ma, Y. Hypoglycemic and hypolipidemic effects of a polysaccharide from Fructus Corni in streptozotocin-induced diabetic rats. *Int. J. Biol. Macromol.* **2019**, *133*, 420–427. [CrossRef]
3. Lee, N.H.; Seo, C.S.; Lee, H.Y.; Jung, D.Y.; Lee, J.K.; Lee, J.A.; Song, K.Y.; Shin, H.K.; Lee, M.Y.; Seo, Y.B.; et al. Hepatoprotective and Antioxidative Activities of *Cornus officinalis* against Acetamino-phen-Induced Hepatotoxicity in Mice. *Evid.-Based Compl. Alt.* **2012**, *2012*, 804924.
4. Nouska, C.; Kazakos, S.; Mantzourani, I.; Alexopoulos, A.; Bezirtzoglou, E.; Plessas, S. Fermentation of Cornus Mas L. Juice for Functional Low Alcoholic Beverage Production. *Curr. Res. Nutr. Food Sci. J.* **2016**, *4*, 119–124. [CrossRef]
5. Kawa-Rygielska, J.; Adamenko, K.; Kucharska, A.Z.; Piórecki, N. Bioactive Compounds in Cornelian Cherry Vinegars. *Molecules* **2018**, *23*, 379. [CrossRef]
6. Xu, M.; Wu, H.Y.; Liu, H.; Gong, N.; Wang, Y.R.; Wang, Y.X. Morroniside, a secoiridoid glycoside from *Cornus officinalis*, attenu-ates neuropathic pain by activation of spinal glucagon-like peptide-1 receptors. *Brit. J. Pharmacol.* **2017**, *174*, 580–590. [CrossRef] [PubMed]
7. Niu, D.; An, S.; Chen, X.; Bi, H.; Zhang, Q.; Wang, T.; Han, B.; Zhang, H.; Kang, J. Corni Fructus as a Natural Resource Can Treat Type 2 Diabetes by Regulating Gut Microbiota. *Am. J. Chin. Med.* **2020**, *48*, 1–23. [CrossRef]
8. Chen, C.C.; Hsu, C.Y.; Chen, C.Y.; Liu, H.K. Fructus Corni suppresses hepatic gluconeogenesis related gene transcription, en-hances glucose responsiveness of pancreatic beta-cells, and prevents toxin induced beta-cell death. *J. Ethnopharmacol.* **2008**, *117*, 483–490. [CrossRef] [PubMed]
9. Feng, G.-D.; Xue, X.-C.; Gao, M.-L.; Wang, X.-F.; Shu, Z.; Mu, N.; Gao, Y.; Wang, Z.-L.; Hao, Q.; Li, W.-N.; et al. Therapeutic Effects of PADRE-BAFF Autovaccine on Rat Adjuvant Arthritis. *BioMed. Res. Int.* **2014**, *2014*, 1–9. [CrossRef]

10. Kayukawa, C.T.M.; Oliveira, M.A.S.; Kaspchak, E.; Sanchuki, H.B.S.; Igarashi-Mafra, L.; Mafra, M.R. Quillaja bark saponin effects on Kluyveromyces lactis beta-galactosidase activity and structure. *Food Chem.* **2020**, *303*, 125388. [CrossRef]

11. Weng, Y.; Yu, L.; Cui, J.; Zhu, Y.-R.; Guo, C.; Wei, G.; Duan, J.-L.; Yin, Y.; Guan, Y.; Wang, Y.-H.; et al. Antihyperglycemic, hypolipidemic and antioxidant activities of total saponins extracted from Aralia taibaiensis in experimental type 2 diabetic rats. *J. Ethnopharmacol.* **2014**, *152*, 553–560. [CrossRef] [PubMed]

12. Xu, J.; Wang, S.; Feng, T.; Chen, Y.; Yang, G. Hypoglycemic and hypolipidemic effects of total saponins from Stauntonia chinensis in diabetic db/db mice. *J. Cell. Mol. Med.* **2018**, *22*, 6026–6038. [CrossRef]

13. Liu, B.; Miao, J.; Peng, M.; Liu, T.; Miao, M. Effect of 3:7 ratio of Astragalus total saponins and Curcumin on the diabetic nephropathy rats model. *Saudi J. Biol. Sci.* **2019**, *26*, 188–194. [CrossRef]

14. Lin, M.-H.; Liu, H.-K.; Huang, W.-J.; Huang, C.-C.; Wu, T.-H.; Hsu, F.-L. Evaluation of the Potential Hypoglycemic and Beta-Cell Protective Constituents Isolated from Corni Fructus To Tackle Insulin-Dependent Diabetes Mellitus. *J. Agric. Food Chem.* **2011**, *59*, 7743–7751. [CrossRef] [PubMed]

15. Huang, J.; Zhang, Y.; Dong, L.; Gao, Q.; Yin, L.; Quan, H.; Chen, R.; Fu, X.; Lin, D. Ethnopharmacology, phytochemistry, and pharmacology of Cornus officinalis Sieb. et Zucc. *J. Ethnopharmacol.* **2018**, *213*, 280–301. [CrossRef] [PubMed]

16. Pu, Z.; Tang, H.; Long, N.; Qiu, M.; Gao, M.; Sun, F.; Dai, M. Assessment of the anti-virulence potential of extracts from four plants used in traditional Chinese medicine against multidrug-resistant pathogens. *BMC Complement. Med. Ther.* **2020**, *20*, 1–10. [CrossRef] [PubMed]

17. Zhong, J.H.; Li, C.S.; Li, D.D. Progress in research of the active chemical components of Cornus officinalis. *Chin. J. New. Drugs* **2001**, *10*, 808–812.

18. Jarzębski, M.; Siejak, P.; Smułek, W.; Fathordoobady, F.; Guo, Y.; Pawlicz, J.; Trzeciak, T.; Kowalczewski, P.Ł.; Kitts, D.D.; Singh, A.; et al. Plant Extracts Containing Saponins Affects the Stability and Biological Activity of Hempseed Oil Emulsion System. *Molecules* **2020**, *25*, 2696. [CrossRef]

19. Fu, S.Y.; Li, Z.J.; Wang, Z.T. Study on the gross saponinsin natural beverage of Cornus officinalis. *Zhejiang Linxueyuan Xuebao* **1998**, *15*, 105–107.

20. Shi, G.Q.; Ma, T.T.; Liu, K.; Liu, Y.Q. Purification of total saponins of Cornus officinalis by membrane separations technology. *Henan Chem. Ind.* **2015**, *9*, 19–23.

21. Zheng, Q.X.; Liu, H.; Li, J.R. The technology of extraction process saponin from Cornus officinalis Sieb. et Zucc. at ultra high pressure. *Food Sci. Tech.* **2011**, *36*, 183–186.

22. You, Q.H.; Yin, X.L.; Zhang, S.N.; Jiang, Z.H. Extraction, purification, and antioxidant activities of polysaccharides from Tricholo-ma mongolicum Imai. *Carbohyd. Polym.* **2014**, *99*, 1–10. [CrossRef] [PubMed]

23. Bezerra, M.A.; Santelli, R.E.; Oliveira, E.P.; Villar, L.S.; Escaleira, L.A. Response surface methodology (RSM) as a tool for optimization in analytical chemistry. *Talanta* **2008**, *76*, 965–977. [CrossRef] [PubMed]

24. Han, B.; Niu, D.; Wang, T.; An, S.; Wang, Y.; Chen, X.; Bi, H.; Xue, X.; Kang, J. Ultrasonic–microwave assisted extraction of total triterpenoid acids from Corni Fructus and hypoglycemic and hypolipidemic activities of the extract in mice. *Food Funct.* **2020**, *11*, 10709–10723. [CrossRef] [PubMed]

25. Matthews, D.R.; Hosker, J.P.; Rudenski, A.S.; Naylor, B.A.; Treacher, D.F.; Turner, R.C. Homeostasis model assessment: Insulin re-sistance and beta-cell function from fasting plasma glucose and insulin concentrations in man. *Diabetologia* **1985**, *28*, 412–419. [CrossRef]

26. Mu, J.; Xin, G.; Zhang, B.; Wang, Y.; Ning, C.; Meng, X. Beneficial effects of Aronia melanocarpa berry extract on hepatic insulin resistance in type 2 diabetes mellitus rats. *J. Food Sci.* **2020**, *85*, 1307–1318. [CrossRef]

27. Tang, W.; Li, S.; Liu, Y.; Huang, M.-T.; Ho, C.-T. Anti-diabetic activity of chemically profiled green tea and black tea extracts in a type 2 diabetes mice model via different mechanisms. *J. Funct. Foods* **2013**, *5*, 1784–1793. [CrossRef]

28. Tian, Y.M.; Liu, Y.; Wang, S.; Dong, Y.; Su, T.; Ma, H.J.; Zhang, Y. Anti-diabetes effect of chronic intermittent hypobaric hypoxia through improving liver insu-lin resistance in diabetic rats. *Life Sci.* **2016**, *150*, 1–7. [CrossRef]

29. Choi, Y.H.; Jin, G.Y.; Li, G.Z.; Yan, G.H. Cornuside suppresses lipopolysaccharide-induced inflammatory mediators by inhibiting nuclear factor-kappa B activation in RAW 264.7 macrophages. *Biol. Pharm. Bull.* **2011**, *34*, 959–966. [CrossRef]

30. Gao, Y.-F.; Zhang, M.-N.; Wang, T.-X.; Wu, T.-C.; Ai, R.-D.; Zhang, Z.-S. Hypoglycemic effect of D-chiro-inositol in type 2 diabetes mellitus rats through the PI3K/Akt signaling pathway. *Mol. Cell. Endocrinol.* **2016**, *433*, 26–34. [CrossRef]

31. Jiang, S.; Ren, D.; Li, J.; Yuan, G.; Li, H.; Xu, G.; Han, X.; Du, P.; An, L. Effects of compound K on hyperglycemia and insulin resistance in rats with type 2 diabetes mellitus. *Fitoterapia* **2014**, *95*, 58–64. [CrossRef] [PubMed]

32. Mayakrishnan, T.; Nakkala, J.R.; Jeepipalli, S.P.; Raja, K.; Chandra, V.K.; Mohan, V.K.; Sadras, S.R. Fenugreek seed extract and its phytocompounds- trigonelline and diosgenin arbitrate their hepatoprotective effects through attenuation of endoplasmic reticulum stress and oxidative stress in type 2 diabetic rats. *Eur. Food Res. Technol.* **2015**, *240*, 223–232. [CrossRef]

Article

Assessment of Fennel Oil Microfluidized Nanoemulsions Stabilization by Advanced Performance Xanthan Gum

Rubén Llinares [1], Pablo Ramírez [1], José Antonio Carmona [1], Luis Alfonso Trujillo-Cayado [2,*] and José Muñoz [1]

[1] Departamento de Ingeniería Química, Facultad de Química, Universidad de Sevilla c/P, García González, 1, E41012 Sevilla, Spain; rllinares@us.es (R.L.); pramirez@us.es (P.R.); joseacarmona@us.es (J.A.C.); jmunoz@us.es (J.M.)

[2] Departamento de Ingeniería Química, Escuela Politécnica Superior, Universidad de Sevilla c/Virgen de África, 7, E41011 Sevilla, Spain

* Correspondence: ltrujillo@us.es; Tel.: +34-95-455-28-45

Abstract: In this work, nanoemulsion-based delivery system was developed by encapsulation of fennel essential oil. A response surface methodology was used to study the influence of the processing conditions in order to obtain monomodal nanoemulsions of fennel essential oil using the microchannel homogenization technique. Results showed that it was possible to obtain nanoemulsions with very narrow monomodal distributions that were homogeneous over the whole observation period (three months) when the appropriate mechanical energy was supplied by microfluidization at 14 MPa and 12 passes. Once the optimal processing condition was established, nanoemulsions were formulated with advanced performance xanthan gum, which was used as both viscosity modifier and emulsion stabilizer. As a result, more desirable results with enhanced physical stability and rheological properties were obtained. From the study of mechanical spectra as a function of aging time, the stability of the nanoemulsions weak gels was confirmed. The mechanical spectra as a function of hydrocolloid concentration revealed that the rheological properties are marked by the biopolymer network and could be modulated depending on the amount of added gum. Therefore, this research supports the role of advanced performance xanthan gum as a stabilizer of microfluidized fennel oil-in-water nanoemulsions. In addition, the results of this research could be useful to design and formulate functional oil-in-water nanoemulsions with potential application in the food industry for the delivery of nutraceuticals and antimicrobials.

Keywords: essential oil; hydrocolloid; microfluidization; nanoemulsion; rheology

Citation: Llinares, R.; Ramírez, P.; Carmona, J.A.; Trujillo-Cayado, L.A.; Muñoz, J. Assessment of Fennel Oil Microfluidized Nanoemulsions Stabilization by Advanced Performance Xanthan Gum. *Foods* **2021**, *10*, 693. https://doi.org/10.3390/foods10040693

Academic Editor: Zisheng Luo

Received: 22 February 2021
Accepted: 20 March 2021
Published: 24 March 2021

Publisher's Note: MDPI stays neutral with regard to jurisdictional claims in published maps and institutional affiliations.

1. Introduction

Essential oils (EOs) are composed of a mixture of volatile hydrophobic compounds produced by aromatic plants which possess interesting properties such as antibacterial, antifungal, antiviral, and antioxidant activity [1,2]. These properties, along with their aromatic properties, give EOs potential applications in a wide variety of fields, such as the food or pharmaceutical industries [2]. Furthermore, their status as safe compounds of biological origin has attracted the attention of researchers and manufacturers. For instance, the use of EOs in food and beverages has risen over the years due to consumer demand for quality and safety in the additives used by the food industry [3]. However, their poor solubility in water, high volatility and undesirable properties when used in high concentrations constrain the direct application of EOs [4,5]. Because of all these limitations, for practical use, they are usually encapsulated as droplets inside an aqueous continuous phase, hence forming emulsions and/or nanoemulsions as a function of the droplet size [6]. The encapsulation of the EOs within the continuous phase prevents the EOs from losing volatiles, avoids oil oxidation and degradation due to environmental stressors, and keeps the biological and physical properties of the EO unchanged [7,8].

Lately, essential oil nanoemulsions have been extensively studied due to their enhanced properties in comparison with conventional emulsions, such as improved stability and the higher bioavailability of lipophilic bioactive compounds [3,9]. These characteristics make nanoemulsions useful tools with great potential in the food sector for the delivery of food ingredients, such as many synthetic and natural antimicrobials, antioxidants, vitamins, flavorings, and colorants [6]. Among all the essential oils studied, several studies have dealt with the formulation and/or application of fennel oil nanoemulsions [10–12]. In some of those studies, fennel oil nanoemulsions have shown potential applications as biocides against food borne pathogens [12] and in transdermal drug delivery for antidiabetic action [11].

The addition of a thickener or rheology modifier to previously formed nanoemulsions can both improve long-time stability and expand their applications in fields such as food, pharmaceuticals, and cosmetics. Hydrocolloids have been used for years in food formulations for their outstanding properties as thickening or gelling agents [13]. They may be used to improve the stability and texture of the product and, in addition, for food formulations, modifying the texture will also change the mouth-feel of the emulsion [14]. Furthermore, gelation in emulsions can be achieved by the addition of hydrocolloids which form a network of cross-linked biopolymer molecules where the emulsions droplets are dispersed. Although the mechanical properties of the emulsion with gel-like behavior obtained by this approach are mainly determined by those of the bulk phase biopolymer, the dispersed droplets containing the surfactant may play a strengthening (active filler) or a weakening (inactive filler) role [9,15].

Xanthan gum has been employed as a stabilizer of emulsions and suspensions since it is able to greatly increase the viscosity of aqueous solutions even at low gum concentrations. Furthermore, the highly ordered structure of the rigid molecules of xanthan gum causes the aqueous xanthan gum solution to have interesting viscoelastic properties typical of weak gel systems. This type of rheological behavior is characterized by a mechanical spectrum in which the elastic moduli, G′, are greater than the loss moduli G″ over the entire frequency range, with the G″/G′ ratio greater than 0.1. In addition, the flow curves show shear-thinning behavior without yield stress, due to the absence of permanent cross-links, and therefore, weak gels flow but they are not irreversibly broken at large shear rates/deformations [16–18]. Advanced performance xanthan gum (APXG) is food-grade xanthan gum (XG) suitable for use in food formulations where fast hydration is required, such as sauces, instant desserts, or beverages. The acetate/pyruvate ratio of conventional XG is 1:1, instead of APXG ratio, which is 10:7. It is stated that the acetate/pyruvate ratio influences the physicochemical and rheological properties of xanthan gum aqueous. It provides high solution viscosity with pseudoplastic properties at low biopolymer concentrations, and presents lower viscous component than the standard xanthan gum, which could be an advantage to stabilize nanoemulsions.

The low water miscibility of several lipophilic bioactive compounds limits their application in aqueous media and their incorporation in food products. Oil-in-water nanoemulsions are in the spotlight of novelty to solve this main problem. The main novelty of this research is the development of stable nanoemulsions formulated with advanced performance xanthan gum and a bioactive volatile oil such as fennel essential oil. For this reason, the present work aims to provide an exhaustive study of the influence of the formulation and processing on microfluidized nanoemulsion properties. In this work, a xanthan gum with boosted rheological properties (advance performance xanthan) e 0.1–0.4/100 g [19–21]. The goal of the present study is to obtain the optimum processing conditions for achieving fennel oil nanoemulsions with the lowest droplet size and narrowest distribution and study the influence of the addition of different concentrations of a hydrocolloid (xanthan gum advanced performance) to the stability and mechanical properties of the previously formed nanoemulsions.

2. Materials and Methods

2.1. Materials

Sweet fennel essential oil (ρ = 0.933 g·cm^{-3}) was provided by Sigma-Aldrich and was used as the dispersed phase in the O/W nanoemulsions. Tween 80 (ρ = 1.064 g·cm^{-3} and HLB = 15) was provided by Sigma-Aldrich and was used as emulsifier. Food-grade xanthan gum "Advanced Performance" was provided by C.P. Kelco (Atlanta, GA, USA).

2.2. Chromatographic Analysis

Samples for GC-MS analysis were prepared by dissolving 100 µL of essential oil in 1 mL of dichloromethane. The chromatographic analysis of fennel oil was carried out following the procedure of Miraldi [22] on a Trace 1300 (Thermo Fisher, Waltham, MA, USA) gas chromatograph to a TSQ 8000 (Thermo Fisher, Waltham, MA, USA) triple quadrupole mass spectrometer. A Zebron ZB-WAXplus column (30 m, 0.25 mm, 0.25 mm) and helium as carrier gas (flow rate 1.0 mL/min) were used. Samples were injected using the split sampling technique, ratio 1:45; sample amount injected was 0.5 mL; injection port temperature, 250 °C. Oven temperature was held at 30 °C for 8 min, then programmed at 30 °C/min to 180 °C, held there for 5 min, and then again programmed at 10 °C/min to 220 °C and held for 10 min. The MS operating parameters were: electron ionization, 70 eV; accelerating voltage, 8 kV; scan range, 30 ± 650 amu. Identification and quantification of the volatile compounds was carried out by a peak-matching library search using the NBS/NIST 11 library.

Table 1 shows the results obtained from the chromatogram of fennel essential oil. The composition of this natural oil is complex since it contains several compounds (up to 12). This composition is similar to other fennel essential oils reported by different authors [23–25]. However, this fennel oil presents a major concentration of anethole. Anethole is a widely used flavoring substance with antimicrobial and antifungal activity [26]. D-limonene was the second major component detected in this fennel oil, followed by fenchone and α-pinene. According to Bilia et al., fenchone concentration in fennel essential oil should be less than 7.5% [27].

Table 1. Percentage composition of the fennel essential oil used in the present study.

Compound	Relative Concentration (%)
α-Pinene	6.8
β-Pinene	0.3
β-Phellandrene	0.6
α-Phellandrene	1.2
β-Myrcene	0.8
D-Limonene	16.7
p-Cymene	0.7
Fenchone	7.4
Estragole	2.3
Anethole	60.6
p-Anisaldehyde	2.3
p-Acetonylanisole	0.3

2.3. Emulsification Process

The emulsification process was performed in two steps. First, a coarse emulsion (200 g) was obtained using a rotor-stator system Silverson L5M (Silverson, East Longmeadow, MA, USA) at 6000 rpm for 30 s. Secondly, the coarse emulsion, previously cooled down to 5 °C, was processed in a high energy homogenization device Microfluidizer M-110P (Microfluidics, Westwood, MA, USA) with two interaction chambers (F12Y and H30Z). The use of two interaction chambers allows better droplet size distributions to be obtained [28]. All nanoemulsions contained 1 g/100 g of EO and a surfactant-to-oil ratio 3:1. The selection of the formulation is based on previous studies so that it results in stable nanoemulsions.

2.4. Gum Addition

The advanced performance xanthan gum was directly added into the previously formed nanoemulsions and gently stirred until complete dissolution. Different samples with a total concentration of 0.05, 0.10, 0.25, 0.40, 0.55, and 0.70 g/100 g were prepared. Samples were stored at 4 °C for 24 h prior to measurement.

2.5. Size Measurement

The droplet size distribution was obtained by using a Zetasizer® ZS (Malvern, Worcestershire, UK). The droplet size of the emulsions was described as the mean diameter (Z-average), whereas the width of the size distribution was indicated by the polydispersity index (PdI). PdI is a measure of the heterogeneity of a sample based on size. This index is a dimensionless number that is calculated from a two-parameter fit to the correlation data. Polydispersity index values smaller than 0.05 are mainly seen with highly monodisperse standards. The refractive index values were 1.538, and 1.343 for the oil phase and the continuous medium, respectively. All measurements were made in triplicate.

2.6. Oscillatory and Flow Curve Measurement

The rheological properties were measured by a controlled-stress AR2000 rheometer (TA Instrument, New Castle, DE, USA) using a serrated plate-plate geometry (60 mm diameter; gap = 1 mm) to avoid slippage of the sample.

Shear stress sweep tests were carried out at a constant frequency of 1 Hz from 0.01 to 50 Pa. The frequency sweep was then carried out at a constant stress value within the linear viscoelastic region in the frequency range from 20 to 0.10 rad/s. The flow curves were measured from 0.01 to 35 Pa, following a step-wise protocol with a maximum time per point of 3 min.

All rheological measurements were repeated at least three times at a controlled temperature of 20 °C and using a solvent trap to prevent solvent evaporation.

2.7. Microstructure of the Emulsions and Continuous Phases

In order to observe the differences between the microstructure of the emulsions and the continuous phases, selected samples were characterized by Cryo-scanning electronic microscopy (Cryo-SEM). A scanning electron microscope Zeiss EVO SEM (Zeiss, Oberkochen, Germany) was used following the procedure described by Carmona et al. [29].

2.8. Statistical Analysis

Experimental data of the study of the influence of the homogenization pressure and number of cycles on nanoemulsion droplet size distributions were fitted to the following quadratic model:

$$Y = \beta_0 + \beta_1 \cdot P + \beta_2 \cdot N + \beta_{12} \cdot P \cdot N + \beta_{11} \cdot P^2 + \beta_{22} \cdot N^2, \tag{1}$$

where Y is the response variable (dependent variable), β_0 is the constant, and P (homogenization pressure) and N (number of cycles) are the coded independent factors. The dependent variables considered were the Z-average and the polydispersity index (PdI) for nanoemulsions aged for 1 day. Statistical analysis of the data was performed with Statistic 10.0 software (StatSoft Inc., Tulsa, OK, USA) to evaluate the analysis of variance (ANOVA). For model construction, terms with $p > 0.05$ were removed and the analysis was recalculated without these terms. The suitability of the models was determined by using the coefficient of determination. All statistical calculations were conducted at a significance level of $p = 0.05$.

3. Results and Discussion

3.1. Influence of Energy Input on Droplet Size Distribution

Energy input in a microfluidizer device is a function of homogenization pressure (P) and the number of cycles (N). Hence, the influence of these variables on droplet size distribution (DSD) for essential oil nanoemulsions has been studied in several works [30,31]. It has been observed that increasing homogenization pressure and the number of cycles reduces the mean droplet sizes and polydispersity of emulsions until a plateau value of both parameters is obtained. Two main variables are commonly used to characterize DSD for Zetasizer devices: Z-average and the polydispersity index (PdI). Figure 1A,B show Z-average and PdI values as a function of P and N, respectively. A clear tendency to decreasing droplet sizes (Z-average) with an increase in homogenization pressure and the number of cycles was observed. It is important to note that nanoemulsions prepared using 12 or more cycles at the highest homogenization pressures showed mean diameters below 10 nm. Nevertheless, results of the ANOVA test demonstrated that there are not significant differences in Z-average for emulsions aged for 24 h prepared in the 12–18 passes range at 140 MPa. Low droplet mean diameters could be a great advantage against some destabilization mechanism, i.e., coalescence and creaming [32]. The lowest value of PdI was obtained for the nanoemulsion processed with 12 passes through the microfluidizer at 140 MPa. However, results of ANOVA test demonstrated that the number of passes in the 12–18 range had no significant effect upon the PdI of the nanoemulsions.

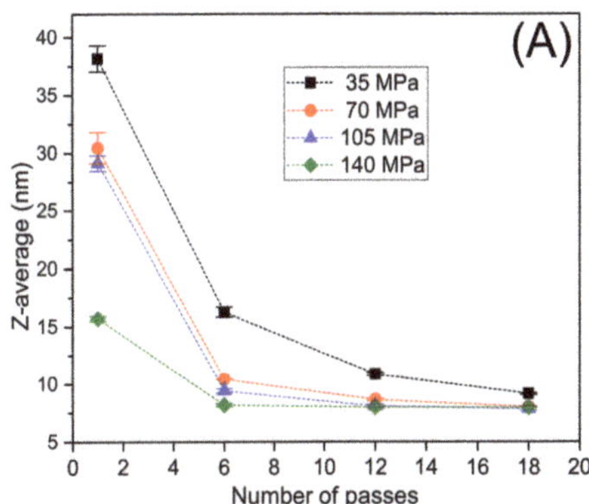
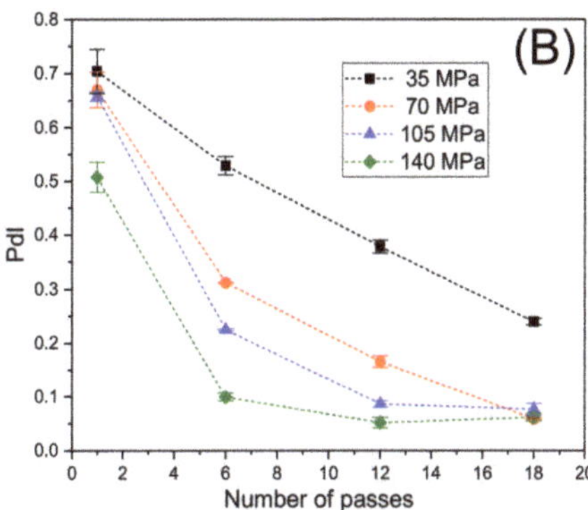

Figure 1. (**A**) Droplet mean diameters and (**B**) polydispersity index values for emulsions aged for 24 h as a function of the number of passes and homogenization pressure. Vertical bars indicate standard deviation of the mean (three replicates).

Figure 2A shows, by way of example, the influence of the number of passes on the droplet size distributions of the nanoemulsions aged for 24 h and processed at a fixed homogenization pressure of 140 MPa. All emulsions in the 6–18 passes range showed monomodal distributions. Conversely, the nanoemulsion developed using only one cycle exhibits a very wide distribution with two peaks centered around 8 and 80 nm, which is probably due to the lack of mechanical energy input during the emulsification process. This fact is also presented in the decrease in polydispersity for nanoemulsions developed at pressure in the range 105–140 MPa comparing with those developed at lower homogenization pressures (see Figure 2B). When the nanoemulsion passed through the high-pressure homogenizer microchannels at higher homogenization pressures, the droplet size distributions become monomodal.

Figure 3A,B illustrates the three-dimension response surface curve of Z-average and polydispersity index for the studied variables (homogenization pressures and number of cycles). This graph provides a visual interpretation of the interaction between the factors analyzed. In addition, response surface methodology is a very powerful tool for the development of new systems. This methodology can link the processing variables with response variables in order to obtain a mathematical model that fits the behavior of nanoemulsions. Z-average fitted a quadratic function of homogenization pressure (P) and number of passes (N):

$$\text{Z-average} = 8.14 - 4.00{\cdot}\text{P} - 8.99{\cdot}\text{N} + 4.59{\cdot}\text{P}{\cdot}\text{N} + 10.02{\cdot}\text{N}^2. \tag{2}$$

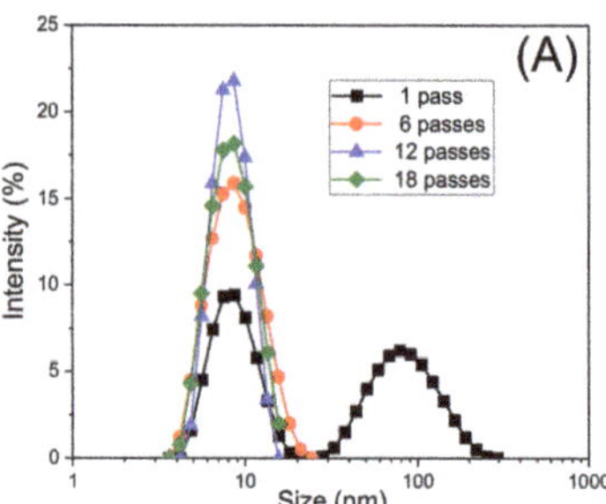

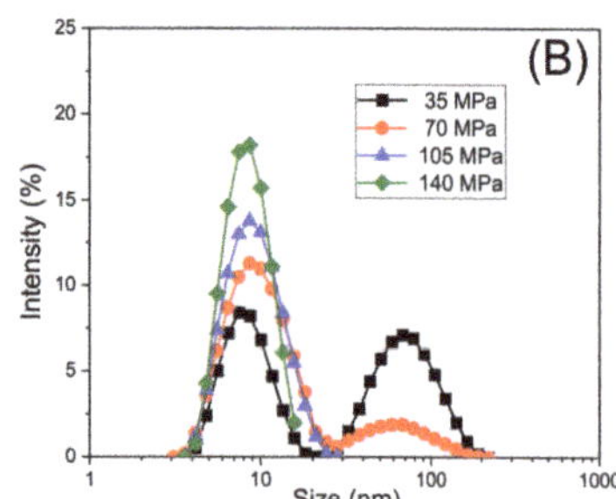

Figure 2. Droplet size distributions for emulsions aged for 24 h as a function of (**A**) the number of passes at a fixed homogenization pressure of 140 MPa and (**B**) homogenization pressure processed at a fixed number of passes of 18.

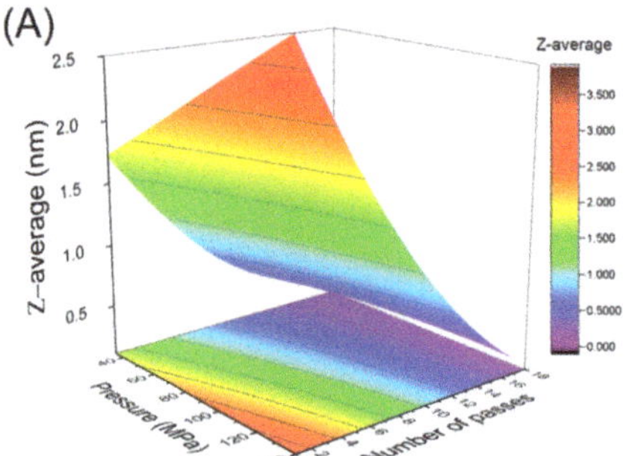

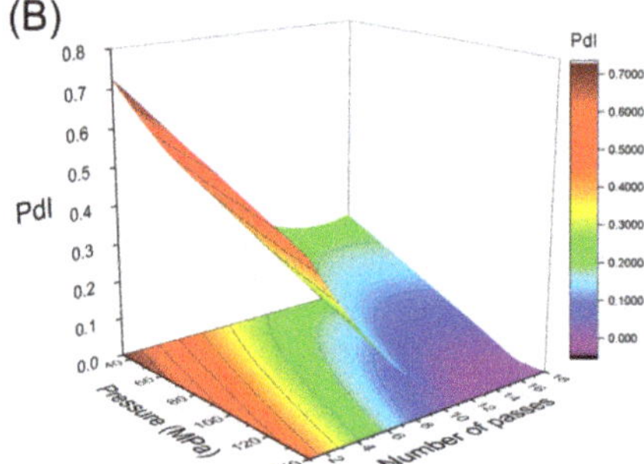

Figure 3. Response surface 3D plot of (**A**) droplet mean diameter and (**B**) polydispersity index as a function of the number of passes and homogenization pressure for emulsions aged for 24 h.

The results indicated that the model employed was adequate, showing a very satisfactory value of $R^2 = 0.917$ and no significant lack of fit ($F_{crit} > F_{lof}$, with $p = 0.05$). The model predicted that Z-average was sensitive to all studied variables. The most significant factor affecting this parameter was the number of passes (N), as supported by their linear and quadratic coefficients (-8.99 and 10.02):

$$\text{PdI} = 0.18 - 0.13{\cdot}\text{P} - 0.25{\cdot}\text{N} + 0.17{\cdot}\text{N}^2. \tag{3}$$

The coefficient of determination (R^2) value of 0.933 indicated a good correlation between the experimental results and predicted responses. It is worth noting the existence of a minimum value of the polydispersity index for the highest value of the number of cycles (N) and homogenization pressure (P). From the model we deduced that the most significant factors affecting PdI in these fennel oil nanoemulsions were the linear term of the homogenization pressure and the linear and quadratic terms of the number of passes.

An optimum process condition can be set for emulsions with minimum droplet sizes and polydispersity. According to the response surface analysis, the predicted minimum Z average mean diameter and polydispersity value are obtained when P = 140 MPa and N = 12 passes. For this reason, these conditions were fixed for the following tests. In this study, minimum initial droplet sizes were observed for all formulations in comparison to those obtained in other studies in which fennel oil nanoemulsions are developed [10].

3.2. Rheological Behavior Influence of Xanthan Gum Concentration

First, the viscosity of fennel oil nanoemulsion with processing conditions of 140 MPa and 12 cycles was studied. All systems presented a Newtonian flow behavior with a viscosity of 1.63 ± 0.05 mPa at 20 °C (data not shown). In order to expand the application of previously obtained nanoemulsions, a natural hydrocolloid (xanthan gum) was added. In this way, the mechanical properties of previously prepared nanoemulsions can be modified as a function of xanthan gum concentration, thus obtaining nanoemulsion gels of tailored viscoelasticity. The rheological properties were studied for xanthan gum concentrations ranging from 0.05 to 0.70 g/100 g.

Most applications of oscillatory tests entail working in linear viscoelastic conditions, that is, in non-destructive conditions. Therefore, a prior step for the determination of viscoelastic properties by this technique consists of selecting a shear stress value that guarantees that the structure of the material is not irreversibly destroyed in the frequency range in which the test is performed. The determination of the linear viscoelastic range was carried out from stress sweep tests, at a fixed frequency of 1 Hz. The critical shear stresses for each concentration were determined as the shear stress value from which the G′ modulus varied more than 2% with respect to the average G′ values in the linear region. G′ values have been used to obtain the shear stress value instead of G″ due to the fact that a departure from a constant value was observed at lower shear stresses for storage moduli for all the concentrations studied. Figure 4 shows the shear stress sweep test for the sample with a xanthan gum concentration of 0.70 g/100 g as an example.

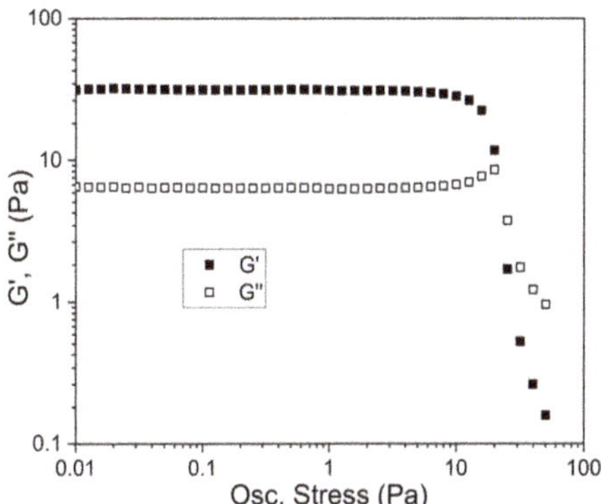

Figure 4. Storage (G′) and loss (G″) moduli at 1 Hz as a function of shear stress for the fennel oil nanoemulsion with 0.70 g/100 g of the AP xanthan gum. Standard deviation of the mean (three replicates) for all the data are below 5%. (T = 20 °C).

Small amplitude oscillatory shear measurements were carried out as shown in Figure 5. Two different viscoelastic behaviors were clearly observed. For xanthan gum concentrations below 0.25 wt.%, a crossover frequency which shifts towards higher values was observed. This crossover is associated with a change of viscoelastic behavior from liquid-like to gel-like. This behavior is typical for diluted solutions of hydrocolloids with low interactions between the polymer chains [33,34]. Nevertheless, for xanthan concentration of 0.25 wt.% and above, the mechanical spectra obtained are completely different. For those systems, the storage modulus (G′) was above the loss modulus (G″) for the whole frequency range, which indicates the occurrence of a more compact structure typical of systems which exhibit weak gel-like behavior [34].

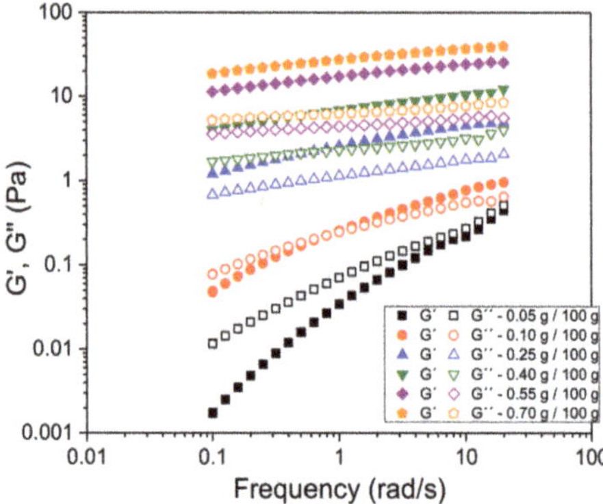

Figure 5. Mechanical spectra for fennel oil nanoemulsions with different xanthan gum concentration at 20 °C. Data shown are the average values of three replicates. The standard deviation for all the experiments was lower than 5%.

With increasing the gum concentration, the system shows a higher weak gel character, with higher values of G′ and G″, increasing differences between G′ and G″ and a reduction in the slope of G′. Thus, the slope of G′ can be used to quantify the solid-like behavior of the system [35]. A marked decrease in the G′ slope is observed with increasing xanthan gum concentration. Comparison of the values of the mechanical spectra from the gelled nanoemulsions with the mechanical spectra of the same xanthan gum network in aqueous solution shows similar value [20]. Hence, it is confirmed that the rheological properties of the weak gelled nanoemulsions are dominated by the biopolymer network which controls the structure of the continuous phase.

When working in the linear viscoelastic domain, the test can be considered non-destructive. Therefore, the results obtained can be related to the structure of the sample and are very sensitive to changes in it [3]. Hence, obtaining mechanical spectra as a function of the storage time of the samples gives us reliable information concerning the possible presence of destabilization processes that cause changes in the structure of the sample. In Figure 6, the evolution over time of the mechanical spectra for the sample of 0.70% by weight of xanthan gum is shown as an example. It was observed that for all the studied concentrations the variations in the values of the viscoelastic modules with respect to the replicates and time of aging were less than 5%.

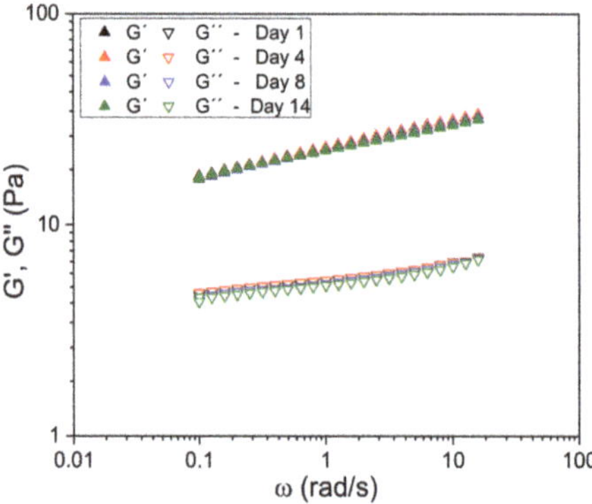

Figure 6. Mechanical spectra as a function of aging time for fennel oil nanoemulsion formulated with advanced performance xanthan gum (0.70 g/100 g) at 20 °C. Data shown are the average values of three replicates. The standard deviation of for all the experiments was lower than 5%.

In order to fulfil the rheological characterization of the gels containing the fennel oil nanoemulsions, flow curves were carried out as a function of the xanthan gum concentration (Figure 7).

Table 2. Fitting parameter values for the Carreau model. The values were obtained by a non-linear regression analysis.

[XG] (g/100 g)	η_0 (Pa·s)	$\dot{\gamma}_c$ (s^{-1})	n
0.05	0.24 ± 0.01	0.11 ± 0.02	0.55 ± 0.01
0.10	2.7 ± 0.1	0.016 ± 0.002	0.43 ± 0.01
0.25	175 ± 9	$3.3 \times 10^{-3} \pm 4 \times 10^{-4}$	0.25 ± 0.01
0.40	660 ± 28	$3.6 \times 10^{-3} \pm 5 \times 10^{-4}$	0.17 ± 0.02
0.55	1730 ± 80	$2.5 \times 10^{-3} \pm 3 \times 10^{-4}$	0.13 ± 0.01
0.70	8900 ± 508	$7 \times 10^{-4} \pm 1 \times 10^{-4}$	0.12 ± 0.01

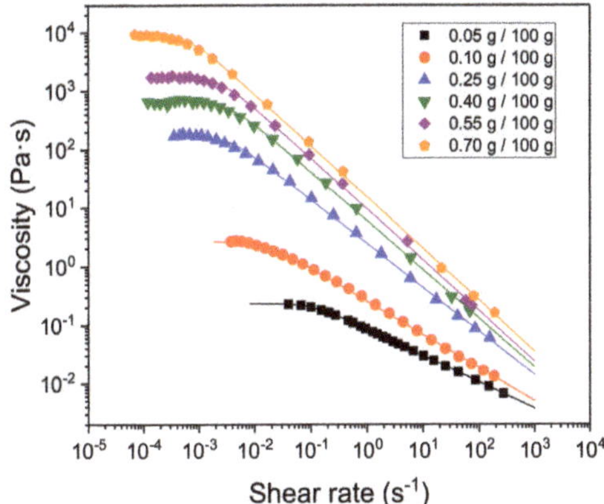

Figure 7. Steady state viscosity as a function of shear rate for the fennel oil nanoemulsion with different AP xanthan gum concentrations at 20 °C. The lines are the best fit to the Carreau model, whose parameters are given in Table 2. Data shown are the average value of three replicates with a standard deviation for all the experiments lower than 10%.

It is observed that all the systems showed shear-thinning behavior characterized by a constant viscosity at low shear rates values (the zero-shear viscosity, η_0) which follows a power-law decay for shear rates below a critical value ($\dot{\gamma}_c$). Therefore, the experimental viscosities (η) can be fitted to the well-known Carreau model:

$$\eta = \frac{\eta_0}{\left(1 + \left(\frac{\dot{\gamma}}{\dot{\gamma}_c}\right)^2\right)^{\left(\frac{1-n}{2}\right)}} \tag{4}$$

where n is the flow index. Solid lines indicating the best fit to the experimental data with the parameter values are given in Table 2. A good agreement between experimental data and the estimated values from the model is obtained. The zero shear viscosity values were compared with those obtained for the same hydrocolloid in aqueous solution [21]. It was observed that for the same concentration, the η_0 values were lower for the nanoemulsion systems (see Table 3). These changes in the zero shear viscosity values between the nanoemulsions and the aqueous solutions of advanced performance xanthan gum are likely to be related to microstructural changes.

Table 3. Comparison of the zero shear viscosity values of the weak gel advanced performance xanthan gum systems containing fennel oil nanoemulsion (* this work) or without fennel oil nanoemulsion (**, Reference [21] by Carmona et al.).

[XG] (g/100 g)	η_0 (Pa·s) *	η_0 (Pa·s) **
0.25	175 ± 9	214 ± 5
0.40	660 ± 28	1638 ± 15

Therefore, in order to establish a relationship between rheological performance and microstructural conformation, cryo-SEM images of an emulsion that contains an advanced performance xanthan gum concentration of 0.25 g/100 g and the correspondent aqueous solutions were obtained (Figures 8A and 8B, respectively). A more structured system is observed in Figure 8B, which corresponds to the aqueous solutions. This structural modification is likely to be related to the fact that nanoemulsion droplets behave as an inactive filler, leading to a slight decrease in the biopolymer network (Figure 8A).

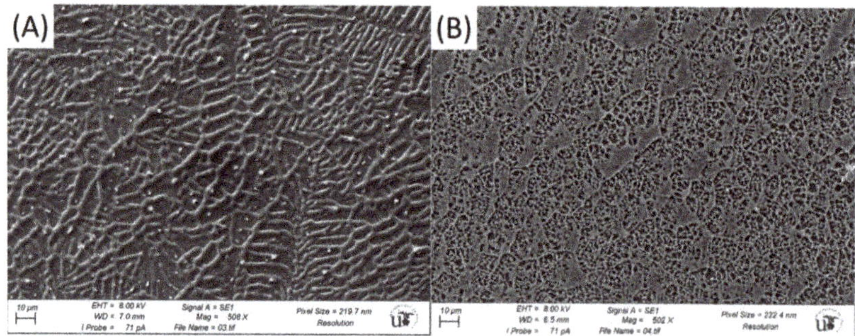

Figure 8. Representative Cryo-SEM micrographs for (**A**) the emulsion formulated with 0.25 g/100 g of advanced performance xanthan gum and (**B**) an aqueous solution that contains 0.25 g/100 g of advanced performance xanthan gum.

4. Conclusions

Stable edible nanoemulsions containing sweet fennel essential oil with tween 80 as emulsifier were processed using a two-step emulsification consisting of a rotor/stator pre-homogenization and microfluidic homogenization. Narrow monomodal distributions (PdI ≈ 0.05) with low average size (Z-average ≈ 8 nm) were obtained and the effect of processing variables on the final nanoemulsion was studied, establishing optimum processing conditions. Xanthan gum was added to the nanoemulsion as a thickening agent to improve the rheological properties. The systems showed viscoelastic properties for every gum concentration used and were characterized by small amplitude oscillatory shear tests and steady flow measurements. Edible nanoemulsion weak gels were obtained with the addition of xanthan gum concentrations above 0.25 g/100 g. The gel strength and zero shear viscosity were functions of the concentration of the gum and, therefore can be conveniently adjusted. Furthermore, from the steadiness of the mechanical spectra over time, the system proved to be stable in the xanthan gum concentration range studied. Comparison of the rheological properties and microstructure of the nanoemulsion weak gels with the ones of the same concentration of xanthan gum aqueous solutions revealed that the mechanical properties are governed by the biopolymer network and the nanoemulsion droplets behave as inactive fillers, slightly decreasing the zero-shear viscosity of the biopolymer network.

Author Contributions: Conceptualization, L.A.T.-C. and P.R.; methodology, J.A.C.; software, J.A.C.; validation, R.L., J.A.C. and P.R.; formal analysis, L.A.T.-C.; investigation, R.L.; resources, J.M.; data curation, J.A.C.; writing—original draft preparation, R.L.; writing—review and editing, L.A.T.-C.; visualization, P.R.; supervision, J.M.; project administration, J.M.; funding acquisition, J.M. All authors have read and agreed to the published version of the manuscript.

Funding: This research was funded by Spanish Ministerio de Economía y Competitividad (MINECO) and FEDER, Grant Number/Project: CTQ2015-70700-P.

Informed Consent Statement: Not applicable.

Data Availability Statement: The data presented in this study are available on request from the corresponding author.

Conflicts of Interest: The authors declare no conflict of interest. The funders had no role in the design of the study; in the collection, analyses, or interpretation of data; in the writing of the manuscript, or in the decision to publish the results.

References

1. Chang, Y.; McLandsborough, L.; McClements, D.J. Fabrication, stability and efficacy of dual-component antimicrobial nanoemulsions: Essential oil (thyme oil) and cationic surfactant (lauric arginate). *Food Chem.* **2015**, *172*, 298–304. [CrossRef] [PubMed]
2. Baser, K.; Buchbauer, G. (Eds.) *Handbook of Essential Oils: Science, Technology, and Applications*; CRC Press: Boca Raton, FL, USA, 2010.
3. Donsì, F.; Ferrari, G. Essential oil nanoemulsions as antimicrobial agents in food. *J. Biotechnol.* **2016**, *233*, 106–120. [CrossRef] [PubMed]
4. Rodríguez-Rojo, S.; Visentin, A.; Maestri, D.; Cocero, M.J. Assisted extraction of rosemary antioxidants with green solvents. *J. Food Eng.* **2012**, *109*, 98–103. [CrossRef]
5. Do, T.K.T.; Hadji-Minaglou, F.; Antoniotti, S.; Fernandez, X. Authenticity of essential oils. *TrAC Trends Anal. Chem.* **2015**, *66*, 146–157. [CrossRef]
6. Salvia-Trujillo, L.; Soliva-Fortuny, R.; Rojas-Graü, M.A.; McClements, D.J.; Martín-Belloso, O. Edible Nanoemulsions as Carriers of Active Ingredients: A Review. *Annu. Rev. Food Sci. Technol.* **2017**, *8*, 439–466. [CrossRef]
7. Choi, M.-J.; Soottitantawat, A.; Nuchuchua, O.; Min, S.-G.; Ruktanonchai, U. Physical and light oxidative properties of eugenol encapsulated by molecular inclusion and emulsion–diffusion method. *Food Res. Int.* **2009**, *42*, 148–156. [CrossRef]
8. Surassmo, S.; Min, S.-G.; Bejrapha, P.; Choi, M.-J. Effects of surfactants on the physical properties of capsicum oleoresin-loaded nanocapsules formulated through the emulsion–diffusion method. *Food Res. Int.* **2010**, *43*, 8–17. [CrossRef]
9. Erramreddy, V.; Ghosh, S. Gelation in nanoemulsion: Structure formation and rheological behavior. In *Emulsions*; Elsevier BV: Amsterdam, The Netherlands, 2016; pp. 257–292.
10. Barradas, T.N.; De Campos, V.E.B.; Senna, J.P.; Coutinho, C.D.S.C.; Tebaldi, B.S.; Silva, K.G.D.H.E.; Mansur, C.R.E. Development and characterization of promising o/w nanoemulsions containing sweet fennel essential oil and non-ionic sufactants. *Colloids Surf. A Physicochem. Eng. Asp.* **2015**, *480*, 214–221. [CrossRef]
11. Mostafa, D.M.; El-Alim, S.H.A.; Asfour, M.H.; Al-Okbi, S.Y.; Mohamed, D.A.; Awad, G. Transdermal nanoemulsions of *Foeniculum vulgare* Mill. essential oil: Preparation, characterization and evaluation of antidiabetic potential. *J. Drug Deliv. Sci. Technol.* **2015**, *29*, 99–106. [CrossRef]
12. Venkadesaperumal, G.; Rucha, S.; Sundar, K.; Shetty, P.H. Anti-quorum sensing activity of spice oil nanoemulsions against food borne pathogens. *LWT* **2016**, *66*, 225–231. [CrossRef]
13. Saha, D.; Bhattacharya, S. Hydrocolloids as thickening and gelling agents in food: A critical review. *J. Food Sci. Technol.* **2010**, *47*, 587–597. [CrossRef]
14. McClements, D.J. *Food Emulsions: Principles, Practices, and Techniques*, 2nd ed.; Series in Contemporary Food Science; CRC Press: Boca Raton, FL, USA, 2005; p. 609.
15. Dickinson, E. Stabilising emulsion-based colloidal structures with mixed food ingredients. *J. Sci. Food Agric.* **2012**, *93*, 710–721. [CrossRef]
16. Rochefort, W.E.; Middleman, S. Rheology of Xanthan Gum: Salt, Temperature, and Strain Effects in Oscillatory and Steady Shear Experiments. *J. Rheol.* **1987**, *31*, 337–369. [CrossRef]
17. Carnali, J.O. Original Contributions Gelation in physically associating biopolymer systems. *Rheol. Acta* **1992**, *412*, 399–412. [CrossRef]
18. Lim, T.; Uhl, J.T.; Prudhomme, R.K. Rheology of Self-Associating Concentrated Xanthan Solutions. *J. Rheol.* **1984**, *28*, 367–379. [CrossRef]
19. Carmona, J.; Ramirez, P.; Calero, N.; Munoz, J.M. Large amplitude oscillatory shear of xanthan gum solutions. Effect of sodium chloride (NaCl) concentration. *J. Food Eng.* **2014**, *126*, 165–172. [CrossRef]
20. Carmona, J.A.; Lucas, A.; Ramírez, P.; Calero, N.; Muñoz, J. Nonlinear and linear viscoelastic properties of a novel type of xanthan gum with industrial applications. *Rheol. Acta* **2015**, *54*, 993–1001. [CrossRef]
21. Carmona, J.A.; Calero, N.; Ramírez, P.; Muñoz, J. Rheology and structural recovery kinetics of an advanced performance xanthan gum with industrial application. *Appl. Rheol.* **2017**, *27*, 1–9. [CrossRef]
22. Miraldi, E. Comparison of the essential oils from ten *Foeniculum vulgare* Miller samples of fruits of different origin. *Flavour Fragr. J.* **1999**, *14*, 379–382. [CrossRef]

23. Marotti, M.; Piccaglia, R.; Giovanelli, E.; Deans, S.G.; Eaglesham, E. Effects of Variety and Ontogenic Stage on the Essential Oil Composition and Biological Activity of Fennel (*Foeniculum vulgare* Mill.). *J. Essent. Oil Res.* **1994**, *6*, 57–62. [CrossRef]
24. Ehsanipour, A.; Razmjoo, J.; Zeinali, H. Effect of nitrogen rates on yield and quality of fennel (*Foeniculum vulgare* Mill.) accessions. *Ind. Crop. Prod.* **2012**, *35*, 121–125. [CrossRef]
25. García-Jiménez, N.; Péerez-Alonso, M.J.; Velasco-Negueruela, A. Chemical Composition of Fennel Oil, *Foeniculum vulgare* Miller, from Spain. *J. Essent. Oil Res.* **2000**, *12*, 159–162. [CrossRef]
26. De, M.; De, A.K.; Sen, P.; Banerjee, A.B. Antimicrobial properties of star anise (*Illicium verum* Hook f). *Phytother. Res.* **2002**, *16*, 94–95. [CrossRef]
27. Bilia, A.; Flamini, G.; Taglioli, V.; Morelli, I.; Vincieri, F. GC–MS analysis of essential oil of some commercial Fennel teas. *Food Chem.* **2002**, *76*, 307–310. [CrossRef]
28. Trujillo-Cayado, L.A.; Santos, J.; Ramirez, P.; Alfaro, M.C.; Muñoz, J. Strategy for the development and characterization of environmental friendly emulsions by microfluidization technique. *J. Clean. Prod.* **2018**, *178*, 723–730. [CrossRef]
29. Carmona, J.; Ramírez, P.; Trujillo-Cayado, L.; Caro, A.; Muñoz, J. Rheological and microstructural properties of sepiolite gels. Influence of the addition of ionic surfactants. *J. Ind. Eng. Chem.* **2018**, *59*, 1–7. [CrossRef]
30. Salvia-Trujillo, L.; Rojas-Graü, M.A.; Soliva-Fortuny, R.; Martín-Belloso, O. Effect of processing parameters on physicochemical characteristics of microfluidized lemongrass essential oil-alginate nanoemulsions. *Food Hydrocoll.* **2013**, *30*, 401–407. [CrossRef]
31. Trujillo-Cayado, L.A.; Santos, J.; Alfaro, M.C.; Calero, N.; Muñoz, J. A further step in the development of oil-in-water emul-sions formulated with a mixture of green solvents. *Ind. Eng. Chem. Res.* **2016**, *55*, 7259–7266. [CrossRef]
32. Izquierdo, P.; Esquena, J.; Tadros, T.F.; Dederen, C.; Garcia, M.J.; Azemar, N.; Solans, C. Formation and Stability of Nano-Emulsions Prepared Using the Phase Inversion Temperature Method. *Langmuir* **2002**, *18*, 26–30. [CrossRef]
33. Cuvelier, G.; Launay, B. Concentration regimes in xanthan gum solutions deduced from flow and viscoelastic properties. *Carbohydr. Polym.* **1986**, *6*, 321–333. [CrossRef]
34. Wyatt, N.B.; Liberatore, M.W. Rheology and viscosity scaling of the polyelectrolyte xanthan gum. *J. Appl. Polym. Sci.* **2009**, *114*, 4076–4084. [CrossRef]
35. Song, K.-W.; Kim, Y.-S.; Chang, G.-S. Rheology of concentrated xanthan gum solutions: Steady shear flow behavior. *Fibers Polym.* **2006**, *7*, 129–138. [CrossRef]

MDPI

St. Alban-Anlage 66

4052 Basel

Switzerland

Tel. +41 61 683 77 34

Fax +41 61 302 89 18

www.mdpi.com

Metals Editorial Office

E-mail: metals@mdpi.com

www.mdpi.com/journal/metals